T0281276

Biological Control
of Crop Diseases

Biological Control
of Crop Diseases

edited by
Samuel S. Gnanamanickam
University of Madras–Guindy
Chennai, Tamil Nadu, India

CRC Press
Taylor & Francis Group
Boca Raton London New York

CRC Press is an imprint of the
Taylor & Francis Group, an **informa** business

First published 2002 by Marcel Dekker, Inc.

Published 2019 by CRC Press
Taylor & Francis Group
6000 Broken Sound Parkway NW, Suite 300
Boca Raton, FL 33487-2742

© 2002 by Taylor & Francis Group, LLC
CRC Press is an imprint of Taylor & Francis Group, an Informa business

First issued in paperback 2019

No claim to original U.S. Government works

ISBN 13: 978-0-367-45497-5 (pbk)
ISBN 13: 978-0-8247-0693-7 (hbk)

Visit the Taylor & Francis Web site at
http://www.taylorandfrancis.com

and the CRC Press Web site at
http://www.crcpress.com

Foreword

In reviewing the chapters in this volume, I am reminded of three separate but distinct responses of three notable colleagues to K. F. Baker's and my first book, *Biological Control of Plant Pathogens*, published in 1974.

The first was the response of Dr. Lilian Fraizer, New South Wales Department of Agriculture (Australia), who, after reviewing an early draft of our manuscript, described it as "defensive," as if we had to *prove* to a skeptical readership that biological control as a science and practice has a place in plant pathology. There was no doubt that our audience at that time consisted of many skeptics, but conveying a defensive posture as the basic message of our book would only reinforce this skepticism. Much to our satisfaction, Dr. Fraizer described our published book as "quietly confident."

I would describe the substance of this book, organized and edited by Samuel S. Gnanamanickam, as exceedingly confident but also pragmatic, based on nearly three decades of lessons from the "school of hard knocks" since K. F. Baker and I wrote our first book. The list of examples of biological control of plant pathogens with introduced microorganisms (Chap. 17) on a global basis is truly remarkable. The pragmatism is apparent in all chapters, at least with respect to biological control with introduced microorganisms, reflecting the experience of the past 30 years that discovery of an antagonist is only the first and relatively easy step on a long path to delivery.

The second response to my first book was from Dr. J. M. Hirst in a review published in *Nature* (252:147) under the title "What Is Biological Control?" Dr. Hirst questioned our broad definition of biological control: "This concept is so broad that it embraces all of microbial ecology and much of agriculture, except the use of chemicals aimed solely against pathogens." Dr. Hirst was half right—

the other exception to our broad concept was physical controls, such as the use of heat, desiccation, or electrocution or the physical separation of host and pathogen, as with quarantine and the production of pathogen-free planting material using meristem culture.

It is worth noting that the two possibly most significant advances in our fundamental understanding of biological control of plant pathogens over the past three decades—suppressive soils and plant–microbe interactions—are outcomes primarily of our understanding of microbial ecology. The descriptions by Sivasithamparam in Chapter 4 of the plant habitats of microbial biocontrol agents and by Mazzola in Chapter 12 of phytomanagement of resident antagonists are prime examples of the application of principles of microbial ecology to understand and achieve biological control.

The third response was from R. R. Baker, who, like K. F. Baker and me, accepted that the principles of microbial ecology are the foundation of biological control of plant pathogens but questioned our inclusion of host plant resistance as part of biological control. He sent a questionnaire to 216 plant pathologists around the world and asked them to select one of three choices for a definition of biological control: (1) the Baker and Cook definition, (2) S. D. Garrett's definition published in the 1965 volume *Ecology of Soil-borne Plant Pathogens*, which focused on agents other than the host and pathogen, or (3) "A 'good' definition has yet to be published." Of 140 respondents, 49% favored the definition of Baker and Cook, 34% favored Garrett's definition, and 17% thought that a good definition was still needed.

Twenty-five years may not be long enough to settle this issue, but clearly the trend is toward acceptance of host–plant resistance as part of biological control. Two chapters in this book illustrate this point—Chapter 3, "Transgenic Plants for the Management of Plant Diseases: Rice, a Case Study," and Chapter 16, "Biocontrol Agents in Signaling Resistance."

Because of advances in molecular biology, all biological control of plant pathogens comes down to delivery of the products of genes and biosynthetic pathways to the right place at the right time. Take-all decline is the result of the antibiotic 2,4-diacetylphloroglucinol produced in adequate amounts in the rhizosphere of wheat by certain genotypes of *Pseudomonas fluorescens* enriched by monoculture wheat. One of the key enzymes in this biosynthetic pathway, chalcone synthase, is the same enzyme used in plants in the biosynthesis of phytoalexins—substances that, if produced by microorganisms, would be called antibiotics. Biological control of fungal pathogens with chitinous cell walls can be achieved by using the right chitinase produced in adequate amounts when and where needed—by *Trichoderma* in the rhizosphere, as a pathogen-related protein induced as part of the systemic acquired resistance in response to a pathogen or nonpathogen, or by a transgene expressed in plants. The best examples, of course, are the family of genes in *Bacillus thuringiensis* (Bt) for production of the crystalline proteins lethal to certain insects: biological control of these insects is the result of the right form of one of these proteins produced in adequate amounts when and where needed, whether by a strain of Bt on the plant or as a product of a Bt transgene expressed in the plant.

The delivery of the products of genes or biosynthetic pathways of microbial biocontrol agents through transformation of the crop plant is compelling because of both the relative simplicity of this approach and the long history of success in breeding crops for resistance to plant diseases. Clearly, this approach is justified. However, there are equally important reasons for delivering these mechanisms of biological control through microorganisms, especially as plant-associated microorganisms. Microbial biocontrol of plant pathogens rarely involves a single mechanism, whereas the transfer of genes from one of these agents to a crop plant will likely focus on just one mechanism, which may be less durable. For a crop such as wheat and a disease such as take-all, the development of a microbial biocontrol seed treatment product that works on multiple cultivars and market classes typical of this crop precludes the need to transform each of the cultivars for each of the environments in which take-all is important.

We must also remember that biological control of plant pathogens with microorganisms includes resident antagonisms managed by cultural practices—by far the most successful method, as illustrated by the examples given for the diseases of the 12 crops discussed in this book. It is easy in this age of sophisticated research to overlook the importance of biological control by resident antagonists—their stimulation by organic amendments, exploitation in suppressive soils, or role in the crop rotation effect. In many respects, biological control in its many variations and applications has been and remains the mainstay of plant disease management, especially for diseases caused by pathogens that are not exclusively seedborne and that therefore have not been subject to control by cleaning up the planting material. Indeed, biological control does embrace much of agriculture. The challenge ahead is how to make greater use of this natural resource. Although the new tools open exciting new possibilities, we must not lose sight of the proven traditional methods.

R. James Cook, Ph.D.
Washington State University
Pullman, Washington

Preface

Agriculture has been man's primary occupation since time immemorial. History shows that agriculture has shaped and dictated the fate of many civilizations and that effective disease management has long been a challenge to mankind. Today, in light of growing concerns about environmental safety, suppression of plant diseases through biological agents is gaining ground as a supplement to traditional disease-management strategies. Several groups are actively engaged in research pertaining to the identification of biocontrol agents against specific plant pathogens, with special focus on their deployment on a commercial scale.

I was fascinated by the elegant early research that emerged from University of California–Berkeley and Washington State University. This led to my involvement in biocontrol research, and today some of my Indian friends in research circles inform me that I was the first one to introduce *Pseudomonas fluorescens* for biocontrol research in India in the early 1980s. The first book on biological control, *Biological Control of Plant Pathogens*, by K. F. Baker and R. J. Cook (published in 1974), and my personal and academic association with Dr. Jim Cook have been sources of great inspiration for developing this volume.

Many books and edited volumes on biological control have been published in recent years, most of them addressing the biological control of a particular nonspecialized pathogen (e.g., *Fusarium*) that causes diseases in different host plants. This volume aims at providing a comprehensive update on principles, a catalogue of the advancements in the application of biocontrol practices for major crops cultivated around the world, and a glimpse of the challenges involved in testing, formulating, and applying them within the context of integrated disease management. It comprehensively presents the agents and methods practiced to control diseases of major crops, each of which is dealt with in a separate chapter.

The chapters contain the experiences of leading experts engaged in biocontrol research. There are twelve such chapters in this volume.

Chapters 15–18 cover very interesting and useful subjects related to biological control: "Implementation of Biological Control of Plant Diseases in Integrated Pest-Management Systems" (Chap. 15), "Comprehensive Testing of Biocontrol Agents" (Chap. 17), and "Formulation of Biological Control Agents for Pest and Disease Management" (Chap. 18). Jo Handelsman has provided in Chapter 19 fine insight into future trends and challenges in biocontrol research and its application.

The reader will be quick to realize that the volume has been designed to accommodate the changing scope of biological control brought on by the advent of biotechnological advances—the definition of biocontrol stands expanded as S. K. Datta (Chap. 3) presents his work on transgenic rices for disease management and L. C. van Loon (Chap. 16) describes signaling mechanisms initiated by rhizobacteria that culminate in resistance to plant pathogens. The first chapter, "Principles of Biological Control," and the Foreword by Dr. Jim Cook recognize this change.

I sincerely hope that this volume will both fill the long-felt need for a textbook on the subject and serve as an advanced treatise for those seeking details on biological management of diseases for any of these 12 crops as well as related topics. I welcome any comments and suggestions for improvement.

I am very grateful for the Foreword by Dr. Jim Cook, whom I consider a living legend on the subject and a staunch supporter of biocontrol research and its application. His contribution definitely adds fillip to this volume. I deeply appreciate the efforts of all the contributors, who have given their best in preparing their chapters. Members of my research team stood by me and assisted in reading and editing the original manuscripts and page proofs: among them, Dr. Preeti Vasudevan deserves special mention. I sincerely thank Marcel Dekker, Inc., and the editorial staff, Ms. Moraima Suarez in particular, for their professionalism and commitment to the development of this volume.

Samuel S. Gnanamanickam

Contents

Contributors

Terry R. Anderson, Ph.D.   Department of Biological Sciences, Greenhouse and Processing Crop Research Centre, Agriculture and Agri-Food Canada, Harrow, Ontario, Canada

Lavanya Babujee, M.Sc., M.Phil.   Center for Advanced Studies in Botany, University of Madras–Guindy, Chennai, Tamil Nadu, India

Swapan K. Datta, Ph.D.   Plant Breeding, Genetics, and Biochemistry Division, International Rice Research Institute, Metro Manila, Philippines

Geneviève Défago, Ph.D.   Department of Phytopathology, Institute of Plant Sciences, Eidgenössische Technische Hochschule, Zurich, Switzerland

Luis del Rio, Ph.D.   Department of Plant Pathology, North Dakota State University, Fargo, North Dakota

Suseelendra Desai, Ph.D.   National Research Centre for Groundnut, Indian Council of Agricultural Research, Junagadh, Gujarat, India

Samir Droby, Ph.D.   Postharvest Science, Agricultural Research Organization, The Volcani Center, Bet Dagan, Israel

Ahmed El Ghaouth, Ph.D.   Faculty of Science and Technology, University of Nouakchott, Nouakchott, Mauritania, Africa

Samuel S. Gnanamanickam, Ph.D.   Center for Advanced Studies in Botany, University of Madras–Guindy, Chennai, Tamil Nadu, India

Jo Handelsman, Ph.D.   Department of Plant Pathology, University of Wisconsin–Madison, Madison, Wisconsin

Barry Jacobsen, Ph.D.   Department of Plant Sciences and Plant Pathology, Montana State University, Bozeman, Montana

S. Kavitha, M.Sc., M.Phil.   Center for Advanced Studies in Botany, University of Madras–Guindy, Chennai, Tamil Nadu, India

G. Krishna Kishore, M.Sc.   Department of Plant Sciences, University of Hyderabad, Hyderabad, Andhra Pradesh, India

Joseph W. Kloepper, Ph.D.   Department of Entomology and Plant Pathology, Auburn University, Auburn, Alabama

Nancy Kokalis-Burelle, Ph.D.   U.S. Horticultural Research Lab, Agricultural Research Service, U.S. Department of Agriculture, Fort Pierce, Florida

Krishnamurthy Konduru, Ph.D.   Department of Entomology and Plant Pathology, Oklahoma State University, Stillwater, Oklahoma

Lise Korsten, Ph.D.   Department of Microbiology and Plant Pathology, University of Pretoria, Pretoria, South Africa

Mark Mazzola, Ph.D.   Agricultural Research Service, U.S. Department of Agriculture, Wenatchee, Washington

D. Mohanraj, Ph.D.   Crop Protection Division, Sugarcane Breeding Institute, Indian Council of Agricultural Research, Coimbatore, Tamil Nadu, India

Kalyan K. Mondal, Ph.D.   Department of Plant Pathology, Regional Research Substation, Chaudhary Sarwan Kumar Himachal Pradesh Krishi Vishwavidyalaya, Sangla, Kinnaur, Himachal Pradesh, India

Bonnie H. Ownley, Ph.D.   Department of Entomology and Plant Pathology, Institute of Agriculture, University of Tennessee, Knoxville, Tennessee

P. Padmanaban, M.Sc.(Ag), Ph.D.   Crop Protection Division, Sugarcane Breeding Institute, Indian Council of Agricultural Research, Coimbatore, Tamil Nadu, India

Corné M. J. Pieterse, Ph.D. Faculty of Biology, Utrecht University, Utrecht, The Netherlands

Appa Rao Podile, Ph.D. Department of Plant Sciences, University of Hyderabad, Hyderabad, Andhra Pradesh, India

V. Brindha Priyadarisini, Ph.D. Center for Advanced Studies in Botany, University of Madras–Guindy, Chennai, Tamil Nadu, India

Munagala S. Reddy, Ph.D. Department of Entomology and Plant Pathology, Auburn University, Auburn, Alabama

Krishnapillai Sivasithamparam, Ph.D. Department of Soil Science and Plant Nutrition, University of Western Australia, Nedlands, Western Australia, Australia

Joseph L. Smilanick, Ph.D. San Joaquin Valley Agricultural Sciences Center, Agricultural Research Service, U.S. Department of Agriculture, Parlier, California

W. Uddin, Ph.D. Department of Plant Pathology, Pennsylvania State University, University Park, Pennsylvania

L. C. van Loon, Ph.D. Faculty of Biology, Utrecht University, Utrecht, The Netherlands

Preeti Vasudevan, Ph.D. Center for Advanced Studies in Botany, University of Madras–Guindy, Chennai, Tamil Nadu, India

Jeevan P. Verma, Ph.D. National Academy of Agricultural Sciences, Indian Agricultural Research Institute, New Delhi, India

G. Viji, Ph.D. Department of Plant Pathology, Pennsylvania State University, University Park, Pennsylvania

R. Viswanathan, Ph.D. Crop Protection Division, Sugarcane Breeding Institute, Indian Council of Agricultural Research, Coimbatore, Tamil Nadu, India

Prem Warrior, Ph.D. Global Research, Valent BioSciences Corporation, Long Grove, Illinois

Charles L. Wilson, Ph.D. Appalachian Fruit Research Station, Agricultural Research Service, U.S. Department of Agriculture, Kearneysville, West Virginia

Michael Wisniewski, Ph.D. Appalachian Fruit Research Station, Agricultural Research Service, U.S. Department of Agriculture, Kearneysville, West Virginia

X. B. Yang, Ph.D. Department of Plant Pathology, Iowa State University, Ames, Iowa

1
Principles of Biological Control

Samuel S. Gnanamanickam and Preeti Vasudevan
University of Madras–Guindy, Chennai, Tamil Nadu, India

Munagala S. Reddy and Joseph W. Kloepper
Auburn University, Auburn, Alabama

Geneviève Défago
Institute of Plant Sciences, Eidgenössische Technische Hochschule, Zurich, Switzerland

I. INTRODUCTION

The term *biological control* was coined by the late Harry Smith of the University of California, who defined it as "the suppression of insect populations by the actions of their native or introduced enemies". There have been debates regarding the scope and definition of biological control ever since, mainly to accommodate the technological advances in the tools available for pest management. The definition presented by Van Drieshce and Bellows [1], "the use of parasitoid, predator, pathogen, antagonist, or competitor populations to suppress a pest population making it less abundant and thus less damaging than it would otherwise be," appears to convey the current thinking on biological control by those researchers involved in insect research.

In their earlier definition of biological control, Baker and Cook [2] described it as "the reduction of inoculum density or disease-producing activities of a pathogen or a parasite in its active or dormant state, by one or more organisms, accomplished naturally or through manipulation of the environment, host or antagonist, or by mass introduction of one or more antagonists." This definition was subsequently reworded to read "the reduction in the amount of inoculum or disease-producing activity of a pathogen accomplished by or through one or more organisms other than man" [3]. These are perhaps the most widely quoted and accepted definitions of biological control.

Some time later, the U.S. National Academy of Sciences introduced some modifications to the definition, referring to biological control as "the use of natural or modified organisms, genes or gene products to reduce the effects of undesirable organisms and to favour desirable organisms such as crops, beneficial insects and microorganisms" [4]. Obviously, due consideration has been given to the advances by the advent of molecular biology to plant pathology and to research in biological control.

There have been other recent definitions, many of which still reflect the basic idea presented in the classical definitions of biological control. According to Shurtleff and Averre [5], biological management or control refers to "disease or pest control through counter balance of microorganisms and other natural components of the environment. It involves the control of pests (bacteria, fungi, insects, mites, nematodes, rodents, weeds, etc.,) by means of living predators, parasites, disease-producing organisms, competitive microorganisms, and decomposing plant material, which reduce the population of the pathogen." Agrios [6] defined biological control as the total or partial destruction of pathogen populations by other organisms.

All these modifications introduced to the basic definition of biological control by different workers indicate its changing scope and perspective. This is largely attributed to the insights obtained by the use of various molecular and biochemical tools for the study of this hitherto poorly understood phenomenon involving interactions within a multicomponent system. Therefore, biological control of plant pathogens has now emerged as a broad concept, evident in the accounts and examples presented by different contributors in this volume, and encompasses several mechanisms. We have, in fact, included the development of transgenic plants and biologically induced systemic resistance (ISR) in hosts under the broad umbrella of biological control.

II. TYPES OF BIOLOGICAL CONTROL

Although biological control broadly refers to the use of living organisms to curtail the growth and proliferation of other, undesirable ones, it may specifically involve one or more of the following strategies.

Importation biological control is often referred to as "classical biological control," reflecting the historical predominance of this approach. It generally involves importation and establishment of a nonnative natural enemy population for suppression of nonnative or native organisms.

Augmentation biological control includes activities in which natural enemy populations are increased through mass culture, periodic release (either inoculative or inundative), and colonization for suppression of native or exotic pests. Inoculative releases are intended to colonize natural enemies early in a crop cycle so that they and their offspring will provide pest suppression for an extended period of time. Inundative releases are conducted to provide rapid pest suppression by the released individuals alone, with no expectation of suppression by their offspring. These two approaches represent extremes on a continuum of activities, with most augmentative releases being a hybrid of the two [7].

Conservation biological control can be defined as the study and modification of human influences that allow natural enemies to realize their potential to suppress pests. While augmentation deals with laboratory-reared natural enemies or microbial antagonists, conservation deals with resident enemy populations. Therefore, it involves (1) identification and remediation of negative influences that suppress natural enemies and (2) enhancement of systems (e.g., agricultural fields) as habitats for natural enemies.

It has to be understood, however, that there are major differences between biological control of weeds (plants) and plant pathogens. While the former attempts to favor a pathogen at the expense of the plant [8], the latter suppresses the pathogen so as to favor the plant. Hence, the concepts of biological control used in entomology can apply to the control of nematodes and parasitic seed plants and not to other pathogens [3].

III. CURRENT CONCEPTS OR APPROACHES TO IMPLEMENT BIOLOGICAL CONTROL

A. Introduction of Microbes in the Phylloplane, Rhizosphere, or Soil

Whereas biological control of insect pests commonly involves foreign exploration and importation of natural enemies of the insect, obtained from the original home of the insect, biological control of plant pathogens and plant parasitic nematodes commonly involves the enhancement or augmentation of antagonists already present on the plant or in the soil but at populations too low to provide adequate or timely biological control.

This practice aims at biological protection against infection. Protection of planting material (crown-gall control) and roots with biological seed treatments (PGPR and root-colonizing *Pseudomonas fluorescens/P. putida* including Pf2-79) as well as the deployment of *Streptomyces* 5406 used in millions of hectares of cotton in China are ideal examples of the successful adoption of this form of plant protection.

Following are some examples in the suppression of major diseases by the introduction of plant-associated antagonists into the rice rhizosphere (as a seed treatment) and phylloplane (as foliar sprays): control of rice blast [9,10] sheath blight [11,12], sheath rot [13], and bacterial blight [14].

Biological protection against infection is accomplished by destroying the existing inoculum, by preventing the formation of additional inoculum, or by weakening and displacement of the existing virulent pathogen population. While the former two are ensured during the augmentation of antagonists, the latter is achieved by reduction of vigor or virulence of the pathogen by agents such as mycoviruses or hypovirulence determinants.

B. Stimulating Indigenous Antagonists

Some agents that are closely related to pathogens such as avirulent or dsRNA-infected hypovirulent strains, epiphytes and endophytes may contribute to making soils naturally disease suppressive. Such microbial agents may be stimulated by the addition of organic amendments such as suppressive compost, thereby increasing the suppressive nature of the soil. The following strategies have been used by different researchers (listed in Ref. 15) to facilitate the enrichment, conservation, and management of resident soil and plant-associated microroganisms by (a) selective elimination of soilborne plant pathogens and enhancement of antagonists by steaming, sublethal fumigation, and soil solarization and (b) management of natural suppression of soilborne pathogens.

A number of mechanisms have been implicated in the suppression of soilborne plant pathogens by naturally suppressive soils or by introduced beneficial microorganisms:

1. Competition for iron: fusarium wilts of flax, carnation, siderophore-mediated control of *Gaeumannomyces graminis* var. *tritici* in wheat and barley.
2. Antibiosis: the production of antibiotics like phenazines, pyoluteorin, pyrrolnitrin, 2,4-diacetylphloroglucinol, which have been reviewed extensively by Defago et al. [16] O'Sullivan and O'Gara [17], Dowling and O'Gara [18], Cook et al. [19], and Handelsman and Stabb [20].
3. Induced resistance (reviewed in Refs. 21,22).
4. LPS derivatives [23].
5. HCN [24].
6. Use of endophytes as biocontrol agents through one or more of the interactions mentioned above.
7. Use of suppressive compost [25,26] in the control of *R. solani*, *Phytophthora*, *F. oxysporum*, and *Pythium*.
8. Ice nucleation: Ice strains of *P. syringae* to suppress ice-nucleation–active strains causing frost injury [27]. Other ice-active strains are found in *P. viridifava*, *P. fluorescens*, *X. campestris* pv. *translucens*, and *E. herbicola* in corn [28].

C. Induced Resistance

Some nonpathogenic microbial agents induce a sustainable change in the plant, resulting in an increased tolerance to subsequent infection by a pathogen. This phenomenon has been described as induced resistance. This emerging concept was discussed extensively in the first international symposium in induced resistance held in Greece in May 2000. Induced systemic resistance (ISR) has been used to describe the systemic resistance induced against pathogens by nonpathogenic or plant growth–promoting rhizobacteria [29,30]. This has been distinguished from systemic acquired resistance (SAR), which essentially refers to the host reaction in response to localized infection by pathogens, manifested as broad range of protection against other pathogens [31,32]. In some cases it has been observed that the elicitation of pathogenesis-related (PR) proteins, which are characteristic of the SAR pathway, are not always associated with ISR [33]. However, since in other cases it does appear that the same suite of genes and gene products are involved in both these phenomena, some workers still believe that the terms can be used interchangeably [34]. (For an update on our knowledge about the role of biocontrol agents in signaling resistance, see Chapter 16.)

D. Biorational Approaches

Mixtures of PGPR strains having identifiable differences in mechanisms (antibiosis, siderophores, HCN production, and induced resistance) may be used to afford increased levels of protection against pathogens [35]. A mixture of six strains of *P. putida*, *P. fluorescens*, *P. aureofaciens*, and *Serratia plymuthica* in cucumber resistance assays against anthracnose caused by *Colletotrichum lagenarium* showed that PGPR could induce systemic resistance to the pathogen that is spatially separated from the root zone.

Use of host resistance and biological control agents is an approach that combines two major strategies, host resistance and biological control. In host-pathogen systems that have a well-documented list of major R genes, it makes sense to combine the use of R genes and biocontrol agents for disease management. Resistance to rice blast caused by *Magnaporthe grisea* is governed by more than 30 major R genes [36] and resistance to bacterial blight of rice (*Xanthomonas oryzae* pv. *oryzae*) is governed by more than 20 major genes [37]. For both systems, near-isogenic rice lines (NILs) have been developed in the genomic backgrounds of CO39 for blast and IR24 for bacterial blight. In preliminary studies, when rice plants carrying the *Pi-1* gene for blast resistance were treated with *P. fluoresccens Pf7-14* and *P. putida V14-I*, they showed enhanced protection against leaf blast development [38]. Similar results were obtained for bacterial blight development when NIL IRBB4 (*Xa4*) were treated with bacterial biocontrol agents. Host resistance and biocontrol agents should complement each other in their activity against pathogens because some major genes are known to be expressed in either the seedling or adult plant stage and not throughout crop growth [37].

Smith et al. [39] established a genetic basis for interactions between recombinant inbred lines (RILs) of tomato that comprised a mapping population to *Pythium torulosum* and disease-suppressive *Bacillus cereus* UW85. Their genetic analysis revealed that three quantitative trait loci (QTL) associated with biological disease suppression explained 38% of the phenotypic variation among the RILs. This research suggests that there are new opportunities to exploit the natural genetic variation governed by major genes or QTL to develop efficient biocontrol strategies.

Data are available about a number of transgenic rice plants that have been generated to express major R genes (e.g., *Xa-21* for bacterial blight resistance) and defense genes (chitinases, glucanases, and thaumatin-like proteins) for the management of rice bacterial blight, sheath blight, and other diseases. The possibility of pyramiding these genes in transgenic plants to afford higher levels of pathogen suppression presents another good opportunity for biological disease management.

IV.   OPPORTUNITIES AND CHALLENGES

There are a range of challenges and opportunities, including discovering and implementation of microorganisms as biological control agents, whether managed as resident communities or introduced as individuals or a mixture of strains. These include scientific and technical challenges and the regulatory processes in different countries that require special reviews and tests as prerequisites for the registration of these agents as microbial pesticides [15].

In many countries as well as the European Union, biological agents are treated on a par with commercial pesticides or fungicides for the purpose of registration. Elaborate toxicological and biosafety data have to be obtained and furnished to regulatory agencies, which ultimately leads to a long waiting period before registration and a very high initial cost for the development of the product.

Different strategies for formulating biocontrol agents and the challenges associated with the process are discussed in Chapter 18. Genetic improvement of biocontrol agents offers many opportunities for their improvement while the transformation of plants with transgenes from these antagonists or other sources of resistance is also an exciting and challenging possibility.

Transgenic rice plants with transient expression of more than one defense gene is an attractive strategy for management of the rice sheath blight pathogen

Rhizoctonia solani [40,41], especially because rice (cultivated or wild) does not have adequate levels of resistance to this devastating disease.

Biological control remains a very vital disease-management strategy for developing countries, where the costs of chemical treatments can be prohibitive.

REFERENCES

1. RG Van Drieshce, TS Bellows, Jr. Pest origins, pesticides and the history of biological control. In: RG Van Drieshce, TS Bellows, Jr., eds. Biological Control. New York: Chapman and Hall, 1996, pp 3–18.
2. KF Baker, RJ Cook. Biological Control of Plant Pathogens. Amer Phytopathol Soc, St. Paul, MN, 1974, pp 35–50.
3. RJ Cook, KF Baker. The Nature and Practice of Biological Control of Plant Pathogens. Amer Phytopathol Soc, St. Paul, MN, 1983.
4. Research briefing. Report of the research briefing panel on biological control in managed ecosystems. Washington, DC: National Academy Press, 1987.
5. MC Shurtleff, CW Averre, III. Glossary of Plant Pathological Terms. APS Press, St. Paul, MN, 1997.
6. GN Agrios. Plant Pathology, 4th ed. San Diego, CA: Academic Press, 1997.
7. DB Orr, CP-C Suh. Parasitoids and predators. In: JE Rechcigl, NA Rechcigl, eds. Biological and Biotechnological Control of Insect Pests. Boca Raton, FL: Lewis Publishers, CRC Press, 2000, pp 3–24.
8. R Charudattan, HL Walker, eds. Biological Control of Weeds with Plant Pathogens. John Wiley and Sons, 1982.
9. SS Gnanamanickam, TW Mew. Biological control of blast disease of rice (*Oryza sativa*, L.) with antagonistic bacteria and its mediation by a *Pseudomonas* antibiotic. Ann Phytopath Soc Jpn 58:380–385, 1992.
10. A Chatterjee, RM Valasubramanian, AK Vachhani, SS Gnanamanickam, AK Chatterjee. Isolation of ant mutants of *Pseudomonas fluorescens* strain Pf7–14 altered in antibiotic production, cloning of ant⁺ DNA and evaluation of the role of antibiotic production in the control of blast and sheath blight of rice. Biol Control 7:185–195, 1996.
11. TW Mew, AM Rosales. Bacterization of rice plants for control of sheath blight caused by *Rhizoctonia solani*. Phytopathology 76:1260–1264, 1986.
12. KV Thara, SS Gnanamanickam. Biological control of rice sheath blight in India: lack of correlation between chitinase production by bacterial antagonists and sheath blight suppression. Plant Soil 160:277–280, 1994.
13. N Sakthivel, SS Gnanamanickam. Evaluation of *Pseudomonas fluorescens* for suppression of sheath rot disease and for enhancement of grain yields in rice, *Oryza sativa* L. Appl Environ Microbiol 53:2056–2059, 1987.
14. P Vasudevan, SS Gnanamanickam. Progress and prospects for biological suppression of rice diseases with bacterial antagonists. In: Proceedings of national symposium on biological control and plant growth promoting rhizobacteria (PGPR) for sustainable agriculture, Hyderabad, India, April 3–4, 2000.

Gnanamanickam et al.

15. EC Tjamos, GC Papavizas, RJ Cook. Biological Control of Plant Diseases: Progress and Challenges for the Future. New York: Plenum Press, 1992.
16. G Defago, D Haas. Pseudomonads as antagonists of soil-borne plant pathogens: modes of action and genetic analysis. In: JM Bollag, G Stotzky, eds. Soil Biochemistry. New York: Marcel Dekker, Inc., 1990, pp 249–291.
17. DJ O'Sullivan, F O'Gara. The traits of fluorescent *Pseudomonas* spp. involved in suppression of plant root pathogens. Microbiol Rev 56:662–676, 1992.
18. DN Dowling, F O'Gara. Metabolites of *Pseudomonas* involved in the biocontrol of plant disease. Trends Biotechnol, 12:133–141, 1994.
19. RJ Cook, LS Thomashow, DM Weller, D Fujimoto, M Mazzola, G Bangera, D Kim. Molecular mechanism of defense by rhizobacteria against root diseases. Proc Natl Acad Sci USA 92:4197–4201, 1995.
20. J Handelsman, EV Stabb. Biocontrol of soilborne plant pathogens. Plant Cell 8: 1855–1869, 1996.
21. J Kuc. Induced immunity to plant disease. Bioscience 32:854–860, 1982.
22. LC Van Loon, PAHM Bakker, CMJ Pieterse. Systemic resistance induced by rhizosphere bacteria. Ann Rev Phytopathol 36:453–483, 1998.
23. M Reitz, K Rudolph, I Schoroder, S Hoffmann-Hergaeten, J Hallmann, RA Sikora. Lipopolysaccharides of *Rhizobium etli* strain G12 act in potato roots as an inducing agent of systemic resistance to infection by the cyst nematode, *Globodera pallida*. Appl Environ Microbiol 66:3515–3518, 2000.
24. C Voisard, C Keel, D Haas, G Defago. Cyanide production of *Pseudomonas fluorescens* helps suppress black root rot of tobacco under gnotobiotic conditions. EMBO J 8:351–358, 1989.
25. HAJ Hoitink, Y Inbar, MJ Boehm. Status of compost-amended potting mixes naturally suppressive to soilborne diseases of floricultural crops. Plant Dis 75:869–873, 1991.
26. HAJ Hiotink. Composted bark, a light weighted growth medium with fungicidal properties. Plant Dis. 66:1106–1112, 1980.
27. SE Lindow, DC Arny, CD Upper. *Erwinia herbicola*: a bacterial ice nucleus active in increasing frost injury to corn. Phytopathology 68:523–527, 1978.
28. SE Lindow, DC Arny, CD Upper. Distribution of ice nucleation active bacteria on plants in nature. Appl Environ Microbiol 36:831–838, 1978.
29. CMJ Pieterse, SCM van Wees, E Hoffland, JA van Pelt, LC van Loon. Systemic resistance in *Arabidopsis* induced by biocontrol bacteria is independent of salicylic acid accumulation and pathogenesis-related gene expression. Plant Cell 8:1225–1237, 1996.
30. E Hoffland, J Hakulinen, JA van Pelt. Comparison of systemic resistance induced by avirulent and non-pathogenic *Pseudomonas* species. Phytopathology 86:757–762, 1996.
31. JA Ryals, UH Neuenschwander, MG Willits, A Molina, H-Y Steiner, MD Hunt. Systemic acquired resistance. Plant Cell 8:1809–1819, 1996.
32. L Sticher, B Mauch-Mani, JP Metraux. Systemic acquired resistance. Annu Rev Phytopathol 35:235–270, 1997.
33. HGM Linthorst. Pathogenesis-related proteins in plants. Crit Rev Plant Sci 10:123–150, 1991.

34. JW Kloepper, MS Reddy. Use of plant growth-promoting rhizobacteria (PGPR) to enhance plant growth and induce systemic disease resistance. In: Proceedings of national symposium on biological control and plant growth promoting rhizobacteria (PGPR) for sustainable agriculture, Hyderabad, India, April 3–4, 2000.

35. JW Kloepper, G Wei, S Tuzun. Rhizosphere population dynamics and internal colonisation of cucumber by plant-growth promoting rhizobacteria which induce systemic resistance to *Colletotrichum orbiculare*. In: ES Tjamos, ed. Biological Control of Plant Diseases. New York: Plenum Press, 1992.

36. L Babujee, SS Gnanamanickam. Molecular tools for the characterization of rice blast pathogen (*Magnaporthe grisea*) population and molecular marker-assisted breeding for disease resistance. Curr Sci 78:248–257, 2000.

37. SS Gnanamanickam, VB Priyadarisini, NN Narayanan, P Vasudevan, S Kavitha. An overview of bacterial blight disease of rice and strategies for its management. Curr Sci 77:1435–1444, 1999.

38. K Krishnamurthy. Biological control of rice blast and sheath blight with *Pseudomonas* spp: survival and migration of the biocontrol agents and the induction of systemic resistance in biological disease suppression. Ph.D. thesis, University of Madras, 1997.

39. KP Smith, J Handelsman, RM Goodman. Genetic basis in plants for interactions with disease suppressive bacteria. Proc Natl Acad Sci USA 96:4786–4790, 1999.

40. W Lin, CS Anuratha, K Datta, I Potrykus, S Muthukrishnan, SK Datta. Genetic engineering of rice for resistance to sheath blight. Bio/Technology 13:686–691, 1995.

41. K Datta, R Velazhagan, N Oliva, T Mew, GS Khush, S Muthukrishnan, SK Datta. Overexpression of cloned rice thaumatin-like protein (PR-5) gene in transgenic rice plants enhances environmental friendly resistance to *Rhizoctonia solani* causing sheath blight disease. Theor Appl Genet 98:1138–1145, 1999.

2
Biological Control of Rice Diseases

Preeti Vasudevan, S. Kavitha, V. Brindha Priyadarisini, Lavanya Babujee, and Samuel S. Gnanamanickam
University of Madras–Guindy, Chennai, Tamil Nadu, India

I. INTRODUCTION

The early 1960s witnessed a dramatic increase in the global population and a mood of despair regarding the world's ability to cope with the food–population balance. Concerned about this impending crisis, several organizations sought to promote rice cultivation in the developing, rice-growing regions of the world. High-yielding, fertilizer-responsive cultivars like IR-8, which could be grown throughout the year, were therefore introduced for cultivation in several parts of Asia, where 92% of the world's rice is grown. Today, 23 countries contribute more than 1 million tons of rice in the global scenario.

This acceleration of growth in rice production has been possible largely due to the replacement of traditional agricultural practices by modern ones. While the deployment of high-yielding cultivars contributed to a direct enhancement of grain yield, the development of effective strategies for pest and disease management minimized crop losses and consequently increased the net availability of rice.

Annual losses of up to 40% are said to be incurred due to biotic stresses like insect pests, pathogens, and weeds, and more than half of the world's rice crop is estimated to be lost [1]. Among several diseases caused by bacterial, fungal, and viral pathogens that devastate rice yields all over the world, *Magnaporthe grisea* (rice blast), *Rhizoctonia solani* (sheath blight), *Sarocladium oryzae* (sheath rot), *Xanthomonas oryzae* pv. *oryzae* (bacterial leaf blight), and the rice tungro virus (tungro disease) are considered serious constraints.

Disease-management strategies aimed at reducing losses and averting outbreak of epidemics have been developed in the past and have been in use ever since. The use of chemicals, especially compounds containing mercury and copper, and diverse antibiotics were widely resorted to in order to achieve high levels of disease suppression. However, the persistent, injudicious use of chemicals has been discouraged owing to their toxic effects on nontarget organisms and due to the undesirable changes they inflict upon the environment. Many of these chemicals are also too expensive for the resource-poor farmers of Asia. Though the exploitation of host resistance and introgression of R-genes into local high-yielding cultivars appear promising, the large-scale and long-term use of resistant cultivars is bound to result in significant shifts in the virulence characteristics of pathogens, culminating in resistance breakdown.

Biological control therefore assumes special significance in being an eco-friendly and cost-effective strategy for disease management, which can also be used in integration with other strategies to afford greater levels of protection and sustain rice yields. Though fungi, viruses, insects, or any organism (other than the damaged host or the pathogen causing the disease) can be used as agents mediating biological control, bacterial antagonists to various plant pathogens have received enormous attention and are considered ideal candidates for biological control for obvious reasons such as rapid growth, easy handling, and aggressive colonization of the rhizosphere. These bacteria may mediate biocontrol by the production of antibiotics, iron-scavenging siderophores, lytic enzymes, and microbial cyanides or by initiating a cascade of events resulting in the induction of systemic resistance in the host [2]. Efficient bacterial antagonists to many rice pathogens affording appreciable levels of disease suppression have been identified, and a few strains demonstrating potential for very high levels of suppression have been studied in detail. Also, the discovery that many of these antagonists may function as plant growth–promoting rhizobacteria (PGPR), contributing directly to the enhancement of plant growth and health, has aroused enormous interest in biological control and favors its deployment as a sound, ecology-conscious strategy for disease management. This chapter aims at highlighting the usefulness of biological control agents in suppressing some major diseases of rice.

II. CANDIDATES FOR BIOLOGICAL CONTROL OF RICE PATHOGENS

Diverse groups of microbes exist in nature. Obviously, biological control agents are not limited to any specific group, but very few groups of microbes have received attention and have been widely acclaimed as ideal candidates for biocontrol. Bacterial antagonists in general, *Pseudomonas* and *Bacillus* in particular, are thought to be the most appealing candidates for biological control [3]. Bacilli are gram-positive endospore-producing bacteria that are tolerant to heat and desiccation, a feature that makes them very attractive for effective deployment. The pseudomonads are gram-negative rods and have simple nutritional requirements. They are known to be excellent colonizers and are widely prevalent in the rice rhizosphere [4]. This chapter describes the use of bacterial biocontrol agents for rice disease management.

Both fluorescent and nonfluorescent bacteria have been implicated in the suppression of rice diseases (Table 1). A number of antagonistic bacteria identified from the rice rhizosphere soils of upland and lowland fields, diseased and healthy plants, and from rice field flood waters [5] have been broadly categorized as fluorescent or nonfluorescent strains. Among them, 91% of the former and 33% of the latter inhibited mycelial growth of *R. solani* in vitro. When used for seed bacterization, these strains reduced rice sheath blight (ShB) severity in greenhouse and field tests.

Rosales et al. [6] have identified different groups of bacterial antagonists for seedborne, foliar, and sclerotium-forming rice pathogens. These antagonists belonged to the genera *Bacillus, Pseudomonas, Serratia,* and *Erwinia.* All of 23 antagonists screened inhibited mycelial growth of *R. solani,* while few of them could inhibit growth of other fungal pathogens like *Sclerotium oryzae, Helminthosporium oryzae, Pyricularia oryzae, Sarocladium oryzae,* and *Fusarium moniliforme.*

Also, studies from our laboratory have revealed that a large number of bacterial strains possess the ability to protect rice plants from diseases such as blast [7], sheath blight [8,9], sheath rot [10], and stem rot [11]. About 40 bacterial isolates antagonistic to the rice bacterial blight pathogen were identified through dual plate assays (Fig. 1). Treatment of susceptible rice plants with these antagonists appears to effect statistically significant reductions in bacterial blight lesion lengths (Table 2), as evident from field assays [12]. Also, there seems to exist a direct correlation between the endophytic survival of a biocontrol agent applied to rice plants through various treatments and the extent of bacterial blight suppression by *Pseudomonas putida* strain V14i [13]. The exploitation of endophytes for biocontrol is therefore an exciting possibility, especially for the control of vascular pathogens [14–16].

A limited number of fungal antagonists have also been reported against rice pathogens, particularly in the control of *R. solani,* among which *Trichoderma* spp. [17], *Penicillium, Myrothecium verrucaria, Chaetomium globosum,* and *Laerisaria arvalis* are known to be efficient [18].

III. SUPPRESSION OF SOME MAJOR RICE DISEASES BY BIOLOGICAL AGENTS AND MODES OF DISEASE REDUCTION

The assessment of the potential of any organism to bring about disease suppression demands that it be applied to the plant system in suitable form. This may

Table 1 Major Rice Diseases: Their Causal Organisms, Symptoms, and Biocontrol Agents

Disease	Causal organism	Symptoms	Biocontrol agents	Ref.
Blast	*Pyricularia oryzae* Teliomorph: *Magnaporthe grisea*	Small, water-soaked, whitish-gray spots that enlarge, forming elliptical lesion with pointed ends and brown margins. In the pannicle, the neck blackens and rots, causing it to fall as the grain sets.	*Pseudomonas fluorescens*	Gnanamanickam and Mew, 1992; Rosales et al., 1993; Valasubramanian, 1994
Brown spot	*Drechslera oryzae/ Helminthosporium oryzae* Teliomorph: *Cochliobolus miyabeanus*	Initiated as a brown, oval, or round spot on leaf or sheath. Spots coalesce in severe cases. Black or brown spots seen on glumes as well.	*Pseudomonas P. aeruginosa Bacillus* sp. *B. subtilis*	Rosales et al., 1993
Bacterial blight	*Xanthomonas oryzae* pv. *oryzae*	Dull green water-soaked lesions on leaf tip or margin that enlarges, forming characteristic yellow lesions with wavy margins that extend up to the sheath. Leaves later turn straw-like and blight.	*Bacillus* sp.	Vasudevan and Gnanamanickam, 2000
Stem rot	*Sclerotium oryzae/ Helminthosporium sigmoideum* Teliomorph: *Nakatara sigmoidea/ Magnaporthe salvinii*	Small black irregular lesions on the outer leaf sheath. Lesions advance to the inner leaf sheath and leaf rots. Internodes rot causing stem to lodge.	*P. fluorescens P. aeruginosa B. subtilis B. pumilus*	Elangovan and Gnanamanickam, 1992; Rosales et al., 1993

Disease	Pathogen	Symptoms	Biocontrol agent	References
Sheath blight	*Rhizoctonia solani* Teliomorph: *Thanatephorus sasaleii*	Ellipsoid to ovoid lesions on leaf sheath with a grayish-white center and brown margin. Lesions enlarge and extend to the leaves, causing severe blight symptoms. Sclerotia formed during advanced stages.	*P. fluorescens* *P. putida* *P. fluorescens* *Bacillus* sp. *B. subtilis* *B. laterosporus* *B. pumilus* *Serratia marcescens* *Pseudomonas* *P. aeruginosa*	Vasantha Devi et al., 1989; Thara, 1994; Rosales et al., 1993; Mew and Rosales, 1986
Sheath rot	*Sarocladium oryzae*	Rotting of the uppermost leaf sheath enclosing young panicles. Lesions begin as oblong, irregular spots with brown margins and gray centers that enlarge and coalesce later. Young panicles remain within sheath or only partially emerge.	*P. fluorescens* *B. subtilis* *P. aeruginosa* *Pseudomonas*	Sakthivel, 1987; Rosales et al., 1993; Sakthivel and Gnanamanickam, 1987
Tungro disease	RTV Vector: *Nephotettix virescens*	Stunting, discoloration of leaves in various shades of yellow to orange. Young leaves mottled; old leaves have rusty specks of various sizes.	*P. fluorescens* (for green leaf hopper vector)	Ganesan, 1999

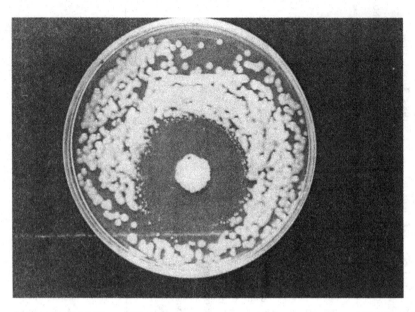

Figure 1 Inhibition of *Xanthomonas oryzae* pv. *oryzae*, the rice bacterial blight pathogen, by bacterial antagonists in dual plate assays.

be achieved either by direct inoculation (dipping seeds in a bacterial culture, aerial spraying, or spreading the bacteria in sowing furrows by drip systems) or by the use of various solid-phase inoculants.

In our laboratory bacterization is achieved by application of biocontrol agents to the seeds before sowing and/or as foliar sprays to the foliage of the plant. Surface-sterilized seeds are dipped overnight in bacterial suspensions (OD at 600 nm = 0.1) prepared in 0.5–1% carboxymethylcellulose for seed bacterization and are then air-dried before sowing. This suspension may also be used to bacterize plants as foliar sprays. Biocontrol agents thus applied have been effective in bringing about significant suppression of several serious foliar rice diseases. Greenhouse and field experiments for sheath rot suppression revealed that *P. fluorescens* treatments applied to the seeds and rice plants prior to inoculation with the pathogen could reduce disease severity by 20–42% in five rice cultivars tested. Bacterization of rice plants was also found to enhance plant height, number of tillers, and grain yields from 3 to 160% [19]. Seed treatments with bacterial antagonists to *Fusarium monoliforme* causing the rice bakane disease could bring about 72–96% disease reduction in seedbed experiments [20].

Similarly, experiments for the biological control of the bacterial blight pathogen revealed that different species of *Bacillus* applied to rice plants as a

Table 2 Suppression of Rice Bacterial Blight
by Treatments with Bacterial Antagonists in Field
Experiments, Conducted at Water Resources Department,
Anna University, Chennai, India

Bacterial isolate	Mean lesion length (cm)	Difference of lesion length from control (cm)	LSD value 1%	5%
Mon#2-16	6.00 R	7.02**	1.20	1.59
M1	7.91 R	5.11**	1.81	2.44
VyI19	6.91 R	6.11**	1.40	1.87
Al23	8.03 R	4.99**	1.47	2.00
Mon5	11.80 S	1.22	1.50	2.01
Alp18	8.73 R	4.29**	1.94	2.61
M9	10.70 S	2.32**	1.72	2.30
Mon13	10.96 S	2.06*	1.76	2.35
VyII17	10.92 S	2.10**	1.31	1.74
Cal9	8.50 R	4.55**	0.99	1.31
Mon#2-17	10.67 S	2.35**	1.23	1.63
M3	10.99 S	2.03*	1.56	2.09
M5	12.05 S	0.97	1.36	1.82
M16	9.56 S	3.46**	2.22	3.00
F1	9.08 R	3.94**	1.94	2.62
VyI18	10.64 S	2.38*	2.16	2.91
Pat8	9.08 R	3.94**	1.12	1.49
M13	11.10 S	1.92*	1.80	2.56
M11	10.58 S	2.44*	1.85	2.50
Nel16	10.56 S	2.46**	1.56	2.09
Control	13.02 S	—	—	—

* Reduction in lesion length significant at 5%; **reduction in lesion
length significant at 1%.
R, Disease reaction resulting in <10 cm bacterial blight lesion length
(resistant); S, a susceptible reaction characterized by lesion length of
>10 cm.
Source: Refs. 12, 21.

seed treatment before sowing, a root dip prior to transplantation, and two foliar
sprays prior to inoculation could afford up to 59% suppression of the disease.
These treatments could also bring about a twofold increase in plant height and
grain yield [21]. Efforts are underway to characterize the mechanism(s) mediating
the biological suppression of bacterial blight disease.

Recent work from our laboratory has demonstrated the insecticidal activity of *P. fluorescens* strains on the rice green leaf hopper (GLH), *Nephotettix virescens*. The GLH is the insect vector of the rice tungro virus (RTV) causing Tungro disease. This is one of the most devastating rice diseases, whose success-ful management has been a challenge to rice growers all over the world. Bacterial strains of *Pf*7–14 and *Pp*V14i showed maximum toxicity to the insect vector, bringing about death of 90% of the insects that were fed on bacteria-treated rice leaves for 7 days. Thin sections of the gut and eyes of the dead insects, however, did not reveal the presence of the bacteria [22]. The death of the insect vector therefore appears to be mediated by certain toxic substances produced by the bacteria. The possibility of using such bacteria to control vectors and thereby reduce incidence of the disease, however, needs to be explored, and intensive studies with regard to the feasibility of such an approach must be carefully as-sessed.

A. Mechanisms of Sheath Blight Suppression

Native bacterial strains, both fluorescent and nonfluorescent, antagonistic to the rice sheath blight pathogen have been identified and analyzed for the mechanism mediating control. Since chitin forms a major component of the cell wall of the pathogen, it was hypothesized that chitinase producers may be efficient antago-nists of *R. solani*. Therefore, as many as 1409 strains were screened for their ability to produce chitinase in the laboratory [9]. Most fluorescent pseudomonads were very poor chitinase producers, while nonfluorescent bacteria like *Bacillus* sp. produced appreciable levels of chitinase [23]. However, seedbed and field experiments revealed that seed bacterization with fluorescent pseudomonads [24], particularly strains of *P. fluorescens* and *P. putida* V14i, were most efficient in reducing ShB severities by 68 and 52%, respectively (Table 3). This lack of correlation between chitinase production and disease suppression suggests that other mechanism(s) may be involved in ShB suppression [25]. A more recent investigation has suggested that the induction of systemic resistance in rice by these fluorescent pseudomonads may also contribute to ShB suppression [26]. Therefore, protection afforded by biocontrol agents may not really be exemplified by just one factor, but is usually a consorted effect of more than one mechanism.

Indeed, antifungal metabolites produced by biocontrol agents have been implicated in the control of the ShB pathogen. Metabolites like phenazines, pyr-rolnitrin, and 2,4-diacetylphloroglucinol were identified in the culture filtrates of different strains of *Pseudomonas* by standard chromatographic methods [27]. These are thought to be responsible for the control of *R. solani* in vitro and for affording limited protection of seedlings against the pathogen [6].

Disease suppression by biocontrol agents is governed by a multitude of factors. These factors influencing biocontrol are bound to vary with the type of biocontrol agent used and the nature of the pathogen targeted for control. There-

Table 3 Suppression of Rice Sheath Blight in IR50 Rice by Bacterial Treatments in the Field, Pattambi, Kerala, Southern India

Strain	R. solani (mm)	Chitinase production	Disease suppression (%)
P. putida V14I	15	0	67.6
P. fluorescens V20d	16	0	35.2
P. fluorescens U113b	25	0	52.1
P. fluorescens U113c	25	0	50.9
P. fluorescens F012	11	0	45.8
P. fluorescens F006	8	2	51.8
Bacillus NF403	5	5	36.3
Validamycin	—	—	26.5
Check	—	—	0
		LSD (0.05) = 6.2	
		LSD (0.01) = 14.5	

Source: Ref. 25.

fore, systematic studies that evaluate the relative importance of various factors in modulating disease suppression by biological agents is essential for the successful deployment of biological control as a disease-management strategy. One such attempt to understand the influence of soil factors and cultural practice on the biological control of ShB with antagonistic bacteria was made by Gnanaman-ickam et al. [28]. The results of this study revealed that acidic soils of pH 5.0 and boron toxicity favored biocontrol. Also, bacterial treatments afforded greater protection in direct-seeded rice than in transplanted rice, which may be attributed to a greater crop canopy in the former, creating an ecological niche that favors colonization and epiphytic survival of the introduced agents.

Trichoderma spp. elicit biocontrol mainly by being mycoparasites and by being aggressive competitors of the pathogen. Several workers have observed growth and coiling of the mycelia of Trichoderma on the host hyphae. The susceptible hyphae usually become vacuolated, collapse, and finally disintegrate. The mycoparasite then grows on the hyphal contents. Species of Trichoderma are also known to produce antibiotics at low pH, which may mediate biocontrol in some cases. Also, T. hamatum and T. harzianum produce lytic enzymes like chitinases and glucanases that attack both the hyphae (resulting in exolysis) and the sclerotia of R. solani. However, the extent of ShB suppression by the fungus is influenced by several parameters, the most important being soil pH. Disease suppression is greatly reduced if the soil pH is about neutral and the maintenance of a threshold population of the fungus in the rhizosphere is impaired to a great extent at higher pH. Antibiotics like viridin and gliotoxin produced by *Gliocladium* spp. are thought to mediate their antagonism against *R. solani* [18].

B. Mechanisms of Blast Suppression

Rice blast is a serious production constraint in most rice-growing regions of the world. Extensive screening of a large number of rice rhizosphere–associated bacteria for antagonism against the rice blast pathogen has been carried out in our laboratory. Some of these strains have demonstrated up to 80% reduction of leaf blast in susceptible rice crops [7].

Strains of *Pseudomonas fluorescens* were probably the first agents of biocontrol identified against the blast pathogen [29]. A strain designated *Pf* 7–14 isolated from the rice rhizosphere afforded significant levels of blast suppression in field experiments. A preliminary effort was also made to characterize the mechanism involved in disease suppression by this promising biocontrol agent. Siderophore production, one of the most important mechanisms known to mediate bacterial antagonism to fungi, could not have been involved here, as the rice soils in which these experiments were carried out were highly acidic (pH 4). Moreover, Fe amendments to the medium did not reverse the antagonism of this strain to the blast fungus. Antibiotics partially purified from the culture filtrate of the bacterium by sephadex column chromatography were implicated in blast suppression These antifungal antibiotics (afa) afforded 70–100% inhibition of condial germination of *P. grisea* at 1.0 ppm concentration [30]. The exact chemical nature of this antibiotic was not elucidated.

A genetic analysis of the genes encoding production of this afa was performed. A genomic library of *Pf* 7–14 was constructed in a cosmid vector pLARF5 [31]. Recombinant cosmids carrying DNA fragments of ca. 20–30 kb were mobilized into mutants deficient in antibiotic production. Three different classes of mutants were generated. The Class I mutants, totally deficient in antifungal antibiotics (ant[-]), were generated by ethyl methyl sulfonate (EMS) or transposon mutagenesis using mini-Tn5-Km. Class II mutants obtained by transposon mutagenesis produced reduced levels of antibiotics (ant[leaky]), while class III mutants generated by mutagenesis with EMS overproduce antibiotics (ant[hyper]). Complementation analysis using the former two classes of mutants revealed that five cosmids restored antibiotic production with different complementation patterns. Two cosmids, one that restored antibiotic phenotype in both classes of mutants (pAKC908) and another that restored the antibiotic phenotype as well as exo-protease activity and the wrinkled colony morphology in their respective mutants (pAKC902), were subsequently analyzed. Mutagenesis of the cosmid pAKC908 with Tn3HoHo1 followed by complementation analyses and Southern hybridization revealed that the genes required for antibiotic production are clustered together [7,32].

*Pf*7–14 is known to produce several antifungal metabolites [6], and it is likely that each of these metabolites requires a different set of genes for its synthesis. This supports the hypothesis that tight ant⁻ mutants could have resulted from a defect in the common regulatory gene responsible for the synthesis of precursors of all these metabolites. *gacA* is one such global regulatory gene known to control antibiotic production in other strains of *Pseudomonas*. However, no homology could be detected between *gacA* and the insert DNAs of the cosmids that restored antibiotic production in mutants upon Southern hybridization even under conditions of high stringency. Moreover, mobilization of *gacA*⁺ plasmid into afa mutants could not restore antibiotic production. Homologs of *lemA* [33] or *phz* [34] could not be detected under similar conditions. It is believed that a comparison of the sequence data of the ant⁺ region of *Pf*7–14 with that of such known global regulators will provide useful information in this regard [32].

The *Pf*7–14 mutants, lacking either totally or partially the ability to produce antifungal metabolites generated by transposon mutagenesis (miniTm5-Km), were tested in field experiments along with the wild strain in an attempt to compare their efficiency in mediating blast suppression (Fig. 2). While the wild-type strain afforded 79 and 82% leaf and neck blast reductions, respectively, the mutants afforded a mere 24–40% and 3–25% suppression of leaf and neck

Figure 2 Suppression of neck blast by *Pseudomonas fluorescens* strain *Pf*7–14 (plants on the right) and lack of protection against neck blast in plants treated with its afa⁻ mutant (plants on the left).

blast (Table 4). Also, the wild-type and mutant strains of $Pf7$–14 controlled ShB by 82 and 10%, respectively. It was also observed that the reductions in blast and sheath blight mediated by $Pf7$–14 in the field was better than those achieved by treatment with a commercial fungicide tricyclazole [7]. Antifungal antibiotics produced by the bacterium were therefore thought to be responsible for biological control of blast by $Pf7$–14.

The direct evidence for the involvement of an antibiotic produced by a biological control agent in disease suppression nevertheless comes with detection of the antibiotic in question in the rhizosphere of the plant system to which the agent has been applied. Perhaps the first of its kind was a high performance liquid chromatography–based assay developed by Thomashow et al. [35]. A similar approach was followed in our laboratory to detect the production of antibiotics produced by *P. fluorescens* strain *Pf*cp in the rice rhizosphere. This strain was originally isolated from citrus leaves [19] and produced phenazine carboxylic acid (PCA). Samples of rice rhizosphere obtained from IR50 rice plants treated with this strain were washed in phosphate buffer and used for isolation of antibiotics by extraction with methylene chloride. The crude antibiotic was subsequently purified by column chromatography. The biologically active fractions thus obtained had spectral properties similar to those of authentic PCA. Also, the severity of blast disease was reduced by up to 60% in plants that received treatments of *P. fluorescens* *Pf*cp [36]. This study provides unequivocal evidence for the production of PCA in vivo and its association in blast suppression.

Many fluorescent pseudomonads and other PGPRs have been reported to induce systemic resistance in different plant systems against their respective pathogens [37–43]. A recent study from our laboratory suggests that induced systemic resistance (ISR) in rice triggered in response to treatments with $Pf7$–14 and PpV14i is an important mechanism in the biological suppression of blast

Table 4 Evaluation of $Pf7$–14 and Its Mini-Tn5 Mutants in the Field for the Suppression of Leaf Blast in IR50 Rice

Bacterial strain	Leaf blast	
	Percent disease incidence	Percent disease control
$Pf7$–14	21.30	78.70
Afa⁻ mutant AC2003	75.30	24.70
Afa^{leaky} mutant AC2007	52.90	47.10
Fungicide (tricyclazole)	31.00	69.00

Source: Ref. 7.

[44]. While an increase in the endogenous levels of salicylic acid was detected, the rice phytoalexin momilactone-A could not be detected in rice plants treated with either strain of bacteria. This increase in SA levels by bacteria-induced ISR was found to contribute to rice blast suppression by 25% [45].

Several workers have attempted the induction of host resistance against rice blast disease. In fact, avirulent strains of *P. grisea* have been used to induce resistance in rice against the highly virulent strains of the pathogen [46–50]. A recent report suggests that treatment of rice plants with strains of avirulent *P. grisea* and isolates of *Bipolaris sorokiniana* could bring about substantial reduction of leaf blast in greenhouse experiments. Application of *B. sorokiniana* to rice plants at the four-leaf stage systemically reduced the disease in leaf 5. A significant reduction in disease severity was also observed in field experiments [51].

IV. MARKING SYSTEMS FOR TRACKING OF INTRODUCED BACTERIAL BIOCONTROL AGENTS IN THE RICE RHIZOSPHERE

Colonization of the rhizosphere and migration to aerial parts of the plant are important prerequisites for the mediation of disease suppression by introduced antagonists. It is therefore imperative to confirm the presence of the introduced antagonist and determine its population dynamics in the rhizosphere in order for a meaningful correlation between the biocontrol agent and the extent of disease suppression to be established. In the past, researchers have relied upon phenotypic traits such as antibiotic resistance as markers for their introduced strains. However, the major disadvantage associated with the use of antibiotic markers is that native populations more often mimic the phenotypic traits of the introduced organisms. This can be overcome by using marker genes that encode unique, easily distinguishable phenotypes in biocontrol agents. The expression of these marker genes will help distinguish introduced strains from the background microflora [52,53]. Such marker genes, apart from detecting introduced organisms in the rhizosphere, will also aid in analyzing their migration, survival, and endophytic presence within the host system [45]. Bioluminescent markers such as *lux* genes [54] and chromogenic markers like *xylE* [55] have been used in the past for this purpose.

The *lacZY* gene constitutes another very useful chromogenic marking system. The stable insertion of the *lacZ* and *lacY* gene from *E. coli* K12 into the chromosome of *Pseudomonas* spp. using a disarmed Tn 7 derivative [56] favors the utilization of lactose as a carbon source, a trait that has not been encountered in the pseudomonads. Also, the *lacZY* gene product (β-galactosidase), cleaves the chromogenic dye X-gal (5-bromo-4-chloro-3-indolyl-β-galactopyranoside), resulting in the development of a blue coloration, and therefore facilitates selection of marked organisms on selection media.

Mageswari and Gnanamanickam [57] developed such a tracking system in *Pp*V14i, an agent for control of *R. solani*. The *lacZY* genes were inserted into the chromosome of this strain using a disarmed transposon (Tn:*lacZY*) by triparental mating. The plasmids pMON7181 and pMON7117 containing the Tn7::*lacZY* genes were mobilized from the donor, *E. coli* DH5α, into *Pp*V14I in the presence of a helper plasmid pRK2013 in *E. coli* HB101. Transconjugants were selected on M9 medium supplemented with 1% lactose, streptomycin, and X-gal and was used to monitor the survival and migration of *Pp*V14I-gal on rice tissues. The insertion of the *lacZY* gene did not affect the biocontrol efficacy of the strain. A similar molecular tracking system was also developed for *Pf*7–14, a potential biocontrol agent of the blast pathogen [45].

V. FORMULATIONS OF BIOCONTROL AGENTS FOR THE CONTROL OF RICE PATHOGENS

The success of a biocontrol program depends largely on the ability of the introduced agent to establish itself in the new environment and maintain a threshold population on the planting material or rhizosphere. Also, the commercial application of biological control and its implementation as a farm-level strategy demands that these agents be preserved in a viable state for long periods of time and be designed to tolerate desiccation and other physiological stresses associated with transport, storage, and application. Therefore, the development of cost-effective formulations of biocontrol agents that are easy to handle and have no adverse effects on seed germination or plant growth is essential.

Studies from our laboratory have revealed that among formulations of *Pp*V14I and *Pf*7–14 with eight different combinations of methylcellulose (mc), talc, and $CaCO_3$ in different proportions, the combination of mc:talc (1:4) emerged as the most satisfactory. The bacteria survived on this formulation for up to 10 months [45]. Formulated *Pp*V14I applied as seed treatment, root dip, and foliar sprays effected ShB suppression of up to 60% [58]. Similarly, a formulation of *Pf*7–14 applied as seed and multiple foliar sprays afforded 60 and 72% suppression of leaf and neck blast, respectively.

Seed treatments with a formulation of the marked *Pf*7–14 strain have also provided useful insights into its survival and migration in rice tissues. Their persistence on rice roots for up to 110 days (almost the entire cropping period) can be correlated with the high levels of disease suppression encountered upon treatment [59]. The limited ability of *Pf*7–14 to migrate to aerial parts of rice plants (until 7–9 days after emergence) suggests that a direct contact between the bicontrol agent and the pathogen may not exist. Disease suppression in such circumstances may be attributed to a systemic resistance induced by the agent in rice [45] or to the production of potent antifungal antibiotics in the rhizosphere and their transport to aerial parts.

VI. TRANSGENIC RICE FOR DISEASE MANAGEMENT

The advent of plant biotechnology and genetic engineering has opened up new vistas for addressing the problem of disease management and the improvement of crop productivity and has facilitated the introduction of agronomically important traits into desirable cultivars. Indeed, several workers have contributed to the establishment of efficient gene-delivery systems [60], which have in turn resulted in the development of disease-resistant transgenic rice lines.

Viral diseases of rice are by and large the most difficult to suppress by use of traditional management practices. The development of transgenic lines for virus resistance may, apart from conventional breeding, be one of the most reliable means of curtailing yield losses. Pathogen-derived resistance (PDR), involving the expression of pathogen-derived transgenes in plants to interrupt the virus infection cycle, is therefore gaining enormous attention. Transgenic rice lines for resistance against several viral rice pathogens like rice stripe virus (RSV), rice hoja blanca virus (RHBV), and rice ragged stunt virus (RRSV) have been developed using this strategy [61]. Rice plants transformed with the coat or capsid protein (CP) genes of RSV, rice dwarf virus (RDV), and the rice brome mosaic virus (RBMV) show increased levels of resistance to the corresponding viral diseases [62–64]. A similar strategy is believed to be useful in generating transgenic plants resistant to the rice tungro virus [65–67]. Viral genes other than the CP may also be used to confer PDR. Transformation of susceptible rice lines with the gene encoding the RNA-dependent RNA polymerase of rice yellow mottle virus (RYMV) renders them highly resistant to attack by viral strains from different African locations [68].

Engineering rice for resistance to fungal pathogens has also been attempted. Indica rice variety IR72 expressing the transformed *bar* gene was resistant to attack by *R. solani* and showed decreased symptoms of rice blast following bialaphos treatment [69]. Transformation of rice plants with a stilbene synthase gene from grape vine under the control of its own promoter allowed the expression of other phytoalexins in addition to the rice phytoalexin momilactone, thereby demonstrating higher levels of resistance to *P. oryzae* [70]. Also, reintroduction of the rice chitinase (*cht-2* and *cht-3*) and the β-1,3- and 1,4-glucanase (*gns1*) genes under the control of an enhanced CaMV 35S promoter into elite japonica cultivars by an *Agrobacterium*-mediated transformation resulted in blast-resistant transgenic plants [71].

Several workers have attempted genetic engineering of rice for resistance to the sheath blight pathogen [72–74]. Transgenic rice lines obtained via the transformation of rice protoplasts with chitinase gene *chiI I* were resistant to *R. solani* [75]. More recently, Datta et al. [76] cloned a 1.1 kb fragment containing the coding region of a thaumatin-like protein (TLP-34) and transformed it into local Indian cultivars. The presence of the 23 kDa TLP in the transgenic lines was confirmed by Western blot. Bioassays of transgenic plants challenged with the ShB pathogen indicated that they overexpressed TLPs and demonstrated enhanced resistance to *R. solani* compared to untransformed control plants.

Rice transformed with the cloned BB R-gene *Xa-21* [77] demonstrated enhanced levels of resistance to the rice bacterial blight pathogen *X. o. oryzae*. This gene confers multi-isolate resistance [78] and is therefore expected to exclude several pathogenic races of the BB pathogen. Recent efforts to transform IR72 rice with this gene have been extremely successful and have resulted in T_1 progeny resistant to BB [79]. Elite indica rice cultivars, CO39 and IR50, have recently been introgressed with genes for blast resistance (*Pi-1* + *Pi-2*) by gene pyramiding through backcross breeding, and the pyramids have shown high levels of blast resistance in the field. An effort is in progress to transform these CO39 and IR50 pyramids with *Xa-21* gene to make them resistant to both blast and bacterial blight (N. N. Narayanan, S. K. Datta, and S. S. Gnanamanickam, unpublished results).

Genetic engineering has therefore contributed enormously to the development of disease-resistant lines and is emerging as an indispensable technique to supplement conventional breeding for resistance to viral, bacterial, and fungal rice pathogens.

VII. CONCLUSIONS AND FUTURE DIRECTIONS

Inconsistent performance of biocontrol agents in the field developed thus far has plagued researchers and their efforts to exploit them for commercial application. Therefore, there is a compelling need to identify efficient and dependable biocontrol agents to be used singly or as mixtures to ensure consistent performance in the farmer's field.

Choice of the correct microbial candidate is indeed one of the most important factors governing the success of biocontrol programs on a commercial basis. It is therefore important to consider biocontrol agents with the ability to reduce the severity of more than one pathogen, as this will make their application cost-effective. Once such agents are selected after stringent testing, it is imperative to have them formulated for high levels of consistent performance. *Bacillus* strains are attractive candidate agents because they withstand desiccation and storage conditions better than fluorescent pseudomonads or other bacteria.

Engineering rice plants with defense genes (chitinases, glucanases, and thaumatin-like proteins), a combination of defense-related or other genes that trigger the signal transduction pathway leading ultimately to the induction of systemic resistance to rice diseases, is an emerging strategy. Work in this direction will gain momentum as more and more genes of the defense pathway become known and are made available. Alternatively, bacterial strains may also be engineered for enhanced biocontrol potential by enabling heterologous expression of antibiotics or other metabolites mediating biological disease suppression.

Recent studies from our laboratory have shown that 1,2,3-benzothiodiazole-7-carbothioic acid S-methyl ester (BTH), a potent chemical inducer of plant defense [37], evokes substantial levels of induced systemic resistance (ISR) to blast, sheath blight, and bacterial blight. The use of BTH in combination with existing disease-management practices for suppressing rice diseases therefore appears very promising.

It needs to be remembered, nevertheless, that most of the world's rice farmers who live in Asia are resource-poor. Therefore, cost-effective formulations of biocontrol agents that perform consistently in their fields when made available, either by themselves or as part of an integrated disease-management (IDM) package, will benefit these low-income group rice farmers with increased grain harvests. In this lies the key to the ultimate success of biocontrol research for rice disease management.

REFERENCES

1. M Hossain. Recent developments in the Asian rice economy: Challenges for the rice research. In: RE Evenson, RW Herdt and M Hossain ed Rice Research in Asia: Progress and Priorities, Wallington, UK: CABI, 1996, 17pp.
2. J Handelsmann, EV Stabb. Biocontrol of soil borne plant pathogens. Plant Cell 8: 1855–1869, 1996.
3. DM Weller. Biological control of soil-borne plant pathogens in the rhizosphere with bacteria. Ann Rev Phytopathol 26:379–407, 1988.
4. N Sakthivel, SS Gnanamanickam. Incidence of different biovars of *Pseudomonas fluorescens* in flooded rice rhizospheres in India. Agriculture, Ecosystems and environment 25:287–298, 1989.
5. TW Mew, AM Rosales. Bacterization of rice plants for control of sheath blight caused by *Rhizoctonia solani*. Phytopathology 76:1260–1264, 1986.
6. AM Rosales, R Vantomme, J Swings, J De Ley, TW Mew. Identification of some bacteria from paddy antagonistic to several rice fungal pathogens. J Phytopathol 138:189–208, 1993.
7. R Valasubramanian. Biological control of rice blast with *Pseudomonas fluorescens* Migula: role of antifungal antibiotics in disease suppression. Ph.D. dissertation, University of Madras, 1994.
8. T Vasantha Devi, R Malarvizhi, N Sakthivel, SS Gnanamanickam. Biological con-

trol of sheath blight of rice in India with antagonistic bacteria. Plant Soil 119:325–330, 1989.

9. KV Thara. Biological control of rice sheath blight by bacterial antagonists: mechanisms of disease suppression. Ph.D. dissertation, University of Madras, 1994.

10. N Sakthivel. Biological control of *Sarocladium oryzae* (Sawada) Gams & Hawksworth, sheath-rot pathogen of rice by bacterization with *Pseudomonas fluorescens* migula. Ph.D. dissertation, University of Madras, 1987.

11. C Elangovan, SS Gnanamanickam. Incidence of *Pseudomonas fluorescens* in the rhizosphere of rice and their antagonism towards *Sclerotium oryzae*. Indian Phytopath 45:358–361, 1992.

12. SS Gnanamanickam, V Brindha Priyadarisini, NN Narayanan, Preeti Vasudevan, S Kavitha. An overview of the bacterial blight disease of rice and strategies for its management. Curr Sci 77:1435–1444, 1999.

13. S Kavitha. Suppression of bacterial blight of rice by *Pseudomonas putida* V14I and its endophytic survival in rice tissues. M. Phil. dissertation, University of Madras, 1999.

14. C Chen, EM Bauske, G Musson, R Rodriguez-Kabana, JW Kloepper. Biological control of *Fusarium* wilt on cotton by use of endophytic bacteria. Biol Control 5:83–91, 1995.

15. CP Chanway. Endophytes: they're not just fungi! Can J Bot 74:321–332, 1996.

16. J Hallmann, A Quadt-Hallmann, WF Mahaffee, JW Kloepper. Bacterial endophytes in agricultural crops. Can J Microbiol 43:895–914, 1997.

17. T Xu, GE Harman, YL Wang, Y Schen. Bioassay of *Trichoderma harzianum*: strains for control of rice sheath blight (abstr). Phytopathology 89(suppl 6): S86, 1999.

18. RJ Cook, KF Baker. The Nature and Practice of Biological Control of Plant Pathogens. St. Paul, MN: Amer Phytopathol Soc, 1983.

19. N Sakthivel, SS Gnanamanickam. Evaluation of *Pseudomonas fluorescens* for suppression of sheath rot disease and for enhancement of grain yields in rice, *Oryza sativa* L. Appl Environ Microbiol 53:2056–2059, 1987.

20. AM Rosales, TW Mew. Suppression of *Fusarium monoliforme* in rice by rice associated antagonistic bacteria. Plant Dis 81:49–52, 1997.

21. P Vasudevan, SS Gnanamanickam. Progress and prospects for biological suppression of rice diseases with bacterial antagonists. Biological control and plant growth promoting rhizobacteria (PGPR) for sustainable agriculture, Hyderabad, India, April 3–4, 2000.

22. P Ganesan. Biological control of fusarium wilts of cotton and tomato with bacterial strains of *Pseudomonas fluorescens* and *P. putida*. Ph.D. dissertation, University of Madras, 1999.

23. M Krishnaveni. Biological control of sheath blight of rice with fluorescent pseudomonads. M. Phil. dissertation, University of Madras, 1991.

24. R Malarvizhi. Biological control of sheath blight disease of rice caused by *Rhizoctonia solani* Kuhn with *Pseudomonas fluorescens*. Masters dissertation, University of Madras, 1987.

25. KV Thara, SS Gnanamanickam. Biological control of rice sheath blight in India: Lack of correlation between chitinase production by bacterial antagonists and sheath blight suppression. Plant Soil 160:277–280, 1994.

26. K Krishnamurthy, SS Gnanamanickam. Biological control of sheath blight of rice: induction of systemic resistance in rice by plant associated *Pseudomonas* spp. Curr Sci 72:331–334, 1997.
27. AM Rosales, LS Thomashow, RJ Cook, TW Mew. Isolation and identification of antifungal metabolites produced by rice associated antagonistic *Pseudomonas* spp. Phytopathology 85:1028–1032, 1995.
28. SS Gnanamanickam, BL Candole, TW Mew. Influence of soil factors and cultural practice on biological control of sheath blight of rice with antagonistic bacteria. Plant Soil 144:67–75, 1992.
29. SS Gnanamanickam, TW Mew. Biological control of rice blast with antagonistic bacteria. Int Rice Res Newslett 14:34–35, 1989.
30. SS Gnanamanickam, TW Mew. Biological control of blast disease of rice (*Oryza sativa* L.) with antagonistic bacteria and its mediation by a *Pseudomonas* antibiotic. Ann Phytopathol Soc Jpn 58:380–385, 1992.
31. NT Keen, S Tamaki, B Kobayashi, D Trollingen. Improved broad-host range plasmids for DNA cloning in gram negative bacteria. Gene 70:191–197, 1988.
32. A Chatterjee, R Valasubramanian, WL Ma, AK Vacchani, SS Gnanamanickam, AK Chatterjee. Isolation of mutants of *Pseudomonas fluorescens* strain *Pf* 7–14 altered in antibiotic production, cloning of ant⁺ DNA and evaluation of the role of antibiotic production in the control of blast and sheath blight of rice. Biol Control 7:185–195, 1996.
33. EM Hrabak, DK Willis. The *lemA* gene required for pathogenicity of *Pseudomonas syringae* pv. *syringae* on bean is a member of a family of two-component regulators. J Bacteriol 174:3011–3020, 1992.
34. LS Pierson III, VD Keppenne, DW Wood. Phenazine antibiotic biosynthesis in *Pseudomonas aureofaciens* 30–84 is regulated by *PhzR* in response to cell density. J Bacteriol 176:3966–3974, 1994.
35. LS Thomashow, DM Weller, RF Bonsall, LS Pierson. Production of the antibiotic phenazine 1-carboxylic acid by fluorescent *Pseudomonas* species in the rhizosphere of wheat. Appl Environ Microbiol 56:908–918, 1990.
36. K Krishnamurthy. Monitoring the production of antifungal antibiotic (afa) by *Pseudomonas fluorescens* in rice rhizosphere. M. Phil. dissertation, University of Madras, 1993.
37. J Ryals, S Uknes, E Ward. Systemic acquired resistance. Plant Physiol 104:1109–1112, 1994.
38. M Maurhofer, C Hase, P Meuwly, JP Metraux, G Defago. Induction of systemic resistance to tobacco necrosis virus by the root-colonizing *Pseudomonas fluorescens* strain CHAO: influence of the *gacA* gene on pyoverdine production. Phytopathology 84:139–146, 1994.
39. M Leeman, JA Van Pelt, FM Den Ouden, M Heinsbroek, PHAM Baker, B Schippers. Induction of systemic resistance against fusarium wilt of radish by lipopolysaccharides of *Pseudomonas fluorescens*. Phytopathology 85:1021–1027, 1995.
40. CMJ Pieterse, SCM Van Wees, E Hoffland, JA Van Pelt, LC Van Loon. Systemic resistance in *Arabidopsis* induced by biocontrol bacteria is independent of salicylic acid accumulation and pathogenesis-related gene expression. Plant Cell 8:1225–1237, 1996.

41. G DeMeyer, M Hofte. Salicylic acid produced by the rhizobacterium *Pseudomonas aeruginosa* 7NSK2 activate the systemic acquired resistance pathway in bean. Mol Plant Microbe Interact 12:450–458, 1997.
42. M Maurhofer, C Reimmann, P Schmidi-Socherer, S Heeb, D Haas, G Defago. Salicylic acid biosynthetic genes expressed in *Pseudomonas fluorescens* strain P3 improve the induction of systemic resistance in tobacco against tobacco necrosis virus. Phytopathology 88:678–684, 1998.
43. J Ton, CMJ Pieterse, LC Van Loon. Identification of a locus in *Arabidopsis* controlling both expression of rhizobacteria mediated induced systemic resistance (ISR) and basal resistance against *Pseudomonas syringae* pv. *tomato*. Mol Plant Microbe Interact 12:911–918, 1999.
44. K Krishnamurthy. Biological control of rice blast and sheath blight with *Pseudomonas* spp.: survival and migration of the biocontrol agents and the induction of systemic resistance in the biological disease suppression. Ph.D. dissertation, University of Madras, 1997.
45. K Krishnamurthy, SS Gnanamanickam. Induction of systemic resistance and salicylic acid accumulation in *Oryza sativa* L. in the biological suppression of rice blast caused by treatments with *Pseudomonas* spp. World J Microbiol Biotechnol 14:935–937, 1998.
46. Y Fujita, K Sonada, H Yaegashi. Leaf blast suppression by per-inoculation of some incompatible lesion-type isolates of *Pyricularia oryzae*. Ann Phytopathol Soc Jpn 56:273–275, 1990.
47. M Iwano. Suppression of rice blast infection by incompatible strain of *Pyricularia oryzae*. Cav Bull Tohoku Natl Agric Exp Stn 75:27–39, 1987.
48. S Kiyosawa, H Fujimaki. Studies on mixture inoculation of *Pyricularia oryzae* on rice. I. Effects of mixture inoculation and concentration on the formation of susceptible lesion in the injection inoculation. Bull Natl Inst Agric Sci 7:1–19, 1967.
49. SK Park, KC Kim. Effects of mixing and reciprocal inoculation with compatible and incompatible races of *Pyricularia oryzae* on the enlargement of disease lesions of the rice blast. Korean J Plant Prot 22:300–306, 1983.
50. G Sun, S Sun. Inducing rice blast (bl) resistance by inoculating with an incompatible race of *Pyricularia oryzae*. Int Rice Res Newslett 17:10–11, 1992.
51. HK Manandhar, HJL Jorgensen, SB Mathur, SV Petersen. Suppression of rice blast by pre-inoculation with avirulent *Pyricularia oryzae* and the non-rice pathogen *Bipolaris sorokiniana*. Phytopathology 88:735–739, 1998.
52. DA Kluepfel. The behavior and tracking of bacteria in the rhizosphere. Annu Rev Phytopathol 31:441–472, 1993.
53. FAAM Deleij, EJ Sutton, JM Whipps, JS Fenlon, JM Lynnch. Impact of field release of genetically modified *Pseudomonas fluorescens* on indigenous microbial population of wheat. Appl Environ Microbiol 61:3443–3453, 1995.
54. GASB Stewart, P Williams. *Lux* genes and the applications of bacterial luminescence. J Gen Microbiol 138:1289–1300, 1992.
55. C Winstanley, AW Morgan, RW Pickup, JR Saunders. Use of a *XylE* marker gene to monitor survival of recombinant *Pseudomonas putida* population in Lake Water by culture on non-selective media. Appl Environ Microbiol 57:1905–1913, 1991.

56. GF Barry. Permanent insertion of foreign genes into the chromosomes of soil bacteria. Biotechnology 4:446–449, 1986.
57. S Mageswari, SS Gnanamanickam. Use of molecular tracking system to study the survival and migration of *Pseudomonas putida*, a biocontrol agent for sheath blight disease of rice. Indian Phytopath 50:469–473, 1997.
58. K Krishnamurthy, SS Gnanamanickam. Biocontrol of rice sheath blight with formulated *Pseudomonas putida*. Indian Phytopath 51:233–236, 1998.
59. K Krishnamurthy, SS Gnanamanickam. Biological control of rice blast by *Pseudomonas fluorescens* strain *Pf* 7–14: evaluation of a marker gene and formulations. Biol Control 13:158–165, 1998.
60. AK Tyagi, A Mohanty, S Bajaj, A Chaudhury, SC Maheshwari. Transgenic rice: a valuable monocot system for crop improvement and gene research. Crit Rev Biotechnol 19:41–79, 1999.
61. NM Upadhyaya, Z Li, MB Wang, S Chen, ZX Gong, PM Waterhouse. Engineering for virus resistance in rice. 4th Int'l Rice Genetics Symposium, IRRI, Philippines, Oct. 22–27, 2000.
62. T Hayakawa, Y Zhu, K Itoh, Y Kimura, T Izawa, K Shimamoto, S Toriyama. Genetically engineered rice resistant to rice stripe virus, an insect-transmitted virus. Proc Natl Acad Sci USA 89:9865–9869, 1992.
63. CC Huntley, TC Hall. Interference with brome mosaic virus replication in transgenic rice. Mol Plant Microbe Interact 9:164–170, 1996.
64. HH Zheng, Y Li, ZH Yu, W Li, MY Chen, XT Ming, R Casper, ZL Chen. Recovery of transgenic rice plants expressing the rice dwarf virus outer coat protein gene (S8). Theor Appl Genet 94:522–527, 1997.
65. CM Fauquet, H Huet, CA Ong, E Sivamani, L Chen, P Viegas, VP Marmey, P Wang, M Daud, A de Kochko, RN Beachy. Control of rice tungro disease by genetic engineering is now a reality! General meeting of the Int'l program on Rice Biotechnol, Malacca, Sept. 15–17, 1997.
66. I Potrykus, PK Burkhardt, SK Datta, J Futterer, GC Gosh Biswas, A Kloti, G Spangenberg, J Wunn. Genetic engineering of indica rice in support of sustained production of affordable and high-quality food in developing countries. Euphytica 85:441–449, 1995.
67. E Sivamani, H Huet, P Shen, CA Ong, A Kochko, C Fauquet, RN Beachy. Rice plant (*Oryza sativa* L.) containing rice tungro spherical virus (RTSV) coat protein transgenes resistant to virus infection. Mol Breeding 5:177–185, 1999.
68. YM Pinto, RA Kok, DC Baulcombe. Resistance to the rice yellow mottle virus (RYMV) cultivated in African rice varieties containing RYMV transgenes. Nat Biotechnol 17:702–707, 1999.
69. Y Tada, M Nakasa, T Adachi, R Nakamura, H Shimada, M Takahashi, T Fijimura, T Matsuda. Reduction to 14 to 16 Kda allergenic proteins in transgenic rice plants by antisense gene. FEBS Lett 391:353–361, 1996.
70. Z Tang, W Tian, L Ding, S Cao, S Dai, S Ye, C Chu, L Li. Transgenic rice with a phytoalexin gene resistant to blast. 4th Int'l Rice Genetics Symposium, IRRI, Philippines, Oct. 22–27, 2000.
71. Y Nishigawa, K Nakazono, M Saruta, M Kamoshita, E Nakajima, M Ugaki, T Hibi. Characterization of blast tolerant transgenic rice constitutively expressing the chi-

tinase or the β-glucanase gene. 4th Int'l Rice Genetics Symposium, IRRI, Philippines, Oct. 22–27, 2000.

72. R Velazhahan, KC Cole, CS Anuratha, S Muthukrishnan. Induction of thaumatin-like proteins (TLPs) in *Rhizoctonia solani*-infected rice and characterization of two new cDNA clones. Plant Physiol 102:21, 1998.

73. K Datta, MF Alam, N Oliva, A Vasquez, E Abrigo, T Mew, S Muthukrishnan, SK Datta. Expression of chitinase and thaumatin-like protein gene in transgenic rice for sheath blight resistance. General meeting of the Int'l Program on Rice Biotechnol, Malacca, Sept. 15–17, 1997.

74. N Baisak. Improvement of rainfed lowland indica rice through in vitro culture and genetic engineering. Ph.D. dissertation, Utkal University, India, 2000.

75. W Lin, CS Anuratha, K Datta, I Potrykus, S Muthukrishnan, SK Datta. Genetic engineering of rice for resistance to sheath blight. Bio/Technology 13:686–691, 1995.

76. K Datta, R Velazhahan, N Oliva, I Ona, T Mew, GS Khush, S Muthukrishnan, SK Datta. Over expression of the cloned rice thaumatin-like protein (PR-5) gene in transgenic rice plants enhances environmental friendly resistance to *Rhizoctonia solani* causing sheath blight disease. Theor Appl Genet 98:1138–1145, 1999.

77. WY Song, GL Wang, LL Chen, HS Kim, LY Pi, T Holsten, J Gardener, B Wang, WX Zhai, LH Zhu, C Fauquet, PC Ronald. A receptor kinase-like protein encoded by the rice disease resistance gene *Xa 21*. Science 270:1804–1806, 1995.

78. GL Wang, WY Song, DL Ruan, S Sideris, PC Ronald. The cloned gene, *Xa 21* confers resistance to multiple *Xanthomonas oryzae* pv. *oryzae* isolates in transgenic plants. Mol Plant Microbe Interact 9:850–855, 1996.

79. J Tu, J Ona, Q Zhang, TW Mew, GS Khush, SK Datta. Transgenic rice variety IR72 with *Xa 21* is resistance to bacterial blight. Theor Appl Genet 97:31–36, 1998.

3

Transgenic Plants for the Management of Plant Diseases: Rice, a Case Study

Swapan K. Datta
International Rice Research Institute, Metro Manila, Philippines

I. INTRODUCTION

The occurrence of plant diseases and their counterattack for self-defense might have evolved simultaneously. Every plant species possesses its own immune system, which signals a cellular response and leads to the death of the attacking pathogen. Plants recognize pathogen-encoded molecules through probable receptors encoded by disease resistance (*R*) genes. A signal transduction study provides an excellent understanding of gene-for-gene resistance, which explains plant pathogen co-evolution in a given environment [1]. Plant diseases cause billions of dollars in crop losses annually. Yield loss in rice alone is enormous. About 20% of total yield is lost due to biotic stresses, including severe diseases such as blast, sheath blight, bacterial blight, tungro, etc. Disease control implies that it is based on the principle of maintaining yield loss below an economic injury level. In most cases, agrochemicals such as fungicides/pesticides and biological control including crop rotation are used to control diseases. Developing varieties with disease resistance will most likely provide the best solution for disease control. This approach is inexpensive and environmentally friendly, and management would be easier than before. The classic *R* gene defined by plant breeders is now isolated and characterized as a cloned gene, and plant biotechnologists can transfer *R* genes along with pathogenesis-related genes into many crop plants, including rice [2–4]. *R* and *PR* genes are listed in Tables 1–3. A few selected areas such as transgenic research with *R* and *PR* genes and expression of those genes in transgenic crop plants leading to strategic management of diseases are stressed here.

Table 1 Classes of Cloned Plant Disease Resistance Genes

Class	R gene	Plant	Pathogen	Avr gene	Structure	Ref.
1	Hm1	Maize	Cochliobolus carbonum (race 1)	None	HC-toxin reductase	13
2	Pto	Tomato	Pseudomonas syringae pv. tomato	avrPto	Serine/threonine protein-kinase	14
3a	RPS2	Arabidopsis	P. syringae pv. tomato	avrRpt2	LZ-NBS-LRR	45
	RPM1	Arabidopsis	P. syringae pv. maculicola	avrRpm1/avrB	LZ-NBS-LRR	46
	Prf	Tomato	P. syringae pv. tomato	avrPto	LZ-NBS-LRR	47
	I2	Tomato	Fusarium oxysporum f. sp. lycopersici	Unknown	LZ-NBS-LRR	48
3b	N	Tobacco	Tobacco mosaic virus	TMV replicase?	TIR-NBS-LRR	49
	L6	Flax	Melamsora lini	AL6	TIR-NBS-LRR	50
	RPP5	Arabidopsis	Peronospora parasitica	AvrPp5	TIR-NBS-LRR	51
3c	Xa1	Rice	Xanthomonas oryzae pv. oryzae	Unknown	NBS-LRR	52
3d	Mlo1	Barley	Erysiphe gramini f. sp hordei		NBS-LRR	53
4	Cf-9	Tomato	Cladosporium fulvum	Avr9	LRR-TM	54
	Cf-2	Tomato	C. fulvum	Avr9	LRR-TM	55
5	Xa21	Rice	X. oryzae pv. oryzae	Unknown	LRR, protein kinase	39
6	Stv-bi	Rice	Rice stripe virus	Unknown	Unknown	56

HC = Helminthosporium; LRR = leucine-rich repeat; LZ = leucine zipper; NBS = nucleotide binding site; TIR = Toll-IL-1R homology region; TM = transmembrane.

Table 2 Pathogenesis-Related Protein Families in Plants (Recognized and Proposed)

Family	Representative plant	Molecular weight (kDa)	Biochemical properties	Gene symbol
PR-1	Tobacco	14,15,17	Antifungal, unknown	*ypr1*
PR-2	Tobacco	31, 33, 35	β-1,3-Glucanase	*ypr2*, [*gns2*]
PR-3	Tobacco	27,28,32,34	Chitinase	*ypr3*, *chia*
PR-4	Tobacco	13,15,20	Similar to potato proteins	*ypr4*, *chid*
PR-5	Tobacco	24,26	Thaumatin-like	*ypr5*
	Rice	23		
PR-6	Tomato	8,13	Proteinase inhibitor	*ypr6*, *pis* ('*pin*')
PR-7	Tomato	69	Endo-proteinase	*ypr*
PR-8	Cucumber	28	Chitinase	*ypr8*, *chib*
PR-9	Tobacco	39,40	Lignin-forming peroxidase	*ypr9*
PR-10	Parsley	17–19	Ribonuclease-like	*ypr10*
PR-11	Tobacco	41,43	Class V chitinase	*ypr11*, *chic*
Others				
Thionins	Barley	5	Antimicrobial	
Plant defensins	Radish	5	Antifungal	

Plants use a variety of strategies to protect against pathogen attack. Plant protection is manifested by a single gene or group of genes to make it work in coordination [5]. Resistance genes are regulatory in nature, whereas defense genes are functional. However, *R* genes regulate the functions of defense genes. Defense genes are usually quiescent in healthy plants but become activated when a pathogen comes in contact with the plant, which then releases a signal [6].

When a plant *R* gene interacts with the corresponding avirulence gene (*Avr*) of the pathogen, this triggers a series of defense responses (Fig. 1). Some common features appear:

1. The cell wall is strengthened to create a barrier between the plant cell and the pathogen.
2. Localized cell death occurs at the infection site, known as hypersensitive response (HR). HR is usually correlated with a transient burst of active oxygen species and accumulation of defense-related gene products [7,8]. HR is particularly associated with increased resistance throughout the plant against subsequent pathogen attacks [9]. This phe-

Table 3 Pathogenesis-Related Proteins and Genes in Grains

PR-protein family	Class/subfamily/ enzyme activity	Name	Protein cDNA/gene	Tissue of expression	Induced by	Ref.
Rice						
PR-1	Acidic	16.5 kDa JIP	P	Roots	Jasmonic acid	57
PR-2	Basic glucanase		P	Grain	Developmental	58
PR-2	Acidic glucanase		P	Bran	Developmental	59
PR-2	Glucanase		P	Leaves	Stress	60
PR-3	Class Ib chitinase		P	Bran	Developmental	61
PR-3	Class I chitinase		P	Leaves	Pathogen	62
PR-3	Class II chitinase		P	Leaves	Pathogen	62
PR-3	Class Ib chitinase	RCH-A, RCH-B	P	Suspension cells	Oligo (NAG)	63
PR-3	Class III chitinase	RCH-C	P	Suspension cells	Oligo (NAG)	63
PR-3	Class Ia chitinase	Chi11	G	Seeds		64
PR-3	Class I chitinase	2-2W	C	Seeds	Developmental	61
PR-3	Class I chitinase	RC-7	C	Leaves	Pathogen	62
PR-5	TLP		P	Leaves	Jasmonic acid, stress	60
PR-5	TLP	pPIR2	C	Leaves	Pathogen	65
PR-5	TLP	C22	C	Leaves	Pathogen	66
PR-5	TLP	D34	C	Leaves	Pathogen	66
PR-6	Cystatin	OC-1	P	Seeds	Developmental	67
PR-6	Cystatin	OC-2	P	Seeds	Developmental	67
PR-6	Cystatin	OC-26	C	Seeds	Developmental	67
PR-6	Cystatin	OC-9b	C	Seeds	Developmental	68
PR-9	Peroxidase	PO-C1	P	Seedlings	Pathogen	69
PR-9	Peroxidase		P	Roots	Jasmonic acid	57
PR-10	RNAse	Osdrr	P	Roots	Jasmonic acid	57

Wheat						
PR-2	Basic glucanase		P	Grain	Developmental	70
PR-2	Subfamily B	LW2	C	Aleurone layer	Developmental	70
PR-2	Glucanase		P	Leaves	Pathogen	71
PR-2	Glucanase	Clone SM289	C	Spikelet	Pathogen	72
PR-2	Glucanase	Clone SM638	C	Spikelet	Pathogen	72
PR-3	Class Ib chitinase		P	Germ	Developmental	73
PR-3	Chitinase		P	Leaves	Aphid infestation	74
PR-3	Chitinase		P	Leaves	Pathogen	71
PR-3	Class IV chitinase	SM383	C	Spikelet	Pathogen	72
PR-3	Class VII chitinase	SM194	C	Spikelet	Pathogen	72
PR-5	TLP	Trimatin	P	Seeds	Developmental	75
PR-5	TLP	WAS-3	P	Suspension cells		76
PR-6	Bowman-Birk	Wali5	C	Roots	Al toxicity	77
PR-6	Trypsin inhibitor		P	Seedlings	Pathogen, salicylic acid	78
PR-9	Peroxidase		P	Leaves	Aphid infestation	74
PR-9	Peroxidase		C	Leaves	Pathogen	79
PR-9	Peroxidase	pox1	C	Roots		80
PR-9	Peroxidase	pox2	C	Roots, leaves	Pathogen (leaves)	80
PR-9	Peroxidase	pox3	C	Leaves		80
PR-9	Peroxidase	pox4	C	Roots		80

Table 3 Continued

PR-protein family	Class/subfamily/ enzyme activity	Name	Protein cDNA/gene	Tissue of expression	Induced by	Ref.
Barley						
PR-1	Basic	pHvPR-1a	C	Leaves	Pathogen	81
PR-1	Basic	pHvPR-1b	C	Leaves	Pathogen	81
PR-1		bpr-1	G		Pathogen	82
PR-2	Basic glucanase	G1-GIII	P	Leaves	Developmental	83
PR-2	Glucanase	BHV-V	C	Roots, leaves	Developmental	84
PR-2	Glucanase		P	Leaves	Pathogen	85
PR-2	Glucanase	GII	P	Grain	Developmental	86
PR-3	Class II chitinase	Chitinase C	P	Grain	Developmental	87
PR-3	Class I chitinase	Chitinase K	P	Grain	Developmental	88
PR-3	Class II chitinase	CH1	P	Flour	Developmental	89
PR-3	Class I chitinase	CH2	P	Flour	Developmental	89
PR-3	Class I chitinase	CH3	P	Flour	Developmental	89
PR-3	Class I chitinase	clone 10	C	Aleurone	Developmental	90
PR-3	Class II chitinase	cht2a	C		Pathogen	81
PR-4	Chitin–binding	BP-R, BP-S	P	Grain, leaves	Pathogen	91
PR-5	TLP	Hv-1	P		Pathogen	91
PR-5	TLP	Bsi1	P	Leaves	Pathogen	92
PR-6	TLP		P		Pathogen	82
PR-13	Thionin		P		Pathogen	82

	Protein	Gene	Tissue		Induction	Ref.
Sorghum						
PR-2	Glucanase		Leaves, sheath	P	Pathogen	93
PR-5	TLP		Seeds	P	Developmental	75
PR-6	Cystatin		Seedlings	C	Developmental	94
PR-10	Peroxidase		Mesocotyls	C	Nonpathogen	95
Maize						
PR-2	Glucanase		Seedlings	P	Pathogen	96
PR-2	Acidic glucanase			C		97
PR-3	Class I chitinase	ChitA, ChitB	Seeds	P	Developmental	98
PR-3	Chitinase		Seeds	G	Developmental	98
PR-3	Chitinase		Seedlings	P	Pathogen	96
PR-3	Class I chitinase		Seedlings	C		33
PR-3	Class I chitinase		Seedlings	C		33
PR-4	Chitin-binding		Grain	P	Developmental	91
PR-4	Chitin-binding		Leaves	P	Stress	91
PR-5	TLP	Zeamatin	Flour	P	Developmental	99
PR-5	TLP		Seeds	P	Developmental	100
PR-5	TLP	Zlp	Seeds	C	Developmental	101
PR-6	Bowman-Birk	WIP1	Coleoptiles	C	Wounding	102
PR-6		MPI	Embryos	C	Pathogen	103

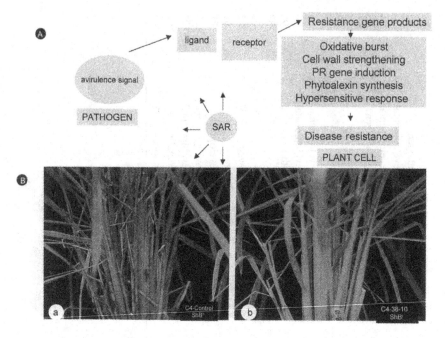

Figure 1   Signal transduction pathway leads to control of plant defense. (A) Model of signal transduction for defense response in plant cell. (B) Sheath blight bioassay showing resistance reaction of transgenic rice to *Rhizoctonia solani* Kuhn.

nomenon is termed systematic acquired resistance (SAR) and is found with the overexpression of *PR* genes and subsequent elevated levels of salicylic acid (SA). The *NPR1* gene cloned from *Arabidopsis*, shown to be independent of the *R* gene, confers resistance to the pathogens *Pseudomonas syringae* and *Peronospora parasitica*. *NPR1* appears to succeed in enhancing plant immunity by overexpressing regulatory induction of the SAR signaling pathway [10].

II. RESISTANCE, AVIRULENCE, AND TRANSGENES

The gene-for-gene hypothesis as proposed [11] can also be explained by the defense response that is often activated by the action of a host resistance (*R*) gene and a pathogen avirulence (*Avr*) gene [12].

   The first cloned *R* gene, *Hm1* from maize, was obtained through transposon tagging. *Hm1* confers resistance to race 1 strains of the fungal pathogen *Cochlio-*

bolus carbonum. Hml encodes an NADPH-dependent reductase that inactivates the potent plant toxin produced by these fungal strains [13]. However, the *Hml R* gene does not involve pathogen *Avr* genes. Nevertheless, this work outlines the first natural or engineered plant disease resistance. Chromosome walking (positional cloning) and heterologous transposon tagging made it possible to clone several *R* genes (Table 1) and enhanced plant resistance against the pathogenic fungus (Table 4). The *R/Avr* approach was reported in tomato [14]. *Pto* cloned from tomato confers resistance against *P. syringae* expressing the *Avr* gene. *Pto*

Table 4 Increased Fungal Resistance in Transgenic Plants

Plant	Transgene(s)	Pathogen	Ref.
Tobacco	PR-1a, SAR 8.2	*Peronospora tabacina, Phytophthora parasitica, Pythium*	104
	Class III chitinase	*Phytophthora parasitica*	104
	Ch-I	*Rhizoctonia solani*	104
	Bean chitinase (CH5B)	*R. solani*	105
	Barley RIP	*R. solani*	106
	Serratia marcescens Chi-A	*R. solani*	106
	Barley Chi + Glu	*R. solani*	107
	Barley Chi + RIP	*R. solani*	107
	Rice Chi + alfalfa Glu	*Cercospora nicotianae*	108
	Radish Rs-AFP	*Alternaria longipes*	109
Carrot	Tobacco Chi-I + Glu-I	*Alternaria dauci, A. radicina*	110
	Tobacco AP24	*Cercospora carotae, Erysiphe heracleï*	110
Tomato	Tobacco Chi-I + Glu-I	*Fusarium oxysporum*	111
Brassica	Bean chitinase	*Rhizoctonia solani*	105
napus	Tomato/tobacco chitinase	*Cylindrosporium* conc.	112
		Sclerotinia sclerotiorum	112
		Phoma lingam	112
Potato	AP24	*Phytophthora infestans*	113
	Glux-ox	*P. infestans*	114
		Verticillium dahliae	114
	Aly AFP	*Verticillium* sp.	115
Rice	Rice-TLP	*Rhizoctonia solani*	16
	Rice-Chi11	*R. solani*	3,18
	Rice Chi, RC7	*R. solani*	4
	Rice-Chi11	*R. solani*	19
	Rhi-Cht2, Cht3	*Magnaporthe grisea*	24
	Rice-*Pi-ta*	*M. grisea*	15
Wheat	Aly AFP	*Fusarium* sp.	116
Sorghum	Chi11	*Rhizoctonia solani*	117

encodes a protein with similarity to serine-threonine protein kinases. Further, a few *R* genes were cloned that encode proteins containing leucine-rich repeat (LRR) domains (Table 1). The gene-for-gene hypothesis has also been demonstrated in rice by Valent et al. [15].

III. *PR* PROTEIN GENES AND THEIR ROLE IN PLANT DEFENSE

Pathogenesis-related proteins were first reported 30 years ago as new protein components induced by tobacco mosaic virus (TMV) in hypersensitively reacting tobacco [2]. Since then a great deal of research has focused on the isolation, characterization, and regulation of expression of this unique class of defense proteins in a variety of plants (Table 2); selected detailed of PR proteins and genes of cereals are shown in Table 3 [16]. The major interest has focused in recent years on the realization that several PR proteins had antimicrobial and insecticidal activity. Several studies led to the conclusion that overexpression of PR proteins in transgenic crops can delay the progression of diseases caused by several patho-

Table 5 Status of Transgenic Rice Plants with Pathogenesis-Related Genes

Cultivar	Gene of interest	Method used	No. of regenerated plants	Analysis (Southern*)	Fertility status (%)
IR72	*Chi11*	B	72	60	70
	RC7	B	20	15	75
IR64	*RC7*	B	3	1	Fertile
CBII	*Chi11*	P	56	30	90
	RC7	P	232	42*	90
	D34	P	141	30*	90
ML7	*Chi11*	B	20	14	85
IRRI-NPT	*Chi11*	B	133	48*	55
Basmati 122	*Chi11*	*Agro*	45	15*	50
Tulsi	*Chi11*	*Agro*	115	111	80
Vaidehi	*Chi11*	*Agro*	64	64	90
Dinorado	*D34*	B	54	5*	80
Swarna	*Chi11*	B			90
IR58	*RC7*	B	35	29	62

B = Biolistic; P = protoplast; *Agro* = *Agrobacterium*; ML = maintainer line; IRRI-NPT = IRRI new plant type.
* All plants were not analyzed.

Table 6 Summary of *Agrobacterium*-Mediated Transformation with Different Rice Cultivars

Cultivar	Explant	*Agro* strain used	No. of explants cocultivated	No. of putative transformed calli after selections	No. of plants in greenhouse	No. of positive plants HPT+	S+
Basmati 122	SC	LBA 4404 (pNO1)	275	52 (12)	45	ND	15/27
Basmati 122	SC	A281 (pNO1)	175	40 (7)	0	—	—
Tulsi	SCM	LBA4404 (pNO1)	400	69 (22)	113	11/16	45/46
Tulsi	SCM	A281 (pNO1)	300	114 (48)	0	—	—
Vaidehi	SCM	LBA4404 (pNO1)	290	41 (15)	64	16/17	27/30

Numbers in parentheses represent the number of primary (independent) calli selected.

SC = Scutellar calli from immature embryo; SCM = scutellar calli from mature seeds; S+ = Southern positive for chitinase gene; HPT+ = Hygromycin phosphotransferase assay positive; ND = not done.

gens belonging to diverse genera (Table 4). *PR* genes are designated as *ypr* followed by the same suffix in accordance with the recommendations of the Commission for Plant Gene Nomenclature (Table 2) [2]. It is necessary to gather information at both the nucleic acid and the protein levels when dealing with stress-related proteins. Newly defined cDNAs may also be added to the existing families when shown to be induced by pathogens or specific elicitors. Defensins and thionins, both families of small, basic cysteine-rich polypeptides, qualify for inclusion as new families of PRs (Table 2). Datta et al. [17], on individual classes of PR protein (PR-1 through PR-11), provide detailed information on the isolation, characterization, and function of individual PR proteins in plants, including a chapter on transgenic research.

There is always some selectivity in the interaction between a PR protein and its intended target pathogens, in that PR proteins represent generalized plant defense responses for broad, albeit incomplete, protection against diverse pathogens (Table 4). Many transgenic plants have now been developed with constitutive inducible expression of PR proteins at effective levels and could be used as a tool to enhance or stabilize yield in areas where pathogens and pests are endemic (Tables 5, 6).

IV. EVALUATION OF *PR* GENES FOR SHEATH BLIGHT RESISTANCE IN CEREALS

Sheath blight disease of rice causes significant yield losses every year and is widespread in all rice-growing countries. Resistance breeding for this disease is not feasible because resistant germplasm is not yet known. It now seems that genetic engineering to manage sheath blight is an attractive and powerful tool by introducing *PR* genes and optimizing the overexpression of PR proteins in transgenic plants. Two different types of *PR* genes, PR3-chitinases (Chi11 and RC7) and PR-5 thaumatin-like protein genes, have been introduced into rice [3,4,16,18–20] (Tables 5, 6). Transformation was done with the biolistic, protoplast, and *Agrobacterium* systems (Fig. 2) described earlier [21–23]. Inheritance was studied by Southern blot analysis (for gene integration) and Western blot analysis with a polyclonal antibody (Fig. 3). The transformants synthesized high levels of PR proteins constitutively and exhibited enhanced resistance when challenged with the sheath blight pathogen (*Rhizoctonia solani*), (Fig. 4). At least 10 rice cultivars have been transformed with several *PR* genes, and they are now at different stages of development, awaiting homozygous status with acceptable levels of PR protein expression (Table 5, 6). A good phenotype with enhanced levels of antifungal activity is now being selected for future field testing and breeding.

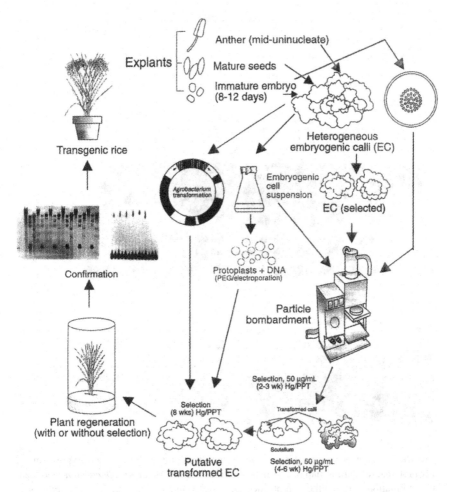

Figure 2 A schematic protocol for production of transgenic rice plants using biolistic-, protoplast-, and *Agrobacterium tumefaciens*–mediated transformation.

V. BLAST RESISTANCE

Rice blast caused by *Magnaporthe grisea* is the most devastating plant disease in Asia, particularly in areas where rice is irrigated or receives high amounts of rainfall and nitrogen fertilizer [24]. Breeders have adopted three methods to suppress blast disease. First, they tried to use varieties with field resistance. Second, they introduced resistance genes into high-yielding cultivars from other varieties.

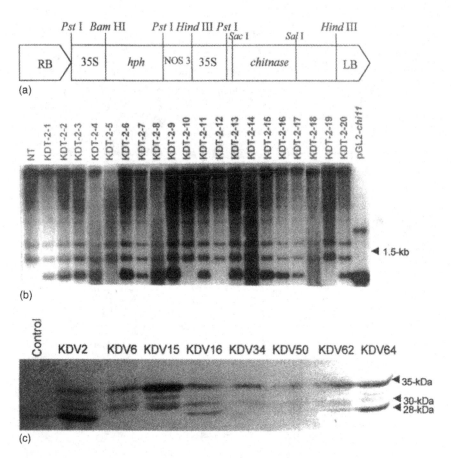

Figure 3 Transgenic rice with *chitinase* gene showing integration and expression of PR proteins. (a) Partial map of reconstructed plasmid PNO1 for *Agrobacterium*-mediated transformation. (b) Southern blot showing monogenic segregation (3:1) of chitinase transgene (*chi11*) in T₁ generation of *Agrobacterium*-derived rice cultivar Tulasi. (c) Western blot showing expression of 35 kDa protein encoded by chitinase transgene (*chi11*) in transgenic rice cultivar Vaidehi.

However, shortly after their release, these cultivars became seriously susceptible to blast disease in many areas because of the appearance of new races of the blast fungus. Thus, this showed clearly that solving the blast problem is more challenging than thought. The third approach adopted was to develop varieties showing a high level of field resistance. Genetic engineering allows shortening breeding time since the genes for a single trait would be transformed without altering the genetic make-up of the adopted cultivar. Two rice *chitinase* genes

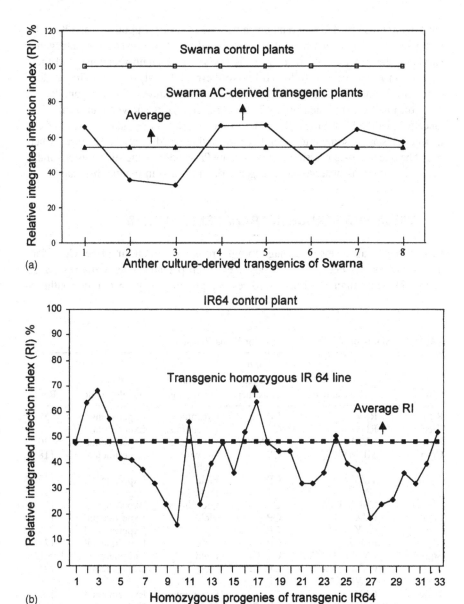

Figure 4 Transgenic rice with *RC7* gene showing enhanced resistance against sheath blight. (a) Bioassay results showing enhanced resistance of homozygous dihaploid transgenics of rice cultivar Swarna carrying chitinase transgene (*Chi11*) to sheath blight fungus under greenhouse conditions. (b) Bioassay results showing enhanced resistance of homozygous transgenics of rice cultivar IR64 carrying chitinase transgene (*RC7*) to sheath blight fungus under greenhouse conditions.

have been transferred into two japonica varieties of rice (Nipponbore and Koshih-ikari) by *Agrobacterium*-mediated transformation. The presence of transgenic *chitinase* genes was confirmed by PCR (polymerase chain reaction), and their expression in leaves was followed by Northern blot analysis [25]. The product of the *cht-2* gene was intracellular, whereas *cht-3* was extracellular in accumulation. The constitutive expression of *cht-2* and *cht-3* chitinases showed enhanced resistance against the blast fungus *M. grisea* compared to control plants. Instead of enhanced resistance, some transgenic lines showed reduced disease resistance, which might be due to co-suppression of endogenous *PR* genes or gene silencing. A similar gene silencing was observed in transgenic rice [26] and in many other plants.

VI. VIRUS RESISTANCE IN TRANSGENIC PLANTS

Plant viruses cause severe damage to numerous crops including rice. Considerable progress has been made in engineering crop plants with virus resistance (Table 7). More than $1 billion in losses is reported yearly for rice in Southeast

Table 7 Status of Transgenic Plants for Virus Resistance

Crop	Resistance	Viral transgene construct	Resistance evaluation	Status or development	Ref.
Rice	RSV	CP	Greenhouse	Experimental	31
Rice	RDV	CP	Greenhouse	Experimental	32
Rice	RYMV	RYMV	Greenhouse	Experimental	30
Corn	MDMV	CP	Growth chamber	Experimental	118
Tomato	TMV	CP	Field	Experimental	119
	CMV	CP	Field	Experimental	120
	CMV, TSWV	CP, NP	Greenhouse	Experimental	121
Potato	PVX	CP	Field	Experimental	122
	PVY	CP	Field	Experimental	123
	PLRV	CP	Field	Experimental	124
Squash	ZYMV, WMV2	CP, CP	Field	Commercial	125,126
	CMV, ZYMV, WMV2	CP, CP, CP	Field	Deregulation	125
Melon	CMV	CP	Greenhouse	Experimental	127
	CMV, ZYMV	CP, CP	Field	Experimental	128
Papaya	PRSV	CP	Field	Deregulation	129,130
Cucumber	CMV	CP	Field	Experimental	131

RSV = Rice stunt virus; RDV = rice dwarf virus; RYMV = rice yellow mosaic virus; MDMV = maise dwarf mosaic virus; TMV = tobacco mosaic virus; CMV = cucumber mosaic virus; TSWV = tomato spotted wilt virus; PVX = potato virus X; PVY = potato virus Y; PLRV = potato leafroll virus; ZYMV = zucchini yellow mosaic virus; WMV2 = watermelon mosaic virus; PRSV = papaya ringspot virus; CP = coat protein; NP = nucleoprotein.

Asia [27]. The coat protein (CP) gene of tobacco mosaic virus (TMV) is the first report of virus-derived resistance in transgenic plants [28]. Transgenic tobacco plants expressing high levels of TMVCP were more resistant to TMV virions than to TMVRNA inocula, suggesting that CP-mediated protection against TMV was through the inhibition of virion disassembly in the initially infected cells [29].

Another approach of antisense RNA in homology-dependent resistance was hypothesized. The posttranscriptional gene silencing and pathogen-derived resistance to viruses was thought to be very effective, but it has yet to be well demonstrated in protecting crops against the viruses, particularly in cereals [30] (J. Futterer, personal communication).

VII. TRANSGENIC RICE RESISTANCE TO YELLOW MOTTLE VIRUS—A CASE STUDY

Rice yellow mottle virus (RYMV) causes major yield losses in African rice production. Though endemic to Africa, RYMV is spreading in newly established, large-scale irrigated rice development schemes and experimental fields of Asian varieties. Control of this disease is difficult because the virus is highly infectious and the epidemiology and role of vectors are not well understood. Natural resistance to RYMV is found in African landraces of rice. However, the resistance is recessive and polygenic, and fertility barriers do not allow the introgression of this trait into cultivated rice. Genetic engineering based on pathogen-derived resistance was applied in this case to disrupt the pathogenesis. A transgene encoding the RNA-dependent RNA polymerase of RYMV, coupled to a 35S promoter and *hpt*, was transferred into susceptible rice cultivars. Fourteen fertile, independent transgenic lines were produced that carried the transgenes. The transformed lines were resistant to RYMV strains from different African locations. One line completely suppressed virus multiplication. Resistance was stable over the last three generations. Further, in the most resistant line, transcription analysis indicated that the resistance derives from an RNA-based mechanism associated with posttranscriptional gene silencing [31]. Some examples also showed transgenic rice conferring resistance to rice stripe tenuvirus [32] and rice dwarf virus [33].

VIII. BACTERIAL RESISTANCE IN TRANSGENIC PLANTS

Few reports are available on transgenic crops conferring resistance to bacterial pathogens. Transgenic potato and tomato plants have been developed with modified apple rootstock. Genetically modified potato containing a gene encoding glucose oxidase from *Aspergillus niger* was highly resistant to soft rot disease incited by *Erwinia carotovora* [34]. Potato plants containing a gene coding for lysozyme, a bacteriolytic enzyme from the bacteriophage T4, showed reduced susceptibility to *E. carotovora* [35]. The first report on map-based cloning and isolation of a resistance gene to combat the bacterial pathogen *Pseudomonas syringae* pv. *tomato* was demonstrated in transgenic tomato plants [14]. Several other genes now available have been shown to be effective against the bacterial pathogens described in Table 7, with great potential for future application [36].

IX. TRANSGENIC RICE DEVELOPED WITH BACTERIAL BLIGHT RESISTANCE

Bacterial blight (BB) caused by *Xanthomonas* pv. *oryzae* (Xoo) is one of the most destructive diseases of rice worldwide. Rice yield losses caused by BB in some areas of Asia can reach 50%. The use of resistant cultivars is the most economical and effective method for controlling this disease [37].

A dominant gene for resistance to BB was transferred from a wild species, *Oryza longistaminata*, to the cultivated variety IR24 [38]. The resulting line with *Xa-21* is called IRBB21. *Xa-21* confers resistance to all the known races of Xoo in India and the Philippines [39]. The molecular structure of *Xa-21* represents an uncharacteristic class of plant disease-resistance genes. From its deduced amino acid sequence, the gene was found to be translated into a receptor kinase–like protein carrying leucine-rich repeats in the putative extracellular domain, a single-pass transmembrane domain, and a serine/threonine kinase intracellular domain. Further, *Xa-21* supports a role for cellular signaling in plant disease resistance [40].

Xa-21 has been transferred to susceptible japonica rice T309, which showed resistance to BB [41]. Because T309 is not a commercial variety, we introduced the gene in elite breeding cultivars, such as IR72, MH63, IR51500, etc. Molecular analysis of transgenic plants revealed the presence of a 3.8 kb *Eco*RV-digested DNA fragment corresponding to most of the *Xa-21* coding region and its complete intron sequence, indicating the integration of *Xa-21* in the genome of rice. Transgenic plants were challenged with two prevalent races (4 and 6) of *Xanthomonas oryzae*. T_0 and T_1 plants positive for the transgene were resistant to bacterial blight [42]. We also observed that the level of resistance to race 4 of Xoo was higher due to pyramiding of *Xa-21* in addition to *Xa-4* already present in IR72. This is a very efficient way to improve BB resistance of rice without genetic dragging, and it requires less than 2 years [3].

X. FIELD EVALUATION OF TRANSGENIC BB RESISTANCE IN IR72

Based on characterization of the resistance phenotype and molecular analysis, several homozygous lines carrying *Xa-21* against the BB pathogen were obtained from previously transformed indica rice IR72. The homozygous line, T103-10, with the best phenotype and seed setting was tested repeatedly under normal field conditions to evaluate its resistance to the BB pathogen in Wuhan, China, in 1998 and 1999. The races of Xoo used in the experiments were PXO61, PXO79, PXO99, and PXO112 isolated from the Philippines, T2 isolated from Japan, and Zhe173 isolated from China. The results demonstrated that the transgenic homozygous line expressed the same resistance spectrum, but with a shorter lesion length to each inoculated race than the lesion length of the *Xa-21* donor line IRBB21 (Table 8) (Fig. 5). The nontransformed control IR72 carrying *Xa-4* was resistant to PXO61, PXO112, Zhe 173, and T2 but susceptible to PXO79 and PXO99. The negative control variety IR24 was susceptible to all isolates under

Table 8 Disease Reaction of 90 Plants of Various Lines to Different Races of *Xoo* Under Field Conditions in 1999, Huazhong Agricultural University, Wuhan, China

Xoo race	Variety	Lesion length (cm)	Reaction
Philippines race 1 (PXO 61)	IR72	1.04 ± 0.12	R
	T103	0.31 ± 0.05	HR
	IR24	16.43 ± 1.32	HS
	IRBB21	0.97 ± 0.24	HR
Philippines race 3 (PXO79)	IR72	9.00 ± 0.86	S
	T103	0.61 ± 0.24	HR
	IR24	14.11 ± 1.46	HS
	IRBB21	0.82 ± 0.40	HR
Philippines race 5 (PXO112)	IR72	0.72 ± 0.13	HR
	T103	0.39 ± 0.09	HR
	IR24	7.60 ± 1.11	S
	IRBB21	1.02 ± 0.62	R
Philippines race 6 (PXO99)	IR72	9.37 ± 1.21	S
	T103	2.43 ± 0.53	R
	IR24	15.69 ± 1.24	HS
	IRBB21	8.00 ± 1.20	MS
Chinese race 4 (Zhe 173)	IR72	1.41 ± 0.53	R
	T103	0.79 ± 0.31	HR
	IR24	11.16 ± 1.89	S
	IRBB21	1.58 ± 0.73	R
Japanese race 2 (T2)	IR72	0.71 ± 0.45	HR
	T103	0.61 ± 0.31	HR
	IR24	20.01 ± 1.89	HS
	IRBB21	1.94 ± 0.91	R

HS = Highly susceptible; S = susceptible; MS = moderately susceptible; R = resistant; HR = highly resistant.

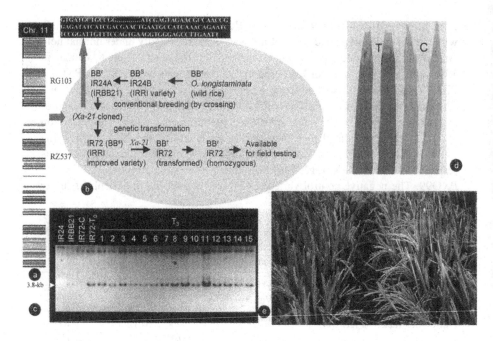

Figure 5   Engineered rice with *Xa21* is resistant to bacterial blight disease.

field conditions. The results demonstrated clearly that the *Xa-21* transgene led to an excellent field performance of the induced bacterial blight resistance trait on the recipient plants [43]. The yield performance of transgenic homozygous line T103-10 is comparable with that of the control under field conditions (Table 9).

We also noticed that an increased level of resistance to the BB pathogen persisted in transgenic plants through several generations, indicating its stable inheritance. The heritable increased level of resistance to the BB pathogen can, in turn, provide an advantage for genetic engineering over classical breeding in cases where the highest levels of resistance are desirable and can be achieved in a short time. It is also noteworthy that various national agricultural research systems in Asia are making efforts to incorporate the other *Xa* genes into popular cultivars through marker-aided selection.

The availability of various cultivars with different resistance genes could significantly decrease the yield loss. Assuming a minimum yield loss of 1% due to this disease, around $32.5 million could be saved over 30 million ha with an average yield of 5.5 t/ha in China, whereas a yield loss of 0.75% covering 132.5 million ha with an average yield of 3.6 t/ha in Asia translates into $715.5 million. Thus, transgenic rice with BB resistance would have a large economic impact.

Table 9 Agronomic Traits of Transgenic IR72 and IR72 Control Under Field Conditions, Huazhong Agricultural University, Wuhan, China, 1999

Variety/line	Days to flowering	Plant height (cm)	Panicles/ plant	Filled seeds/ panicle	Empty grains/ panicle	Total seeds/ panicle	Seed- setting rate (%)	1000-seed weight (g)	Yield (t/ha)
IR72 transgenic	98	85.6	16.2	67.9	41.1	108.9	62.4	20.8	4.89
IR72 control	96	93.6	17.1	65.0	29.8	94.8	68.5	21.3	4.97

This study shows that conventional and molecular breeding techniques could be a powerful combination in rice breeding. Genetic transformation is a one-step process of introducing novel genes into a desirable genetic background of important crops. Because it is a fast and efficient gene integration tool, it could well be the answer to catching up with the pathogen's ability to mutate quickly and render once-resistant plants susceptible. For instance, rice cultivars carrying the *Xa-4* gene for resistance, which were widely deployed in the Philippines in the early 1970s, became susceptible to the predominant race of Xoo within 5 years [44]. Transformation techniques could help to develop transgenic plants in less than 2 years to minimize the effects of a breakdown in resistance in the host plant. With the availability of resistance genes from other sources, the strategic deployment of transgenic rice with gene pyramiding may provide durable resistance in rice breeding.

XI. CONCLUSIONS

Disease resistance will be an exciting and challenging research area in the coming decades. The cloning of many specific and broad-spectrum resistance genes can be anticipated in the near future. Functional analysis will allow the dissection of molecular specificity, leading to the *explanta* generation of new transgenic plants with resistance-gene specificities. Disease-resistance genes control pathogens at a low phenotypic cost by inducing programmed cell death by hypersensitive reaction due to the response of the pathogens. A signal transduction study now provides an excellent understanding of gene-for-gene resistance, which explains plant pathogen co-evolution in a given environment. The recent breakthrough in

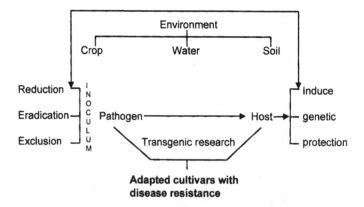

Figure 6   Integrated disease management.

cloning several *R* genes, such as *Xa-21*, shows that conventional and molecular breeding techniques could be a powerful combination in rice breeding. Genetic transformation is a one-step process of introducing novel genes into a desirable genetic background of important crops. Because it is a fast and efficient system, it will create improved varieties faster than the pathogen can overcome the resistance. Overexpression of PR genes and the combination of more than one gene will delay disease symptoms and protect plants in a sustainable manner. However, a disease is the outcome of interactions among the host, pathogen, and environment, and therefore, using all possible approaches including genetic engineering in an adopted elite variety with enhanced disease resistance is the only way to achieve durable resistance (Fig. 6).

ACKNOWLEDGMENTS

Financial support from the BMZ (Germany) and the Rockefeller Foundation are gratefully acknowledged. Thanks are due to Dr. Bill Hardy for comments and editorial assistance and to Dr. Karabi Datta, Michelle Viray, N. N. Narayanan, and Niranjan Baisakh for artwork and general assistance.

REFERENCES

1. AF Bent. Plant disease resistance genes: function meets structure. Plant Cell 8: 1757–1771, 1996.
2. LC Van Loon. Occurrence and properties of plant pathogenesis-related proteins. In: SK Datta, S Muthukrishnan, eds. Pathogenesis-Related Proteins in Plants. Boca Raton, FL: CRC Press, 1999, pp. 1–19.
3. SK Datta, S Muthukrishnan. Pathogenesis-Related Proteins in Plants. Boca Raton, FL: CRC Press, 1999.
4. SK Datta. Transgenic rice breeding: protection against bacterial blight, sheath blight, and stem borer. In: SA Leong, PS Teng, N Cattlin, eds. A Colour Handbook of Pests, Diseases, and Disorders of Rice. London: Manson Pub., 2000.
5. RP Purkayastha. Disease resistance and induced immunity in plants. Indian Phytopath 51(3):211–221, 1998.
6. P Vidhyasekaran. Molecular biology of pathogenesis and induced systematic resistance. Indian Phytopath 51(2):111–120, 1998.
7. JL Dangl, RA Dietrich, MH Richberg. Death don't have no mercy: cell death programs in plant-microbe interactions. Plant Cell 8:1793–1807, 1996.
8. KE Hammond-Kosack, JDG Jones. Resistance gene-dependent plant defense responses. Plant Cell 8:1771–1791, 1996.
9. JL Dangl. Plants just say no to pathogens. Nature 394:525–527, 1998.
10. H Cao, X Li, X Dong. Generation of broad-spectrum disease resistance by over-

expression of an essential regulatory gene in systematic acquired resistance. Proc Natl Acad Sci USA 95:6531–6536, 1998.

11. H Flor. Current status of the gene-for-gene concept. Annu Rev Phytopathol 9:275–296, 1971.

12. D Gabriel, B Rolfe. Working models of specific recognition in plant-microbe interactions. Annu Rev Phytopathol 28:365–391, 1990.

13. GS Johal, SP Briggs. Reductase activity encoded by the HM1 disease resistance gene in maize. Science 258:985–987, 1992.

14. GB Martin, SH Brommonschenkel, J Chunwongse, A Frary, MW Ganal, R Spivey, T Wu, ED Earle, SD Tanksley. Map-based cloning of a protein kinase gene conferring disease resistance in tomato. Science 262:1432–1436, 1993.

15. Y Jia, SA AcAdams, GT Bryan, P Hershey, B Valent. Direct interaction of resistance gene and avirulence gene products confers rice blast resistance. EMBO J 19(15):4004–4014, 2000.

16. SK Datta. Transgenic cereals: *Oryza sativa* (rice). In: IK Vasil, ed. Molecular Improvement of Cereal Crops. Dordrecht: Kluwer Academic Publisher, 1999, pp. 149–187.

17. K Datta, S Muthukrishnan, SK Datta. Expression and function of PR-protein genes in transgenic plants. In: SK Datta, Muthukrishnan S, eds. Pathogenesis-Related Proteins in Plants. Boca Raton, FL: CRC Press, 1999, pp. 261–277.

18. W Lin, CS Anuratha, K Datta, I Potrykus, S Muthukrishnan, SK Datta. Genetic engineering of rice for resistance to sheath blight. Bio/Technology 13:686–691, 1995.

19. N Baisakh, K Datta, N Oliva, I Ona, GJN Rao, TW Mew, SK Datta. Rapid development of homozygous transgenic rice using anther culture harboring rice *chitinase* gene for enhanced sheath blight resistance. Plant Biotechnology 18:101–108, 2001.

20. K Datta, R Velazhahan, N Oliva, T Mew, GS Khush, S Muthukrishnan, SK Datta. Over expression of cloned rice thaumatin-like protein (PR-5) gene in transgenic rice plants enhances environmental friendly resistance to *Rhizoctonia solani* causing sheath blight disease. Theor Appl Genet 98:1138–1145, 1999.

21. SK Datta, A Peterhans, K Datta, I Potrykus. Genetically engineered fertile Indica-rice plants recovered from protoplasts. Bio/Technology 8:736–740, 1990.

22. SK Datta, K Datta, N Soltanifar, G Donn, I Potrykus. Herbicide-resistant Indica rice plants from IRRI breeding line IR72 after PEG-mediated transformation of protoplasts. Plant Mol Biol 20:619–629, 1992.

23. SK Datta, L Torrizo, J Tu, N Oliva, K Datta. Production and molecular evaluation of transgenic rice plants. IRRI Discussion Paper Series No. 21. International Rice Research Institute, Manila, Philippines, 1997.

24. RS Zeigler, SA Leong, PS Teng. Rice Blast Disease. Wallingford, United Kingdom: CAB International, 1994.

25. Y Nishizawa, Z Nishio, K Nakazono, M Soma, E Nakajima, M Ugaki, T Hibi. Enhanced resistance to blast (*Magnaporthe grisea*) in transgenic japonica rice by constitutive expression of rice chitinase. Theor Appl Genet 99:383–390, 1999.

26. S Chareonpornwattana, VT Krishnarajapuran, L Wang, SK Datta, W Panbangred, S Muthukrishnan. Inheritance, expression, and silencing of a chitinase transgene in rice. Theor Appl Genet 98:371–378, 1999.

27. RW Herdt. Research priorities for rice biotechnology. In: GS Khush G Toenniessen, eds. Rice Biotechnology. Wallingford, United Kingdom: CAB International, 1991.
28. P Powell-Abel, RS Nelson, B De, N Hoffmann, SG Rogers, RT Fraley, RN Beachy. Delay of disease development in transgenic plants that express the tobacco mosaic virus coat protein gene. Science 232:738–743, 1986.
29. RN Beachy, S Loesch-Fries, NE Tumer. Coat protein-mediated resistance against virus infection. Annu Rev Pytopathol 24:451, 1990.
30. DC Baulcombe. Mechanism of pathogen-derived resistance to viruses in transgenic plants. Plant Cell 8:1833–1844, 1996.
31. YM Pinto, RA Kok, DC Baulcombe. Resistance to rice yellow mottle virus (RYMV) in cultivated African rice varieties containing RYMV transgenes. Nat Biotechnol 17:702–706, 1999.
32. T Hayakawa, Y Zhu, K Itoh, Y Kimura, T Izawa, K Shimamoto, S Toriyama. Genetically engineered rice resistant to rice stripe virus, an insect-transmitted virus. Proc Natl Acad Sci USA 89:9865–9869, 1992.
33. HH Zheng. Recovery of transgenic rice plants expressing the rice dwarf virus outer coat protein gene (S8). Theor Appl Genet 94:522–527, 1997.
34. S Wu, AL Kriz, MJ Widholm. Molecular analysis of two cDNA clones encoding acidic class I chitinase in maize. Plant Physiol 105:1097–1105, 1994.
35. K Düring, P Porsch, M Fladung, H Lörz. Transgneic potato plants resistant to the phytopathogenic bacterium *Erwinia carotovora*. Plant J 34:587, 1993.
36. M Fuchs, D Gonsalves. Genetic engineering. In: NA Rechcigl, JE Rechcigl, eds. Environmentally Safe Approaches to Crop Disease Control. Boca Raton, FL: CRC Press, 1997, pp. 333–368.
37. T Ogawa. Methods and strategy for monitoring race distribution and identification of resistance genes to bacterial leaf blight (*Xanthomonas campestris* pv. *oryzae*) in rice. JARQ 27:71–80, 1993.
38. GS Khush, E Bacalangco, T Ogawa. A new gene for resistance to bacterial blight from *O. longistaminata*. Rice Genet Newslett 7:121–122, 1990.
39. R Ikeda, GS Khush, R Tabien. A new resistance gene to bacterial blight derived from *O. longistaminata*. Jpn J Breed 40(suppl 1):280–281, 1990.
40. WY Song, GL Wang, LL Chen, HS Kim, LY Pi, T Holsten, J Gardner, B Wang, WX Zhai, LH Zhu, C Fauquet, P Ronald. A receptor kinase-like protein encoded by the rice disease resistance gene *Xa-21*. Science 270:1804–1806, 1995.
41. GL Wang, WY Song, DL Ruan, DL, S Sideris, PC Ronald. The cloned gene *Xa21* confers resistance to multiple *Xanthomonas oryzae* pv. *oryzae* isolates in transgenic plants. Mol Plant-Microbe Interact 9:850–855, 1996.
42. J Tu, I Ona, Q Zhang, TW Mew, GS Khush, SK Datta. Transgenic rice variety IR72 with *Xa21* is resistant to bacterial blight. Theor Appl Genet 97:31–36, 1998.
43. J Tu, K Datta, GS Khush, Q Zhang, SK Datta. Field performance of *Xa21* transgenic indica rice (*Oryza sativa* L.), IR72. Theor Appl Genet 101:15–20, 2000.
44. TW Mew, CM Vera Cruz, ES Medalla. Changes in race frequency of *Xanthomonas oryzae* pv. *oryzae* in response to the planting of rice cultivars in the Philippines. Plant Dis 76:1029–1032, 1992.
45. BJ Staskawicz, FM Ausubel, BJ Baker, JG Ellis, DG Jones. Molecular genetics of plant disease resistance. Science 268:661–667, 1995.

46. MR Grant, L Godlard, E Straube, T Ashfield, J Lewald, A Sattler, RW Inner, JL Dangl. Structure of the Arabidopsis *RPM1* gene enabling dual specificity disease resistance. Science 269:843–846, 1995.

47. JM Salmeron, SJ Barker, FM Carland, AY Mehta, BJ Staskawics. Tomato mutants altered in bacterial disease resistance provide evidence for a new locus controlling pathogen recognition. Plant Cell 6:511–520, 1994.

48. G Simons, R Flur. Resistance gene-dependent plant defense responses. Plant Cell 8:1771–1791, 1996.

49. S Whitham, SP Dinesh-Kumar, D Choi, R Hehl, C Corr, B Baker. The product of the tobacco mosaic virus resistance gene N: similarity to Toll and the interleukin-1 receptor. Cell 78:1011–1015, 1994.

50. GJ Lawrence, EJ Finnegan, MA Ayliffe, JG Ellis. The *L6* gene for flax rust resistance is related to the *Arabidopsis* bacterial resistance gene RPS2 and the tobacco viral resistance gene, *N*. Plant Cell 7:1195–1206, 1995.

51. J Parker, M Coleman, V Szabo, J Daniels, J Jones. Resistance gene-dependent plant defense responses. Plant Cell 8:1771–1791, 1997.

52. S Yoshimura, SR McCouch, TW Mew, RJ Nelson. Tagging and combining bacterial blight resistance genes in rice using RADP and RFLP markers. Mol Breed 1: 375–387, 1995.

53. R Büschges, K Hollricher, R Panstruga, G Simons, M Wolter, A Frijters, R van Daelen, T vander Lee, P Diergaarde, J Groenendijk, S Töpsch, P Vos, F Salamini, P Schulze-Lefert. The barley *Mlo* gene: a novel control element of plant pathogen resistance. Cell 88:695–705, 1997.

54. DA Jones, CM Thomas, KE Hammond-Kosack, PJ Balint-Kurti, JDG Jones. Isolation of the tomato *Cf-9* gene for resistance to *Cladosporium fulvum* by transposon tagging. Science 266:789–793, 1994.

55. MS Dixon, DA Jones, JS Keddie, CM Thomas, K Harrison, JDG Jones. The tomato *Cf-2* disease resistance locus comprises two functional genes encoding leucine-rich repeat proteins. Cell 84:451–459, 1996.

56. Y Hayano-Saito, T Tsuji, K Fujii, K Saito, M Iwasaki, A Saito. Localization of the rice tripe disease resistance gene, *Stv-b*, by graphical genotyping and linkage analyses with molecular markers. Theor Appl Genet 96(8):1044–1049, 1998.

57. A Moons, E Prinsen, G Bauw, M Van Montague. Antagonistic effects of abscisic acid and jasmonates on salt stress-inducible transcripts in rice roots. Plant Cell 9: 2243–2259, 1997.

58. T Akiyama, H Kaku, N Shibuya. Purification and properties of a basic endol 1-3-β-glucanase from rice (*Oryza sativa* L.). Plant Cell Physiol 37:702–705, 1996.

59. T Akiyama, N Shibuya, M Hrmova, GB Fincher. Purification and characterization of a (1-3)-beta-D-glucan endohydrolase from rice (*Oryza sativa*) bran. Carbohydr Res 14:365–374, 1997.

60. R Rakwal, GK Agarwal, M Yonekura. Separation of proteins from stressed rice (*Oryza sativa* L.) leaf tissues by two-dimensional polyacrylamide gel electrophoresis: induction of pathogenesis-related and cellular protectant proteins by jasmonic acid, UV irradiation and copper chloride. Electrophoresis 20:3472–3478, 1999.

61. CS Anuratha, JK Huang, A. Pingali, Muthukrishnan S. Isolation and characteriza-

tion of a chitinase and its cDNA clone from rice. J Plant Biochem Biotech 1:5–10, 1992.

62. CS Anuratha, KC Zen, KC Cole, T Mew, S Muthukrishnan. Induction of chitinases and β-glucanases in *Rhizoctonia solani*-infected rice plants: isolation of an infection-related chitinase cDNA clone. Physiol Plant 97:39–46, 1996.

63. H Inui, Y Yamaguchi, Y Ishigami, S Kawaguchi, T Yamada, H Ihara, S Hirano. Three extracellular chitianses in suspension-cultured rice cells elicited by N-acetylchitooligosaccharides. Biosci Biotechnol Biochem 60:1956–1961, 1996.

64. JK Huang, L Wen, M Swegle, HC Tran, TH Thin, HM Naylor, S Muthukrishnan, GR Reeck. Nucleotide sequence of a rice genomic clone that encodes a class I endochitinase. Plant Mol Biol 16:479–480, 1991.

65. RR Dudler. cDNA cloning and sequence analysis of a pathogen-induced thaumatin-like protein from rice (*Oryza sativa*). Plant Physiol 101:1113–1114, 1993.

66. R Velazhahan, KC Cole, CS Anuratha, S Muthukrishnan. Induction of thaumatin-like protein (TLPs) in *Rhizoctonia solani*-infected rice and characterization of two new cDNA clones. Physiol Plant 102:21–28, 1998.

67. K Abe, Y Emori, H Kondo, K Suzuki, S Arai. Molecular cloning of a cysteine protease inhibitor of rice (*Oryza cystatin*): homology with animal cystatins and transient expression in the ripening process of rice seeds. J Biol Chem 262:16793–16797, 1987.

68. MJ Chen, B Johnson, L Wen, S Muthukrishnan, KJ Kramer, TD Morgan, GR Reeck. Rice cystatin: bacterial expression, purification, cysteine proteinase inhibitory activity, and insect growth suppressing activity of truncated form of the protein. Protein Expr Purif 3:41–49, 1992.

69. PJ Reimers, A Guo, JE Leach. Increased activity of a cationic peroxidase associated with an incompatible interaction between *Xanthomonas oryzae* pv. *oryzae* and rice (*Oryza sativa*). Plant Physiol 99:1044–1050, 1992.

70. DM Lai, PB Hoj, GB Fincher. Purification and characterization of (1-3, 1-4)-beta-glucan endohydrolases from germinated wheat (*Triticum aestivum*). Plant Mol Biol 22:847–859, 1993.

71. S Munch-Garthoff, JM Neuhas, T Boller, B Kemerlin, KH Kogel. Expression of a β-1,3-glucanase and chitinase in healthy, stem-rust-affected and elicitor-treated near-isogenic wheat lines showing Sr5- or Sr24-specified race-specific ruts resistance. Planta 201:235–244, 1997.

72. WL Li, JD Faris, JM Chittoor, JE Leach, DJ Liu, PD Chen, BS Gill. Genomic mapping of defense response genes in wheat. Theor Appl Genet 98:226–233, 1999.

73. J Molano, I Polacheck, A Duran, E Cabib. An endochitinase from wheat germ. Activity on nascent and performed chitin. J Biol Chem 254:4901–4907, 1979.

74. AJ Van der Westhuizen, XM Qian, AM Botha. Differential induction of apoiplastic peroxidase and chitinase activities in susceptible and resistant wheat cultivars by Russian wheat aphid infestation. Plant Cell Rep 18:132–137, 1998.

75. AJ Vigers, WK Roberts, CP Selitrennikoff. A new family of plant antifungal proteins. Mol Plant-Microbe Interact 4:315–323, 1991.

76. C Kuwabara, K Arakawa, S Yoshida. Abscisic acid-induced secretory proteins in suspension-cultures cells of winter wheat. Plant Cell Physiol 40:184–191, 1999.

77. KC Snowden, KD Richards, RC Garner. Aluminum-induced genes. Induction by

toxic metals, low calcium, and wounding and pattern of expression in root tips. Plant Physiol 107:341, 1995.

78. OO Molodchenkova, AP Levitskii, IA Levitskii, VG Adamovskaia, Dymokovskaia. Trypsin inhibitors of wheat seedlings infected and treated with salicylic acid. Ukr Biokhim Zh 70:30–37, 1998.

79. G Rebmann, F Mauch, R Dudler. Sequence of a wheat cDNA encoding a pathogen-induced thaumatin-like protein. Plant Mol Biol 17:283–285, 1991.

80. M Baga, RN Chibbar, KK Kartha. Molecular cloning and expression analysis of peroxidase genes from wheat. Plant Mol Biol 29:647–662, 1995.

81. T Bryngelsson, J Sommer-Knudsen, PL Gregersen. Purification, characterization, and molecular cloning of basic PR-1-type pathogenesis-related proteins from barley. Mol Plant-Microbe Interact 7:267–275, 1994.

82. C Stevens, E Titarenko, JA Hargreaves, SJ Gurr. Defense-related gene activation during an incompatible interaction between *Staganospora* (*Septoria*) *nodorum* and barley (*Hordeum vulgare* L.) coleoptile cells. Plant Mol Biol 31:741–749, 1996.

83. M Hrmova, GB Fincher. Purification and properties of three (1-3)-β-D-glucanase isoenzymes from young leaves of barley (*Hordeum vulgare*). Biochem J 289:453–461, 1993.

84. P Xu, AJ Harvey, GB Fincher. Heterologous expression of cDNAs encoding barley (*Hordeum vulgare*) (1-3)-β-glucanase isozyme GV. FEBS Lett 348:206–210, 1994.

85. W Jutidamrongphan, JB Andersen, G Mackinnon, JM Manners, RS Simpson, KJ Scott. Induction of β-1,3-glucanases in barley in response to infection by fungal pathogens. Mol Plant-Microbe Interact 4:234–238, 1991.

86. KM Kragh, S Jacobsen, JD Mikkelsen, KA Nilesen. Purification and characterization of three chitinases and one β-1,3-glucanases accumulating in the medium of cell suspension cultures of barley (*Hordeum vulgare* L.). Plant Sci 76:65–77, 1991.

87. R Leah, H Tommerup, J Mundy, I Svendsen. Biochemical and molecular characterization of three barley seed proteins with antifungal properties. J Biol Chem. 266(3): 1564–1573, 1991.

88. S Jacobsen, JD Millelsen, J Hejgaard. Characterization of two antifungal endochitinases from barley grain. Physiol Plant 79:554–562, 1990.

89. M Swegle, KJ Kramer, S Muthukrishnan. Properties of barley seed chitianses and release of embryo-associated isoforms during early stages of imbibition. Plant Physiol 99(3):1009–1014, 1992.

90. MS Swegle, JK Huang, G Lee, S Muthukrishnan. Identification of an endochitinase cDNA clone from barley aleurone cells. Plant Mol Biol 12:403–412, 1989.

91. J Hejgaard, S Jacobsen, SE Bjorn, KM Kragh. Anti-fungal activity of chitin-binding PR-4 type proteins from barley grain and stressed leaf. FEBS Lett 307:389–392, 1992.

92. T Bryngelsson, B Green. Characterization of a pathogenesis-related, thaumatin-like protein isolated from barley challenged with an incompatible race of mildew. Physiol Mol Plant Pathol 35:45–52, 1989.

93. S Krishnaveni, S Muthukrishnan, GH Liang, G Wilde, A Manickam. Induction of chitinases and β-1,3-glucanases in resistant and sensitive cultivars of sorghum in response to insect attack, fungal infection and wounding. Plant Sci 144:9–16, 1999.

94. QK Huynh, JR Borgemeyer, JF Zobel. Isolation and characterization of a 22kDa

protein with antifungal properties from maize seeds. Biochem Biophys Res Commun 182:1–5, 1992.

95. SC Lo, JD Hipskind, RL Nicholsen. cDNA cloning of a sorghum pathogenesis-related protein (PR-10) and differential expression of defense-related genes following inoculation with *Cochliobolus heterostrophus* or *Colletotrichum sublineolum*. Mol Plant-Microbe Interact 12:479–489, 1998.

96. MJ Cordero, D Raventos, B San Segundo. Differential expression and induction of chitinases and β-1,3-glucanases in response to fungal infection during germination of maize seeds. Mol Plant-Microbe Interact 7:23–31, 1994.

97. S Wu, AL Kriz, JM Widholm. Nucleotide sequence of a maize cDNA for a class II, acidic β-1,3-glucanase. Plant Physiol 106:1709–1710, 1994.

98. QK Huynh, CM Hironaka, EB Levine. Antifungal proteins from plants—purification, molecular cloning, and antifungal properties of chitinases from maize seed. J Biol Chem 267(10):6635–6640, 1992.

99. WK Roberts, CP Selitrennikoff. Zeamatin, an antifungal protein from maize with membrane-pemeabilizing activity. J Gen Microbiol 136:1771–1778, 1990.

100. QK Huynh, JR Orgemeyer, JF Zobel. Isolation and characterization of a 22 kDa protein with antifungal properties from maize seeds Biochem Biophys Res Commun 182:1–5, 1992.

101. DE Malehorn, JR Borgmeyer, CE Smith, DM Shah. Characterization and expression of an antifungal zeamatin-like protein (Zlp) gene from *Zea mays*. Plant Physiol 106:1471–1481, 1994.

102. T Rohrmeier, L Lehle. WIP1, a wound-inducible gene from maize with homology to Bowman-Brik proteinase inhibitors. Plant Mol Biol 22:783, 1993.

103. MJ Cordero, D Raventos, B San Segundo. Expression of a maize protease inhibitor is induced in response to wounding and fungal infection: systematic wound response of a monocot gene. Plant J 6:141–150, 1994.

104. D Alexander, RM Goodman, M Gut-Rella, C Glascock, K Weymann, L Friedrich, D Maddox, P Ahl-Goy, T Luntz, E Ward, J Ryals. Increased tolerance to two oomycete pathogens in transgenic tobacco expressing pathogenesis-related protein 1a. Proc Natl Acad Sci USA 90:7327–7331, 1993.

105. K Broglie, I Chet, M Holliday, R Cressman, P Biddle, S Knowlton, CJ Mauvis, R Broglie. Transgenic plants with enhanced resistance to the fungal pathogen *Rhizoctonia solani*. Science 254:1194–1197, 1991.

106. J Logemann, G Jach, H Tommerup, J Muncy, J Schell. Expression of a barley ribosome inactivating protein leads to increased fungal protection in transgenic tobacco plants. Bio/Technology 10:305–308, 1992.

107. G Jach, B Gornhardt, J Mundy, J Logemann, E Pinsdorf, R Leaf, J Schell C Maas. Enhanced quantitative resistance against fungal disease by combinatorial expression of different barley antifungal proteins in transgenic tobacco. Plant J 8:97–109, 1995.

108. Q Zhu, EA Maher, S Masoud, RA Dixon, CJ Lamb. Enhanced protection against fungal attack by constitutive co-expression of chitinase and glucanase genes in transgenic tobacco. Bio/Technology 12:807–812, 1994.

109. FRG Terras, K Eggermont, V Kovaleva, NV Raikhel, RW Osborn, A Kester, SB Rees, S Torrekens, F van Leuven, J Venderleyden. Small cysteine-rich antifungal

proteins from radish (*Rhaphanus sativus* L.): their role in host defense. Plant Cell 7:573–588, 1995.

110. LS Melchers, M Stuiver. Novel genes for disease-resistance breeding. Curr Opin Plant Biol 3:147–152, 2000.

111. E Jongedjik, H Tigelaar, JSC van Roekel, SA Bres-Vloemans, I Dekker, PJM van den Elzen, BJC Cornelissen, LS Melchers. Synergistic activity of chitinases and β-1,3 glucanases enhances fungal resistance in transgenic tomato plants. Euphytica 85:173–180, 1995.

112. R Grison, B Grezes-Besset, M Schneider, N Lucante, L Olsen, JJ Leguay, A Toppan. A field tolerance to fungal pathogens of *Brassica napus* constitutively expressing a chimeric chitinase gene. Nat Biotechnol 14:643–646, 1996.

113. D Liu, KG Raghothama, PM Hasegawa, RA Bressan Osmotic over-expression in potato delays development of disease symptoms. Proc Natl Acad Sci USA 91: 1888–1892, 1994.

114. G Wu, BJ Shortt, EB Lawrence, EB Leveine, C Fitzsimmons, DM Shah. glucose oxidase in transgenic potato plants. Plant Cell 7:1357–1368, 1995.

115. J Liang, Y Wu, C Rosenberger, S Hakimi, S Castro, J Berg. AFP gene confer disease resistance to transgenic potato plants expressing sense and antisense genes for an osmotin-like protein. Planta 198:70–77, 1998.

116. WP Chen, PD Chen, DJ Liu, RJ Kynast, B Friebe, R Velazhahan, S Muthukrishnan, BS Gill. Development of wheat scab symptoms is delayed in transgenic wheat plants that constitutively express a rice thaumatin-like protein gene. Theor Appl Genet 99:755–760, 1999.

117. H Zhu, S Krishnaveni, GH Laing, S Muthukrishnan. Biolistic transformation of sorghum using a rice chitinase gene. J Genet Breed 52:243–252, 1998.

118. LE Murray, LG Elliott, SA Capitant, JA West, KK Hanson, L Scarafia, S Johnston, C De Luca-Flaherty, S Nichols, D Cunanan, PS Dietrich, IJ Mettler, S Dewald, DA Warnick, C Rhodes, RM Sinibaldhi, KJ Brunke. Transgenic corn plants expressing MDMV strain B coat protein are resistant to mixed infections of maize dwarf mosaic virus and maize chlorotic mottle virus. Bio/Technology 11:1559, 1993.

119. RS Nelson, SM McCormick, X Delannay, P Dube, J Layton, EJ Anderson, M Kaniewska, RK Proksch, RB Horsch, SG Rogers, RT Fraley, RN Beachy. Virus tolerance, plant growth, and field performance of transgenic tomato plants expressing coat protein from tobacco mosaic virus. Bio/Technology 6:403, 1988.

120. M Fuchs, R Provvidenti, JL Slightom, D Gonsalves. Evaluation of transgenic tomato plants expressing the coat protein gene of cucumber mosaic virus strain WL under field conditions. Plant Dis 80:270, 1996.

121. C Gonsalves, B Xue, SZ Pang, R Provvidenti, JL Slightom, D Gonsalves. Breeding multiple virus-resistant transgenic tomatoes. Proceedings of 6th Int Congr Pathology, Montreal, July 28 to August 6, 1993, p. 190.

122. E Jongedijk, AAJM de Schutter, T Stolte, PJM van den Elzen, BJC Cornelissen. Increased resistance to potato virus X and preservation of cultivar properties in transgenic potato under field conditions. Bio/Technology 10:422, 1992.

123. P Malnöe, L Farinelli, GF Collet, W Reust. Small-scale field tests with transgenic potato, cv. Bintje, to test resistance to primary and secondary infections with potato virus Y. Plant Mol Biol 25:963, 1994.

124. H Barker, KD Webster, CA Jolly, B Reavy, A Kumar, MA Mayo. Enhancement of resistance to potato leafroll virus multiplication in potato by combining the effects of host genes and transgenes. Mol Plant-Microbe Interact 7:528, 1994.

125. D Tricoli, KJ Carney, PF Russell, JR McMaster, DW Groff, KC Hadden PT Himmell, JP Hubbard, ML Boeshore, JF Reynolds, HD Quemada. Field evaluation of transgenic squash containing single or multiple virus coat protein gene constructs for resistance to cucumber mosaic virus, watermelon mosaic virus 2, and zucchini yellow mosaic virus. Bio/Technology 13:1458, 1995.

126. M Fuchs, D Gonsalves. Resistance to transgenic hybrid squash ZW-20 expressing the coat protein genes of zucchini yellow mosaic virus and watermelon mosaic virus 2 to mixed infections by both potyviruses. Bio/Technology 13:1466, 1995.

127. C Gonsalves, B Xue, M Yepes, M Fuchs, K Ling, S Namba, P Chee, JL Slightom, D Gonsalves. Transferring cucumber mosaic virus-white leaf strain coat protein gene into *Cucumis melo* L. and evaluating transgenic plants for protection against infections. J Am Soc Hortic Sci 119:345, 1994.

128. GH Clough, PB Hamm. Coat protein transgenic resistance to watermelon mosaic and zucchini yellow mosaic virus in squash and cantaloupe. Plant Dis 79:1107, 1995.

129. S Lius, RM Manshardt, MMM Fitch, JL Slightom, JC Sanford, D Gonsalves. Pathogen-derived resistance provides papaya with effective protection against papaya ringspot virus. Mol Breed 3:161–168.

130. MMM Fitch, RM Manshardt, D Gonsalves, JL Slightom, JC Sanford. Virus resistant papaya plants derived from tissues bombarded with the coat protein gene of papaya ringspot virus. Bio/Technology 10:1466, 1992.

131. D Gonsalves, P Chee, R Provvidenti, R Seem, JL Slightom. Comparison of coat protein-mediated and genetically-derived resistance in cucumber to infection by cucumber mosaic virus under field conditions with natural challenge inoculations by vectors. Bio/Technology 10:1562, 1992.

4
Biological Control of Wheat Diseases

Krishnapillai Sivasithamparam
University of Western Australia, Nedlands, Western Australia, Australia

I. INTRODUCTION

As with most other plant diseases, the extent of success in biological control of wheat diseases, especially those caused by soilborne fungi, is determined by the environment. This is especially true in rain-fed crops, especially in Mediterranean-type environments. The low-input agriculture in these regions implies that there is usually little or no attempt made to employ cultural practices to change moisture, pH, or in some instances even the fertility of the soils. These are, however, some of the soil conditions that determine the level of success of biocontrol methods. This becomes a major issue when one extrapolates research results from biocontrol studies of relatively and naturally nutrient-rich soils of the cool temperate, warm tropical, or subtropical regions to the microbially and nutrient-impoverished soils of regions such as southern Australia. Much of the coverage on the biocontrol of wheat diseases in this chapter deals with take-all caused by the fungus *Gaeumannomyces graminis* var. *tritici* and *Rhizoctonia* root rots, with which I am familiar. Most, if not all, major soilborne fungal diseases of wheat are caused by necrotrophs, which tend to be more damaging to crops in nutrient-impoverished soils [1].

Difficulties encountered with the biocontrol of pathogens in such disease-conducive soils are not dissimilar to those experienced elsewhere with excessively large disease pressures. This is further complicated by the fact that the harsh environment to which the host and pathogen have naturally adapted may be inhospitable to a biocontrol agent that requires soil conditions with favorable soil moisture, temperature, and nutrient supply. The success and failure of a biocontrol agent, therefore, may be determined by its behavior in bulk soil, rhizosphere, rhizoplane, and/or the root cortices. In this scenario one must also include the growth stage and the tissue of the host affected and the site and phase of activity of the pathogen.

In southern Australia, which enjoys a Mediterranean-type climate, the wheat crop is essentially a rain-fed winter crop. Following seeding in late autumn, the winter and spring rains support crop growth. The winters are mild in these regions, and the summers are generally hot and dry [2]. Seedling diseases under these conditions are affected by conditions which are very different from, for instance, the winter wheat of the temperate regions of the Northern Hemisphere. Wheat seedlings in the tropics, such as in Thailand, however, can be affected severely by *Sclerotium rolfsii*, which is active only under conditions associated with temperatures over 25°C, significantly higher than temperatures experienced at germination in either the temperate or Mediterranean environments. The spectrum of pathogens infecting wheat and the severity of disease they cause in any region can vary significantly depending on the soil mineralogy and soil conditions such as nutrient and moisture levels and temperature. It is noteworthy that biocontrol agents can be similarly affected by variations in soil conditions. In Washington State, *Fusarium* root rot can be serious in a rain-fed crop of wheat, while take-all (caused by *Gaeumannomyces graminis* var. *tritici*) may replace it as the major pathogen if the crop was irrigated [3]. While take-all may be favored by moist soils, flooding of soil such as that practiced in rice production, could reduce the inoculum of the take-all fungus. Thus, although the take-all fungus has been recorded in tropical or subtropical regions of Asia, such as India and Japan [4], the disease is not considered to be a serious threat due to periodic flooding between wheat crops. Soil pH may have differential effects, *Fusarium* may be favored by moderately low pH, while liming soil may render the soil conducive to the take-all fungus, which is a serious threat to wheat in neutral to moderately alkaline soils [5]. Not all biocontrol agents are likely to be effective under such varying soil conditions. Targeting a single pathogen in fields where there is a complex of pathogens can also create new problems. Control of a *Fusarium* root rot by fluorescent pseudomonads [6] or of the take-all fungus by a sterile fungus [7] could lead to emergence of *Rhizoctonia solani* AG8, the causal agent of the bare-patch disease of cereals, which is not suppressed by these antagonists.

The biotic and abiotic environments can play a significant role not only in the pathogenicity of the disease-causing organisms, but also in the antagonistic activities of the biocontrol agents. Biological control of a disease can be affected by "pathogen suppression" or "disease suppression" [3,8]. Pathogen suppression is brought about by the antagonistic activities of soil microflora occurring outside the influence of the host plant, invariably in the "bulk soil." Disease suppression, on the other hand, is considered to be directly or indirectly mediated by the plant. Much of the biocontrol work on root disease of wheat in the past two decades has concentrated on fluorescent pseudomonads, which are thought to bring about disease suppression essentially through their activities both in the rhizosphere/rhizoplane region as well as in the root tissues [6]. Their suppression of the take-all fungus on the root surface is favored by the ectotrophic growth habit of the take-all fungus, which occurs not only before initial penetration [9] but also in the subsequent spread of the pathogen up the root axes. This suppression of spread facilitates the abortion on the root axis of lesion extension towards the crown. It is caused by the antagonistic fluorescent pseudomonads, which are considered to be natural occupants of the rhizosphere and rhizoplane [6]. It is likely that in soils subjected to high temperature and drying cycles, spore-forming bacteria (e.g., *Bacillus* spp.) are likely to be more active as antagonists, especially in the bulk soil.

The nature of bacterial and fungal flora of the root surface affects the ectotrophic habit of the take-all fungus on wheat roots [5]. The growth of the pathogen towards the root and the ectotrophic growth on the root is critical for the establishment of its parasitic phase. It is therefore possible that the root-infecting pathogen could be disoriented in its trophic growth by the presence in the rhizosphere/rhizoplane region of bacterial [10] and fungal [11] flora that are typically occupants mainly of the bulk soil. This is certainly a novel approach in the biocontrol of root-invading pathogenic fungi, the management of which conventionally has been mainly through the introduction onto the roots of rhizosphere-competent strains mainly belonging to *Pseudomonas* spp. [6].

II. ABIOTIC ENVIRONMENT

The activities of microorganisms are environment dependent. In relation to biocontrol this would apply not only to the pathogen but also to its antagonists. In addition, environmental factors may also affect the susceptibility of the host plant to disease [2]. In field trials carried out to screen strains of microbial antagonists, these complex interactions are often overlooked with only the response of the pathogen to the abiotic conditions taken into consideration. Very often the variability between sites and between seasons in the performance of biocontrol agents in the field could also be attributed to the heterogeneous and/or inhospitable environments provided for the introduced biocontrol agent. A good example of this is the acidification of bulk and rhizosphere soil resulting from the continued use of ammoniacal nitrogen (N). The shift, although slight, in the soil pH is adequate to render the soil suppressive to the take-all fungus. This shift is also sufficient to render the soil conducive to the native populations of *Trichoderma koningii* (considered to be the cause of soil suppressiveness) but not to the take-all fungus. Continued use of the ammoniacal form of N eventually results in the soil becoming unsuitable for the wheat crop. The remedial application of lime, however, renders the soil suppressive to *T. koningii* but conducive to the disease [8]. This is an interesting example of the abiotic component of the environment affecting disease through the mediation of soil microflora.

Abiotic factors affect biocontrol activities of microorganisms through a variety of direct and indirect effects. For instance, Leggett et al. [13] found that strains of fluorescent pseudomonads they tested on wheat roots reduced ecto-trophic growth of the take-all fungus and lesion development only when no N was added to a nutrient-impoverished sandy soil. They suggested that the N effect in this case might be related to other soil characteristics. They also demonstrated that nutrient supply affected root growth, colonization by the take-all fungus, colonization of the root by the antagonistic pseudomonad, and the interaction of the three.

It is important to note that take-all, like many other diseases caused by necrotrophic root-infecting fungi, is severe in soils that are nutrient impoverished and poorly buffered microbially. Any soil amendment to improve the fertility of such soils invariably reduces the severity of the disease. Manganese (Mn) as an element has received considerable attention [14], with cereals that are adequately supplied with Mn and genetically efficient in its utilization usually less severely affected by take-all. The availability of Mn to the plant, however, can be affected by soil pH and the activity of manganese-reducing and -oxidizing bacteria, in addition to the genetically determined ability of the host to be efficient in the requirement and utilization of Mn. At least in theory, any improvement in the ability of the host roots to support a large population of manganese reducers should render the host tolerant to the disease. In this case the manganese-reducing bacteria could effectively function as biocontrol agents, although their effect on the pathogen is largely indirect and host mediated.

Crown rot of wheat caused by *Fusarium roseum* complex appears to be exacerbated by moisture stress [3]. Biocontrol agents selected to manage this disease should also be capable of being active and effective under these conditions. Take-all, on the other hand, is most severe in alkaline sandy soils that are cool and moist [5]. This may explain the relative success of a variety of soil bacteria screened for the biocontrol of take-all in the cool temperate wheat-growing regions worldwide [6].

The abiotic factors most critical for the activity of both pathogen and antagonists are temperature and moisture. Early work [15] showed that take-all in field soil declined at temperatures of 25°C and above. This effect of temperature was not evident in sterilized soil infested with the pathogen, indicating the role of soil microflora in amelioration of the disease under warm conditions. Thus, although the temperature optima for in vitro agar cultures of the pathogen may be close to 23°C, the disease caused by the take-all fungus in the glasshouse may peak around 15°C [16]. *Rhizoctonia solani* AG8, however, may be most damaging in relatively warm and dry soils [17]. It is therefore likely that in fields where the inocula of both the take-all fungus and *R. solani* are present, as is the case in some wheat field soils in southern Australia, the ecological attributes of the strains successful in the biocontrol of the two diseases may well be different and reflect the conditions that favor each disease.

Temperature ranges in soil at seed germination may be critical for the beneficial activities of the biocontrol agents. Temperature and moisture affect both the multiplication of microorganisms as well as the production of antibiotics in soil by biocontrol agents [3]. Temperature can also change the behavior of a biocontrol agent. For instance, *Phialophora graminicola*, which causes significant reductions in the root rot caused by the take-all fungus [18,19] under relatively cool conditions, has been found to cause serious root rot of grasses under warm (>30°C) conditions [20].

Pythium spp., which affect the early stages of growth in wheat, may need to be challenged by biocontrol strains capable of rapidly establishing themselves in the spermosphere. This is necessary because certain pathogenic strains of *Pythium* spp. invade the seed within 28 hours of sowing [3]. Because *Pythium* spp. are also favored by relatively high soil moisture levels, the antagonists chosen to control these pathogens should be capable of being effective under these conditions. Aspects such as these are overlooked or underrated because of the "optimal" conditions that are generally used in greenhouse biocontrol trials, which always form the initial and accepted basis of selection preceding field tests. Natural selection may also favor bacteria that produce antifungal compounds in lesions caused by a virulent pathogen. Charigkapakorn and Sivasithamparam [21] found that a greater proportion of fluorescent pseudomonads antagonistic to the take-all fungus in vitro were isolated from the wheat rhizosphere with each successive planting of wheat in the greenhouse in natural soil infested with the take-all fungus. It is interesting that several "effective" biocontrol strains are isolated from fields in which the disease is widespread and sometimes rampant. It would be tempting to assume that the presence of such strains indicates the enrichment of the soil by these strains associated with lesioned roots and may relate to the potential for a future "decline" in the disease at that site [3]. It is also probable that the infrequent presence in nonsuppressive soils of strong antagonists may indicate that these strains in such situations are incapable of causing an effect because environmental conditions at that point of time do not favor the large-scale multiplication required to build up an effective population of the antagonist. This may also explain the existence of suppressive and nonsuppressive soils in close proximity [22].

III. BIOCONTROL OF WHEAT DISEASES—WHOLE PLANT

A. Root Systems

Fungal pathogens target one or more parts of the wheat plant, the root system generally being the most affected, especially in Mediterranean environments (Fig. 1). The seminal roots, which number around five in most commercial varieties of wheat, are the most important, being responsible for the uptake of moisture from the depths of the soil profile [23]. The secondary roots are less involved in

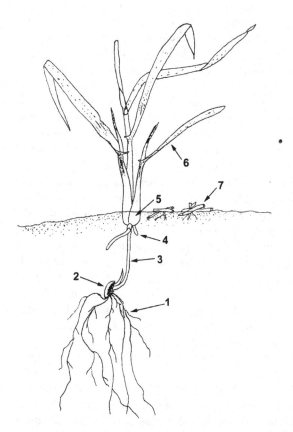

Figure 1 Niches associated with the wheat plant normally targeted for the activities of biocontrol agents: 1, seminal roots; 2, the seed; 3, the subcrown internode; 4, nodal and tiller roots; 5, crown; 6, leaves, flowers, and seeds; 7, residues around the plant. Note that the abiotic and biotic environment for each of these niches may be different.

seeking moisture from such storage, especially under Mediterranean-type environments, where surface soil usually explored by these roots retains little moisture as the cropping season progresses. Thus, take-all and *Rhizoctonia* bare-patch cause greater crop damage under these conditions, where early disfunction of the seminal roots results in the stunting of seedlings or premature senescence of adult plants. The infected adult plants suffer from deprivation of moisture and nutrients following cessation of seasonal rains. Unlike the take-all fungus, which can invade plants almost until the "booting" stage of plant growth, much of the root invasion by *R. solani* AG8 occurs on wheat seedlings in the first 2 weeks of growth. Therefore, while protection by biocontrol agents of *Rhizoctonia* root rot may be needed only for the early seedling stages of wheat, protection against take-all may need to be extended until flowering, even though attack of seedling roots may cause the most severe crop loss. Thus, it would require biocontrol organism(s) to be adaptable enough to be active not only on and in seminal roots with live cortices, but also on older seminals, as well as young and mature secondary (nodal and tiller) roots. Relatively superior performance of microbial mixtures used biocontrol agents against the take-all fungus [6,24] may indicate the value of introducing a variety of microorganisms that could succeed with changing age and type of roots as well as the variation in the physical environments of the soil that could be expected with the progression of the growing season.

With the varieties of wheat sown in Western Australia, the nodal roots do not appear until day 21 after germination. Therefore, any antagonist introduced on seed has to survive on or around the seminal roots for a period of time to be able to colonize the secondary roots. Although the first internode in shallow sown wheat is telescoped and indistinguishable, deep-sown crops tend to have a pronounced subcrown internode (first internode), which would separate the seminals from the secondaries not only by time but also by space. The subcrown internode is not usually favored for infection by the take-all fungus, *Rhizoctonia solani* AG8 or *Pythium* spp., but the common root rot fungus (*Cochliobolus sativus*) and sometimes certain fusaria colonize it and are often diagnosed by their activity on it. Barley differs from wheat in its ability to produce large numbers of compensatory roots, which helps the host, to some extent, cope with infections by the take-all fungus [25] but not against *Rhizoctonia solani* AG8 infections under Western Australian conditions. This ability appears to be absent in wheat. These aspects of root formation and function in wheat need to be considered in relation to the survival and spread of the introduced biocontrol agent.

B. Stem Base/Crown

These regions are commonly affected by fusaria and *C. sativus*. *Fusarium culmorum* is associated with foot and root rot in the northwestern United States [26], and the crown rot or dryland rot of wheat in Australia and South Africa is caused by *Fusarium graminearum* group 1, which persists in or on the tillage layer, predominantly as hyphae in crop residues. Infection occurs mainly at the coleoptile, subcrown internode and crown, depending on the nature of the inoculum [27]. Current practices related to minimum tillage and the retention of stubble have resulted in the increased incidence of these trashborne diseases in rain-fed crops. Under these conditions, infection could even occur via senescent leaf sheaths [28]. Thus, the variety of infection courts involved would require a biocontrol agent capable of being effective at all the sites of infection or a selection of agents capable of being active in as many of these niches involved would need to be employed. In addition, the moisture stress and high temperatures associated with disease severity would restrict the nature of antagonists that could be effective. Pathogenic fusaria are in general active in conditions that are not favorable to bacterial activity [3]. It has, therefore, been proposed that fusarial pathogens are active mainly in soils that are not conducive to competition from soil bacterial flora [3]. This would indicate that choice of antagonists for the control in the field of crown rot fungi would need to be made in the knowledge of conditions that favor infection and development of these diseases.

C. Foliar and Glume Diseases

Necrotrophic foliar and glume diseases are caused by trashborne inoculum of pathogens such as *Stagonospora* (Septoria) *nodorum*. These diseases have become a greater threat to wheat production since the use of minimum tillage, which favors conservation of inoculum based in trash. Although greenhouse studies have identified several phyllosphere organisms with noticeable potential to reduce the parasitic activity of these pathogens on foliage, their value as field-scale applications still remain to be proven.

The management of diseases through the manipulation of trash is currently more promising. Any practice that helps to degrade the stubbleborne inoculum will help to reduce the disease hazard. This is likely to be ineffective in back-to-back wheat crops, especially in Mediterranean regions with dry hot summers, where the inoculum is conserved relatively intact between the cropping seasons. Although management of inoculum in host residues has been considered in horticultural crops [29], little has been done in the field to develop methods to enhance field degradation of trashborne inoculum of necrotrophic foliar pathogens of wheat, either through introduced antagonist(s) or by encouraging microbiological breakdown of the stubbleborne inoculum through appropriate cultural practices. Complex interactions, however, may be involved in the succession of parasitic fungi by saprophytic fungi in crop residues in soil [30].

IV. PHASES OF ACTIVITY OF PATHOGEN TARGETED

A. Saprophytism

All major wheat root–infecting pathogenic fungi have necrotrophic habits. The two major root pathogens in southern Australia are the take-all fungus and *R. solani* AG8. Although both pathogens are capable of producing spores, their spores are involved in neither the spread nor the survival of the pathogens. The inoculum of these two pathogens exists as mycelia. The take-all fungus inoculum is essentially the mycelium in the stem base and roots ("saprophytic survival" [5]), colonized during its parasitic phase, while the *R. solani* inoculum exists as mycelia within and on residues as well those growing in bulk soil ("saprophytic growth" [5]). *R. solani*, and to a much lesser extent the take-all fungus, also depends on saprophytic colonization [5] to establish a mycelial network in soil (Fig. 2). Sandy soil favors soil growth of the take-all fungus [2] and *R. solani* [17]. The Mediterranean environments are conductive to the saprophytic activity of these pathogens when soil moisture and temperature are favorable [2].

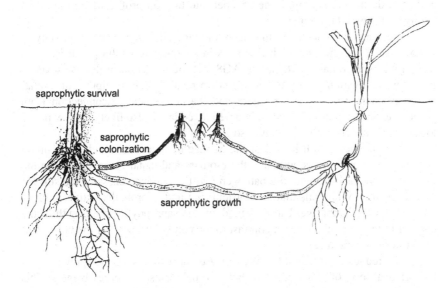

saprophytic survival

saprophytic colonization

saprophytic growth

Figure 2 Aspects of soil saprophytism relevant to both the pathogens and the biocontrol agents. Left to right: saprophytic survival involving the survival of the pathogen in residues of tissues colonized during its parasitic phase; saprophytic colonization of soil organic matter by the pathogen preceding parasitic activity in the cropping season; saprophytic growth of the mycelium of the pathogen through soil, from the substrate where it survived, to the young root. (Concepts from Ref. 5.)

When the biological suppression of a pathogen occurs in the bulk soil or in any niche outside the influence of the plant host, it is considered to operate through pathogen suppression [3,8]. Pathogen suppression therefore is involved in the suppression of the pathogen in its saprophytic phase (during saprophytic survival, saprophytic growth, and/or saprophytic colonization [5]). Pathogen suppression can also be effective against plant parasitic nematodes [31].

Much of the attack on the saprophytic phase of pathogens, by biological means or otherwise, can only be successful under field conditions if appropriate cultural practices are employed. Although the saprophytic survival phase is relatively sensitive to microbial attack [3], such attacks are particularly difficult under field conditions in Mediterranean environments where the soil is too hot and dry during summer for microbial activity. Unseasonal (i.e., summer when the inoculum is resting under Mediterranean conditions) rains can result in the destruction of the inoculum between crops. This is evident when delayed sowing in May/ June has been shown to result in reduced inoculum density and disease [2]. Deep ploughing can not only break the stubble-based inoculum into smaller and less infective units, but could also render the inoculum less infective if the infective units are, during cultivation, buried deeper into the soil profile where the availability of O_2 may be limited.

Saprophytic colonization in soil by the take-all fungus may occur only in impoverished and disease-conducive soils in southern Australia [2]. It is, however, a widespread habit of R. solani AG8, but the reduction of disease severity following cultivation of soil is considered to result in the disruption of hyphal networks connecting organic residues and the consequent disconnection from particulate food sources [32]. The disrupted mycelia are also likely to be exposed to microbial attack in the cultivated soil.

Saprophytic growth in soil of the hyphae or mycelial strands is common with most root pathogens, although the biomass and distance traveled from the food base would differ with the pathogen and the conduciveness of the soil for hyphal growth. With soils that are nutrient rich and microbially active, such as those found in Cambridge, United Kingdom, the saprophytic soil growth of the take-all fungus is limited [5], in contrast to the conductive soils of the wheat belt of Western Australia [2].

The reduction of take-all in Western Australia following the application of ammoniacal form of N has been attributed to induced soil suppressiveness. This suppressiveness resulting from the shift in soil pH has been shown [8] to be related to the increased activity of antagonistic strains of Trichoderma koningii in treated soil. A strain from this study has shown considerable promise in field crops as a biocontrol agent of take-all in the United States [33] and China [34]. The study of Simon and Sivasithamparam [8] also showed that the reduction of take-all in soil amended with ammonium N was essentially a response to pathogen suppression.

The biological suppression of saprophytic soil growth of the take-all fungus is considered to result in the reduction in the number of hits by the pathogen on the roots of seedlings of wheat and further interference of the pathogen in the rhizosphere.

B. Parasitism

Both the take-all fungus and R. *solani* AG8 have an ectotrophic phase preceding invasion of the cortex. The ectotrophic growth of the take-all fungus is, however, comparatively more elaborate [9] than that of R. *solani* AG8. The activity on the root of the runner hyphae of the take-all fungus occurs not only before penetration but also during the spread of the fungus along and between root axes. This habit lent to successful testing of a wide variety of rhizosphere organisms for the biocontrol of take-all. The rhizosphere habit of R. *solani* AG8 is different. The growth of the pathogen in wheat rhizosphere, probably at the expense of the root exudates, consists of loosely branched and detached runner hyphae, while the infective hyphae, more adpressed to the root, consist of pigmented hyphae that branch heavily at right angles (J. S. Gill, unpublished). Such branching is associated with infection cushions typical of R. *solani* on other hosts. These mycelia in the rhizosphere/rhizoplane region are easily targeted or outcompeted by antagonists that inhabit these niches. The major differences between the two pathogens is that while the infections can continue to be established till flowering with the take-all fungus, colonization by R. *solani* AG8 appears to be restricted mainly to seedling roots, and thus the antagonists have only a short period of time to interact with and suppress R. *solani* in the rhizosphere. This period is still shorter with *Pythium* spp. With dryland *Fusarium* rot, and to a lesser extent with C. *sativus*, the infection court activity of the pathogen can last from early infections of young roots and coleoptiles to senescing leaf sheaths of mature plants.

The final frontier for the defense of stelar elements of roots against root-infecting pathogenic fungi is the root cortex [7]. In addition to mycorrhizal fungi [35], a variety of nonmycorrhizal fungi [2] and bacteria [36] inhabit the live cortices of roots and succeed in defending the stele through mechanisms that involve direct and indirect interactions. These interactions have been recognized for a considerable period of time [3] and have been, with the exception of few fine studies (e.g., Refs. 37, 38), largely unexplored. As endophytic fungi they inhabit live cortices and cause little or no damage to the host. Instead, some of these fungi (such as the sterile red fungus [7]) produce antifungal compounds and promote plant growth. The sterile red fungus is capable of preemptively colonizing the seminal roots in the presence of the take-all fungus and masks the damaging effects of the pathogen by producing compensatory roots. The sterile red fungus also appears to compete with the take-all fungus for thiamine, which they both require, and cause hyphal lysis in the pathogen. This antagonist also causes induced resistance of roots to the take-all fungus [7]. It is, however, not clear whether or when one or more of the mechanisms observed are responsible for the disease reduction observed in limited field trials. Although this antagonist is endophytic, it offered no protection against *R. solani*. It also requires a relatively moist soil to be active, indicating that it is adapted only to a limited set of environmental conditions. This study, however, points to the fact that such fungi, which in all probability are cosmopolitan in distribution, deserve more attention than they have received to date.

Our work also showed that a wide variety of soil fungi invade the cortices of young wheat roots, many of which enhance plant growth and/or reduce root rot caused by the take-all fungus [2]. These fungi invariably invade up to the endodermis of the root and appear to do little or no harm to the host plant. Whether or not all of them could be classified as endophytes is not clear and certainly warrants investigation. The main advantage of these dwellers of root cortices is that if they are introduced to seed at sowing, their survival and activity in and along the root is extended in time and space. As cortical inhabitants they are also protected from harsh environments on and around the root, such as with drought-affected or waterlogged soils.

Disease suppression is considered to occur where the biocontrol agent reduces the activity of the pathogen on and inside the roots [2,8]. These interactions are directly or indirectly influenced by the host plant, the indirect action usually mediated by the host plant (induced resistance). Although by definition rhizosphere relates to the soil as influenced by host roots, it is currently accepted that rhizosphere activities of biocontrol agents could extend into the cortical cells of the root [39]. Thus, a population of antagonistic bacteria, for instance, could interact with the pathogen outside in the rhizosphere and could continue its activities inside the cortex (see Ref. 40). The activities of such bacteria inside the live root cortices is considered to facilitate the induction of resistance in the host plant.

V. BIOLOGICAL CONTROL AGENTS

A. Bacteria and Actinomycetes

Bacteria have predominantly been used for the control of cereal diseases. This is partly because of the ease of multiplication and application and also because of the advantage of having the background technology already developed for *Rhizobium* inoculants. Actinomycetes, unfortunately, have received relatively less attention. Numerically, bacteria dominate the rhizosphere, and many of them produce a wide variety of antifungal compounds. In general, the rhizosphere has been recognized as the frontline defense in protection of roots against invading root pathogens [6]. Constant phenotypic changes in the biocontrol bacteria, the heterogenous soil environment, and unpredictable climatic conditions contribute to the variation in the performance of many of the strains in the field.

Bacteria are relatively easy to screen in the greenhouse against the take-all fungus and *Pythium* spp. where average yield increases of wheat of up to 17% have been realized [41]. The pseudomonad strain 2–79, for instance, can be an aggressive root colonizer that could be reisolated through the cropping season, forming 50% of total fluorescent pseudomonads reisolated from seminal roots [42]. Kloepper et al. [43] demonstrated the importance of spermosphere competence, the level of importance of which, however, could vary from test to test. Root colonization ability is most critical for protection against root pathogens. The dose of *P. fluorescens* 2–79 applied to seed was reflected in its population of the strain on roots, and there was an inverse relationship between the population of the strain 2–79 on roots and incidence and severity of take-all lesions on wheat roots. Populations of these strains on individual roots, however, varied up to a 1000-fold [6]. Following inoculation with the strain, 20–40% of the roots were not colonized 4 weeks after sowing. It is, therefore, critical to ensure that colonization of the roots by the protectant bacterium is complete at the time the susceptible region of the root is exposed to the root pathogen resident in the surface organic layer soil. Fortunately, reports (see Ref. 6) indicate that the bacterial colonization by competent strains can be very high at the region near the inoculum layer of the pathogen in the soil, declining towards the root tip. The colonization of the root was considered to take place at the rate of 2–9 cm/d. The successful strains are supposed to attach themselves to the roots, often displacing ineffective resident strains, the growth being determined by the soil environment, which also affects root growth. In this matter their behavior is similar to the competitiveness of effective *Rhizobium* strains on legume roots.

P. fluorescens 2–79 was reported [6] to be active in the range of -0.3 to -0.7 bars in which O_2 availability and turgor potential of the cells and/or nutrient availability was optimum, as was the temperature range of 25–30°C. *Pseudomonas putida* prefers <20°C for root colonization and a pH of 6–6.5 rather than 7. These requirements may be critical for the application of these bacteria in soils where aeration, temperatures, and moisture may not be conducive for their multiplication. It may therefore be advisable to consider application of strains suitable for each environment or, if possible, consider application of a mixture of compatible strains from which appropriate strains could be naturally selected by the unpredictable microenvironment around the germinating wheat seed. In addition, the activity of the individual strain could also be affected by the phenotype of the host plant [6] associated with it.

It is also probable that different biocontrol strains may be required to target different pathogens. The strain Q2-79-80, for instance, could suppress a *Pythium* sp., while increasing or having no effect on *R. solani* AG8 [6]. Where a certain strain lacks ecological competence in a specific environment, other introduced and better adapted strains in the mixture could take over the antagonistic activities to bring about the suppression of root rot caused by the take-all fungus [6,24]. It is also possible that some useful strains have specific site(s) of activity. *P. chloraphis* MA342, for instance, appears to have strong spermosphere but poor rhizosphere colonizing ability [45]. This strain was shown to have strong potential for control of seedborne plant pathogens such as *Drechslera teres*, *Tilletia caries*, *Microdochium nivale*, and *Stagonospora nodorum* [45]. This strain produces the metabolite 2,3-deepoxy-2,3-didehydrorhizoxin (DDR), which was considered to be involved in disease control [45]. A variety of bacterial metabolites with involvement in disease suppression have been reported. These include siderophores (e.g., pyroverdin, pseudobactin, pyochelin), antibiotics (e.g., acetylphloroglucinols, 2,4-diacethylphloroglucinol, phenazines, phenazine-1-carboxylic acid, pyrroles, pyoluteorin, pyrrolnitrin, macrocyclic latones, DDR), lytic enzymes (e.g., β-1,3-glucanase, chitinases, proteases) biosurfactants (e.g., rhamnolipids), and HCN. A variety of plant growth–promoting substances are also reported to be produced by bacterial strains and include plant hormones such as auxins, indole-3-acetic acid, and gibberellins [46–48]. Certain hormones can help the host to cope with the disease by producing compensatory roots.

The ability of many of these strains to produce bioactive compounds in vivo has been demonstrated in several studies. It is, however, accepted that the production of these compounds in effective quantities is determined by the soil environment, including nutrient supply. Increased production of these compounds by biocontrol agents, however, is not always related to increased disease suppression.

Because many of these biocontrol bacterial strains produce growth-promoting substances in addition to antifungal compounds and, in some cases, can also cause induced resistance [36,49], it is not clear whether all three mechanisms are employed by the producers in an environment at a given time.

B. Fungi

Several monographs (e.g., Ref. 3) and reviews (e.g., Ref. 39) have comprehensively covered the potential of fungi as biocontrol agents of fungal pathogens in cereal crops. Essentially, various fungi (including yeasts) are capable of antagonizing pathogens in all niches associated with the wheat plant illustrated in Figure 3. In soil they can be operative in the bulk soil (e.g., *Trichoderma koningii* [8]) affecting pathogen suppression. Such strains can also be operative in the rhizosphere, including the root cortex [11]. Effectiveness of such strains are certainly related to their ability to suppress the pathogen at several of the niches referred to in Figures 1 and 2.

Ideally, antagonists of each pathogen should be screened in relation to the soil behavior and spreading habits on and in roots by the pathogen. For instance,

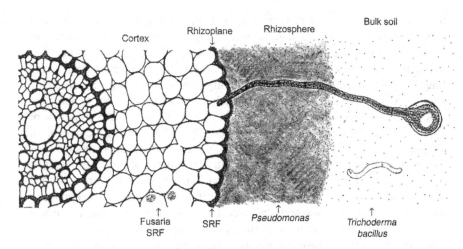

Figure 3 Regions in the soil and root where the pathogen could be suppressed. The pathogens including nematodes could be suppressed by biocontrol agents (e.g., *Bacillus* spp. [3], *Trichoderma* spp. [8], and *Verticillium chlamydosporium* [31]) in the bulk soil. In the rhizosphere and/or rhizoplane it could be suppressed by biocontrol agents such as *Pseudomonas* spp. [6]. The rhizoplane and cortices (up to endodermis) could be the sites of activities of biocontrol fungi such as *Fusarium* spp. [37], *G. graminis* var. *graminis* [55], *Phialophora graminicola* [19], and the sterile red fungus (SRF) [7].

while the soil saprophytic growth and soil colonization habit of the take-all fungus may be limited, these habits are significantly elaborate and extensive in *R. solani* AG8, the pathogenicity of which is therefore determined to a large extent by the establishment of its soil phase preceding infection. In bare patch disease of wheat caused by *R. solani* AG8, the severity of disease depends heavily on the integrity of the hyphal inoculum of the pathogen in soil.

The mechanisms by which fungal antagonists have been proposed to suppress diseases caused by fungal plant pathogens are many and varied. Unlike with bacteria, only a few genetic studies of them have been conducted in order to establish the basis of disease suppression. Competition, antibiosis, predation, and mycoparasitism have been proposed as mechanisms [3] but have generally been difficult to prove exclusively. Harman [50], discussing mechanisms of suppression by antagonistic *Trichoderma*, suggested that where a certain mechanism employed by a biocontrol agent fails or is ineffective, it is likely that another mechanism could come into play and compensate.

While most studies with fungal antagonists of fungal pathogens involve in-depth investigations into mechanisms of suppression, relatively few examine the niche behavior in soil and root of the antagonists and their role in the determination of the suitability of the biocontrol agent. Fungi, although numerically smaller than bacteria, have an advantage in their ability to grow towards a stimulus or target and also tolerate a greater range of moisture and/or pH stresses. The nutrient status of the niche is also critical for all forms of biocontrol agents, including fungi. In the bulk soil, in the absence of a high level of competitive saprophytic ability, the biocontrol agent may opt to rely to a large extent on its ability to sustain itself as a mycoparasite or to live off the exudates from the propagule or hyphae of the pathogen (e.g., through externally coiling around the hyphae of the pathogen). The hyphae of the pathogen in these nutrient-impoverished soils low in organic matter can be a relatively attractive source of nutrients attracting "hyphasphere" bacteria, actinomycetes, and fungi. It has been proposed [51] that strains of *Trichoderma* spp., which are naturally attracted to the hyphae of the take-all fungus in these soils, produce antibiotics that could help to weaken and predispose the hyphae to successful mycoparasitism by the antagonist. The root exudates in the rhizosphere support a large spectrum of bacteria and fungi that are able to compete and utilize them. The biocontrol agents in the rhizosphere/rhizoplane region are therefore expected to have a certain level of rhizosphere competency to be sufficiently active to suppress the ectotrophic activities of the pathogen. The outer cortical cells of wheat roots progressively die and are invaded by soil microflora [52,53]. This region of roots sometimes called the "endorhizosphere" (see Ref. 40) is to both the antagonists and pathogen ecologically indistinguishable from the natural rhizosphere. Whether rhizosphere inhabiting biocontrol agents invariably invade live cortical cells needs verification, especially where they cause induced resistance against the invading pathogens. Another approach has been to seek antagonists among fungi which are taxonomically and ecologically similar to pathogens [54–56].

A variety of fungal invaders and inhabitant of root cortices, some mycorrhizal and others not [7], have some level of parasitic competency and are effective in protecting the vascular elements from fungal pathogens. This is a promising area for exploration, at least in annuals such as wheat where the protection and growth promotion of young seminal roots confers significant boost to the early growth, which could lead to improved yield from the crop.

C. Microfauna

Activities of microfauna in the organic layer of wheat field have been largely ignored or underrated, especially in relation to pathogen suppression. Chakraborty and Warcup [57] reported the ability of soil protozoan populations to parasitize or feed on runner hyphae of the take-all fungus. The role of microfauna in the development of "general suppression" [58] to root disease such as take-all and *Rhizoctonia* bare-patch warrants detailed studies. Following the report of "decline" of root rot of wheat caused by *R. solani* AG8 in South Australia by

Roget [59], Gupta et al. [60] found that protozoan predation and the subsequent decomposition of damaged *Rhizoctonia* hyphae was one of the main reasons for the reduced survival of *Rhizoctonia* inoculum from one season to another. Grazing by microfauna of the fungal network of *Rhizoctonia*, which form the large component of the fungal inoculum in soil, could be expected to have a significant effect on the inoculum potential of the pathogen. Although it may be difficult to culture and introduce individual species of beneficial micro- or mesofauna on a field scale, there is certainly scope to develop cultural practices, which may promote the activity of these organisms. There is currently considerable interest in the use in Australia of brassica residues for the "biofumigation" of wheat field soils [61]. The glucosinolates in these plant tissues are hydrolyzed to isothiocyanates in soil, which has been proposed to reduce the inoculum of the take-all fungus. Although this approach may lead to the emergence of diseases not responding to the amendment, it certainly deserves further evaluation. It is likely that certain microfauna populations could be enhanced following such "green-manure" residues enhancing soil fertility and inducing "general suppression" [58].

VI. CONCLUSION

Biological control of wheat diseases, mainly those caused by soilborne fungal pathogens, have been studied worldwide. Certain diseases, such as the take-all caused by *Gaemannomyces graminis* var. *tritici*, have been useful "models" to study the mechanisms of biocontrol in depth [47]. Despite the knowledge that has accrued on this subject, there is no indication to date of a reliable biological means of field control of cereal root rots on a large scale. The only exception is, however, the "take-all decline" that occurs following continuous cropping of wheat in fields infested with the take-all fungus [62,63]. While this method has wide acceptance in Europe [64] and the United States [3], it has not been widely successful in Australia. Similar "decline" has also been reported with *R. solani* AG8 in Australia [59] and the United States [63]. The nature of agriculture in Australia, however, may not accommodate a practice that requires continuous cropping with wheat. Studies on this suppressiveness phenomena may yield information that could help to develop new strategies for the cultural control of cereal root rots.

Specific-suppression strategies have attracted most of the attention of those working on the biocontrol of wheat root diseases. Although these studies have yielded exciting results, many of the trials using this strategy have, in general, resulted in inconsistent results [64], mainly due to unpredictable and variable environmental conditions. There are several instances where combinations of specific and general suppression could operate together and yield more reliable results (e.g., Refs. 8, 37). Manipulation of the environment through cultural practices to enhance the activities of specific and/or general microflora clearly holds great promise.

Another aspect receiving inadequate attention currently is the potential hazard of a nontarget pathogen replacing a pathogen that is specifically suppressed following application of an effective biocontrol agent. A complex of pathogens is commonly resident in wheat field soils [2]. Even within a pathogenic species, variation in sensitivity to a biocontrol agent may occur. For example, anastomoses groups of *Rhizoctonia solani* may vary in their sensitivity to gliotoxin [65]. Strategies to manage the major diseases simultaneously may eventually become the solution. This is again the reason why, even where specific suppression of a disease is promising, a management strategy that includes practices enhancing "general suppression" of two or more major diseases may still be necessary and agronomically useful.

ACKNOWLEDGMENTS

I thank M. J. Barbetti and B. Gaskell for reviewing the manuscript and Alanna Wood and Janet King for their immense patience and help in the preparation of the manuscript.

REFERENCES

1. RD Graham. Effects of nutrient stress on susceptibility of plants to disease with particular references to the trace elements. Adv Bot Res 10:221–276, 1983.
2. K Sivasithamparam. Ecology of root infecting pathogenic fungi in mediterranean environments. Adv Pl Pathol 10:245–280, 1993.
3. RJ Cook, KF Baker. The Nature and Practice of Biological Control of Plant Pathogens. St. Paul, MN: The American Phytopathological Society, 1983.
4. MJC Asher, PJ Shipton. Biology and Control of Take-All. London: Academic Press, 1981.
5. SD Garrett. Pathogenic Root-Infecting Fungi. London: Cambridge University Press, 1970.
6. DM Weller. Biological control of soil-borne plant pathogens in the rhizosphere with bacteria. Annu Rev Phytopathol 26:379–407, 1988.
7. K Sivasithamparam. Root cortex—the final frontier for the biocontrol of root-rot with fungal antagonists: a case study on a sterile red fungus. Annu Rev Phytopathol 36:439–452, 1998.
8. A Simon, K Sivasithamparam. Pathogen suppression: a case study in biological suppression of *Gaeumannomyces graminis* var. *tritici* in soil. Soil Biol Biochem 21:331–337, 1989.

9. ME Brown, D Hornby. Behaviour of *Ophiobolus graminis* on slides buried in soil in the presence or absence of wheat seedlings. Trans Br Mycol Soc 56:95–103, 1971.

10. J Handelsman, GS Gilbert. Root camouflage and disease control. Phytopathology 84:222–225, 1994.

11. SK Dyer. The role of colonisation of soil and wheat roots by *Trichoderma koningii* in biological control of *Gaeumannomyces graminis* var. *tritici*. Ph.D. dissertation, Adelaide University, Adelaide, 2000.

12. K Sivasithamparam. The effect of soil nutrients on microbial suppression of soil borne diseases. In: RS Utkhede, VK Gupta, eds. Management of Soilborne Diseases. New Delhi: Kalyani Publishers, 1996, pp 123–145.

13. ME Leggett, K Sivasithamparam, MJ McFarlane. Effect of nitrogen supply on rhizosphere interactions and take-all disease of wheat. Can J Microbiol 37:42–51, 1991.

14. DM Huber. The role of nutrition in the take-all disease of wheat and other small grains. In: AW Engelhard, ed. Soilborne Plant Pathogens: Management of Diseases with Macro- and Microelements. St. Paul, MN: The American Phytopathological Society, 1996, pp 46–74.

15. SD Garrett. Factors affecting the severity of take-all II. Soil temperature. J Dep Agric S Aust 37:799–805, 1934.

16. MJ Grose, CA Parker, K Sivasithamparam. Growth of *Gaeumannomyces graminis* var. *tritici* in soil: effects of temperature and water potential. Soil Biol Biochem 16:211–216, 1984.

17. JS Gill, K Sivasithamparam, KRJ Smettem. Soil types with different texture affects development of *Rhizoctonia* root rot of wheat seedlings. Pl Soil 221:113–120, 2000.

18. PR Scott. *Phialophora radicicola*, an avirulent parasite of wheat and grass roots. Trans Br Mycol Soc 55:163–167, 1970.

19. JW Deacon. Ecological relationships with other fungi: competitors and hyperparasites. In: MCJ Asher, PJ Shipton, eds. Biology and Control of Take-all. London: Academic Press, 1981, pp 75–101.

20. RW Smiley, MC Fowler, RT Kane. Temperature and osmotic potential effects on *Phialophora graminicola* and other fungi associated with patch diseases of *Poa pratensis*. Phytopathology 75:1160–1167, 1985.

21. N Charigkapakorn, K Sivasithamparam. Changes in composition and population of fluorescent pseudomonads on wheat roots inoculated with successive generations of root-piece inoculum of the take-all fungus. Phytopathology 77:1002–1007, 1987.

22. HX Peng, K Sivasithamparam, DW Turner. Chlamydospore germination and *Fusarium* wilt of banana plantlets in suppressive and conducive soils are affected by physical and chemical factors. Soil Biol Biochem 31:1363–1374, 1999.

23. A Troughton. The Roots of Temperate Cereals (Wheat, Barley, Oats and Rye). Hurley: Commonwealth Agricultural Bureaux, 1962.

24. K Sivasithamparam, CA Parker. Effects of certain isolates of bacteria and actinomycetes on *Gaemannomyces graminis* var. *tritici* and take-all of wheat. Aust J Bot 26:773–782, 1978.

25. MJC Asher. Effect of *Ophiobolus graminis* infection on the growth of wheat and barley. Ann Appl Biol 70:215–223, 1972.

26. RJ Cook. Fusarium root rot of wheat and its control in the Pacific Northwest. Pl Dis 64:1061–1066, 1980.

27. LW Burgess, CM Liddell, BA Summerell. Laboratory Manual for Fusarium Research. Sydney: University of Sydney, 1988.
28. BA Summerell. Wheat stubble decomposition and stubble-borne diseases. Ph.D. dissertation, University of Sydney, Sydney, 1988.
29. JH Andrews. Comparative Ecology of Microorganisms and Macroorganisms. New York: Springer-Verlag, 1991.
30. GW Bruehl. Management of food resources by fungal colonists of cultivated soils. Annu Rev Phytopathol 14:247–264, 1976.
31. BR Kerry, DH Crump, LA Mullen. Studies of the cereal cyst nematode, *Heterodera avenae* under continuous cereals, 1975–1978. II Fungal parasitism of nematode females and eggs. Ann Appl Biol 100:489–499, 1982.
32. HJ McDonald, AD Rovira. Development of inoculation technique for *Rhizoctonia solani* and its application to screening cereals for resistance. In: CA Parker, AD Rovira, KJ Moore, PTW Wong, JF Kollmorgen, eds. Ecology and Management of Soilborne Plant Pathogens. St. Paul, MN: The American Phytopathological Society, 1985, pp 174–176.
33. BK Duffy, A Simon, DM Weller. Combination of *Tricoderma koningii* with fluorescent pseudomonads for control of take-all on wheat. Phytopathology 86:188–194, 1996.
34. M Ryder, unpublished data.
35. DH Marx. Ectomycorrhizae as biological deterrents to pathogenic root infections. Annu Rev Phytopathol 10:429–454, 1972.
36. G Wei, JW Kloepper, S Tuzun. Internal and external colonisation of cucumber by bacteria which induce systemic resistance to *Colletotrichum orbiculare*. Phytopathology 82:1108, 1992.
37. C Alabouvette. Biological control of *Fusarium* wilt pathogens in suppressive soils. In: D Hornby, ed. Biological Control of Soilborne Plant Pathogens. Wallingford: CAB Int., 1990, pp 27–43.
38. L Gasoni, BG de Gurfinkel. The endophyte *Cladorrhinum foecundissimum* in cotton roots: phosphorus uptake and host growth. Mycol Res 101:867–870, 1997.
39. JM Whipps. Developments in the biological control of soil-borne plant pathogens. Adv Bot Res 26:1–134, 1997.
40. JW Kloepper, B Schippers, PAHM Bakker. Proposed elimination of the term endorhizosphere. Phytopathology 82:726–727, 1992.
41. DM Weller, RJ Cook. Suppression of take-all of wheat by seed treatments with fluorescent pseudomonads. Phytopathology 73:463–469, 1983.
42. DM Weller. Colonisation of wheat roots by a fluorescent pseudomonad suppressive to take-all. Phytopathology 73:1548–1553, 1983.
43. JW Kloepper, RM Scher, M Laliberte, I Zaleska. Measuring the spermosphere colonising capacity (spermosphere competence) of bacterial inoculants. Can J Microbiol 31:926–929, 1985.
44. DM Weller. Effect of wheat genotype on root colonisation by a take-all suppressive strain of *Pseudomonas fluorescens*. Phytopathology 76:1059, 1986.
45. M Hokeberg. Seed bacterization for control of fungal seed-borne diseases in cereals. Doctoral thesis, Swedish University of Agricultural Sciences, Uppsala, 1998.
46. G Defago, D Haas. Pseudomonads as antagonists of soilborne plant pathogens.

Modes of action and genetic analysis. In: JM Bollag, G Stotzky, eds. Soil Biochemistry 6. New York: Marcel Dekker, 1990, pp 249–289.

47. RJ Cook, LS Thomashow, DM Weller, D Fujimoto, M Mazzola, G Bangera, DS Kim. Molecular mechanisms of defense by rhizobacteria against root disease. Proc Natl Acad Sci USA 92:4197–4201, 1995.

48. JJ Borowicz. Traits of biologically interacting pseudomonads. Doctoral thesis, Swedish University of Agricultural Sciences, Uppsala, 1998.

49. LC van Loon, PAHM Bakker, CMJ Pieterse. Systemic resistance induced by rhizosphere bacteria. Annu Rev Phytopathol 36:453–483, 1998.

50. GE Harman. Myths and dogmas of biocontrol: changes in perceptions derived from research on *Trichoderma harzianum* T-22. Pl Dis 84:377–393, 2000.

51. EL Ghisalberti and K Sivasithamparam. Antifungal antibiotics produced by *Trichoderma* spp. Soil Biol Biochem 23:1011–1020, 1991.

52. J Holden. Use of nuclear straining to assess rates of cell death in cortices of cereal roots. Soil Biol Biochem 7:333–334, 1975.

53. K Sivasithamparam, CA Parker, CS Edwards. Rhizosphere microorganisms of seminal and nodal roots of wheat grown in pots. Soil Biol Biochem 11:155–160, 1979.

54. K Sivasithamparam. *Phialophora* and *Phialophora*-like fungi occuring in the root region of wheat. Aust J Bot 23:193–212, 1975.

55. PTW Wong, RJ Southwell. Field control of take-all by avirulent fungi. Ann Appl Biol 94:41–49, 1980.

56. B Sneh. Use of non-pathogenic or hypovirulent fungal strains to protect plants against closely related fungal pathogens. Biotechnol Adv 16:1–32, 1998.

57. S Chakraborty, J Warcup. Soil amoebae and saprophytic survival of *Gaemannomyces graminis tritici* in a suppressive pasture soil. Soil Biol Biochem 15:181–185, 1983.

58. M Gerlagh. Introduction of *Ophiobolus graminis* into new polders and its decline. Meded Lab Phytopath No. 241, 1968.

59. DK Roget. Decline in root rot (*Rhizoctonia solani* AG8) in wheat in a tillage and rotation experiment at Avon, South Australia. Aust J Exp Agric 35:1009–1013, 1995.

60. VVSR Gupta, R Coles, S McClure, SM Neate, DK Roget, CE Pankhurst. The role of micro-fauna in the survival and effectiveness of plant pathogenic fungi in soil. Aust Microbiol A64, 1996.

61. JA Kirkegaard, M Sarwar. Biofumigation potential of brassicas. I. Variation in glucosinolate profiles of diverse field-grown brassicas. Pl Soil 201:71–89, 1998.

62. PJ Shipton. Monoculture and soilborne plant pathogens. Annu Rev Phytopathol 15: 387–407, 1977.

63. P Lucas, RW Smiley, HP Collins. Decline of Rhizoctonia root rot on wheat in soils infested with *Rhizoctonia solani* AG8. Phytopathology 83:260–265, 1993.

64. D Hornby. Suppressive soils. Annu Rev Phytopathol 24:65–85, 1983.

65. RW Jones, R Pettit. Variations in sensitivity among anastomosis groups of *Rhizoctonia solani* to antibiotic gliotoxin. Pl Dis 71:34–36, 1987.

5
Biological Control of Cotton Diseases

Kalyan K. Mondal
Regional Research Substation, Chaudhary Sarwan Kumar Himachal Pradesh Krishi Vishwavidyalaya, Sangla, Kinnaur, Himachal Pradesh, India

Jeevan P. Verma
National Academy of Agricultural Sciences, Indian Agricultural Research Institute, New Delhi, India

I. INTRODUCTION

Cotton (*Gossypium* spp.) is an important fiber crop cultivated in almost every tropical country as well as in many parts of the subtropical world. In India, cotton has been cultivated since ancient times, playing a key role in the national economy. Cotton fields in India occupy an area of 8.9 m ha, representing the largest cotton-growing area in the world. However, it is only fourth place in production (11.1 m bales, each bale = 170 kg lint), and the mean productivity is 292 kg lint/ha, which is just 50% of the world average of 583 kg lint/ha [1]. Lack of suitable disease-management programs is considered to be one of the most important factors limiting the productivity of this crop [2]. The present chapter will discuss the changing disease scenario of cotton and the biocontrol approaches so far taken in combating such diseases, keeping in view the fundamental research and its ecofriendly application in integrated disease/crop management.

II. COTTON DISEASE SCENARIO

A. Diseases of Major Concern

Several diseases afflict the cotton crop, but bacterial blight, root rot, new wilt, and cotton leaf curl are of major concern. Bacterial blight of cotton, caused by *Xanthomonas campestris* pv. *malvacearum*, was of minor importance until 1947–48 primarily because 97% of the total area was planted with diploid, indigenous cottons (*G. arboreum* L. and *G. herbaceum* L.), which are naturally resistant to the disease [3–5]. The disease assumed severe proportions during 1948–52, when a campaign was started to replace the rainfed, indigenous, diploid cottons with the high-yielding, good-quality, tetraploid cottons (*G. hirsutum* and *G. barbadense*), which were susceptible to disease [4].

During the last decade a new problem of wilt has emerged, which affects all the four cultivated *Gossypium* spp., causing much damage in the central and southern parts of India In affected plants a loss of 14.76–94.5% has been observed, depending upon the genotype and the time of manifestation of wilting [2]. Direct and indirect evidence suggests that the disease is not caused by any fungus, bacterium, virus, or phytoplasma; neither is it transmitted through grafting nor through infected debris or soil [2,6]. According to Sahay et al. [7], excess production of ethylene may be responsible for this wilting.

Recent surveys in 10 villages in East Nimar (Khandwa) during 1989–90 and 1990–91 to assess the incidence of new wilt revealed that on average the incidence ranged between 3.0 and 4.0% in 1989 and 3.2, and 4.8% in 1990 in the four most commonly cultivated hybrids/varieties; the maximum incidence was found, however, at Burhanpur on Maljari (6.2 and 5.2% in 1989 and 1990, respectively). The incidence on both hybrids and varieties was almost same, contradicting the earlier view that the disease is more severe in or restricted to hybrids [8]. The relative incidence of the disease varies from season to season, even in the same genotype. The observation that some genotypes are relatively disease-free in one season but others are affected and that some seasons are wilt-free while others show a high incidence is suggestive of the environment influencing the expression of the disease. The susceptibility to new wilt was found to be distributed in all three species widely cultivated in India (*G. hirsutum*, *G. arboreum*, and *G. herbaceum*, including inter- and intrahirsutum hybrids). It was concluded that new wilt is a genetically controlled physiological disorder expressed under strong influence of the environment.

Root rots caused by *Rhizoctonia solani* Kuhn and *R. bataticola* (Taub.) Butler have been reported from all the cotton-growing areas but are serious problems only in Punjab, Rajasthan, and Gujarat. A loss of about 3% in general and 90% in certain cases have been reported [9].

Cotton leaf curl virus (CLCuV) has become the most important disease problem in the northern cotton zone of India and Pakistan. The virus belongs to the Gemini virus group [10]. In India, the disease was noticed in the Sriganganagar, Abohar, and Fazilka areas of Rajasthan and Punjab [11].

B. Diseases of Minor Importance

With the change in cotton species complex, the disease situation also changed. *Fusarium* wilt became less important because American (New World) cottons are immune to the Indian race of *Fusarium* wilt pathogen, and these upland cottons now occupy about 70% of the cotton area. So far no change in the virulence of this pathogen has been noticed, and American cotton varieties continue to maintain immunity to this wilt. Of late *Verticillium* wilt appeared on cotton in Tamil Nadu, which is, however, restricted to southern areas [2]. Grey mildew, which was a serious problem for diploid cottons, especially in Maharashtra and Gujarat, has almost disappeared due to large-scale cultivation of tetraploid cottons immune to this disease. Other diseases such as *Alternaria, Myrothecium, Cercospora, Helminthosporium* leaf spots and rust occur only sporadically. Diseases of viral origin, e.g., leaf crumple [12], anthocyanosis [13], and viral wilt [14], have been restricted in the Marathwada area of Maharashtra, where these diseases were first reported. Bacterial wilt [15] and stenosis [16] are mainly restricted to Karnataka.

III. MANAGEMENT PRACTICES

The existing disease-management approaches are primarily chemical based and are, therefore, not efficient in terms of energy use or ecological concerns. Further, soilborne cotton pathogens, especially root-rot pathogens, are very difficult to manage with chemicals. Soil fumigation is effective but not affordable. The development of resistance to some soil-treatment chemicals (brassicol and vitavax) was also evident in *R. bataticola*, the inducer of root rot [4]. Further, the injudicious use of pesticides is becoming a global concern, mainly because of its negative impact on environment, ecology, and society. Resistance against widely used chemicals is gradually becoming more common. The continuous exposure of bacterial blight pathogens to antibiotics has resulted in the development of antibiotic-resistant mutants [17,18]. Verma et al. [5] suggested different need-based integrated practices for the eradication of bacterial blight of cotton, depending on the availability of chemicals, date of sowing, disease symptoms, and other agronomic practices; they also demonstrated natural biological control through the presence of less virulent/avirulent races of *X. c.* pv. *malvacearum* and phylloplane bacteria with preinoculative protective ability [3,17]. Agronomic practices including tillage to manage pathogens present in crop residues has led to loss of soil organic matter and soil erosion. Resistant varieties are known for a large number of bacterial [19], fungal [20,21], and viral diseases [22]. But the ecological sustainability of the resistance breeding essentially depends on a constant supply of new resistant sources in the germplasm, which is unlikely in view of the genetic diversity of pathogens, particularly when several pathogens are involved. Thus, the life of a resistant variety is short.

Genetic engineering approaches towards developing transgenics have offered hope because of wider access to useful genes, but it may take some time for this to become socially and ecologically acceptable.

IV. NEED FOR BIOLOGICAL CONTROL

Environmental and related considerations have led to a progressive decline in the availability and use of pesticides, stressing the need to develop dependable biological control. Although complete replacement of chemicals in disease management is not anticipated in the near future, the considerable progress already made in the development of eco-friendly biocontrol agents gives hope that there may someday be a remarkable reduction in the use of agro-chemicals. Thus, a rapid shift from a chemical-intensive to a biological agent–intensive integrated cropping system [ICS, including the appropriate IPM (Integrated Pest Management) menu] is clearly visible. In the near future, the IPM menu, based on environment, agricultural practices, and pest scenario (including forecast network), will be available to farmers so that they can choose the best locally and ecologically sustainable menu in the ICS.

V. USE OF SAPROPHYTIC/UNRELATED MICROORGANISMS IN BIOCONTROL OF COTTON DISEASES

A. Phylloplane Microorganisms

Biological control using unrelated saprophytic microorganisms has been demonstrated for bacterial blight of cotton [16,23–25]. Spraying with several phylloplane bacteria resulted in the reduction of disease; these phylloplane bacteria (species of *Aeromonas*, *Flavobacterium*, and *Pseudomonas*) were not antagonistic to *X. c.* pv. *malvacearum* but possessed a preinoculative protective ability [i.e., when these phylloplane bacteria were preinoculated 8–20 hours before the challenge dose of *X. c.* pv. *malvacearum*, there was no (reduced) disease development], perhaps due to the production of a protection factor (something similar to elicitors) and faster use of nutrients. Some strain specificity was also observed, and the cultivars Bikaneri Nerma and BC-131-2 were most responsive to phylloplane bacteria spray for the control of bacterial cotton blight. This area requires more research to develop a phylloplane bacterium or a mixture of phylloplane bacteria/microorganisms for the biological control of cotton diseases.

The phylloplane bacteria possessing this preinoculative protective effect belonged to the genera *Flavobacterium*, *Pseudomonas*, and *Aeromonas* and represented 14.2% of the total phylloplane bacterial population. Greater protection was afforded to a less virulent race of *X. c.* pv. *malvacearum* than to a highly virulent race. Field trials have already indicated that a phylloplane bacteria spray could effectively reduce the severity of bacterial cotton blight and could be used effectively in the integrated management of the disease [25].

B. Endophytes as Biocontrol Agents

The interaction of bacterial endophytes on cotton with other plant-associated bacteria has been critically reviewed [26]. *Pseudomonas fluorescens* 89B-61 is a root colonist that has been shown to reduce the incidence of *Fusarium* wilt of cotton (causal agent *Fusarium oxysporum* f. sp. *vasinfectum*) [27]. Endophytic bacteria have shown significant control in model systems of *R. solani* on cotton [28] Chen et al. [29] suggested that the biocontrol effect of endophytic bacterial strains resulted mainly from enhanced host defense rather than from bacterial metabolites. Bacterial endophytes are also thought to play a role in the resistance observed in the multiple adversity-resistant (MAR) cotton lines [30]. Biswas isolated five bacterial endophytes from cotton plants showing antagonistic activity towards *X. c.* pv. *malvacearum*, *Sclerotium rolfsii*, and *R. solani* in vitro; all of these endophytes belonged to the genus *Pseudomonas* (B. Bhowmik, personal communication, 2000).

C. Rhizosphere Microorganisms Including Plant-Growth–Promoting Rhizobacteria

Fluorescent pseudomonads are unique rhizobacteria possessing the ability to promote plant growth and suppress plant diseases [31] (Table 1). Laha and Verma [35] demonstrated that 18.75% of fluorescent pseudomonad isolates were effective against *X. c.* pv. *malvacearum* (angular leaf spot phase of bacterial blight of cotton), *R. solani* (root rot), and *S. rolfsii* [38]; seed bacterization with fluorescent pseudomonads effectively increased seed germination and reduced seedling infection by *R. solani*, *S. rolfsii*, and seedling blight by *X. c.* pv. *malvacearum*.

A strain of *Pseudomonas fluorescens* antagonistic to *R. solani* was isolated from rhizosphere of cotton seedlings [33], which produced an antibiotic (pyrrolnitrin) that was strongly inhibitory to *R. solani*. Seed treatment with this bacterium or pyrrolnitrin increased seedling survival from 30 to 79% and from 13 to 70%, respectively. Pyrrolnitrin persisted for 30 days in moist nonsterile soil with no measurable loss in activity. Verma et al. [39] also observed more than 80% im-

Table 1 Cotton Diseases Controlled by Fluorescent Pseudomonads

Disease	Pathogen	Ref.
Bacterial blight	*Xanthomonas campestris* pv. *malvacearum*	3,32
Seedling blight	*Rhizoctonia solani*	33–35
Seedling rot	*Pythium ultimum*	84,86
Wilt	*Fusarium oxysporum* f. sp. *vasinfectum*	28,37

proved seed germination on seed treatment of various cotton cultivars with fluorescent pseudomonads. Sezgin et al. [40] identified *Penicillium* sp., *P. petuletum*,
Aspergillus sp., *A. terreus*, *A. fumigatus*, and *Chaetomium* sp. as antagonistic to
R. solani. Seed coating with *Trichoderma* spp. reduced disease (*R. solani*) incidence up to 83% in greenhouse [41]. In field experiments, disease severity was
reduced to 47–60%, which was equal to that obtained with PCNB. *Pseudomonas
fluorescens* CRb-26, a cotton rhizobacterium, when applied on *X. c.* pv. *malvacearum*–inoculated cotton seeds, significantly improved [32] germination (by
35.71 and 82.07% under sterilized and unsterilized soil, respectively, over the
nonbacterized *X. c.* pv. *malvacearum*–inoculated seeds). The isolate CRb-26,
which was highly effective in in vitro antibiosis, also showed the highest ability
to colonize upon seed bacterization along the young cotton seedling, especially
on the cotyledon. At the same time it significantly reduced the cotyledon infection
(71.59 and 58.51% in sterile and nonsterile soils, respectively) [32]. Further, the
rhizosphere resident CRb-26 was also effective on aerial parts. The isolate CRb-
26 caused significant reduction in bacterial blight disease intensity on susceptible
cotton line Acala 44 when applied 8 and 24 hours after or before challenging
with *X. c.* pv. *malvacearum* [32]. This observation suggests that a viable cotton
bacterial blight management strategy can be developed with isolate CRb-26 as
presowing seed treatment to control seedborne infection and as postemergence
foliar spray to control secondary spread of the disease during the growing season.
The effectiveness of fluorescent pseudomonads/PGPR against seedling infection
of cotton caused by fungal pathogens including *R. solani*, *Pythium ultimum*, *Fusarium* spp., *Verticillium* sp., *Sclerotinia* sp., etc. has also been reported in other
crops [42–46]. An isolate of *Burkholderia cepacia* (D₁) recovered from cotton
bolls in Arizona proved to be an extremely effective control agent (in the field
as soil drench) against *Aspergillus flavus*–induced cotton boll decay [47] and *R.
solani*–induced cotton seedling damping-off [48] in the field.

D. Bacteriophages

Bacteriophages have been reported to act as biocontrol agents [49]. A field experiment in 1936 in Tashkent showed that bacteriophage treatment of the cotton seed
in combination with vernalization lowered the infection of seeding diseases by
74%; treatment of the seeds shortly before sowing showed very little effect [50].
Phages of *X. c.* pv. *malavacearum* are widely distributed and could be isolated
from cotton soil and fresh and stored diseased leaves [51]. Lysogeny was also
demonstrated [18]. A detailed study was made of the uptake and translocation
of phage lysate and longevity of phages in cotton seedlings with a view to resolve
the potentials of phages as biocontrol agents [52]. It was evident that cotton
seedlings absorbed phages when root tips of 72-hour-old cotton seedlings were

immersed in phage lysate, and subsequently the phages reached the cotyledon within 12 hours. It was concluded that phages could persist in cotton seedlings at least 48 hours, even in the absence of *X. c.* pv. *malvacearum*. When phage lysate (90 × 10^7 pfu/mL) was spray-inoculated on leaves, the phages were detected for only 24 hours, while in case of leaf infiltration phages were detected for 48 hours. Verma et al. [52] also noted that the preinoculation of phage 1 hour before *X. c.* pv. *malvacearum* challenge afforded maximum protection for susceptible reaction of cv. Acala-44 (40%) and hypersensitive reaction of 101–102 B (80%). It was found that at the 0 hour challenge, better control was achieved when phage was applied first. It appears that when the bacterial pathogen is applied first, the bacterium occupies the infective sites in the host within the intercellular spaces, which makes them relatively inaccessible to the phages.

VI. USE OF AVIRULENT STRAINS IN BIOCONTROL

A. Cross-Protection

A protective effect of preinoculation with an avirulent strain or heat-killed races of *X. c.* pv. *malvacearum* was demonstrated when the challenge dose of virulent strain of *X. c.* pv. *malvacearum* was given at 8–24 hours [17,24]. It has been observed that the virulence of the mixture of races may be synergistic, intermediate, or even less [3,53]. When race 2 and race 32 of *X. c.* pv. *malvaceanum* were co-inoculated or when leaves preinoculated with race 2 were challenged with race 32 or vice versa, an additional (synergistic) effect was observed on cultivars susceptible to both the races. Acala-44 is a variety susceptible to both races, and when they were co-inoculated, the disease increase (measured in terms of lesions area) was 72.7% greater than race 2 (less virulent and attacks only polygenes) and 48.8% greater than race 32 (more virulent and attacks polygenes and at least five major genes, e.g., B_7, B_4, B_2, B_{In}, and B_N). Further, it was observed that veinal infection started in race 32 in 12 days, while in co-inoculation the veinal infection started on the 9th day or even earlier [53]. It has been concluded that resistant reaction of cotton hosts towards race 2 was not changed upon mixing or challenging with race 32 (the highly virulent genotypes). But when race 2 was challenged on leaves pre inoculated with race 32, the susceptible reaction of the host was expressed at a 48 h challenge but not at a 0 h challenge. Thus, in a mixed inoculation, a synergistic effect was obtained on cultivars susceptible to both the races, whereas on cultivars resistant to the less virulent race, a hypersensitive response (HR) was observed. It appears that in a mixed inoculation of two races on a cultivar that is resistant to one race, the incompatible reaction, HR, started first and the product of this reaction inhibited the compatible reaction, and accordingly instead of a susceptible reaction (SR), an HR was obtained. The incompatible reaction, i.e., rapid browning or HR, was dominant, started earlier, and probably inhibited the compatible (disease) reaction, but it did not kill the pathogen.

As many as 13 races could be differentiated from 19 isolates of *X. c.* pv. *malvacearum* collected from four different lesions on one leaf of Acala-44, which is the most susceptible line of differentials. This was taken to indicate that most of the races colonized this highly susceptible variety very easily and could grow, perhaps, individually. It was concluded [54] that multiple races were present in the same host, in the same leaf, and even in the same lesion, at least on highly susceptible cultivars in natural infection. Preliminary studies have indicated [53] that new races possessing more virulent genotype could be generated in nature (perhaps by a process similar to recombination) from mixed infection of *X. c.* pv. *malvacearum* genotypes on susceptible cultivars. Future work may demonstrate the significance of cross-protection and recombination in the evolution of the interaction of host-pathogen systems and how these could be exploited in integrated biointensive disease management.

B. Microbial Metabolites

The secondary metabolites of fluorescent pseudomonads have received considerable attention, probably due to the abundance of this bacterium in the rhizosphere and also because of their ability to produce a range of secondary metabolites that are inhibitory to other microflora including the plant pathogen [31]. Siderophores, metabolites with antimicrobial properties, and other substances like cyanide are the main groups of metabolites known to be involved in the pathogen suppression [55–60].

Siderophores are low molecular weight compounds excreted under low iron conditions, which selectively chelate iron with very high affinity. Kloepper et al. [61] were the first to demonstrate the importance of siderophores in the biological control of plant pathogens. The function of siderophores is to supply iron to the cell. The concentration of soluble ferric ion at pH 7 needed to sustain microbial growth is very low (10^{-17} M). Soilborne microorganisms active in low iron conditions are capable of producing these iron-chelating compounds. In the field, siderophores produced by plant growth–promoting rhizobacteria (PGPR) such as fluorescent pseudomads are thought to deprive the deleterious rhizosphere microorganism (DRMO) of iron, a limiting essential nutrient for metabolism, presumably due to the higher affinity for iron of the siderophores produced by PGPR than those produced by the pathogens or DRMOs. Further. PGPR also have the ability to utilize the ferric-siderophore complex produced by DRMO. In contrast, DRMO and pathogens lack the receptor protein for the ferric-siderophore complex of PGPR [61,62]. Mondal et al. [64] demonstrated production of several antimicrobial secondary metabolites including HCN and siderophores by five cotton rhizobacteria antagonistic to race 32 of *X. c.* pv. *malvacearum*, the most prevalent and virulent race of bacterial blight pathogen found in India. The most efficient rhizobacterium was *Pseudomonas fluorescens* CRb-26, which produced four major phenolic metabolites, two of which were fluorescent and two nonfluorescent; one of the fluorescent metabolites, which was produced maximally, was identified as 2,4-diacetylphloroglucinol. In vitro efficacy of these metabolites indicated that 2,4-diacetylphloroglucinol and a nonfluorescent compound suppressed the growth of bacterial blight pathogen more effectively than the other two; these two metabolites also protected cotton plants (*Gossypium hirsutum* and *G. barbadense*) when applied simultaneously (along with pathogen) or 8 hours after the pathogen inoculation. Thus these metabolites were identified as candidates for exploitation in the management of bacterial blight of cotton. Further, the iron tolerance level of CRb-26 for siderophore production was higher, 10^{-2} M $FeCl_3$, in comparison to other rhizobacterial isolates with tolerance levels of 10^{-3} M $FeCl_3$. This again suggested that strain CRb-26 has great potential as a biocontrol agent due to its ability to produce siderophores even at higher iron levels. Such iron-tolerant strains would be better suited as biocontrol agents in iron-toxic areas. Besides siderophores, fluorescent pseudomonads produced several other secondary metabolites with antimicrobial properties such as cyanide, acetylphloroglucinols, pyoluteorin, and pyrrolnitrin. Pyrrolnitrin and pyoluteorin produced by *P. fluorescens* were the key antagonistic factors against *Rhizoctonia solani* and *Pythium* spp. that infect cotton [33,65].

The induction of a defense response in plants by biotic elicitors is well known. Certain resistance-inducing compounds (resistance elicitors) are known today that are involved in systemic acquired resistance (SAR), e.g., oxalate, phosphate, 2,6-dichloroisonicotinic acid and its ester derivatives, salicylic acid and ethylene. Systemin is a likely endogenous signal in the tobacco and cucumber resistance response, i.e., they act as inducers of pathogenesis-related protein in plants. Several chemicals are known to inhibit the development of bacterial blight symptoms in cotton [66]. SAR can be effective against infection by a broad range of pathogens totally unrelated to the inducer. For example, in cucumber or tobacco, a first infection by fungus/bacteria/virus/abiotic stress (e.g., wounding, UV exposure, etc.) protects the plants against subsequent infection by fungus/bacteria/virus in both infected and uninfected parts of plant [67,68]. Several studies have revealed that fluorescent pseudomonads also induce systemic resistance against many pathogens and control disease caused by them [69]. Pea root bacterization with *P. fluorescens* or *Bacillus pumilus* triggered a set of plant defense reactions [70]. Seed bacterization of rice with fluorescent pseudomonads also induced systemic resistance [71]. Marked host metabolic changes culminating in a number of structural (accumulation of callose or lignin) and biochemical responses (synthesis of chitinases) occurred at the onset of bacterial antagonist–induced resistance in several plants [69]. An increase in peroxidase activity as well as an increase in the level of mRNAs encoding for PAL and chalcone synthase could be seen in bean roots colonized by bacterial antagonists [72]. Fluorescent pseudomonads are known to produce salicylic acid and several other SAR inducers and inhibitors of pathogens [31,73]. Molecular cloning of genetic determinants for inhibition of *Pythium ultimum* (inducer of damping-off of cotton) by a fluorescent pseudomonad was done, and at least five genes were required for fungal inhibition [74]. Biotic and abiotic stress-related expression of SAR genes is also known [75]. *P. fluorescens* is known to control a large number of diseases in various crops [76], and it is hoped that this bacterium will be used in biointensive management programs, because data on effective dosage of the bacterial inoculant, survivability, SAR activity, microbial inhibitors, proper strain mixtures, time of application, and suitable delivery systems are already available in several cases.

VII. BIOCONTROL IN INTEGRATED CROP/PEST MANAGEMENT

Integrated crop/pest management (ICM/IPM) involves need-based use of pesticides and adoption of farming techniques to keep the pest incidence below economic threshold levels, thereby promoting the build-up of many biocontrol agents in the crop ecosystems. Thus, this approach, a greener alternative to the conventional use of chemicals, is an attempt to promote natural, economic, and sociological farming methods through the most effective combination of farming techniques and judicious and limited use of fungicides. Biological control is an important component of the ICM/IPM program, in which the pathogen activity is reduced through the use of other living organisms, resulting in a reduction of disease incidence and severity.

In this context it is imperative to discuss the MAR (Multiple Adversity Resistance) system developed in cotton by Bird [30,63], which makes it easier to produce cultivars with stable resistance to two or more types of pest. The techniques and procedures that evolved into the MAR program originated in information gathered on seed conditioning, bacterial blight resistance, and genetic interrelationships for resistance to several diseases. Interrelationships among genes conferring resistance to several diseases of cotton suggested one of four possible situations: (1) genes causing resistance to two or more pathogens; (2) close linkages among genes conditioning resistance to several pathogens; (3) genes conditioning a mechanism effective against pathogens causing several diseases; or (4) a rather broad pleiotropic system. Further, the genetic interrelationship pathways indicated a strong association between resistance to bacterial blight and the *Fusarium* wilt/root-knot nematode complex. Lesser but significant associations existed between blight resistance and resistance to *Verticillium* wilt,

Phymatotrichum root rot, and seed coat resistance to mold. The measured associations with bacterial blight were stronger when at least four races of *X. c.* pv. *malvacearum* were used to evaluate and select for resistance. Similar associations existed between resistance to the *Fusarium* wilt/root-knot nematode complex, *Verticillium* wilt, and *Phymatotrichum* root rot. A reduced rate of seed germination at low temperatures (13–18°C) was associated with resistance to *Verticillium* wilt and seedling pathogens. Seed coat resistance to mold was strongly associated with high yield potential and early maturity. Small seed size was associated with rapid germination at reduced temperatures and susceptibility to *Verticillium* wilt. These associations formed a complex model of interrelationship that have been measured in formal genetic studies and have held up under 15 years of use in the MAR genetic improvement system [30,77]. Other correlations that may complement the system were between the B_2, B_3, and B_6 genes for bacterial blight resistance and resistance to a boll-rotting fungus (*Colletotrichum* sp.) and the cotton stainer, a boll- and seed-puncturing insect [78]. Three hybrid pools have been processed from the MAR program. Two varieties, Tamcots SP 21 and SP 37, released in 1972 were first pool hybrids comprising progenies in 1967–1968. Replacement of older varieties by the new Tamcot SP varieties doubled the average yield per hectare in the Coastal End of Texas. The third hybrid pool represented by three varieties (Tamcots SP 21S, SP 37H, and CAMD-E) released in 1975 and cultivars LEBO and CPPS were the progenies during 1976–1978. Comparisons made in the Coastal End area among cultivars representing each hybrid pool showed that each sequence of MAR improvement resulted in higher yield and earliness potentials. Resistance to diseases and insects has improved simultaneously with increased yield and earliness. The greatest improvements have been in resistance to various problems that were low or nonexistent during 1967–1968. The rate of progress slowed as a high level of resistance was approached. As more and more MAR genes were accumulated, resistance to some problems increased while resistance to other adversities occurred where more could be measured earlier and production potential increased. This indicated that a weak MAR gene complex was adequate for some adversities but that a stronger one was necessary to influence others [79–81]. Within a plant species, genes for viability, adaptation, fitness, and some morphological traits tend to be linked or associated [82]. Many of these genes are not likely dealt with in domestic plant improvement. If this is the case, the MAR system may have evolved a bioassay procedure that identified little-used but important genes for survival and fitness. Such genes, having a small genetic but a broad biological effect for resistance, may be accumulated under natural selection. But these genes may be bypassed or diluted by domestic procedures that deal more with specific traits. It is believed [77] that the explanation for genetic events recurring in the MAR program lies within this general concept. The program offers a system in which quantitative genes are easier to use than qualitative ones in developing cultivars with resistance to several diseases. The nature of resistance conferred by the MAR system is based on the fact that cotton has the genetic potential to alter its natural symbiotic microflora, which occur on seed and root surfaces and in tissues, to organisms that are unfavorable to pathogens and insects. This system is under the control of MAR genes. Further, the change is accomplished by genetic alteration of the quality and quantity of components of exudates from seed coat, seedlings, or plant roots and fluids in tissues. The altered fluids are nutritionally unfavorable for pathogens and insects and selectively favorable for bacteria, actinomycetes, and fungi that are highly competitive with pathogens and insects. Unfavorable nutrition and microorganisms function together to provide a mechanism of multiadversity resistance. Initial investigations revealed that bacteria and fungi (*Bacillus* spp., *Pseudomonas* spp., and *Fusarium* spp.) from seedling roots of MAR cultivars when used as treatments reduced damage by root pathogens. The bacteria and fungi together were more effective than either alone. Two bacteria (*Bacillus* spp.) types from leaves and squares of Tamcot CAMD-E (high resistance to bacterial blight and partial resistance to boll weevil) induced resistance to the bacterial blight pathogen in leaves of a susceptible cotton. The same bacteria when applied to a susceptible cotton cultivar made it as resistant as Tamcot CAMD-E to the boll weevil. Thus, bacteria that can influence host responses to diseases and insects do exist in tissues of MAR cottons [83,84]. These investigations emphasize that more meaningful results may be obtained in the future by focusing research planning not only on the complex interaction of host-pathogen epiphyte/saprophyte/antagonist/symbiont environment, but also on their components, rather than only on the simpler cases of classic host-pathogen-environment interaction.

VIII. SPECIFIC RECOMMENDATIONS FOR COTTON DISEASE BIOCONTROL AGENTS

Some specific recommendations made for the biocontrol of cotton diseases are listed in Table 2. Of these, the use of fluorescent pseudomonads in the biocontrol of seedling diseases has been approved by the U.S. Environmental Protection Agency [85]. *Trichoderma, Gliocladium, Bacillus,* and *Paecilomyces* have been approved for use in other crops [86–90]. Fluorescent pseudomonads are known to control more than 46 diseases in 23 crops [76]. A mixture of *P. fluorescens* and *B. subtilis* was effective against *Pythium ultimum, Fusarium oxysporum,* and *Thielaviopsis basicola,* the causal pathogens of seedling diseases in cotton [91]. Fluorescent pseudomonads were also effective against *X. c.* pv. *malvacearum* [32,35]. The application of *Burkholderia cepacia* (isolate D1) as soil drench significantly increased numbers of emerged seedlings in two field trials in Safford and Tucson, Arizona, conducted in 1996 [87]. Although biological agents have been suggested for the management of CLCuV (through vector control), certain

Table 2 Biocontrol of Cotton Diseases

Diseases/pathogens	Biocontrol agents identified	Recommendations	Ref.
Bacterial blight (*Xanthomonas campestris* pv. *malvacearum*)	*P. fluorescens* CRb26	Seed bacterization followed by foliar spray with bacterial suspension	32
Cotton leaf curl (Gemini virus)	*Paecilomyces* spp. (fungi) *Eretmocerus mundus* (aphelinid) *Chrysoperla cama* (lacewing)	All these are natural enemies of whitefly, the vector of cotton leaf curl; therefore, the recommendation is not to use increased and early insecticides in cotton	84
Damping-off (*Rhizoctonia* spp., *Pythium* spp.)	*P. fluorescens*	Seed bacterization followed by seedling drench with bacterial suspension	85
	Trichoderma spp. *Gliocladium virens*	Seed coating	86
	Bacillus cereus		22
	Burkholderia cepacia	Drenching furrow shortly after sowing cotton seed with bacterial suspension	48,87
	Bacillus subtilis	Seed bacterization	
	Nonpathogenic binucleate *Rhizoctonia*	Seed coating before sowing	88
Root rot (*Macrophomina phaseolina*, *Rhizoctonia solani*, *Fusarium solani*)	*Pseudomonas aeruginosa, Paecilomyces lilacinus, Trichoderma koningii*	Seed dressing	89
Wilt (*Fusarium oxysporum* f. sp. *vasinfectum*)	*P. fluorescens*	Seed bacterization with bacterial suspension	29,37
	Gliocladium virens Trichoderma hamatum	Seed dressing	90

other nonchemical recommendations have also been made by the Indian Council of Agricultural Research (ICAR) Committee. The main recommendations are:

1. Seed production of CLCuV-resistant variety like RG 8, HD 107, LD 327 of *G. arboreum*, RS 810, RS 875, LRA 5166, LRK 516 of *G. hirsutum* and hybrid LHH 144, LK 515, RS-2013, HHH-223;
2. Creation of Buffer Zone 1 (consisting of entire belt adjoining the international border) of *G. arboreum* cotton up to 20 km and Buffer Zone II' of resistant/tolerant varieties of *G. hirsutum* in next 20 km in all the three affected states (Rajasthan, Punjab, and Haryana);
3. Rogueing and destruction of diseased plants, alternate hosts/weeds in the off-season as well as during crop season;
4. Avoid cultivation of *G. hirsutum* cotton in and around orchards.

Since there is obviously no single solution to this highly complex and widespread problem of CLCuV, a multidisciplinary integrated disease-management approach is required. Both short- and long-term strategies involve delaying *Bemisia tabaci* infestation combined with use of early-maturing varieties and crop sanitation, reducing whitefly population and virus reservoirs, and developing tolerant and resistant crop varieties will go a long way towards management of cotton leaf curl disease.

IX. PROBLEMS IN BIOLOGICAL CONTROL

There appears to be a nongeneral correlation between the in vitro ability of the antagonists and their ability to suppress diseases in the field; accordingly, the strains/chemicals producing the largest zones of inhibition on agar media do not always make the best biocontrol agents [85,92–94]. It is, therefore, necessary to develop suitable in vitro assay procedures, which closely simulate natural conditions. Colonization and rapid multiplication of the biocontrol agent in the natural environment of the pathogens (i.e., seed, root, stem, and leaf) are considered very important attributes for effective biocontrol [31,95].

Biological control essentially depends upon maintaining a threshold population of the antagonist on planting material or in soil; any drop in the viable count below this critical level may render biological control ineffective. A large number of edaphic and environmental factors affect the viability of antagonists. Gram-positive spore-forming species of *Bacillus* have the advantage of producing endospores (for longer survival/viability). But gram-negative formulations face the same difficulties as rhizobia, which do not produce endospores, and their vegetative cells are, therefore, sensitive to drying and heat. A granular peat formulation of *P. fluorescens* [96] has shown promise in controlling seedborne pathogens of cotton. It, therefore, appears necessary to develop appropriate techniques for suitable formulations for biological control so that inconsistencies in their performance are eliminated. Ecological data also must be generated, particularly on the viability of antagonists and optimum conditions for their multiplication on plant surfaces or in soil and the production/secretion of the secondary metabolite involved in biocontrol. Currently, little is known about the bacterial traits that contribute to their ecological competence. Another complication is the effect of biocontrol agents on pathogens other than the target pathogen; thus, when one pathogen is controlled, another may become predominant [96]; thus, an understanding of the pathogens in the agro-ecosystem and conditions that favor each is essential.

X. RECENT PROGRESS IN THE IMPROVEMENT OF BIOCONTROL ACTIVITY

In recent years efforts have been made to increase the efficiency of *P. fluorescens*. For example, introduction of gene(s) *phlx* encoding a monoacetylphloroglucinol acetyl transferase into a wild-type strain (M114) of *Pseudomonas* sp., which is unable to synthesize the more active antifungal metabolite 2,4-diacetylphloroglucinol, has resulted in an enhanced biocontrol ability of strain M114 against *Pythium ultimum* both in the laboratory and in greenhouse experiments [97]. Likewise, a chitinase biosynthesis gene from *Serratia marcescens* was transferred to *P. fluorescens* to achieve a wider protection of plant by the transformed strain against fungal pathogens that have chitin in their cell wall [98–101]. Further, the development of a constitutively siderophore-producing mutant improved siderophore-mediated biocontrol under condition of high iron in vitro [102]. Increasing the copy number of gene(s) in a wild-type strain also led to the overproduction of 2,4-diacetylphloroglucinol and pyoluteorin and resulted in enhanced disease suppression. The biocontrol activity of a genetically manipulated strain of *Trichoderma virens* was enhanced against cotton seedling disease incited by *R. solani* (as compared with wild-type strain) due to the overexpression of a chitinase gene (Cht 42) [103]. However, the transformant showed patterns similar to the wild-type strain with respect to other characteristics like growth rate, sporulation, antibiotic production, colonization of cotton roots, and growth/survival in soil [103].

   An ecofriendly approach to the management of CLCuV would be to investigate the molecular mechanism by which the virus is transmitted by *B. tabaci* in the hope of designing genetically engineered plants that express an "antitransmission factor" that will effectively neutralize the vector's ability to transmit the virus. Some of the approaches reported [104] for the development of transgenic resistance to CLCuV include (1) expression of antisense RNA against complete or fragments of *AC1* gene, (2) overexpression of *AC1* in transgenic plants, and (3) expression of a virus-induced cytotoxin gene in transgenic plants.

XI. FUTURE CHALLENGES

The challenges of biological control are to select suitable naturally occurring strains of biocontrol agents or create them through genetic engineering. This could be problematic because production of many antimicrobial metabolites by superior strains is not governed by a single gene [105]. Further, it may take time to develop the procedures of risk assessment of the genetically modified microorganisms and make them ecologically sustainable as well as socially acceptable. The ultimate aim must be to develop disease-resistant transgenic plants by incorporating useful genes.

It is also necessary to develop inexpensive mass production of antagonists and easily applicable formulations that remain viable under less than optimal conditions. Because it is known that *P. fluorescens* can penetrate host epidermal cells [69] and trigger a set of ultrastructural changes that lead to plant defense [69,70], it is necessary to generate information on the mechanism of interaction and the ideal time and place for application of antagonists. For example, soil application of fluorescent pseudomonads just before sowing of seeds effectively controlled cotton seedling diseases [34,35]. One strain of *Bacillus subtilis* (JM 339) attained the highest endophytic populations in cotton following a 2-hour seed soak, while another strain (CC-90-471) had maximum population with foliar spray [106]. If seed treatments offer only transient protection (by inducing defense genes), then seed treatment must be followed by foliar spray applications of the antagonists. The final challenge is the modification of farm equipment and practices to accommodate biological treatment as one component in integrated disease/crop management. The quality control of commercial biocontrol agents should be also strictly enforced. Efforts should also be made to see that these bioformulations are not treated as pesticides for registration purposes.

XII. CONCLUSIONS

The major challenges for plant pathologists are to develop techniques for easy, early, and fast detection of pathogens and to postulate appropriate management practices for diseases considering the environmental, social, economic, and sustainable values. A single disease-management approach/agro-technique cannot meet these challenges. Therefore, an integrated approach incorporates all the available strategies of disease management as well as crop production techniques, blending them in such a way so that the detrimental effects of individual components can be overcome. We call this the "cafeteria" approach, where the user, with the help of a computer, can select the best "menu." Data need to be generated so that the appropriate software can be developed. Biological control of plant pathogens play a major role in this effort. Though a total replacement of chemicals for disease control is not feasible, the IPM approach would at least reduce their use. There will be in the future a shift from fungicides, which have a direct mode of action on the target pathogen, to chemicals that induce systemic resistance (e.g., 2,6-dichloro-isonicotinic acid, which protects cucumber and tobacco against pathogens). However, the successful deployment of a biocontrol agent would heavily depend on the time, dosage, and method of its application and, above all, on the information available regarding its consistent and good performance. The success of future programs will be information based and will depend on advances in computer technologies.

REFERENCES

1. Economic survey 1998–99, Government of India. Government of India Press, New Delhi, 1999.
2. JP Verma, Sheo Raj. Blight and wilt of cotton. In: AN Mukhopadhyay, HS Chaube, J Kumar, US Singh, eds. Plant Diseases of National Importance Vol. IV. Diseases of Sugar, Forest and Plantation Crops. Englewood Cliffs, NJ: Prentice Hall, 1992, pp. 138–160.
3. JP Verma. Bacterial Blight of Cotton. Boca Raton, FL: CRC Press, 1986.
4. Sheo Raj, JP Verma. Diseases of cotton in India and their management. Rev Trop Pl Path 5:207–254, 1988.
5. JP Verma, RP Singh, ML Nayak. Bacterial blight of cotton and steps for its eradication. Cotton Dev 4:23–27, 1974.
6. CD Mayee. Etiology and epidemiology of parawilt of cotton. Proceedings of Vasantrao Naik Memorial National Seminar, PVK, Nagpur, 1992, pp. 116–126.
7. RK Sahay, R Pundarikakshudu, MC Gawande. New wilt in cotton. Proceedings of Group Discussion on New Wilt in Cotton, Nagpur, India, 1984, pp. 33–34.
8. CD Mayee, AB Choulwar, HS Acharya. Distribution pattern of parawilt of cotton (*Gossypium* species). Indian J Agri Sci 60:777–778, 1990.
9. Sheo Raj, NK Taneja. Diseases of cotton and their management. In: M Veerbhadra Rao, S Sithanantham, eds. Plant Protection in Field Crops. Plant Protection Association of India, 1987, pp. 301–317.
10. A Mohsin, E Haq, AA Hasmi, S Hamid, S Khalid. Dilemma of virus disease in cotton. PAPA Bulletin, 1992, pp. 23–25.
11. A Varma, CG Malathi, A Handa, M Aiton, BD Harrison, JP Verma, RP Singh, M Singh, M Srivastava, J Singh. Occurrence of lead curl of cotton and okra in northern India. 6th Int Congr Pl Pathol, Montreal, Canada, July 28–August 6, 1993.

12. VR Mali. Cotton leaf crumple virus disease—a new record for India. Indian Phyto-
     path 30:326–329, 1977.
13. VR Mali. Anthocyanosis virus disease of cotton—a new record for India. Curr Sci
     47:235–237, 1978.
14. VR Mali. Viral wilt—a new disease hitherto unrecorded on cotton. Curr Sci 48:
     687–688, 1979.
15. VV Sulladamath, RK Hegde, BGP Kulkarni, PC Hiremath. A new bacterial disease
     on 'Varalaxmi' a hybrid cotton. Curr Sci 44:286, 1975.
16. GL Kottur, MK Patel. Malformation of the cotton plant leading to sterility. Agri
     J India 15:640–643, 1920.
17. JP Verma, RP Singh. Races of *Xanthomonas malvacearum*, loss in their virulence
     and the protective effect of avirulent strains, heat-killed cells and phylloplane bacte-
     ria. Z Pflkrankh Pflschutz 83:748–757, 1976.
18. SK Duttamajumder, JP Verma. Lysogeny in *Xanthomonas campestris* pv. *malva-
     cearum*. Indian Phytopath 41:617–619, 1988.
19. JP Verma, RP Singh. Control of bacterial diseases through resistance breeding.
     In: A Hussain, K Singh, BP Singh, VP Agnihotri, eds. Recent Advances in Plant
     Pathology. Lucknow, India: Print House, 1983, pp. 20–47.
20. YL Nene. Opportunities for research on diseases of pulse crops. Indian Phytopath
     39:333–342, 1986.
21. JS Grewal. Diseases of pulse crops—an overview. Indian Phytopath 41:1–14,
     1988.
22. VV Chenulu. Plant virology in India—past present and future. Indian Phytopath
     37:1–20, 1984.
23. P Singh, N Singh. Antagonistic effect of cotton phylloplane bacteria against *Xan-
     thomonas campestris* pv. *malvacearum* in vitro and in vivo. Indian Phytopath 34:
     116–117, 1981.
24. JP Verma, HD Chowdhury, RP Singh. Interaction between phylloplane bacteria
     and *Xanthomonas malvacearum*. Proceeding 4th International Conference on Plant
     Pathogenic Bacteria, Angers, France, 1978, pp. 795–802.
25. JP Verma, RP Singh, HD Chowdhury, PP Sinha. Usefulness of phylloplane bacteria
     in the control of bacterial blight of cotton. Indian Phytopath 36:574–577, 1983.
26. A Quadt-Hallmann, J Hallmann, JW Kloepper. Bacterial endophytes in cotton:loca-
     tion and interaction with other plant-associated bacteria. Can J Microbiol 43:254–
     259, 1997.
27. C Chen, EM Banske, G Musson, R Rodriguez-Kabana, JW Kloepper. Biological
     control of Fusarium wilt on cotton by use of endophytic bacteria. Biol Control 5:
     83–91, 1995.
28. S Pleban, F Ingel, I Chet. Control of *Rhizoctonia solani* and *Sclerotium rolfsii* in the
     greenhouse using endophytic *Bacillus* spp. Euro J Plant Pathol 101:665–672, 1995.
29. C Chen, EM Banske, G Musson, JW Kloepper. Biological control potential and
     population dynamics of endophytic bacteria in a cotton/Fusarium wilt system. In:
     MH Ryder, PM Stephans, GD Bowan, eds. Improving Plant Productivity with Rhi-
     zosphere Bacteria. Adelaide, Australia: Graphic Services, 1994, pp. 191–193.
30. LS Bird. The MAR (multiadversity-resistance) system for genetic improvement of
     cotton. Plant Dis 66:172–176, 1982.

31. KK Mondal, JP Verma. Role of secondary metabolites of *Pseudomonas fluorescens* in the biocontrol of plant pathogens. In: SP Singh, SS Hussaini, eds. Biological Suppression of Plant Diseases—Phytoparasitic Nematodes and Weeds. Bangalore, India: Project Directorate of Biological Control, 1998, pp. 70–80.

32. KK Mondal, RP Singh, JP Verma. Beneficial effects of indigenous cotton rhizobacteria on seed germinability, growth promotion and suppression of bacterial blight disease. Indian Phytopath 52:228–235, 1999.

33. CR Howell, RD Stipanovic. Suppression of *Pythium ultimum* induced damping off of cotton seedlings by *Pseudomonas fluorescens* and its antibiotic, pyoluteorin. Phytopathology 70:712–715, 1980.

34. C Hagedorn, WD Gould, TR Bardinelli. Field evaluation of bacterial inoculants to control seeding disease pathogens on cotton. Plant Dis 77:278–282, 1993.

35. GS Laha, JP Verma. Role of fluorescent pseudomonads in the suppression of root rot and damping off of cotton. Indian Phytopath 51:275–278, 1998.

36. WJ Howie, TV Suslow. Role of antibiotic biosynthesis in the inhibition of *Pythium ultimum* in cotton spermosphere and rhizosphere by *Pseudomonas fluorescens*. Mol Plant-Microbe Interact 4:393–399, 1991.

37. A Gamliel, J Katan. Suppression of major and minor pathogens by fluorescent pseudomonads in solarized and non-solarized soils. Phytopathology 83:68–75, 1993.

38. GS Laha, RP Singh, JP Verma. Biocontrol of *Rhizoctonia solani* in cotton by fluorescent pseudomonads. Indian Phytopath 45:412–415, 1992.

39. JP Verma, RP Singh, HD Chowdhury. Relationship of epiphytic bacteria to bacterial diseases of plants. In: KS Bilgrami, KM Vyas, eds. Recent Advances in Biology of Microorganisms. Dehra Dun, Bishan Singh Mahendra Pal Singh, 1980, pp. 95–106.

40. E Sezgin, A Karcilioghe, U Yemiscioglu. Investigation on the effects of some cultural application and antagonistic fungi on *Rhizoctonia solani* Kuhn and *Verticillium dahliae* Kleb in the Aegean region. II. Effects of herbicides and antagonistic fungi. J Turkish Phytopathol 11:79–91, 1982.

41. Y Elad, A Kalfon, I Chet. Control of *Rhizoctonia solani* in cotton by seed-coating with *Trichoderma* spp. spores. Plant Soil 66:279–281, 1982.

42. G Berg, C Knaapa, G Ballin, D Seidel. Biological control of *Verticillium dahliae* Kleb. by naturally occurring rhizosphere bacteria. Arch Phytopathol Pl Prot 29:249–262, 1994.

43. JM Expert, B Digat. Biocontrol of sclerotinia wilt of sunflower by *Pseudomonas fluorescens* and *Pseudomonas putida* strains. Can J Microbiol 41:685–691, 1995.

44. XY Liang, HC Huang, LJ Yanke, GC Kozuls. Control of damping-off of safflower by bacterial seed treatment. Can J Plant Pathol 18:43–49, 1996.

45. CS Nautiyal. Selection of chickpea-rhizosphere competent *Pseudomonas fluorescens* NBRI 1303 antagonistic to *Fusarium oxysporum* f. sp. *ciceris*, *Rhizoctonia bataticola* and *Pythium* sp. Curr Microbiol 35:52–58, 1997.

46. CS Nautiyal. Plant beneficial rhizosphere competent bacteria. Proceedings of National Academy of Science, India, 70(8):107–123, 2000.

47. IJ Misaghi, PJ Cotty, DM Decianne. Bacterial antagonists of *Aspergillus flavus*. Biocont Sci Technol 5:387–392, 1995.

48. K Zaki, IJ Misaghi, A Heydari. Control of cotton seedling damping-off in the field by *Burkholderia* (*Pseudomonas*) *cepacia*. Plant Dis 82:291–293, 1998.

49. AK Vidaver. Prospects for control of phytopathogenic bacteria by bacteriophages and bacteriocins. Annu Rev Phytopathol 14:451–465, 1976.

50. NP Lebedeva. Abstracts of the investigation of the plant protection stations of the Pan-Soviet Institute for scientific research on cotton. Plant Prot Leningr 15:99, 1937.

51. RP Singh, JP Verma. The distribution and lysis pattern of the bacteriophages of *Xanthomonas malvacearum*, the incitant of bacterial blight of cotton. Proc Indian Natl Sci Acad 40B:363–368, 1974.

52. JP Verma, GN Dake, RP Singh. Interaction between bacteriophages, *Xanthomonas campestris* pv. *malvacearum* and *Gossypium hirsutum*. Indian Phytopath 47:27–33, 1994.

53. JP Verma, HD Chowdhury, RP Singh. Interaction between different races of *Xanthomonas malvacearum* in leaves of *Gossypium hirsutum*. Z Pflkrankh Pflschutz 86:460–464, 1979.

54. HD Chowdhury, RP Singh, JP Verma. Presence of more than one race of *Xanthomonas malvacearum* in lesions on leaves of *Gossypium hirsutum*. Indian Phytopath 32:110–112, 1979.

55. J Leong. Siderophore: their biochemistry and possible role in the biocontrol of plant pathogens. Annu Rev Phytopathol 24:187–209, 1986.

56. L Thomashow, DM Weller. Role of a phenazine antibiotic from *Pseudomonas fluorescens* in biological control of *Gaeumannomyces graminis* var. *tritici*. J Bacteriol 170:3499–3508, 1988.

57. G Defago, D Haas. Pseudomonads as antagonists of soilborne plant pathogens: modes of action and genetic analysis. Soil Biochem 6:246–291, 1990.

58. C Keel, P Wirthner, TH Oberhansli, C Voisard, U Berger, D Haas, G Defago. Pseudomonads as antagonists of plant pathogen in rhizosphere: role of antibiotic 2,4-diacetylphloroglucinol in the suppression of black root rot of tobacco. Symbiosis 9:327–341, 1990.

59. JE Loper, JS Buyer. Siderophores in microbial interaction on plant surfaces. Mol Plant-Microbe Interact 4:5–13, 1991.

60. Z Piotrowska-Seget. The effect of cyanogenic strain of *Pseudomonas fluorescens* on plant growth. Acta Microbiol Polonica 44:161–170, 1995.

61. JW Kloepper, J Leong, M Teintza, MN Schroth. *Pseudomonas* siderophores: a mechanism explaining disease suppressive soils. Curr Microbiol 4:317–320, 1980.

62. SA Leong, JB Neilands. Siderophore production by phytopathogenic microbiol species. Arch Biochem Biophys 218:315–359, 1982.

63. LS Bird, C Liverman, P Thaxton, RG Percy. Evidence that microorganisms in and on tissues have a role in a mechanism of multi-adversity resistance in cotton. Proceedings of Beltwide Cotton Production Research Conference, National Cotton Council, Memphis, TN, 1980, pp. 283–285.

64. KK Mondal, RP Singh, P Dureja, JP Verma. Secondary metabolites of cotton rhizobacteria in the suppression of bacterial blight of cotton. Indian Phytopath 53:22–27, 2000.

65. WF Pfender, J Kraus, JE Loper. A genomic region from *Pseudomonas fluorescens*

Pf-5 required for pyrrolnitrin production and inhibition of *Pyrenophora tritici-repentis* in wheat straw. Phytopathology 83:1223–1228, 1993.

66. SG Borkar, JP Verma. Chemical inhibition of compatible/incompatible reaction in cotton. Indian Phytopath 42:236–240, 1989.

67. H Kessmann, T Sraub, C Hofmann, E Ward, S Uknes, J Ryals. Induction of systemic acquired resistance in plants by chemicals. Annu Rev Phytopathol 32:439–459, 1994.

68. JP Verma, Anupam Verma. Current questions in plant pathology. Proceedings of the National Academy of Sciences, India, 1995, pp. 1–21.

69. P M'Piga, RR Belanger, TC Paulitz, N Benhamou. Increased resistance to *Fusarium oxysporum* f. sp. *radicis lycopersici* in tomato plants treated with the endophytic bacterium *Pseudomonas fluorescens* strain 6328. Physiol Mol Plant Pathol 50:301–320, 1997.

70. N Benhamou, JW Kloepper, A Quadt Hallmann, S Tuzun. Induction of defense-related ultrastructural modifications in pea root tissues inoculated with endophytic bacteria. Plant Physiol 112:919–929, 1996.

71. K Krishnamurthy, SS Gnanamanickam. Biological control of sheath blight of rice: induction of systemic resistance in rice by plant associated *Pseudomonas* spp. Curr Sci 72:331–334, 1997.

72. RE Zdor, AJ Anderson. Influence of root colonizing bacteria on the defense response of bean. Plant Soil 140:99–107, 1992.

73. DN Dowling, F O'Gara. Metabolites of *Pseudomonas* involved in the biocontrol of plant disease. TIBTECH 12:133–141, 1994.

74. NI Gutterson, TJ Layton, JS Ziegle, GS Warren. Molecular cloning of genetic determinants for inhibition of fungal growth by a fluorescent pseudomonad. J Bacteriol 165:696–703, 1986.

75. YS Kim, D Choi, MM Lee, SH Lee, WT Kim. Biotic and abiotic stress-related expression of 1-amino-cylopropapane-1-carboxylate oxidase gene family in *Nicotiana glutinosa* L. Plant Cell Physiol 39:565–573, 1998.

76. P Vidhyasekaran. Biological suppression of major diseases of field crops using bacterial antagonists. In: SP Singh, SS Hussaini, eds. Biological Suppression of Plant Diseases—Phytoparasitic Nematodes and Weeds. Project Directorate of Biological Control, Bangalore, India, 1998, pp. 81–95.

77. LS Bird. Breeding for disease and nematode resistance in cotton. In: MK Harris, ed. Biology and Breeding for Resistance to Arthropods and Pathogens of Agricultural Plants. Tex Agric Exp Stn MP-1451, 1980.

78. J Cauquil. Cotton Boll Rot; New Delhi, India. Washington, DC: Amerind Publishing Co., for USDA and NSF, 1975.

79. LS Bird, DL Bush, FM Bourland, RG Percy. Performance of multi-adversity resistant cottons in the presence of adversity-progress for insect resistance. Proceedings of Beltuide Cotton Production Research Conference, National Cotton Council, Memphis, TN, 1976, pp. 28–30.

80. LS Bird, RG Percy. Advancements in developing multi-adversity (diseases, insects and stress) resistant glandless and glanded okra leaf, frego bract and glabrous cotton. Proceedings of Beltwide Cotton Production Research Conference, National Cotton Council, Memphis, TN, 1979, pp. 230–232.

81. L Reyes, LS Bird, P Thaxton, RG Percy, G Sparniel, N Vestal, D Pawlik, H Hoermann. Performance of Texas A & M multi-adversity resistant (TAM-MAR) cottons in the Texas Coastal Plains. Tex Agric Exp Stn PR-3757, 1980.
82. JR Harlan. Diseases as a factor in plant evolution. Annu Rev Phytopathol 14:31–51, 1976.
83. P Lemanceau, C Alabourette, JM Muyer. Production of fusarinine and iron assimilation by pathogenic and non-pathogenic *Fusarium*. In: TR Swinburne, ed. Iron, Siderophore and Plant Disease. NATO Advanced Science Institute Series: Series A: Life Science. USA, 1986, vol. 117, pp. 251–259.
84. IIBC. Technical assistance completion report for RETA 5514 (Integrated Pest Management in Cotton), Unpublished report, IIBC Malaysia Regional Station, 1997.
85. RJ Cook. Making greater use of introduced microorganisms for biological control of plant pathogens. Annu Rev Phytopathol 31:53–80, 1993.
86. JA Lewis, GC Papavizas. Biocontrol of cotton damping-off caused by *Rhizoctonia solani* in the field with formulation of *Trichoderma* spp. and *Gliocladium virens*. Crop Prot 10:396–402, 1991.
87. A Heydari, IJ Misaghi. Biocontrol activity of *Burkholderia cepacia* against *Rhizoctonia solani* in herbicide-treated soils. Plant Soil 202:109–116, 1998.
88. AR Villajuan, K Kageyama, M Hyakumachi. Biocontrol of Rhizoctonia damping-off of cucumber by non-pathogenic binucleate *Rhizoctonia*. Eur J Plant Pathol 102:227–285, 1996.
89. S Shamim, N Ahmad, Atta Ur Rahaman, S Ehteshamul Haque, A Chaffer. Efficacy of *Pseudomonas aeruginosa* and other biocontrol agents in the control of root rot infection in cotton. Acta Agrobotanica 50:5–10, 1997.
90. RP Larkin, Dr Fravel, P Dugger, D Richter. Biological control of wilt pathogens with fungal antagonists. Proceedings Beltwide Cotton Conference, San Diego, California, Vol. 1, 1998, pp. 125–127.
91. T Aqil, WE Batson. Evaluation of a radicle assay for determining the biocontrol activity of rhizobacteria to selected pathogens of the cotton seeding disease complex. Pakistan J Phytopathol 11:30–40, 1999.
92. JP Verma, RP Singh, ML Nayak. Laboratory evaluation of chemicals against *Xanthomonas malvacearum*, the incitant of bacterial blight of cotton. Indian Phytopath 28:170–174, 1975.
93. ML Nayak, RP Singh, JP Verma. Effective chemical sprays to control bacterial blight of cotton. Z Pflkrankh Pflschutz 83:407–415, 1976.
94. MN Schroth, JG Hancock. Selected topics in biological control. Annu Rev Microbiol 35:453–476, 1981.
95. ME Juhnke, DE Mathre, DC Sands. Identification and characterisation of rhizosphere-competent bacteria of wheat. Appl Environ Microbiol 53:2793–2799, 1987.
96. DM Weller. Biological control of soilborne plant pathogens in the rhizosphere with bacteria. Annu Rev Phytopathol 26:379–407, 1988.
97. AM Fenton, PM Stephens, J Crowley, M O'Callaghan, F O'Gara. Exploitation of gene(s) involved in 2,4-diacetylphloroglucinol biosynthesis to confer a new biocontrol capability to a *Pseudomonas* strain. Appl Environ Microbiol 58:3873–3878, 1992.
98. JDG Jones, KL Grady, TV Suslow, JR Bedbrook. Isolation and characterisation of

genes encoding two chitinase enzymes from *Serratia marcescens*. EMBO J 5:467–473, 1986.

99. RL Fuchs, SA Mc Pherson, DJ Drahos. Cloning and expression of a *Serratia marcescens* gene encoding chitinase. In: EL Civerolo, A Colliner, RE Davis, A Gillespic, eds. Plant Pathogenic Bacteria. Proceeding of 6th International Conference on Plant and Pathogenic Bacteria, Martinus, Nijhoff, 1987.

100. L Sundheim. Biocontrol of *Fusarium oxysporum* with a chitinase encoding gene from *Serratia marcescens* on a stable plasmid in *Pseudomonas*. J Cell Biochem 13A (suppl):171, 1989.

101. S Koby, H Schickler, I Chet, AB Oppenheim. The chitinase encoding Tn7-based ChiA gene endows *Pseudomonas fluorescens* with the capacity to control plant pathogens in soil. Gene 147:81–83, 1994.

102. DJ O'Sullivan, F O'Gara. Genetic improvement of siderophore production aimed at enhancing biocontrol in *Pseudomonas* strains. In: DL Keister, Cregan, eds. The Rhizosphere and Plant Growth. Beltsville Symposia in Agricultural Research, 1991, PB No. 14.

103. JM Back, CR Howell, CM Kenerley. The role of an extracellular chitinase from *Trichoderma virens* GV 29-8 in the biocontrol of *Rhizoctonia solani*. Curr Genetics 35:41–50, 1999.

104. Y Zafar, A Bashir, S Mansoor, M Saeed, S Asad, NA Saeed, R Briddon, PG Markham, CM Fauquet, KA Malik. Cotton leaf curl virus epidemic in Pakistan: virus characterisation, diagnosis and development of virus resistant cotton through genetic engineering. Selected publications of Plant Biotech Division, NIBGE, Pakistan, 1998, pp. 33–39.

105. A Kerr. The impact of molecular genetics on plant pathology. Annu Rev Phytopathol 25:87–110, 1987.

106. G Musson, JA Meinroy, JW Kloepper. Development of delivery systems for introducing endophytic bacteria into cotton. Biocont Sci Technol 5:407–416, 1995.

6

Biological Control of Tobacco Diseases

Bonnie H. Ownley
University of Tennessee, Knoxville, Tennessee

I. INTRODUCTION

Currently, disease management in tobacco is based on appropriate cultural practices, with host resistance and chemical applications being used to varying degrees depending upon the specific problems and their severity [1]. Although there are many research reports on microbial biological control of tobacco diseases, as outlined in this chapter, commercial products are generally unavailable. A search yielded only one product, Trieco, listed for use on tobacco [2]. Trieco is a formulation of *Trichoderma viride* for control of various soilborne pathogens. Commercially available biological control products have not targeted the tobacco market. This is unfortunate because fungicide residues on tobacco continue to pose a major problem on the world market. This review is by no means exhaustive, but it attempts to cover some of the more recent work on biological control of diseases in tobacco.

II. ROOT AND STEM DISEASES CAUSED BY FUNGI

A. Black Root Rot

The soilborne fungus *Thielaviopsis basicola* causes black root rot. This disease occurs in most major tobacco-growing regions of the world and is most severe in cool climates [3]. The roots of infected plants are black and rotted, resulting in a greatly reduced root system. However, the fungus is restricted to the cortex of the root and does not enter the vascular tissues. Control is based on a combination of cultural practices, genetic resistance, and chemical applications [3].

Biological control of black root rot of tobacco with fluorescent pseudomonads has been studied extensively [4–19]. *Pseudomonas fluorescens* strain CHA0 was originally isolated from tobacco roots grown in a soil from Morens, Switzerland, that was naturally suppressive to black root rot [4]. In initial pot experiments, addition of strain CHA0 (10^7 cfu/cm^3 soil) to soils that were conducive to black root rot rendered most of them suppressive (36 of 39 soils) [4]. In order to understand the mechanisms of disease suppression, most of the subsequent work with *P. fluorescens* CHA0 was conducted in a controlled environment gnotobiotic system [6–8,10–14,16]. In natural soil artificially infested with *T. basicola*, in both field plots and pot experiments, biocontrol with CHA0 has been less successful [5].

Strain CHA0 produces several secondary metabolites that are involved in its ability to suppress plant diseases. Fluorescent siderophores (iron-chelating compounds), cyanic acid, and antibiotics are produced by this pseudomonad [6–11]. Ahl et al. [6] reported that iron-free siderophores (pyoverdine type [9]) reduced the production of endoconidia by *T. basicola* in vitro but did not inhibit germination of endoconidia or chlamydospores, nor did it inhibit mycelial growth of the fungus. However, Fe^{3+}-bound siderophores strongly inhibited both mycelial growth and spore germination in culture [6]. Ahl et al. [6] suggested that the bacterial siderophores had increased iron to the point where it became highly toxic to *T. basicola*. Iron competition is not the mechanism of black root rot suppression. However, sufficient iron is required for *P. fluorescens* CHA0 to effectively suppress black root rot of tobacco [7]. The iron requirement is most likely related to production of hydrogen cyanide (HCN) by CHA0 [7–9].

Voisard et al. [8] tested the importance of cyanide in a gnotobiotic system containing an iron-rich vermiculite soil. A cyanide-negative mutant, CHA5, was constructed by gene replacement. The mutant was less effective than the wild-type CHA0 in protecting tobacco from black root rot. Complementation of CHA5 by the cloned wild-type hcn^+ genes of CHA0 restored the strain's ability to suppress disease [8]. Production of HCN by *P. fluorescens* CHA0 and restored mutants was also highly correlated with root hair formation [9].

Pseudomonas fluorescens CHA0 produces the antibiotics 2,4-diacetylphloroglucinol (Phl), pyoluterorin, and monoacetylphloroglucinol [9]. The importance of 2,4-diacetylphloroglucinol in suppression of black root rot has been demonstrated. This metabolite has antifungal, antibacterial, and phytotoxic activity [10]. A Phl-negative mutant, CHA625, obtained by Tn5 insertion, suppressed black root rot to a lesser extent than the wild-type CHA0 under gnotobiotic conditions [10–12]. The mutant and the parental strain did not differ in root colonization [10]. The ability of CHA625 to suppress disease and produce 2,4-diacetylphloroglucinol was restored with a cosmid obtained from a genomic library of strain CHA0 [10–12]. Since Phl has herbicidal activity, it has been suggested that, like other herbicides, Phl may induce plant defense mechanisms against pathogens [10,12]. It has also been hypothesized that Phl synthesized in the rhizosphere might locally inhibit pathogens on plant roots [10,12]. Strain CHA0 also suppresses take-all of wheat, and Phl has been recovered from the rhizosphere of wheat colonized by CHA0 in a gnotobiotic system [10].

A global regulator gene, *gacA*, has been identified which regulates production of HCN, Phl, and pyoluteorin in *P. fluorescens* during restricted growth and/or limited nutrient supply [13]. The *gacA* mutants of *P. fluorescens* CHA0 were greatly reduced in their ability to suppress black root rot of tobacco [13].

In an effort to enhance the biocontrol capacity of *P. fluorescens* CHA0, a recombinant strain (CHA0/pME3090) that overproduced pyoluteorin and Phl three- to fivefold in vitro was evaluated in a gnotobiotic system [14]. The antibiotic-overproducing strain protected tobacco roots significantly better against black root rot than the parental strain CHA0, but growth of tobacco plants was greatly reduced [14], suggesting that the amounts of pyoluteorin and Phl produced in the rhizosphere of tobacco by the overproducing mutant were great enough to be toxic to the plants [14].

To elucidate the physical relationship between CHA0 and *T. basicola*, Troxler et al. [15] conducted immunofluorescence microscopy studies on tobacco roots. Strain CHA0 was introduced into sterile soil microcosms, and within 4–7 days of planting tobacco seeds cells of CHA0 were observed between and inside cells in the epidermis and cortex and in xylem vessels of the roots. The presence of CHA0 delayed colonization of the root interior by *T. basicola*, but CHA0 was seldom in contact with the mycelium of the fungus, suggesting that direct colonization of the mycelium of *T. basicola* by CHA0 was not required for protection against black root rot [15].

Recently, the genetic basis of hydrogen cyanide synthesis in *P. fluorescens* CHA0 was investigated [16]. Under microaerophilic conditions, HCN is produced by *P. fluorescens* from glycine. The structural gene cluster encoding HCN synthase in CHA0 was expressed in *Escherichia coli*, resulting in HCN production. The gene encoding an anaerobic regulator protein, ANR, was cloned from CHA0 and sequenced [16]. An *anr* mutant of CHA0 produced little HCN and, similar to an *hcn* deletion mutant, was impaired in its ability to suppress black root rot of tobacco. The anaerobic regulator ANR is required for synthesis of HCN by *P. fluorescens* CHA0, suggesting that ANR-mediated cyanogenesis contributes to biocontrol of black root rot, especially under limited oxygen conditions, such as a poorly aerated, water-saturated soil [16].

Pseudomonas fluorescens CHA0 suppresses black root rot of tobacco through multiple mechanisms. In addition to any direct effects of the various toxic metabolites produced by CHA0 on the fungus, these compounds may also affect tobacco plants, resulting in the induction of stress-related defense responses [15,17].

In addition to *P. fluorescens* CHA0, there are other reports of black root rot suppression by a fluorescent pseudomonad. Reddy and Patrick isolated a pseu-domonad designated as RD:1 from rye residues. Root bacterization with RD:1 increased seedling growth and reduced severity of black root rot [18,19]. Strain RD:1 became established in the rhizosphere of tobacco seedlings and reduced populations of other fluorescent and nonfluorescent pseudomonads and total aerobic bacteria [19]. Vegetative growth of *T. basicola* in vitro also was inhibited by RD:1. The authors indicated that the biocontrol activity of RD:1 might be due to the production of diffusible antibiotics and to its inhibitory effects on indigenous soil microflora [18,19].

In Italy, a reduction in the number of propagules of *T. basicola* has been reported in soils with tobacco roots colonized by the vesicular-arbuscular mycorrhizal fungus, *Glomus microcarpum* [20]. However, in the burley tobacco–growing regions of Kentucky, mycorrhizal fungi, specifically *Glomus macrocarpum* and *G. microcarpum*, are responsible for a common and widespread disease known as tobacco stunt [21].

B. Stem Rot

Stem rot of tobacco is caused by *Sclerotium rolfsii* (teleomorph *Athelia rolfsii*) [22]. The disease is not a major problem in tobacco but may cause losses of up to 10% in some fields. Stem rot is widely distributed in warmer regions of the temperate zone and in the tropics [22]. Overwintering sclerotia of the fungus germinate to produce fanlike mats of hyphae that contact and infect plant stems.

In the Philippines, various isolates of the biocontrol fungi *Trichoderma harzianum*, *T. hamatum*, and *T. aureoviridae* were evaluated for control of *S. rolfsii* on tobacco seedlings and for parasitism of sclerotia of *S. rolfsii* [23]. In greenhouse experiments with soil artificially and naturally infested with *S. rolfsii*, seedling survival was significantly increased by *Trichoderma* treatments. The number of transplantable seedlings was fourfold greater from soil with *Trichoderma* than from the infested control soil. Two of the more effective isolates (*T. harzianum* no. 1 and *T. hamatum* no. 1) were originally recovered from parasitized sclerotia of *S. rolfsii* from tobacco field soil. In a controlled experiment, 76 and 82% of recovered sclerotia were parasitized by these isolates, respectively [23].

C. Black Shank

Phytophthora parasitica var. *nicotianae* causes black shank, a very destructive root and stem disease of tobacco. The disease is worldwide in distribution but is most destructive in warmer climates [24]. The roots and lower portions of the stem are primarily affected by the pathogen, but all parts of the plant can be infected. Symptoms of black shank vary with plant age and environment. In the final stages of the disease, the stem or shank becomes black, hence the name black shank [24]. Control of black shank is based mainly on the use of resistant cultivars, crop rotation, and the fungicide metalaxyl. However, there are reports on efforts to control black shank biologically [25–28].

English and Mitchell [25] isolated fungi and bacteria that colonized developing tobacco roots rapidly in natural field soil. These isolates included *Trichoderma harzianum*, *Aspergillus carbonarium*, *Aspergillus terreus*, *Penicillium steckii*, and *Pseudomonas putida*. In a glasshouse study, a composite of these organisms was added to soil infested with *P. parasitica* var. *nicotianae*. During the first 14 days the number of root infections was not affected by the microbial composite amendment. However, after 90 days of plant growth there was a significant decrease in mortality of tobacco in the soil amended with the microbial composite [25].

The biocontrol bacterium *Bacillus cereus* strain UW85 was evaluated for antagonism against *P. parasitica* var. *nicotianae* in a laboratory microassay [26]. Previously, *Bacillus cereus* UW85 had been shown to lyse zoospores of *Phytophthora* [29] and to protect alfalfa from infection by *P. megasperma* f. sp. *megasperma* [30]. In the microassay, one 7-day-old tobacco seedling was placed in each well of a 96-well microtiter plate and inoculated with 500 zoospores of *P. parasitica* var. *nicotianae*. Sporangia of *Phytophthora* did not develop on seedlings that were inoculated simultaneously with zoospores and either 5 µL of filtrate of a sporulated culture of *B. cereus* UW85 or 1 µg/mL of the fungicide metalaxyl. The authors suggested that the microassay might be useful for rapid screening of potential biological and chemical control agents and for studying mechanisms of infection and control of *Phytophthora* spp. under hydroponic conditions [26].

Three isolates (P9023, L9125, KC94J) of nonpathogenic binucleate *Rhizoctonia* fungi (BNR) controlled black shank on greenhouse-grown seedlings in Styrofoam float trays [27]. The BNR were incorporated into soilless mix on colonized, pulverized, sifted rice particles; colonized whole rice grains; or pelleted tobacco seeds with 0.5% methylcellulose. The degree of protection varied with application method and was greatest with BNR applied as rice inocula. The mechanism(s) of protection may involve root colonization and competition for attachment sites or systemic induced resistance [27].

Trichoderma harzianum and the VA mycorrhizal fungus *Glomus fasciculatum* were evaluated singly and in combination for control of black shank in a naturally infested tobacco nursery in India [28]. The soil was infested with both *P. parasitica* var. *nicotianae* and *Pythium aphanidermatum*. *Trichoderma* (100 g produced in wheat bran and sawdust, 2×10^3 propagules/g) and *G. fasciculatum* (250 g, 8×10^2 propagules/g) were spread on the surface of 1×1 m seedbeds and covered with a layer of topsoil. Tobacco seed was mixed with fine sand and broadcast over the seedbed. Seed germination and the number of transplantable seedlings were greatest and seedling mortality was lowest with the combination of *T. harzianum* and *G. fasciculatum*. This treatment was equal to treatment with the fungicide metalaxyl (Ridomil MZ-72 WP 0.2% in 500 mL water applied as a soil drench at sowing and as a spray 20 days after sowing). Both the dual inoculation of *T. harzianum* and *G. fasciculatum* and the metalaxyl treatment were significantly better than the untreated control [28].

D. *Pythium* Diseases

Pythium diseases occur wherever tobacco is grown and include damping-off of seedlings, stem and root rot of young plants, and feeder root necrosis of field plants [31]. *Pythium* does generally not attack mature tissues, and therefore infections in older plants are often limited to the root tips [31]. Damping-off can cause severe losses in outdoor seedbeds and of greenhouse-grown plants seeded in trays floating in water reservoirs. *Pythium* diseases occur over a wide range of temperature. The prevailing temperature determines the species of *Pythium* that is active in tobacco tissues. Some of the more frequently isolated species of *Pythium* from tobacco roots include *P. aphanidermatum*, *P. ultimum* var. *ultimum*, and *P. myriotylum*.

Trichoderma harzianum has been evaluated for control of damping-off caused by *P. aphanidermatum* [32,33] and *P. myriotylum* [33]. Application of *T. harzianum* (wheat bran sawdust medium) to artificially infested soil significantly reduced damping-off compared with control and metalaxyl-treated seeds [32]. In pot tests, *T. harzianum* (on wheat bran) was added to natural and sterile soils infested with *P. aphanidermatum* or *P. myriotylum* [33]. Control of seedling damping-off was achieved in sterile soil but not in natural soil. The authors suggested that effective control of *Pythium* could be achieved in seedbeds if they were fumigated first, then amended with *T. harzianum* to prevent reinfestation by *Pythium* spp. [33].

An experimental system was developed [34] to determine the potential for combining systemic acquired resistance (SAR) and microbial biocontrol with *Bacillus cereus* UW85 for control of damping-off diseases caused by *Pythium torulosum*, *P. aphanidermatum*, and *Phytophthora parasitica*. *Bacillus cereus* UW85 produces the antibiotics zwittermicin A [35–37] and kanosamine [36,38]. In the presence of plant roots, cultures of UW85 or either of the antibiotics it produces can inhibit zoospore movement, reduce zoospore encystment on roots, and delay germination of zoospore cysts and the elongation rate of germ tubes [39].

Systemic acquired resistance occurs in many plant species in response to pathogen infection and certain chemicals. SAR is associated with expression of plant defense genes. It can last for several weeks to months after induction and is effective against a broad range of pathogens. To induce SAR, 7-day-old tobacco seedlings were treated with 0.5 mM salicylic acid or 0.1 mM 2,6-dicholoroisonicotinic acid. Induction of SAR suppressed damping-off and did not affect growth of UW85 on tobacco roots. Strain UW85 did not induce SAR. The combination of induced SAR and treatment with UW85 resulted in greater disease suppression than did either alone [34].

E. *Rhizoctonia* Sore Shin and Damping-Off

Rhizoctonia solani is a broadly distributed soilborne pathogen with an extensive host range. The fungus causes stem cankers (sore shin) and damping-off of tobacco. In the field, losses are usually small, but damage can be extensive to seedlings produced in outdoor seedbeds or in the float system in the greenhouse [40].

Several isolates of *Trichoderma* were collected from tobacco field soils in Zimbabwe and tested for pathogenicity against *R. solani* in vitro [41]. In sterilized soil, isolates of *T. harzianum* reduced the build-up of populations of *R. solani*, and to a lesser extent *Fusarium solani*. Biocontrol of *R. solani* and *F. solani* in tobacco transplants was achieved by addition of *T. harzianum* T77 to seedbeds that had been fumigated with methyl bromide, then infested with *R. solani* before seed was sown [41]. Plant growth and yield were also increased. Integration of the fungicide triadimenol with *Trichoderma* treatment enhanced disease control.

Another mycoparasite, *Verticillium biguttatum*, has been isolated from strains of *R. solani* recovered from tobacco plants with symptoms of sore shin in Italy [42]. *Verticillium biguttatum* isolates have been widely studied for their ability to suppress black scurf of potato caused by *R. solani* [43–45]. However, the study from Italy is the first report of *V. biguttatum* on tobacco isolates of *R. solani* [42]. In a greenhouse test, *V. biguttatum* isolates reduced severity of sore shin in a sterile soil artificially infested with *R. solani*.

B. cereus strain BA55 [46] also has potential as a biological control for sore shin of tobacco. Strain BA55 was originally isolated from a tobacco field soil and inhibits mycelial growth of *R. solani* in vitro [47]. Cultures of BA55 produced in a minimal medium were more effective than inoculum grown in a nutrient-rich medium in reducing disease severity caused by sore shin [46].

III. FOLIAR DISEASES CAUSED BY FUNGI

A. Target Spot

The basidiospores of *Thanatephorus cucumeris* (anamorph of *Rhizoctonia solani*) cause a leaf spot disease of tobacco termed target spot. The disease has been reported from mild humid climates where tobacco is grown. Symptoms begin as small water-soaked lesions about 2–3 mm in diameter [48]. When temperatures are moderate and relative humidity is high, the lesions expand, becoming light green with irregular margins and chlorotic halos [48]. The lesion tissue becomes necrotic and will drop out, leaving a ''shot-hole'' effect in the leaf. Environmental conditions within greenhouses for transplant production by floating Styrofoam trays in water reservoirs can be highly conducive to development of target spot [48].

Bacillus cereus strain BA55 has been tested for control of target spot in a transplant production greenhouse [47,49]. An aqueous suspension of BA55 cells, applied as a foliar spray, was as effective as foliar fungicides (fluazinam, mancozeb, iprodione) on direct-seed and seed-and-transfer tobacco, on both burley (DF485) and dark-fire (TR Madole) cultivars. In a subsequent test, with direct-seed plants only, BA55 was effective only on DF485 [47,49]. The causes of variability in disease control with BA55 include method of production and rate of application [46].

B. Brown Spot/*Alternaria* Leaf Spot

Alternaria alternata causes brown spot, a major foliar disease of tobacco [50]. It occurs on tobacco grown worldwide but is more severe in warm climates. Damage is greatest when older leaves are infected, leading to premature ripening and death of large areas of leaf tissue [50]. Under favorable environmental conditions, infections can result in premature abscission of leaves and dark brown sunken lesions on stems of suckers, petioles, seed capsules, and stalks [50].

Most isolates of *Alternaria* spp. from tobacco are nonpathogenic [50]. Spurr [1977] applied nonpathogenic isolates of *Alternaria* to tobacco leaves and reduced brown spot by 60% in laboratory tests and 65% in artificially induced field infections [51]. In controlled environment studies, *Bacillus cereus* subsp. *mycoides* effectively controlled development of tobacco brown spot lesions by inhibiting germination of *A. alternata* conidia [52]. In a field study conducted in China, the percentage of plants with *Alternaria* brown spot disease was decreased, and yield and quality of tobacco were increased with biological control applications composed of attenuated cucumber mosaic virus and its satellite RNA [53].

C. Gray Mold

Botrytis cinerea causes gray mold on seedlings wherever tobacco is grown [54]. Gray mold usually appears when seedlings are large enough for transplant. During rainy periods wet rot lesions form and the surface of the lesions are covered with *Botrytis*. If infected seedlings are transplanted to the field, cankers may develop on the stem and the plants may die. If dry conditions prevail, the seedlings may recover [54].

Trichoderma harzianum T39 applied at sites (soil or leaves) spatially separated from inoculation with *B. cinerea* significantly reduced severity of gray mold and caused a delay in spreading lesion formation in tobacco [55]. The spatial separation of the pathogen and biocontrol agent suggested that biocontrol was at least partially attributed to induction of systemic resistance by T39 [55]. Competition for nutrients and suppression of *B. cinerea* pathogenicity enzymes are also potential biocontrol mechanisms by T39 [55].

D. Blue Mold

Blue mold is a devastating foliar disease of tobacco caused by the oomycete *Peronospora tabacina*. Destruction of leaf tissue, systemic infection, and stunting of plants can be extensive when the environment is cool and wet [56]. The systemic fungicide metalaxyl is commonly used in seedbeds and field tobacco for control of blue mold.

The concept of immunization of plants or induction of SAR has come largely from early work with the blue mold pathogen of tobacco. In 1960, Cruickshank and Mandryk [57] reported high foliar resistance to blue mold when tobacco plants were stem-infected with a spore suspension of *P. tabacina*, and the foliage was subsequently inoculated with the same pathogen. Subsequently, it was shown that application of *P. tabacina* spores to the soil surface around stems could restrict infection in the stem and induce systemic resistance [58]. Induction of resistance by either method was associated with premature senescence, smaller leaves, dwarfing, and symptoms of nitrogen deficiency [57,58].

Later, another technique for immunization was developed that involved injection of *P. tabacina* into tobacco stem tissue external to the cambium [59–61]. With this technique, plants were 95–99% protected. At flowering, height and dry weight were increased by 40%, fresh weight was increased by 30%, and immunized plants had four to six more leaves than control plants [59–61]. SAR is characterized by an accumulation of salicylic acid and various pathogenesis-related proteins, including β-1,3-glucanases, chitinases, and peroxidases in the immunized plant [17]. In field experiments in Mexico, stem injections with *P. tabacina* protected tobacco against metalaxyl-tolerant strains of *P. tabacina* [63]. Technology that induces SAR is still in development and most recently has led to commercialization of chemicals that induce SAR in tobacco.

IV. ROOT DISEASE CAUSED BY BACTERIA

Bacterial wilt is caused by the soilborne pathogen *Ralstonia solanacearum* (formerly *Pseudomonas solanacearum*). The disease has been reported from tropical, semitropical, and warm temperate regions where tobacco is grown [64]. Wilting, stunting, and yellowing of the foliage are symptomatic of bacterial wilt and may occur at any stage of plant development. Effective control of bacterial wilt includes crop rotation, nematode management, resistant cultivars, and sanitation.

Avirulent bacteriocin-producing strains of *R. solanacearum* provided protection of tobacco seedlings against a virulent strain of *R. solanacearum* in greenhouse and field experiments [65,66]. A combination of avirulent *R. solanacearum* and a bacteriophage of *R. solanacearum* were tested for control of bacterial wilt [67]. Tobacco roots were placed in a suspension of the avirulent *R. solanacearum* for one hour, transplanted to pots, and drenched with culture filtrate containing the bacteriophage. Equal protection against a phage-susceptible pathogenic strain of *R. solanacearum* was provided by the combined treatment even when the interval between the two treatments was varied from 1 hour to 7 days. Treatment with the combination of avirulent *R. solanacearum* and the bacteriophage was more efficacious than either treatment alone in reducing wilting. The combined treatment was not effective against a phage-tolerant pathogenic strain of *R. solanacearum* [67].

More recently [68], mutant Hrp- strains of *R. solanacearum* were selected for their ability to aggressively colonize tobacco root tissue. These mutants do not elicit a hypersensitive response and are not pathogenic. Root inoculation of tobacco with an Hrp- mutant strain of *R. solanacearum* gave protection against subsequent inoculation by a pathogenic strain. The more invasive the Hrp- strain, the greater protection it provided against pathogenic *R. solanacearum* [68].

V. FOLIAR DISEASES CAUSED BY BACTERIA

A. Wildfire

Pseudomonas syringae pv. *tabaci* causes a foliar disease of tobacco commonly called wildfire. The primary symptom is leaf spot surrounded by a yellow halo [69]. Wildfire is known in almost all tobacco-producing regions of the world. Losses from wildfire have largely been eliminated with development of resistant cultivars [69].

Within the past 10 years there have been numerous reports of induced systemic resistance (ISR) to fungal, bacterial, and viral pathogens in different host plants by nonpathogenic rhizobacteria. The disease resistance induced by ISR is phenotypically similar to chemical- and pathogen-induced SAR [17]. Bacterial determinants of ISR include siderophores, lipopolyssaccharides, and salicylic acid [17]. The rhizobacterium *Serratia marcescens* strain 90–166 can induce resistance in tobacco to *P. syringae* pv. *tabaci* [70]. Research on the bacterial determinants responsible for ISR by strain 90–166 is underway. However, unlike other rhizobacteria-host-pathogen interactions [17], bacterial salicylic acid does not appear to be involved in disease resistance induced by 90–166 [70].

VI. DISEASES CAUSED BY VIRUSES

More than 20 viruses are known to infect tobacco naturally [71]. There are reports of biological control of tobacco mosaic virus (TMV), tobacco necrotic virus (TNV), and cucumber mosaic virus (CMV) on tobacco. These viruses occur worldwide. CMV is an important pathogen of tobacco in Asia and has caused significant losses in Europe. TMV is economically important but can be controlled effectively with crop rotation, sanitation, and resistant cultivars. TNV is soilborne and may kill seedlings. Its economic importance is unknown [71].

Foliar inoculations with TMV can induce SAR to challenge inoculations with TMV and the blue mold pathogen *P. tabacina* [17,72,73]. The rhizobacterium *Pseudomonas aeruginosa* strain 7NSK2 can mediate ISR in tobacco to TMV [17,74,75]. In this interaction, bacterially produced salicylic acid contributes to induction of systemic resistance [74,75].

Tobacco plants can be immunized against TNV with prior TNV inoculations [77]. *Pseudomonas fluorescens* CHA0 can also induce resistance to TNV in tobacco. Tobacco plants grown for 6 weeks in autoclaved natural soil inoculated with *Pseudomonas fluorescens* CHA0 showed resistance in leaves to infection with TNV [77]. The response was accompanied by an increase in salicylic acid in the leaves. A siderophore-negative mutant of CHA0 only induced partial resistance to TNV, thus implicating a role for the pyoverdin siderophore of CHA0 in the induction of resistance against TNV [77].

Compounds purified from fungi [78,79] and plants [80] can inhibit TMV infection in tobacco. Mycolaminaran, a β-1,3-glucan purified from the cytoplasm of *Phytophthora megasperma*, significantly reduced the number of necrotic lesions caused by TMV when co-inoculated with the virus, but it was ineffective when applied 4 hours after TMV inoculation [78]. Mycolaminaran is thought to induce a rapid general resistance in certain plants to local viral infection on treated leaf surfaces, rather than a systemic resistance [78]. A polysaccharide derived from the basidiomycete *Fomes fomentarium* (designated BAS-F) induced systemic resistance in tobacco to TMV [79]. A basic protein derived from the leaves of *Clerodendrum aculeatum* induced a high level of systemic resistance against TMV infection in tobacco [80].

In a field study conducted in China, tobacco plants were treated with preparations of attenuated cucumber mosaic virus and its satellite RNA [53]. Following inoculation with virulent CMV, the percentage of attenuated CMV-treated plants with CMV was significantly less than control plants.

VII. DISEASES CAUSED BY NEMATODES

Root-knot nematode is a major disease problem in tobacco production worldwide. Root-knot is most severe on light sandy or sandy loam soils in warm climates [81]. Root galling by nematodes leads to symptoms of drought stress and mineral deficiency that result from reduced efficiency of the root system. The quality and yield of tobacco is reduced, resulting in severe losses [81]. Four species of root-knot nematode are found on tobacco, including *Meloidogyne incognita* (southern root-knot nematode), *M. javanica* (Javanese root-knot nematode), *M. arenaria* (peanut root-knot nematode), and *M. hapla* (northern root-knot nematode) [81]. *Meloidogyne javanica* is the most common in subtropical and tropical regions, and *M. incognita* is the most common in temperate tobacco-growing regions [81]. Management of root-knot in tobacco is based on crop rotation, cultural practices, resistant cultivars, and fumigant and nonfumigant chemical pesticides [81].

There are several reports on biological control of root-knot with fungal and bacterial parasites of nematodes [82–90]. *Pasteuria* (formerly *Bacillus*) *penetrans* is an obligate, endospore-forming bacterial parasite of root-knot nematodes. In a field experiment in Florida, addition of *P. penetrans* to the soil of field plots reduced yield losses in tobacco caused by *M. incognita* by 23% [82]. In another field study in Florida [83], *P. penetrans* was suppressive to a mixed population of *M. incognita* and *M. javanica*. The population density of *Meloidogyne* spp. second-stage juveniles (J2) in soil was negatively correlated with both the numbers of *P. penetrans* endospores attached per J2 and the percentage of J2 with endospores attached [83].

A field site in Florida was identified as suppressive to *Meloidogyne* spp. [84]. The field had a mixed population of *M. incognita* and *M. javanica* and had been in tobacco monoculture for 7 years. Nematode management practices had not been applied to the site during this time. Initially yield losses were severe, but production recovered to the point that there was little difference in the growth of nematode-susceptible and resistant tobacco cultivars. The underlying basis of root-knot disease suppression was investigated, and it was concluded that reductions in root galling, numbers of egg masses, and eggs observed in the suppressive field soil were caused by infection of both nematode species with *P. penetrans* [84].

Paecilomyces lilacinus is a fungal parasite of nematodes and has been associated with egg masses of *M. incognita* and *M. javanica* from a Florida tobacco field soil [85]. *Pasteuria penetrans*, in combination with *Paecilomyces lilacinus*, a fungal parasite of nematodes, was evaluated for control of *M. incognita* in tobacco [86]. In greenhouse experiments, the combination of *P. penetrans* and *P. lilacinus*, or either agent applied alone, provided significant control of *M. incognita* [86]. Field tests on tobacco with applications of *P. lilacinus* have not been as successful. The fungus did not control *M. javanica* [87] or *M. incognita* [88] on tobacco.

Incorporation of the endomycorrhizal fungus *Glomus fasciculatum* into soil infested with *M. incognita* resulted in better growth and higher yield of tobacco than in nonmycorrhizal soil [89]. The number of root-knot galls, endoparasites, and egg masses per infested seedling was reduced by 61–89% as a result of mycorrhizal inoculation [89].

DiTera, a new biological nematicide derived from the fermentation of a nematode-parasitic isolate of the fungus *Myrothecium verrucaria*, has been registered as a microbial nematicide in several countries. The product has activity against several nematode species, including species of root-knot nematode [90]. Currently, it is available for cole crops and turf in the United States and table grapes in Mexico [90]. Although the fungus is cosmopolitan in distribution, it has been reported only once from cysts of the soybean cyst nematode *Heterodera glycines* [90,91]. The isolate of *M. verrucaria* used to produce DiTera was derived through subisolate selection from the soybean cyst nematode isolate [90]. The fermentation mass is utilized in the production of DiTera, and the organism is killed during the processing steps, resulting in a nonliving microbial product [90]. Field evaluations indicate that this product has potential for nematode control in tobacco [90].

VIII. DISEASES CAUSED BY PARASITIC HIGHER PLANTS

Broomrapes (*Orobanche* spp.) are the most important parasitic seed plants that cause disease in tobacco [92]. Plants attacked early in the season are stunted, while quality is reduced in plants attacked late in the season [92]. Effective management of broomrape is very difficult and relies on hand weeding and selective herbicides [92]. Fungal pathogens of *Orobanche* are being studied for their potential as biological controls of broomrape [93–96].

Fusarium oxysporum, isolated from naturally infected *Orobanche* in Iran, reduced broomrape by 75% and increased dry weight of tobacco by 81% in a field trial [93]. The fungus was not pathogenic on tobacco. Similarly in greenhouse studies in Germany, *F. oxysporum* f. sp. *orthoceras* was highly pathogenic on *Orobanche* from tobacco [94]. Studies on *Fusarium lateritium* indicate that it also has good potential as a mycoparasite of *Orobanche* spp. on tobacco. In a field study in Bulgaria, protection of tobacco by *F. lateritium* against broomrape ranged from 62 to 68% [95]. Application of the fungus in irrigation water at the time of transplant provided long-lasting protection [95].

IX. TRANSGENIC TOBACCO

In the last few years there have been hundreds of reports on development of transgenic tobacco with disease resistance. A review of this topic is beyond the scope of this work. However, the production of transgenic tobacco with a gene encoding a strongly antifungal endochitinase from the mycoparasitic biocontrol fungus *Trichoderma harzianum* [96] serves as an example of future directions in biological control of tobacco diseases. Expression of the *Trichoderma* endochitinase in tobacco resulted in a high level of resistance to soilborne and foliar fungal pathogens. The degree of disease resistance represented a major improvement in comparison to transgenic expression of other chitinase genes from plants or bacteria [96]. The authors suggested that this probably results from the stronger and wider spectrum of antifungal activity by *Trichoderma* chitinases compared with those from plants and bacteria [97]. Incorporation of microbial genes responsible for biological control into plants is one possible solution to the problem of inconsistent performance that can occur when biocontrol microorganisms are introduced into a variety of environments [96].

X. CONCLUSION

As illustrated by these numerous research studies, there is great potential for biological control of tobacco diseases. As with many other crops, inconsistent performance and multiple disease problems in the field have hampered commercial development. In the last 10 years our understanding of microbial biocontrol mechanisms and plant responses has increased tremendously. Continued advances will lead to a greater understanding of the reasons for variability in disease control and enhance our capacity to manipulate biocontrol mechanisms in order to optimize biological disease control.

REFERENCES

1. NT Powell. Approaches for management of tobacco diseases. In: HD Shew, GB Lucas, eds. Compendium of Tobacco Diseases. St. Paul, MN: APS Press, 1991, pp 55–58.
2. D Fravel. Commercial biocontrol products for use against soilborne crop diseases. *http://www.barc.usda.qov/psi/bpdlprod/bioprod.html.* Updated Sep 20, 2000. Accessed Nov 28, 2000.
3. HD Shew. Black root rot. In: HD Shew, GB Lucas, eds. Compendium of Tobacco Diseases. St. Paul, MN: APS Press, 1991, pp 21–23.
4. EW Stutz, G Défago, H Kern. Naturally occurring fluorescent pseudomonads in-

volved in suppression of black root rot of tobacco. Phytopathology 76:181–185, 1986.

5. B Wüthrich, G Défago. Suppression of wheat take-all and black root rot of tobacco by *Pseudomonas fluorescens* strain CHA0: results of field and pot experiments. Bulletin—SROP 14:17–22, 1991.

6. P Ahl, C Voisard, G Défago. Iron bound-siderophores, cyanic acid, and antibiotics involved in suppression of *Thielaviopsis basicola* by a *Pseudomonas fluorescens* strain. J Phytopathol 116:121–134, 1986.

7. C Keel, C Voisard, CH Berling, G Kahr, G Défago. Iron sufficiency, a prerequisite for the suppression of tobacco black root rot by *Pseudomonas fluorescens* strain CHA0 under gnotobiotic conditions. Phytopathology 79:584–589, 1989.

8. C Voisard, C Keel, D Haas, G Défago. Cyanide production by *Pseudomonas fluorescens* helps suppress black root rot of tobacco under gnotobiotic conditions. EMBO J 8:351–358, 1989.

9. G Défago, CH Berling, U Burger, D Haas, G Kahr, C Keel, C Voisard, P Wirthner, B Wüthrich. Suppression of black root rot of tobacco and other root diseases by strains of *Pseudomonas fluorescens*: potential applications and mechanisms. In: D Hornby, ed. Biological Control of Soil-borne Plant Pathogens. Wallingford, UK: CAB International, 1990, pp 93–106.

10. C Keel, U Schnider, M Maurhofer, C Voisard, J Laville, U Burger, P Wirthner, D Hass, G Défago. Suppression of root diseases by *Pseudomonas fluorescens* CHA0: importance of the bacterial secondary metabolite 2,4-diacetylphloroglucinol. Mol Plant-Microbe Interact 5:4–13, 1992.

11. C Keel, PH Wirthner, T Oberhänsli, C Voisard, U Burger, D Haas, G Défago. Pseudomonads as antagonists of plant pathogens in the rhizosphere: role of the antibiotic 2,4-diacetylphloroglucinol in the suppression of black root rot of tobacco. Symbiosis 9:327–341, 1990.

12. D Haas, C Keel, J Laville, M Maurhofer, T Oberhänsli, U Schnider, C Voisard, B Wüthrich, G Défago. Secondary metabolites of *Pseudomonas fluorescens* strain CHA0 involved in the suppression of root diseases. In: H Hennecke, DPS Verma, eds. Advances in Molecular Genetics of Plant-Microbe Interactions, Vol. 1. Dordrecht, Netherlands: Kluwer Academic Publishers, 1991, pp 450–456.

13. J Laville, C Voisard, C Keel, M Maurhofer, G Défago, D Haas. Global control in *Pseudomonas fluorescens* mediating antibiotic synthesis and suppression of black root rot of tobacco. Proc Natl Acad Sci 89:1562–1566, 1992.

14. M Maurhofer, C Keel, D Haas, G Défago. Influence of plant species on disease suppression by *Pseudomonas fluorescens* strain CHA0 with enhanced antibiotic production. Plant Pathol 44:40–50, 1995.

15. J Troxler, C-H Berling, Y Moënne-Loccoz, C Keel, G Défago. Interactions between the biocontrol agent *Pseudomonas fluorescens* CHA0 and *Thielaviopsis basicola* in tobacco roots observed by immunofluorescence microscopy. Plant Pathol 46:62–71, 1997.

16. J Laville, C Blumer, C von Schroetter, V Gaia, G Défago, C Keel, D Haas. Characterization of the hcnABC gene cluster encoding hydrogen cyanide synthase and anaerobic regulation by ANR in the strictly aerobic biocontrol agent *Pseudomonas fluorescens* CHA0. J Bacteriol 180:3187–3196, 1998.

17. LC van Loon, PAHM Bakker, CMJ Pieterse. Systemic resistance induced by rhizo-sphere bacteria. Annu Rev Phytopathol 36:453–483, 1998.

18. MS Reddy, ZA Patrick. Effect of a fluorescent pseudomonad on growth of tobacco seedlings and suppression of black root rot caused by *Thielaviopsis basicola*. Bulle-tin—SROP 14:23–29, 1991.

19. MS Reddy, ZA Patrick. Colonization of tobacco seedling roots by fluorescent pseu-domonad suppressive to black root rot caused by *Thielaviopsis basicola*. Crop Prot 11:148–154, 1992.

20. L Tosi, M Giovannetti, A Zazzerini, G Della Torre. Influence of mycorrhizal tobacco roots, incorporated into the soil, on the development of *Thielaviopsis basicola*. J Phytopathol 122:186–189, 1988.

21. HS Modjo, JW Hendrix. The mycorrhizal fungus *Glomus macrocarpum* as a cause of tobacco stunt disease. Phytopathology 76:688–691, 1986.

22. HD Shew. Stem rot. In: HD Shew, GB Lucas, eds. Compendium of Tobacco Dis-eases. St. Paul, MN: APS Press, 1991, pp 23–24.

23. HX Truong, MD Salinas, AS Obien, RC Carasi. Biological control of *Sclerotium rolfsii* and *Pythium aphanidermatum* damping off on tobacco with *Trichoderma har-zianum* culture. Brighton Crop Protection Conference—Pests and Diseases, Brigh-ton, 1988, pp 1189–1194.

24. HD Shew. Black shank. In: HD Shew, GB Lucas, eds. Compendium of Tobacco Diseases. St. Paul, MN: APS Press, 1991, pp 17–20.

25. JT English, DJ Mitchell. Influence of an introduced composite of microorganisms on infection of tobacco by *Phytophthora parasitica* var. *nicotianae*. Phytopathology 78:1484–1490, 1988.

26. J Handelsman, WC Nesmith, SJ Raffel. Microassay for biological and chemical con-trol of infection of tobacco by *Phytophthora parasitica* var. *nicotianae*. Curr Micro-biol 22:317–319, 1991.

27. DK Cartwright, HW Spurr Jr. Biological control of *Phytophthora parasitica* var. *nicotianae* on tobacco seedlings with non-pathogenic binucleate *Rhizoctonia* fungi. Soil Biol Biochem 14:1879–1884, 1998.

28. KR Sreeramulu, T Onkarappa, HN Swamy. Biocontrol of damping off and black shank disease in tobacco nursery. Tob Res 24:1–4, 1998.

29. GS Gilbert, J Handelsman, JL Parke. Role of ammonia and calcium in lysis of zoo-spores of *Phytophthora cactorum* by *Bacillus cereus* UW85. Exp Mycol 14:1–8, 1990.

30. J Handelsman, S Raffel, EH Mester, L Wunderlich, CR Grau. Biological control of damping-off of alfalfa seedlings with *Bacillus cereus* UW85. Appl Environ Micro-biol 56:713–718, 1990.

31. HD Shew. Pythium diseases. In: HD Shew, GB Lucas, eds. Compendium of Tobacco Diseases. St. Paul, MN: APS Press, 1991, pp 20–21.

32. AB Brahmbhatt, AN Mukhopadhyay, KK Patel. *Trichoderma harzianum*, a potential bio-control agent for tobacco damping-off. PKV Res J 13:170–172, 1989.

33. NS Devaki, SS Bhat, SG Bhat, KR Manjunatha. Antagonistic activities of *Tricho-derma harzianum* against *Phythium aphanidermatum* and *Pythium myriotylum* on tobacco. J Phytopathol 136:82–87, 1992.

34. J Chen, LM Jacobson, J Handelsman, RM Goodman. Compatibility of systemic ac-

quired resistance and microbial biocontrol for suppression of plant disease in a laboratory assay. Mol Ecol 5:73–80, 1996.

35. H He, L Silo-Suh, J Handelsman, J Clardy. Zwittermicin A, an antifungal and plant protection agent from *Bacillus cereus*. Tetrahedron Lett 35:2499–2502, 1994.

36. LA Silo-Suh, BJ Lethbridge, SJ Raffel, H He, J Clardy, J Handelsman. Biological activities of two fungistatic antibiotics produced by *Bacillus cereus* UW85. Appl Environ Microbiol 60:2023–2030, 1994.

37. LA Silo-Suh, EVS Stabb, SJ Raffel, J Handelsman. Target range of zwittermicin A, an aminopolyol antibiotic from *Bacillus cereus*. Curr Microbiol 37:6–11, 1998.

38. JL Milner, L Silo-Suh, JC Lee, H He, J Clardy, J Handelsman. Production of kanosamine by *Bacillus cereus* UW85. Appl Environ Microbiol 62:3061–3065, 1996.

39. H Shang, J Chen, J Handelsman, RM Goodman. Behavior of *Pythium torulosum* zoospores during their interaction with tobacco roots and *Bacillus cereus*. Curr Microbiol 38:199–204, 1999.

40. HD Shew. Sore shin and damping-off. In: HD Shew, GB Lucas, eds. Compendium of Tobacco Diseases. St. Paul, MN: APS Press, 1991, pp 24–25.

41. JS Cole, JS Zvenyika. Integrated control of *Rhizoctonia solani* and *Fusarium solani* in tobacco transplants with *Trichoderma harzianum* and triadimenol. Plant Pathol 37:271–277, 1988.

42. R Nicoletti, E Lahoz. *Verticillium biguttatum* recovered from *Rhizoctonia solani* isolates pathogenic on tobacco. Petria 8:145–150, 1998.

43. G Jager, H Velvis, JG Lamers, A Mulder, J Roosjen. Control of *Rhizoctonia solani* in potato by biological, chemical, and integrated measures. Potato Res 34:269–284, 1991.

44. RAC Morris, JR Coley-Smith, JM Whipps. The ability of the mycoparasite *Verticillium biguttatum* to infect *Rhizoctonia solani* and other plant pathogenic fungi. Mycol Res 99:997–1003, 1995.

45. TJ Wicks, B Morgan, B Hall. Chemical and biological control of *Rhizoctonia solani* on potato seed tubers. Aust J Exp Agric 35:661–664, 1995.

46. JP McMahan. Biological control of target spot and sore shin of tobacco seedlings with biological seed treatments and foliar sprays. M.S. thesis, The University of Tennessee, Knoxville, TN, 1996.

47. BA Ashby Jr. Biological and chemical control of target spot of float-grown dark fire-cured tobacco seedlings. M.S. thesis, The University of Tennessee, Knoxville, TN, 1995.

48. HD Shew. Target spot. In: HD Shew, GB Lucas, eds. Compendium of Tobacco Diseases. St. Paul, MN: APS Press, 1991, pp 13–15.

49. BH Ownley, RL Ashby Jr, BB Reddick, DO Onks. Effect of biological and chemical controls, cultivar, host age, and planting method on severity of target spot, caused by *Thanatephorus cucumeris*, on float-grown tobacco seedlings. Tob Sci 40:137–146, 1996.

50. HW Spurr. Brown spot. In: HD Shew, GB Lucas, eds. Compendium of Tobacco Diseases. St. Paul, MN: APS Press, 1991, pp 10–12.

51. HW Spurr Jr. Protective applications of conidia of nonpathogenic *Alternaria* sp. isolates for control of tobacco brown spot disease. Phytopathology 67:128–132, 1977.

52. DR Fravel, HW Spurr Jr. Biocontrol of tobacco brown-spot disease by *Bacillus cereus* subsp. *mycoides* in a controlled environment. Phytopathology 67:930–932, 1977.
53. B Qin, X Zhang, G Wu, P Tien. Plant resistance to fungal diseases induced by the infection of cucumber mosaic virus attenuated by satellite RNA. Ann Appl Biol 120: 361–366, 1992.
54. GB Lucas, HD Shew. Gray mold and dead-blossom leaf spot. In: HD Shew, GB Lucas, eds. Compendium of Tobacco Diseases. St. Paul, MN: APS Press, 1991, p 15.
55. G De Meyer, J Bigirimana, Y Elad, M Höfte. Induced systemic resistance in *Trichoderma harzianum* T39 biocontrol of *Botrytis cinerea*. Eur J Plant Pathol 104:279–286, 1998.
56. CE Main. Blue mold. In: HD Shew, GB Lucas, eds. Compendium of Tobacco Diseases. St. Paul, MN: APS Press, 1991, pp 5–9.
57. IAM Cruickshank, M Mandryk. The effect of stem infestation of tobacco with *Peronospora tabacina* Adam on foliage reaction to blue mold. J Aust Inst Agricul Sci 26:369–372, 1960.
58. Y Cohen and J Kuć. Evaluation of systemic resistance to blue mold induced in tobacco leaves by prior stem inoculation with *Peronospora hyoscyami tabacina*. Phytopathology 71:783–787, 1981.
59. J Kuć, S Tuzun. Immunization for disease resistance in tobacco. Recent Adv Tob Sci 9:179–213, 1983.
60. S Tuzun, J Kuć. A modified technique for inducing systemic resistance to blue mold and increasing growth of tobacco. Phytopathology 75:1127–1129, 1985.
61. S Tuzun, W Nesmith, RS Ferriss, J Kuć. Effects of stem injections with *Peronospora tabacina* on growth of tobacco and protection against blue mold in the field. Phytopathology 76:938–941, 1986.
62. J Kuć. Immunization for the control of plant disease. In: D Hornby, ed. Biological Control of Soil-borne Plant Pathogens. Wallingford, UK: CAB International, 1990, pp 355–373.
63. S Tuzun, J Juarez, WC Nesmith, J Kuć. Induction of systemic resistance in tobacco against metalaxyl-tolerant strains of *Peronospora tabacina* and the natural occurrence of the phenomenon in Mexico. Phytopathology 82:425–429, 1992.
64. E Echandi. Bacterial wilt. In: HD Shew, GB Lucas, eds. Compendium of Tobacco Diseases. St. Paul, MN: APS Press, 1991, pp 33–35.
65. W Chen, E Echandi, H Spurr Jr. Protection of tobacco plants from bacterial wilt avirulent bacteriocin-producing strains of *Pseudomonas solanacearum*. In: JC Lozano, ed. Proceedings of the Fifth International Conference on Plant Pathogenic Bacteria. Centro Internacional de Agricultura Tropical, Colombia, 1981, pp 482–492.
66. WY Chen, E Echandi. Effects of avirulent bacteriocin-producing strains of *Pseudomonas solanacearum* on the control of bacterial wilt of tobacco. Plant Pathol 33: 245–253, 1984.
67. H Tanaka, H Negish, H Maeda. Control of tobacco bacterial wilt by an avirulent strain of *Pseudomonas solanacearum* M4S and its bacteriophage. Ann Phytopath Soc 56:243–246, 1990.
68. A Trigalet, D Trigalet-Demery, R Feuillade. Aggressiveness of French isolates of

Ralstonia solanacearum and their potential use in biocontrol. OEPP/EPPO Bull 28: 101–107, 1998.

69. PB Shoemaker. Wildfire and angular leaf spot. In: HD Shew, GB Lucas, eds. Compendium of Tobacco Diseases. St. Paul, MN: APS Press, 1991, pp 30–32.
70. CM Press, M Wilson, S Tuzun, JW Kloepper. Salicylic acid produced by *Serratia marcescens* 90–166 is not the primary determinant of induced systemic resistance in cucumber or tobacco. Mol Plant-Microbe Interact 10:761–768, 1997.
71. GV Gooding. Diseases caused by viruses. In: HD Shew, GB Lucas, eds. Compendium of Tobacco Diseases. St. Paul, MN: APS Press, 1991, pp 41–47.
72. XS Ye, SQ Pan, J Kuć. Activity, isozyme pattern, and cellular localization of peroxidase as related to systemic resistance of tobacco to blue mold (*Peronospora tabacina*) and to tobacco mosaic virus. Phytopathology 80:1295–1299, 1990.
73. XS Ye, SQ Pan, J Kuć. Pathogenesis-related proteins and systemic resistance to blue mould and tobacco mosaic virus induced by tobacco mosaic virus, *Peronospora tabacina* and aspirin. Physiol Mol Plant Pathol 35:161–175, 1989.
74. G De Meyer, M Höfte. Induction of systemic resistance by the rhizobacterium *Pseudomonas aeruginosa* 7NSK2 is a salicylic acid dependent phenomenon in tobacco. In: B Duffy, U Rosenberger, G Défago, eds. Molecular Approaches in Biological Control. Delemont, Switzerland: Bulletin OILB-SROP 1998, pp 117–121.
75. G De Meyer, K Audenaert, M Höfte. *Pseudomonas aeruginosa* 7NSK2-induced systemic resistance in tobacco depends on in planta salicyclic acid accumulation but is not associated with PR1a expression. Eur J Plant Pathol 105:513–517, 1999.
76. M Maurhofer, C Hase, P Meuwly, Métraux, G Défago. Induction of systemic resistance of tobacco to tobacco necrosis virus by the root-colonizing *Pseudomonas fluorescens* strain CHA0: influence of the *gacA* gene and of pyoverdine production. Phytopathology 84:139–146, 1994.
77. M Maurhofer, C Reimmann, P Schmidli-Sacherer, S Heeb, D Haas, G Défago. Salicylic acid biosynthetic genes expressed in *Pseudomonas fluorescens* strain P3 improve the induction of systemic resistance in tobacco against tobacco necrosis virus. Phytopathology 88:678–684, 1998.
78. TM Zinnen, CM Heinkel, MES Hudspeth, R Meganathan. The role of cytoplasmic mycolaminaran in inhibiting initial viral infection of certain *Nicotiana* species. Phytopathology 81:426–428, 1991.
79. M Aoki, M Tan, A Fukushima, T Hieda, S Kubo, M Takbayashi, K Ono, Y Mikami. Antiviral substances with systemic effects produced by basidiomycetes such as *Fomes fomentarius*. Biosci Biotech Biochem 57:278–282, 1993.
80. HN Verma, S Srivastava, Varsha, D Kumar. Induction of systemic resistance in plants against viruses by a basic protein from *Clerodendrum aculeatum* leaves. Phytopathology 86:485–492, 1996.
81. SM Schneider. Root-knot nematodes. In: HD Shew, GB Lucas, eds. Compendium of Tobacco Diseases. St. Paul, MN: APS Press, 1991, pp 37–40.
82. SM Brown, JL Kepner, GC Smart Jr. Increased crop yields following application of *Bacillus penetrans* to field plots infested with *Meloidogyne incognita*. Soil Biol Biochem 17:483–486, 1985.
83. S Chen, DW Dickson, EB Whitty. Response of *Meloidogyne* spp. to *Pasteuria penetrans*, fungi, and cultural practices in tobacco. J Nematol 26:620–625, 1994.

84. E Weibelzahl-Fulton, DW Dickson, EB Whitty. Suppression of *Meloidogyne incognita* and *M. javanica* by *Pasteuria penetrans* in field soil. J Nematol 28:43–49, 1996.
85. SY Chen, DW Dickson, EB Whitty. Fungi associated with egg masses of *Meloidogyne incognita* and *M. javanica* in a Florida tobacco field. Nematropica 26:153–157, 1996.
86. B Dube, GC Smart Jr. Biological control of *Meloidogyne incognita* by *Paecilomyces lilacinus* and *Pasteuria penetrans*. J Nematol 19:222–227, 1987.
87. TE Hewlett, DW Dickson, DJ Mitchell, ME Kannwischer-Mitchell. Evaluation of *Paecilomyces lilacinus* as a biocontrol agent of *Meloidogyne javanica* on tobacco. J Nematol 20:578–584, 1988.
88. PM Yuen, SA Karim, OA Shokri. Comparison of the efficacy of *Paecilomyces lilacinus* and fenamiphos in controlling *Meloidogyne incognita* on tobacco seedlings. Proceedings of the 3rd International Conference on Plant Protection in the Tropics, Genting Highlands, Pahang, Malaysia, 1992, pp 38–41.
89. KSK Prasad. Influence of a vesicular mycorrhiza on the development and reproduction of root-knot nematode affecting flue cured tobacco. Afro-Asian J Nematol 1:130–134, 1991.
90. P Warrior, LA Rehberger, M Beach, PA Grau, GW Kirfman, JM Conley. Commercial development and introduction of DiTera™, a new nematicide. Pestic Sci 55:376–379, 1999.
91. B Gintis, G Morgan-Jones, R Rodriguez-Kabana. Fungi associated with several developmental stages of *Heterodera glycines* from an Alabama field soil. Nematropica 13:181–200, 1983.
92. GB Lucas. Broomrape. In: HD Shew, GB Lucas, eds. Compendium of Tobacco Diseases. St. Paul, MN: APS Press, 1991, p 47.
93. A Mazaheri, N Moazami, M Vaziri, N Moayed-Zadeh. Investigations on *Fusarium oxysporum* a possible biological control of broomrape (*Orobanche* spp.). Proceedings of the Fifth International Symposium on Parasitic Weeds, Nairobi, Kenya, 1991, pp 93–95.
94. H Thomas, J Sauerborn, D Müller-Stöver, A Ziegler, JS Bedi, J Kroschel. The potential of *Fusarium oxysporum* f. sp. *orthoceras* as a biological control agent for *Orobanche cumana* in sunflower. Biol Control 13:41–48, 1998.
95. H Bozoukov, I Kouzmanova. Biological control of tobacco broomrape (*Orobanche* spp.) by means of some fungi of the genus *Fusarium*. In: AH Pieterse, JAC Verkleij, SJ ter Borg, eds. Proceedings of the Third International Workshop on *Orobanche* and related *Striga* research. Amsterdam, The Netherlands, 1994, pp 534–538.
96. M Lorito, SL Woo, I Garcia, G Colucci, GE Harman, JA Pintor-Toro, E Filippone, S Muccifora, CB Lawrence, A Zoina, S Tuzun, F Scala. Genes from mycoparasitic fungi as a source for improving plant resistance to fungal pathogens. Proc Natl Acad Sci 95:7860–7865, 1998.
97. M Lorito, GE Harman, CK Hayes, RM Broadway, A Tronsmo, SL Woo, A Di Pietro. Chitinolytic enzymes produced by *Trichoderma harzianum*: antifungal activity of purified endochitinase and chitobiosidase. Phytopathology 83:302–307, 1993.

7

Biological Control of Peanut Diseases

Appa Rao Podile and G. Krishna Kishore
University of Hyderabad, Hyderabad, Andhra Pradesh, India

I. INTRODUCTION

The peanut, or groundnut (*Arachis hypogaea* L.), is a very important legume crop of tropical and subtropical areas of the world. The importance of peanut as a food and oil crop and as a cash source in the semi-arid tropics is well known. Peanut is grown in many countries throughout the world, with ~80% of the crop grown in developing countries. Peanut is grown on ~3.1×10^7 ha throughout the world with a production of ~3.8×10^7 metric tons [1]. Average yields, ~825 kg/ha, of peanut in developing countries are low in comparison with the 2650 kg/ha or higher obtained in developed countries. An important factor contributing to low yield is disease attack by different microorganisms, nematodes, and insect pests. Peanut is prone to attack by more than 55 pathogens, including fungi, bacteria, viruses, mycoplasma, nematodes, and parasitic flowering plants, among which fungal diseases cause the majority of economic losses of yield. The most significant diseases of peanut are listed in Table 1 [2]. This chapter summarizes the efforts to develop biological control of the major fungal and nematode diseases of peanut. Biological control of bacterial and viral diseases of peanut using fungal and bacterial agents has not been studied in detail. The experience of the authors' laboratory, including unpublished results, and findings of various research groups as to biological control of fungal diseases of peanut using fungal and bacterial agents are presented here.

Table 1 Important Diseases of Peanut and Their Pathogens

Disease	Causal organism
Fungal diseases	
Foliar diseases	
Alternaria leaf blight	*Alternaria tenuis* Auct.
	A. tenuissima (Kunze ex Pers.) Wiltshire
	A. arachidis Kulk.
Alternaria leaf spot	*A. arachidis* Kulk.
	A. alternata (Fr.) Keissler
	Alternaria sp.
Cercospora leaf blight	*Cercospora canescens* Ellis & Martin
Choanephora leaf spot, wet blight	*Choanephora cucurbitarum* (Berk. & Rav.) Thaxt;
	Choanephora sp.
Drechslera leaf blight	*Bipolaris spicifera* (Bainier) Subramanian
Early leaf spot	*Cercospora arachidicola* Hori
Late leaf spot	*Phaeoisariopsis personata* (Berk. & Curt.) v. Arx
Pepper spot and leaf scorch	*Leptosphaerulina crassiasca* (Sechet) Jackson & Bell
Pestalotiopsis leaf spot	*Pestalotiopsis arachidis* Satya
	P. adusta (Ell. & Ev.) Steyaert
	P. versicolor (Speg.) Steyaert
Phoma leaf blight	*Phoma arachidis*
	P. microspora Balasubramaniam & Narayanasamy sp nov.
	P. sorghina (Sacc.) Boerema, Dorenbosch & V. Kest
Rust	*Puccinia arachidis* Speg.
Scab	*Sphaceloma arachidis* Bitancourt & Jenk.
Web blotch, net blotch, ascochyta leaf spot, phoma leaf spot, spatselviek, muddy spot	Didymella arachidicola (Chock.) Taber, Pettit & Philley
Zonate leaf spot	*Cristulariella pyramidalis* Waterman & Marshall
Seed and seedling diseases	
Aspergillus crown rot, crown rot, collar rot, black mold	*Aspergillus niger* van Tieghem
	Aspergillus pulverulentus (McAlpine) Thom.
Collar rot, diplodia collar rot, diplodia blight	*Lasiodiplodia theobromae* (Pat.) Grif. & Maub.

Table 1 Continued

Disease	Causal organism
Damping-off and foot rot	*Rhizoctonia solani* Kuhn
	Fusarium solani f. sp *phaseoli* (Burk.) Sny. & Hans.
	F. oxysporum Schl. emend Sny. & Hans.
	Fusarium spp.
	Pythium myriotylum Dreschsler
	Pythium debaryanum Hesse
	Pythium irregulare Buisman
	Pythium ultimum Trow
	Macrophomina phaseolina (Tassi.) Goid.
	Sclerotium rolfsii Sacc.
Preemergence seed and seedling rots	*Rhizopus arrhizus* Fischer
	R. stolonifer (Ehr. ex Fr.) Vuillemin
	R. oryzae Went & Gerlings
	Fusarium solani f. sp *phaseoli* (Burk.) Sny. & Hans.
	F. oxysporum Schl. emend Sny. & Hans.
	Fusarium spp.
	Pythium myriotylum Dreschsler
	P. debaryanum Hesse
	P. irregulare Buisman
	P. ultimum Trow
	P. butleri Subramaniam
	Rhizoctonia solani Kuhn
	Macrophomina phaseolina (Tassi.) Goid.
	Sclerotium rolfsii Sacc.
	Aspergillus niger van Tieghem
	A. flavus Link ex Fries
	Botryodiplodia theobromae Pat.
	Cochliobolus bicolor Paul & Par.
	Penicillium citrinum Thom.
	P. funiculosum Thom.
Yellow mold, aflaroot	*Aspergillus flavus* Link ex Fries
	A. parasiticus Speare
Stem, root, and pod diseases	
Blackhull	*Thielaviopsis basicola* (Berk. & Br.) Ferr.
Blacknut, charcoal rot, ashy stem	*Macrophomina phaseolina* (Tassi.) goid.
Blight, dry rot, dry wilt	=*Rhizoctonia bataticola* (Taub.) Butler
Blue damage	*Sclerotium rolfsii* Sacc.
Concealed damage	*Diplodia gossypina* Cooke
	Aspergillus niger van Tieghem
	Macrophomina phaseolina (Tassi.) Goid.

Table 1 Continued

Disease	Causal organism
Cylindrocladium black rot	*Cylindrocladium crotalariae* (Loos) Bell & Sobers (anamorph)
Fusarium wilt	*Fusarium oxysporum* Schl. emend Sny. & Hans.
Pythium vascular wilt	*Pythium myriotylum* Dreschsler
Root rot, pod rots, pod breakdown	*Rhizoctonia solani* Kuhn
	Fusarium solani f. sp *Phaseoli* (Burk.) Sny. & Hans.
	F. oxysporum Schl. emend Sny. & Hans.
	Fusarium spp.
	Pythium myriotylum Dreschsler
	P. debaryanum Hesse
	P. irregulare Buisman
	P. ultimum Trow
	Macrophomina phaseolina (Tassi.) Goid.
	Sclerotium rolfsii Sacc.
Sclerotinia blight	*Sclerotinia minor* Jagger
	S. sclerotiorum (Lib.) de Bary
Stem rot, white mold, southern blight, Sclerotium blight, Sclerotium rot, Sclerotium wilt	*Sclerotium rolfsii* Sacc.
Verticillium wilt, floury rot	*Verticillium albo-atrum* Reinke & Bert.
	V. dahliae Kleb.
Bacterial diseases	
Bacterial wilt	*Pseudomonas solanacearum* (Smith) Smith
Witches' broom	Mycoplasma-like organism
Rugose leaf curl	Rickettsia-like organism
Viral diseases	
Peanut mottle	Peanut mottle virus (potyvirus group)
Groundnut rosette, chlorotic rosette, green rosette, and mosaic rosette	Groundnut rosette virus and its satellite RNA
	Groundnut rosette assistor virus (luteovirus group)
Peanut clump	Peanut clump virus (furovirus group)
Peanut stripe	Peanut stripe virus (potyvirus group)
Bud necrosis, spotted wilt, bud blight	Tomato spotted wilt virus (tomato spotted wilt virus group)
Peanut stunt	Peanut stunt virus (cucumovirus group)
Peanut green mosaic	Peanut green mosaic virus (potyvirus group)

Table 1 Continued

Disease	Causal organism
Nematode diseases	
Chlorosis and stunting	*Scutellonema cavenessi* Sher
Kalahasti malady	*Tylenchorhynchus brevilineatus* Williams
Peanut rot	*Ditylenchus destructor* Thorne
Root-knot	*Meloidogyne arenaria* (Neal) Chitwood
	M. hapla Chitwood
	M. javanica (Treub) Chitwood
Phanerogamic parasites	
Dodder	*Cuscuta campestris* Yunck.
Witch weed	*Alectra vogelii* Benth.

II. BIOLOGICAL CONTROL OF FUNGAL DISEASES

A. Leaf Spots (Early and Late)

1. Fungal Biocontrol

Dicyma pulvinata, Fusarium spp., *Penicillium* spp., and *Verticillium lecanii* parasitize *Cercospora arachidicola* and *Phaeoisariopsis personata*, the causal agents of early and late leaf spot diseases of peanut. *D. pulvinata* appears as a peculiar whitish, downy growth on diseased spots and parasitizes hyphae and spores of *P. personata* on peanut [3–5]. *D. pulvinata* penetrates stromatic cells, and the hyphae can be both inter- and intracellular. The growth of the mycoparasite is mainly confined to the peripheral region of the lesion, indicating a correlation with the actively growing pathogen. Microscopic observations of the mycoparasite on the lesion appear as extensively branched, septate, hyaline mycelia, conidioiphores, and conidia.

The use of *D. pulvinata* offers a model system for studying the effects of different environmental factors relevant to the development of an effective foliar biocontrol system and to provide a better understanding of the pathogen-mycoparasite interaction. Mitchell and Taber [6] determined the survival and/or growth of *D. pulvinata* in culture media and on host tissue invaded by *C. personatum* and tested the potential carriers for spray formulations of the mycoparasite. Variables tested included temperature, pesticide concentration, pH, and relative humidity. *D. pulvinata* exhibited a broader growth range in terms of temperature (23–28°C) on peanut leaflets infected with *C. personatum* compared with growth in broth culture (23–25°C). On detached leaves sporulation occurred at constant temperatures up to 30°C, and no sporulation occurred at 31.5°C. *D. pulvinata* grew over a broad pH range with maximal growth in the range of 3.3–

7.7. The germination and growth of *D. pulvinata* was completely inhibited by pesticides benomyl, mancozeb, and triphenyltin hydroxide during commercial field applications, and slight inhibition of growth was observed with carbofuran, quintozene, and carboxin. *D. pulvinata* therefore could be incorporated with the insecticides chloropyrifos and carbofuran onto peanut plants without any adverse effects. A chemically induced mutant of *D. pulvinata* (BR 30), tolerant of benomyl, was selected to incorporate into an existing pest-management system that uses benomyl.

Spores of *D. pulvinata* survived for 29 days in carriers consisting of H_2O, 0.1–0.2% carboxy methyl cellulose (CMC), 0.2–0.4% citrus pectin, or 0.25% ghatti gum incubated at 25°C or 60 days at 6°C, with little loss of viability [7]. Viability of *D. pulvinata* spores was higher after 29 days in the above carriers when incubated at 25°C compared with 6°C.

Mitchell et al. [8] further established that at 26°C, conidia of *D. pulvinata* close to both hyphae and conidia of *C. personatum* germinate within 11–17 hours. Visible signs of colonization of lesions of *C. personatum* by *D. pulvinata* appear within 58–65 hours. *D. pulvinata* was an effective protectant only when plants were exposed to continuous leaf wetness at 26°C for 5 days, which limits its use as a biocontrol agent. Lesions of *C. personatum* were visibly colonized by *D. pulvinata* within 4 days after applying their conidial suspensions under 40 hours of leaf wetness, 60 hours at 23–28°C, and 17.31 cm rainfall. *D. pulvinata* probably could not be used as a protectant of peanut plants against infection by *C. personatum* but possibly to control its secondary spread. It is, therefore, necessary to increase biocontrol efficacy through selection of superior agents, improved formulation techniques, or more appropriate application schedules.

Mycoparasites and their culture filtrates have a significant role in the reduction of leaf spot diseases. *D. pulvinata* and *V. lecanii* and their culture filtrates inhibit the in vitro conidial germination of *P. personata*, with *V. lecanii* culture filtrate being more efficacious than *D. pulvinata* [9]. That the mycoparasites were better than their respective culture filtrate, in inhibiting spore germination of *P. personata*, suggests nutrient competition or elaboration of fresh inhibitory substances by the spores of mycoparasites in the germinating medium. The mycoparasites and their culture filtrates reduced the in vivo development of late leaf spot to a significant level when compared with control. *V. lecanii* and its culture filtrate completely suppressed the development of late leaf spot even 12 days after inoculation. Preinoculation with *V. lecanii* 48 hours before the pathogen gave better disease control when compared with simultaneous inoculation [10]. In most cases, *V. lecanii* established contact directly through the spore wall, and in a few cases through germspores. Mycelia were observed inside the spores of leaf spot pathogens, but sporulation was not evident. Lysing of the pathogen spores was common. Bursting of spores due to extensive growth of the fungus was also observed occasionally. *V. lecanii* has the potential to control peanut late leaf spot disease, but significant work has not been done on the formulation of *V. lecanii* in suitable carriers for field application of this biocontrol agent. Intercropping peanut with pigeonpea and two sprays of cell-free culture filtrate of *P. islandicum* at 55 and 70 days after planting protected groundnut crop from leaf spot diseases [11]. A few species of *Fusarium* and *Penicillium* commonly present on the phyllosphere of peanut effectively reduced late leaf spot [12].

2. Bacterial Biocontrol

Epiphytic bacteria have the potential to control foliar plant diseases through antagonistic activity against pathogens [13]. A chitinolytic *Bacillus cereus* was isolated from the foliar chitin amended leaves and when applied again to chitin amended leaves survived better than on nonamended leaves [14]. Significant reductions in the severity of early leaf spot were obtained by chitin amendment on leaves, and further reductions were obtained by chitin and *B. cereus* amendments. Scanning electron microscopy revealed chitin deposits, fungal hyphae, and spores colonized by bacilliform bacteria. Colonized fungal hyphae and spores were pitted and distorted, indicating a potential for biological control of chitin-containing fungal pathogens by chitinolytic bacterial antagonists.

Four antagonistic bacterial strains, *B. thuringiensis* (HD-1), *B. thuringiensis* (HD-521), *B. cereus* var. *mycoides* (Ox-3), and *Pseudomonas cepacia* (Pc 742), were formulated as wettable powders and dusts [15]. Bacterial preparations were applied to the plants at biweekly intervals as aqueous suspensions of wettable powders or as dusts. Survival of *Bacillus* spp. formulated as wettable powder was less variable than survival of *P. cepacia* formulated as wettable powder or dust, and mean log populations were higher for *Bacillus* spp. However, *P. cepacia* controlled the disease more effectively.

Efforts are being made in our laboratory to identify a potent bacterial strain(s) for control of late leaf spot. A large number of rhizobacterial strains isolated from the rhizosphere of peanut [16] were tested for in vitro antagonistic activity against *P. personata*. Bacterial strains that inhibited the germination of *P. personata* conidia by more than 90% were further tested for control of late leaf spot. The bacteria were used as a foliar spray at different time intervals 48 hours before, 24 hours before, simultaneously, 24 hours after, and 48 hours after the pathogen inoculation. Two bacterial strains that appeared promising in control of late leaf spot in the greenhouse [17] are being field tested. Development of suitable formulations for field application and designing an effective spray schedule for late leaf spot control are also underway.

B. Rust

1. Fungal Biocontrol

Verticillium lecanii and *Penicillium islandicum* and their culture filtrates inhibit in vitro germination of urediniospores of *Puccinia arachidis* and significantly reduce in vivo development of rust [10,18]. *V. lecanii* establishes inside the spores of *P. arachidis*, and the growth of the fungus causes bursting of the spores [10]. *V. lecanii* has a high potential for use in biological control of rust and early and late leaf spot diseases of peanut, since it can parasitize all three pathogens, which normally occur together.

Treatment of peanut leaves with *Acremonium obclavatum* resulted in significant reductions in the number of pustules and uredospores, delay in maturity and opening of uredosori, and reduced viability of uredospores [19]. The pathogenicity of spores collected from infected sori was reduced, and *A. obclavatum* reappeared on most of the sori developed from infected uredospores. *A. obclavatum* (mycelium or conidia) survives until the time of rust development and are carried along with the rust fungal spores, when they are liberated from the pustule, reducing the infection capacity of rust fungus and leading to biocontrol of rust disease. Cell-free culture filtrates of *A. obclavatum* also reduced the number and size of rust pustules both when applied along with the pathogen and one day before the pathogen. This indicates a possible role of cell-free culture filtrate of *A. obclavatum* in induced systemic resistance. A water-soluble glucan isolated from the culture filtrate of *A. obclavatum* inhibited the germination of uredeospores of *P. arachidis*. Prior treatment of peanut leaves with glucan prolonged the incubation period and decreased the number of pustules and uredospores/sorus. Increased levels of endogenous salicylic acid, intercellular chitinase, and β-1,3-glucanase activities were found in glucan-treated peanut leaves [20].

Mycoparasites of rust fungi produce lytic enzymes capable of degrading the cell wall polymers of pathogenic fungi. Chitinase is a key enzyme involved in cell wall lysis of higher fungi and plays a vital role in biological control. *Fusarium chlamydosporum*, a mycoparasite to rust isolated from uredosori of *P. arachidis*, produces extracellular chitinase and has potential as a biological control agent for peanut rust [21,22]. Purified chitinase inhibited the germination of uredospores and also lysed the walls of uredospores and germ tubes [23]. An antifungal metabolite inhibiting the germination of uredospores of *P. arachidis* was also isolated from the culture filtrate of *F. chlamydosporum* [24].

Trichoderma harzianum inhibits the germination and germ tube growth of *P. arachidis* when uredospore suspensions were mixed with *T. harzianum* conidial suspensions either with or without prior incubation of the conidial suspension. Treatment with *T. harzianum* before or at the same time as inoculation with *P. arachidis* decreases the number of rust pustules and number of uredospores/pustule. *T. harzianum* colonizes pustules formed before application of the conidial suspension better than those forming after application of conidial suspension. A phenol-like antifungal compound inhibitory to *P. arachidis* was isolated from the germination fluid of *T. harzianum* [25].

2. Bacterial Biocontrol

Bacillus subtilis AF 1, a plant growth–promoting biocontrol rhizobacterium, isolated from soils nonconductive to pigeon pea wilt was used as a foliar spray to control rust in a detached leaf bioassay. AF 1 effectively reduced the severity of rust. Partially purified chitinase of AF 1 inhibits the germination of uredospores and development of rust (unpublished data from authors' laboratory).

C. *Aspergillus* Crown Rot

1. Fungal Biocontrol

Trichoderma harzianum and *T. viride* compete with organisms that cause crown or pod rot disease in peanuts and damping-off on other hosts [26,27]. The linear growth of *Aspergillus niger* considerably decreased in the presence of *T. harzianum*. A decrease in disease incidence occurred in *T. harzianum* soil treated at both the seedling and vegetative growth stages. The percentage of rotted pods was greatly reduced. The number and weight of pods or seeds significantly increased in soil infested with *A. niger* in combination with *T. harzianum*. Seed dressing with *T. harzianum* resulted in a decrease in crown rot infection at different *Aspergillus* inoculum levels [28]. Similarly, soil application of *T. harzianum* was effective in controlling seed and collar rot [29].

2. Bacterial Biocontrol

Two fluorescent pseudomonad strains, FPC 32 and FPO 4, applied as seed treatment significantly protected peanut against *A. niger* infection and increased the yield [30]. Significant control of crown rot was obtained by bacterization of peanut seeds with *Bacillus subtilis* AF 1 in *A. niger*–infested soil [31]. In dual cultures, AF 1 lysed *A. niger* cell wall, suppressed >90% fungal growth, and inhibited sporulation of *A. niger*. An extracellular protein precipitate from the filtrate of *B. subtilis* AF 1 culture exerted a growth-retarding effect on *A. niger*. The mycelial preparation of *A. niger*, as principal carbon source, supported the growth of AF 1 as much as chitin. When treated with *B. subtilis* AF 1, lipoxygenase levels in peanut seedlings increased earlier than they did when treated with *A. niger* [32]. 13-Hydroperoxyoctadecadienoic acid and 13-hydroperoxy-octadecatrienoic acid formed upon incubation of peanut lipoxygenase with linoleic acid or α-linolenic acid inhibited the growth of *A. niger* in vitro. Toxic phytoalexin accumulated in peanut in response to *A. niger* infection is converted to nontoxic or less toxic lower homologs [33]. However, in dual inoculation with *A. niger* together with *B. subtilis* AF 1, the phytoalexin remains unchanged, to the benefit of the plant. The multiplication and survival of *B. subtilis* AF 1 in different formulations in calcium alginate, peat, mushroom spent compost, peat amended with chitin, and peat amended with *A. niger* mycelium was determined at different temperatures and over a period of 6 months of storage and used for control of crown rot in greenhouse experiments. Of these different formulation products, seed bacterization with chitin supplemented peat formulation of *B. subtilis* AF 1 showed better disease control than freshly grown cells of AF 1 [34]. Two rhizobacterial strains exhibited antagonism towards growth, and a few strains inhibited sporulation of *A. niger* when a large number of rhizobacteria from peanut were tested (authors' unpublished data).

D. *Rhizoctonia* Damping-Off

1. Fungal Biocontrol

The majority of *Trichoderma* isolated from soil, pine bark, and other bark were antagonistic to *Rhizoctonia solani*, and the isolates from pine bark were more aggressive against *R. solani* than those from other sources [35]. ANF-777, a fungus with an exceptionally strong inhibitory effect against *R. solani* was selected from a collection of 7500 microbial colonies [36].

The hyperparasites *Botryotrichum piluliferum*, *Coniothyrium sporulosum*, *Dicyma olivacea*, *Gliocladium catenulatum*, *Stachybotrys chartarum*, *Stachybotrys elegans*, *Stachylidium bicolor*, *Trichothecium roseum*, *Verticillium chlamydosporium*, *V. tenerum*, and *V. bigguttatum* parasitize the hyphae of *R. solani* [37,38]. Profuse coiling of the parasitic hyphae around the host hyphae result in the cytoplasm of the host cells becoming granulated, disintegrated, and the contents disappear. In advanced stages of parasitism, the host hyphae finally die.

The growth of *R. solani* significantly decreased in the presence of *Trichoderma harzianum* Th008 culture filtrates [39]. Scanning electron microscopic observations of parasitism of *T. harzianum* and *T. hamatum* on *R. solani* revealed that the hyphae of *Trichoderma* coil around the host. *T. harzianum* attached to host mycelium by forming hooks, and *T. hamatum* produces appressoria at the tips of short branches. *Trichoderma* spp. penetrated the host by partial degradation of cell wall [40]. In the later stages of parasitism pronounced collapse and loss of turgor of *R. solani* hyphae occurs besides cell wall breakdown and occasional hyphal disintegration. Continuous production of chitinases by the antagonist resulted in gradual breakdown of chitin. Disorganization of the cell wall structure of *R. solani* appears to be an early event that promotes internal osmotic balance, which, in turn, triggers intracellular disorders, such as retraction of the plasma membrane and cytoplasm aggregation [41]. Recognition of the host by the parasite is the first step in the fungus-fungus interaction leading to mycoparasitism. *R. solani* contains a lectin, which binds to O but not A and B erythrocytes, and this attachment is prevented by galactose and fucose. A lectin present in *R.*

solani hyphae binds to galactose residues on *Trichoderma* cell walls and plays a role in prey recognition by the predator [42]. Culture filtrates of *S. elegans* possessing β-1,3-glucanase and chitinase activities were capable of degrading *R. solani* mycelium. Depending on the carbon source used, different isoforms of chitinases and β-1,3-glucanases were detected in the culture filtrate, suggesting a role for these enzymes in mycoparasitism [43].

R. *solani* failed to form sclerotia in the presence of *Gliocladium virens*, which colonizes both mycelia and sclerotia of *R. solani*. *G. virens* formed appressoria and penetrated the host hyphal cells. The parasitized hyphae began to collapse and shrink and finally collapsed completely and died. Mycelia of *G. virens* were found in the sclerotial cells of *R. solani*, demonstrating the intracellular parasitism of sclerotia by *G. virens* [44]. Two mycoparasitic strains of *G. virens* and their mutants with no mycoparasitic activity on *R. solani* showed similar efficacy as biocontrol agents of cotton seedling disease induced by *R. solani*, indicating that mycoparasitism is not the major mechanism in the biological control of *R. solani* by *G. virens* [45].

T. *harzianum* prevented reinfestation of the fumigated soil by *R. solani* and *S. rolfsii* up to 88% under field conditions. The combined treatment of fumigation and *T. harzianum* applications caused almost total mortality of sclerotia in soil in the laboratory and in the field [46]. *T. harzianum* in wheat bran/peat preparation applied at the time of sowing was used for the control of *Rhizoctonia* damping-off [47]. Infestation of soil with antagonists of *R. solani*—binucleate *Rhizoctonia* spp., *Laetisaria arvalis*, and an unidentified orange basidiomycetous fungus—had no effect on the yield of peanut grown in rotation with corn or corn and snapbean [48]. *Coniothyrium minitans* applied as seed treatment significantly reduced pre- and postemergence damping-off [49].

2. Bacterial Biocontrol

Pseudomonas fluorescens 2-79, *P. aeruginosa* EP 7, and *Bacillus subtilis* AF 1 inhibited the growth of *R. solani* [50]. Culture filtrates from *B. megaterium* B153-2-2 reduced the growth of *R. solani* by >90%. *B. megaterium* produced a relatively large amount of Ca^{2+}-dependent endoproteinase, inactivating pectinase and pectin lyase from *R. solani*, which are detrimental to plant cell wall. *B. megaterium* produced several other extracellular enzymes including phospholipase A, glucanase, and endochitinase [39]. *P. fluorescens*, with strong inhibitory action against *R. solani* and the ability to promote plant growth, reduced the average lengths of *R. solani*–induced lesions under greenhouse conditions [51]. The bacterium inhibits the growth of *R. solani* by the production of siderophores, and the germination of *R. solani* sclerotia was completely inhibited by the bacterium. *P. lindbergii*, when applied as seed treatment, reduced pre- and postemergence damping-off more effectively than the fungicides benomyl and carboxin [49].

   B. subtilis when added as a seed treatment reduced levels of root cankers caused by *R. solani* AG-4, consistently colonized the roots, increased germination and emergence, increased nodulation by *Rhizobium* spp., enhanced plant nutrition, increased root growth, and increased the yield of peanut [52]. A crude preparation of the antibiotic(s) from the culture filtrate of *B. subtilis*, obtained from *Stylosanthes guianensis* showed antifungal effects against *R. solani* and other fungi such as *Thanatephorus cucumeris, Phytophtohora parasitica*, and *Pyricularia oryzaze*. The inhibitory effect of a cell-free culture filtrate of *B. subtilis* was often comparable with or better than that obtained with benomyl [53].

E. Charcoal Rot

1. Fungal Biocontrol

Trichoderma harzianum was effective in control of charcoal rot [54]. Distillary effluents had no effect on *T. harzianum*, but dairy effluents caused reduced growth of *T. harzianum* indicating that it can be used for the biological control of *R. bataticola* even in the presence of dairy and distillary effluents.

   The influence of soil moisture levels on root rot diseases caused by *Rhizoctonia bataticola* and the rhizosphere population of *R. bataticola* and antagonist *T. viride* is one of the major factors affecting the distribution, survival, proliferation, and subsequent establishment of *T. viride* in soil and in the rhizosphere [55]. At 40% moisture holding capacity, the population of *R. bataticola* in the groundnut rhizosphere was highest and the disease incidence was maximum. Disease reduction positively correlated with increasing moisture levels, and the rhizosphere population of *T. viride* was best at 40–60% moisture levels. The growth and sporulation of *T. viride* was highest at 50% moisture. However, high moisture levels were unfavorable for both the pathogen and the antagonist due to increased activity of bacteria and also near anaerobic conditions that are not favorable for either the pathogen or the antagonist. There was also a significant increase in *T. viride* population in the rhizosphere of peanut with an increase in the age of the crop, with maximum population at 75 days after sowing.

2. Bacterial Biocontrol

One hundred and fifty rhizobacterial strains from the rhizosphere of groundnut were evaluated in vitro for antagonistic activity against *R. bataticola* on potato dextrose agar medium. Nine bacterial strains exhibit potent antagonistic activity against *R. bataticola* with an inhibition zone of above 15 mm (authors' unpublished data). *Pseudomonas cepacia* UPR 5C was found to be a strong inhibitor to *R. bataticola* [56].

F. *Sclerotinia* Blight

1. Fungal Biocontrol

Sporidesmium sclerotivorum, a mycoparasite proliferates throughout the cortex and medulla of sclerotia of *Sclerotinia minor* [57]. The colonizing hyphae were most abundant in the extracellular matrix but were occasionally seen in the lumen of sclerotial hyphae that were highly vacuolated or empty. These intracellular structures were considered to be haustoria of the mycoparasite. Fungicides, benomyl, chlorothalonil, vinclozoline, iprodiene, and procymidone were not toxic to *S. sclerotivorum* in soil at concentrations likely to be encountered in the field [58]. Hence, these pesticides can be used in conjunction with *S. sclerotivorum* in an integrated approach to disease control. *Teratosperma oligocladum* effectively reduces the survival of *S. minor* sclerotia in soil [59], both when *S. minor* sclerotia were added to soil after soaking in antagonist spore suspension and when the spores of antagonist in water were mixed into soil containing sclerotia of *S. minor*.

   Gliocladium virens colonized the sclerotia of *S. minor*, *S. sclerotiorum*, *S. rolfsii*, and *R. bataticola* [60]. Optimum conditions for parasitism of *S. sclerotiorum* sclerotia by *G. virens* were 25–35°C and pH 4.0–5.6. Germination of conidia and infection of sclerotia took place only at 100% relative humidity and only when free water was present. In soil, within a pH range of 4.5–7.8 and at moisture content of 80% field capacity, the infection of sclerotia by *G. virens* was maximum.

   G. virens inhibited the formation of sclerotia by *S. sclerotiorum*, parasitized both mycelia and sclerotia of the host fungus, and sporulated profusely on sclerotia when introduced to the culture of *S. sclerotiorum* after formation of sclerotia. *G. virens* produced appressoria at the tips of short branches at various points of contact between the two fungi. These appressoria gave rise to infection hyphae, which penetrated the host cell wall and initiated intracellular parasitism. After penetration of the host hypha the appressorium shrank and the parasitized host hypha also shrank gradually with time. *G. virens* penetrated the host sclerotia but failed to sporulate internally and the parasitized sclerotia failed to germinate [61]. Crude culture filtrates of *Coniothyrium minitans* and *T. viride* grown on autoclaved crushed sclerotia of *S. sclerotiorum* lyse wall isolated from hyphal cells or the inner pseudoparenchymatous cells of the sclerotia, in which a branched β-(1,3)-β-(1 $\rightarrow$ 6)-glucan, sclerotan, is a major constituent. Endo- and exo-β(1 $\rightarrow$ 3)-glucanases produced by *C. minitans* act together to degrade the glucan completely [62].

   T. harzianum (TH-88) applied as seed treatment and foliar spray did not affect the incidence of sclerotinia blight [63]. *S. minor* inhibits *Penicillium citrinum* isolated from sclerotia of *S. minor* recovered from field soil planted with peanuts [64]. Growth of *S. minor* was significantly inhibited with 10% (v/v) filtrate from 2- to 3-week-old cultures of *P. citrinum*. The inhibitor(s) was active against *S. minor* even after autoclaving. The molecular weight of the active compound(s) was ≤ 1000 Da. The active component against *S. minor* in the culture filtrate of *P. citrinum* was identified as citrinin. Sclerotia of *S. minor* soaked in a conidial suspension of *P. citrinum* leads to colonization of sclerotia by *P. citrinum* [65]. Similar treatments when incubated in pasteurized and nonpasteurized soils resulted in destruction by *P. citrinum*, suggesting the potential use of *P. citrinum* as a biocontrol agent for *S. minor*. Biological control of sclerotinia blight using various rates and formulations of *Sporicidium* spp. and a *Trichoderma* sp. was comparable with standard fungicide rovral [66].

G. Stem Rot

1. Fungal Biocontrol

Trichoderma spp. parasitizes the mycelium and sclerotia of *Sclerotium rolfsii*. The mechanisms of parasitism were well studied in this interaction. Hyphae of *Trichoderma* penetrate the rind and cortex of sclerotia without affecting the host cells. Upon reaching the medulla, it ramified, lysed the medullar tissue, and produced chlamydospores and finally underwent autolysis. Degraded sclerotia became dark, soft, and empty and disintegrated under slight pressure [67]. Observation of cross sections of sclerotia parasitized by *T. harzianum* revealed that fungal growth mainly was intracellular in the rind layer, and cell invasion occurred through localized host wall penetration. Incubation of ultrathin sections of parasitized sclerotia with wheat germ agglutinin/ovomucoid-gold complex for localization of chitin monomers revealed that, except in the area of hyphal penetration, the chitin component of the host cell walls was structurally preserved. The host cytoplasm had undergone complete disorganization at a time when β-1,3-glucans were still distributed evenly over the rind cell walls, indicating that the production of cell wall–degrading enzymes by *T. harzianum* probably was not the first event involved in sclerotial decay [68].

Isolates of *T. harzianum*, *T. hamatum*, and *T. koningii* kill the sclerotia of *S. rolfsii*. The penetration of sclerotia by *Trichoderma* spp. and multiplication inside the sclerotium was dependent on the ability of the biocontrol agent to attack the sclerotia and establish on the rind of the sclerotium [69]. Isolates of *T. harzianum*, *T. hamatum*, and *T. koningii* were antagonistic to *S. rolfsii* by production of diffusable and volatile metabolites, lysis, hyphal interference, and mycoparasitism. Vacuolation, granulation, coagulation, and disintegration of *S. rolfsii* cytoplasm and lysis of the host cells were observed [70]. Detachment of a coiled hypha of *T. harzianum* and *T. hamatum* from around *S. rolfsii* revealed a digested area and the penetration sites on the host mycelium. Fluorescence microscopic observations of parasitism of *T. harzianum* on *S. rolfsii* and *R. solani* indicated the presence of *N*-acetyl-D-glucosamine oligomers at coiling zones, suggesting that chitin fibrils are exposed in the cell walls of *S. rolfsii* and *R. solani* as a result of extracellular β-1,3-glucanase excreted by *Trichoderma* at the contact sites. High β-(1,3)-glucanase and chitinase activities were detected in dual agar cultures when *T. harzianum* parasitized *S. rolfsii* compared with either fungus grown alone, and these enzymes are responsible for the degradation of the host cell wall [40].

Talaromyces flavus parasitized both hyphae and sclerotia of *S. rolfsii*. *T. flavus* penetrated the thick cell walls of *S. rolfsii* sclerotia and subsequently the host cell organelles disintegrated and the cytoplasmic content disappeared. Extracellular chitinase production by *T. flavus* positively correlated with mycoparasitism of *S. rolfsii* [71]. The expression of various *N*-acetylglucosaminidases and endochitinases during mycoparasitism was under a finely tuned regulation that was affected by the host. In dual culture, when *T. harzianum* was antagonizing *S. rolfsii*, a 102 kDa *N*-acetyl-glucosaminidase (CHIT 102) was the first to be induced. As soon as 12 hours after contact, its activity diminished, and another 73 kDa *N*-acetyl-glucosaminidase (CHIT 73) was expressed, and none of the *T. harzianum* endochitinases were detected during the parasitic interaction with *S. rolfsii*. When *T. harzianum* was antagonizing *R. solani*, 12 hours after contact, CHIT 102 activity was elevated, and the activities of three additional endochitinases of 52, 42, and 33 kDa were detected. As the antagonistic interaction proceeded, CHIT 102 activity decreased and the activity of the endochitinases gradually increased. The activity of CHIT 73, which was highly expressed during the parasitic action of *T. harzianum* towards *S. rolfsii*, was not detected during its parasitic action towards *R. solani*. This differential expression of *T. harzianum* chitinases may influence the overall antagonistic ability of the fungus against a specific host [72].

S. rolfsii produces two agglutinins with molecular weights of 60 and 55 kDa. The crude agglutinin agglutinated certain gram-negative bacteria and yeasts. A positive correlation between the ability of isolates of the mycoparasite *Trichoderma* spp. to attack *S. rolfsii* and the agglutination of conidia of *Trichoderma* by *S. rolfsii* agglutinin suggests a role for the agglutinin in the recognition of *S. rolfsii* by the *Trichoderma* spp. [73]. A biomimetic system based on the binding of lectins to the surface of nylon fibers simulating the host hyphae that enables examining the role of lectins in mycoparasitism was developed [74]. *T. harzianum*, when allowed to grow on nylon fibers treated with concovalin A or crude *S. rolfsii* agglutinin, coiled around the nylon fibers and produced hooks in a pattern similar to that observed with the real host hyphae. The incidence of interaction was significantly higher with agglutinin-treated fibers than with control. In contrast to the lectin isolated by Barak et al. [73], a lectin of molecular mass 45 kDa was isolated from the culture filtrates of *S. rolfsii* and its agglutination activity was not inhibited by any of the mono- or disaccharides tested. Incubation of lectin with trypsin, chymotrypsin, and β-1,3-glucanase totally inhibited lectin activity, and protease inhibited 70% activity. Incubation of lectin with either chitinase or β-glucuronidase had no effect on agglutination activity. Both the protein and β-1,3-glucan are necessary for agglutination of the lectin. The presence of the purified agglutinin on the surface of the fibers specifically induced mycoparasitic behavior in *T. harzianum*. *Trichoderma* formed tightly adhering coils around the purified agglutinin-treated fibers. Other mycoparasite-related structures, such as appressorium-like bodies and hyphal loops, were only observed in the interaction between *T. harzianum* and the purified agglutinin-treated fibers [75].

Correlation studies on the infectivity of sclerotia of *S. rolfsii* on bean hypocotyls and sclerotial colonization and ability to germinate after incubation in nonsterile soil amended with *G. virens* suggested that, in addition to mycoparasitism, antibiosis may be involved in the degradation of sclerotia by *G. virens* in soil and in reducing the infectivity of germinable sclerotia [76].

T. harzianum was the most effective biocontrol agent against *S. rolfsii* when compared with potential biocontrol agents *Aspergillus flavus*, *A. niger*, *Bacillus subtilis*, *Penicillium chrysogenum*, *Streptomyces* spp., and *T. viride* [77]. A moderate 36% control of stem rot was obtained using *G. virens* and *T. longibrachiatum* as biocontrol agents [78]. UV-induced mutants of *T. viride* exhibit increased biocontrol abilities. Two mutants, M1 and M2, increased the zone of inhibition in dual plate cultures and decreased *S. rolfsii* sclerotial production and germination [79]. The growth rate of *T. harzianum* in culture was greater than that of *S. rolfsii* and invaded its mycelium under growth conditions adverse to *S. rolfsii*, e.g., high pentachloronitrobenzene concentrations, high pH levels, and low temperatures. The improved antagonism under conditions adverse for the pathogen emphasizes the potential of integrating various means of disease control [80].

A diatomaceous earth granule impregnated with a 10% molasses solution was suitable for growth and delivery of *T. harzianum* to peanut fields due to its low bulk and no residues. Viability of *Trichoderma* on air-dried infested granules was excellent, with virtually 100% of granules displaying mycelial growth. Reduction in number of dead plants with signs of *S. rolfsii* by application of 140 kg/ha *Trichoderma* granules was equivalent to that achieved using 10% PCNB granules at 112 kg/ha [81]. *T. harzianum* added to soil on a wheat-bran formulation reduced the viability of sclerotia in the soil. A combination of urea and *T. harzianum* further reduced the viability of sclerotia. *T. harzianum* and calcium ammonium nitrate applied together was most effective in reducing the disease severity [82]. A wheat bran and biogas manure mixture (1:1) that stimulated the growth and multiplication of *T. harzianum* was suitable for formulation of the biocontrol agent. This formulation product suppressed *S. rolfsii* and increased seedling emergence of peanut [83]. It was recommended that *T. harzianum* could be multiplied in the wheat bran and biogas manure mixture for continued survival of the biocontrol agent as a soil amendment.

Seed treatments with *Rhizobium* and a fungicide are included in the package of practices for peanut. The effect of superimposition of antagonist *T. harzianum* along with these treatments on plant growth and the incidence of root rot caused by *S. rolfsii* was studied to develop an integrated method of management [84]. Application of *T. harzianum* inoculum to soil at sowing was better than other treatments including seed treatment. Seed treatment with *T. harzianum* and application of the antagonist to soil on the sixth day after sowing gave equal results as with application of *T. harzianum* to soil at the time of sowing. These results are similar to that obtained with carbendazim seed treatment and drenching 0.1% carbendazim on the sixth day after inoculation. However, seed treatment with *T. harzianum* and *Rhizobium* plus *T. harzianum* inoculum added to the soil on the sixth day after sowing plus 0.1% carbendazin soil drenching on 30 days after sowing was superior in disease control. Also, *T. harzianum* colonized well in the rhizosphere of peanut and its population increased with increasing time, reaching a maximum at 75 days after sowing. Integration of *T. harzianum* with carbendazim treatment offers great hope in the management of the disease under field conditions. These results are in contrast with the earlier reports on management of *S. rolfsii* using *T. harzianum* in combination with pentachloronitrobenzene (PCNB) or carboxin [85]. None of the *Trichoderma* treatments alone or in combination with PCNB or carboxin gave better disease control. However, increases in yield were observed in plots treated with *T. harzianum* and PCNB applied on demand. *T. harzianum* mixed with wheat middlings was also ineffective in increasing yield. This could be due to the survival of *T. harzianum* in the soil for only 5–8 days and its sensitivity to foliar fungicides applied to peanut. These contrasting results support the concept that colonization of the rhizosphere is an essential prerequisite for the success of a biocontrol agent in the control of a soilborne pathogen.

2. Bacterial Biocontrol

Rhizobium used for increase in nodulation and growth of peanuts was also found to reduce the population of *S. rolfsii* in the rhizosphere of peanut [86]. Similarly soil inoculation with *Rhizobium* reduced the population of *S. rolfsii* in the rhizosphere. Disease incidence significantly reduced and the yields increased by 13 and 48% in two successive field trials when both *S. rolfsii* and *Rhizobium* were present. Four bacterial isolates were inhibitory to the mycelial growth of *S. rolfsii* among a collection of 150 rhizobacterial isolates (authors' unpublished data). *P. fluorescens* 2-79, *P. aeruginosa* EP 7, and *B. subtilis* AF 1 were antagonistic to *S. rolfsii*. *P. aeruginosa* completely inhibited the growth of *S. rolfsii* by producing a siderophore [50].

Sclerotial germination is an important stage in the life cycle of *S. rolfsii*. Inhibition of sclerotial germination and reduction of their viability decrease both the inoculum potential of the pathogen and disease severity [87]. An antagonistic strain of *Serratia marcescens* inhibited the germination percentage of *S. rolfsii* sclerotia dipped in the bacterial suspension. The incidence of stem rot was reduced by 60% when *S. marcescens* suspension was mixed with soil [88]. Direct attack of the biocontrol strain on the pathogen and not the interaction with the soil microflora was thought to be the mechanism of biocontrol, as determined by testing for disease control in sterilized and unsterilized soils. This direct antagonism may be due to the chitinolytic ability of *S. marcescens* by the production of extracellular chitinase. Culture filtrate of *S. marcescens* grown on medium containing colloidal chitin as a sole source of carbon degraded the cell wall or mycelium of *S. rolfsii*. Scanning electron microscopy showed direct degradation of hyphae when *S. rolfsii* mycelium served as the sole carbon source for the bacterium. Crude chitinase of *S. marcescens* caused lysis of hyphal tips [89]. Among the hydrolytic enzymes produced by the antagonists, chitinase plays a major role in the antagonistic activity against *S. rolfsii*. Hyphal tips, septa, and branches of hyphae of *S. rolfsii* contain oligomers of β-glucan and *N*-acetyl glucosamine. The chitinase produced by the biocontrol agent attacks these sites and causes a release of *S. rolfsii* β-glucanase, which, together with the chitinase, completely degrades the hyphae.

P. *fluorescens* restricted *S. rolfsii* mycelial growth and sclerotial germination. Germination of sclerotia was inhibited after the sclerotia were immersed in *P. fluorescens* cell suspension. The antagonistic activity was mediated by the production of siderophores. In greenhouse experiments, 99% of peanut plants were protected from *S. rolfsii* infection after inoculation with *P. fluorescens*. Fresh weights of tops of plants inoculated with bacteria and infected with *S. rolfsii* were significantly higher than weights of the wilted control plants [90]. Bacterization of peanut seeds with native strains of *P. fluorescens* might prove very valuable for efficient management of soilborne plant pathogens *S. rolfsii* and *R. solani. C. minitans* showed noticeable activity against pod rot caused by *S. rolfsii* and is more effective than benomyl and carboxin [49].

H. Mycotoxins

1. Fungal Biocontrol

The potential for biological control of preharvest aflatoxin contamination of peanut was demonstrated using an atoxigenic strain of *Aspergillus parasiticus* as a biocompetitive agent [91]. Edible peanuts from the treated soil contained aflatoxin concentrations of 11, 1, and 40 ppb for three consecutive crop years in comparision with 531, 96, and 241 ppb in untreated soil. Reductions of aflatoxin B_1 levels in peanut kernels at maturity were also obtained by simultaneous inoculation of root regions of 1- to 2-week-old peanut plants with toxigenic and atoxigenic strains of *A. flavus* when compared with plants inoculated with the toxigenic strains alone [92]. Preinoculation of atoxigenic strains one day earlier resulted in greater inhibition of aflatoxin production, whereas toxin level was not much reduced when the atoxigenic strain was introduced 1 day after the toxigenic strain. Inoculation of an atoxigenic strain at fivefold higher spore concentrations within 12 hours of toxigenic strain inoculation led to a significant reduction in aflatoxin B_1. Thus, the potential of atoxigenic strains of *A. flavus* in biological control against pre-harvest aflatoxin contamination of developing peanuts was established subject to the practicability of the approach. The effect of different inoculum rates of nontoxigenic color mutants of *A. flavus* and *A. parasiticus* on preharvest aflatoxin contamination of peanuts was studied. A stronger relationship was found between inoculum rate and aflatoxin concentrations. It is possible to achieve a higher degree of control when plots or fields are retreated with biocontrol agents in subsequent years [93]. The hypothesis behind these studies was that the introduced atoxigenic isolates would exclude or outcompete the toxigenic strain. A procedure was developed to encapsulate mycelia of an atoxigenic strain of *A. flavus* in alginate pellets with corn cob grits as filler and wheat gluten as adjuvant for seedling into agricultural fields in order to reduce aflatoxin contamination via competitive exclusion [94].

In dual cultures, *A. shirousamii* decreased the production of aflatoxin by *A. flavus* [95]. Application of *T. harzianum* to the soil reduced the colonization of peanuts by *A. flavus*, and application of gypsum further enhanced the reduction [96].

2. Bacterial Biocontrol

The soil, rhizosphere, and geocarposphere constitute three distinct ecological niches within the vicinity of the peanut root system. Specific bacterial taxa are preferentially adapted to colonize the geocarposphere, rhizosphere, and root-free soil as identified by analysis of fatty acid methyl esters of the bacterial isolates [97]. The geocarposphere microbial community is the last barrier for *A. flavus* prior to pod colonization. Bacteria that colonize the geocarposphere were examined as potential biological control agents for pod-invading fungi such as the toxigenic strains of *A. flavus* and *A. parasiticus*. Selected bacterial strains isolated from the geocarposphere significantly reduced pod colonization by *A. flavus* in both greenhouse and field conditions [98]. *Flavobacterium odortum* decreased the production of aflatoxins by *A. flavus* in peanut kernels, and *Bacillus megaterium*, *B. laterosporeus*, *Pseudomonas aurofaciens*, *Xanthomonas maltophila*, *Cellulomonas carate*, and *Phyllobacterium rubiacearum* increased aflatoxin production on peanut kernels [99]. *P. cepacia* completely inhibited the growth of *A. flavus* [100].

Treatment of viable peanuts with a combination of chitosan or *Bacillus* reduced the growth of *A. flavus*. Peanuts treated with chitosan + *Bacillus* sp. resulted in the greatest induction of chitosanase. The introduction of a bioinducer into the plant-pathogen interaction is an effective method for induction of plant resistance against pathogenic and toxigenic fungi [101].

III. DISEASES CAUSED BY NEMATODES

A. Root-Knot

1. Fungal biocontrol

Paecilomyces lilacinus, parasitizing *Meloidogyne* spp., was used as biocontrol agent in a wide range of crops [102]. *P. lilacinus*, grown on neem cake, significantly reduced the root-knot on peanut plants in *M. javanica*–infested soils [103]. *P. lilacinus* and two entamopathogenic fungi, *Metrarhizium anisopliae* and *Beauveria brongniartii*, applied as soil amendment were effective in reducing *M. javanica* infection [104]. Dilute culture filtrate of *A. niger* killed >90% of the juveniles and also affected hatching of eggs significantly [105].

2. Bacterial Biocontrol

Pasteuria penetrans is an obligate mycelial and endospore-forming bacterial parasite of root-knot nematodes [106] and has great potential as a biocontrol agent of root-knot nematodes [107]. The endospore germinates after attaching to the nematode cuticle, the germ tube extends through the nematode cuticle and forms vegetative thalli in the nematode body cavity, which enlarge, branch, and spread throughout the nematode body producing sporangia that endogenously form single spores [108].

Application of endospores of *P. penetrans* effectively reduced the damaging effects of *M. arenaria* on peanuts [109]. Soil inoculation reduced root gall index and pod galls. The major suppressive mechanism of *M. arenaria* by *P. penetrans* on peanut is the initial endospore infestation of second-stage juveniles at planting. Suppression was proportional to *P. penetrans* infestation levels, regardless of the initial population densities of root-knot nematodes [110]. The chance of *P. penetrans* infection of subsequent generations of nematodes during the cropping season was negligible because of the minor endospore build-up in soil, and the J2 that hatched from egg masses on root surfaces quickly enter adjacent roots without being exposed to endospores in soil [111].

IV. CONCLUSIONS

Biological control is rapidly developing as an important component of integrated disease management (IDM) because of its cost-effectiveness and the focus on environmental pollution and residual toxicity caused by fungicides. In the past two decades, several successful attempts were made to achieve biological control

Table 2 Antagonistic Activity and Control of Peanut Pathogens by Extracts of Commonly Available Nonhost Plants

Plant	Type of extract	Targeted pathogens	Disease control	Ref.
Azadirachta indica	Aqueous leaf extracts	Aspergillus flavus	In vitro incidence was less than Dithane M-45	112
		Aspergillus niger	Control of in vitro incidence of A. niger was higher than that obtained with commercial fungicides thiran and dithane M-45	112
		Phaeoisariopsis personata	Effective in control of late leaf spot under field conditions	113
		Puccinia arachidis	Effective in control of rust under field conditions	113
	Ethanolic leaf extract	A. flavus	Highly significant in reducing in vitro incidence of A. flavus and reducing seed rotting	114
		A. niger	Significant reduction in seed and seedling rotting was obtained in in vitro studies	114
		Rhizoctonia bataticola	R. bataticola infection was reduced significantly in vitro	114
	Neem oil	P. personata	50% disease reduction was obtained in field conditions	115
		P. arachidis	Significant reduction in rust incidence was obtained in field conditions	115
	Seed kernel extract	Cercospora arachidicola	Effective in disease control as a foliar spray along with potash	116
		P. personata	Disease control was effective and comparable to fungicides	117
		P. arachidis	Disease control was better than carbendazim + mancozeb spray	117

Table 2 (Continued)

Plant	Type of extract	Targeted pathogens	Disease control	Ref.
Curcuma longa	Turmeric powder	*A. flavus*	In vitro reduction in *A. flavus* incidence was significant	112
		A. niger	Reductions in in vitro incidence of *A. niger* was comparable to thiram and dithane M-45	112
Lawsonia inermis	Aqueous leaf extract	*A. flavus*	Reduced in vitro incidence of *A. flavus*	112
		A. niger	Reduced in vitro incidence of *A. niger*	112
		P. personata	Reduced disease incidence in field conditions	113
		P. arachidis	Effective in disease reduction in field	113
Nerium odorum	Aqueous leaf extract	*P. personata*	Effective in disease control	115
		P. arachidis	Effectively reduced the disease incidence	115
Ocimum gratissimum	Ethanolic leaf extract	*A. flavus*	Marginal reduction in in vitro incidence of *A. flavus*	114
		A. niger	Marginal reduction in in vitro incidence of *A. niger*	114
Ocimum sanctum	Ethanol extract and essential oils	*A. niger*	Increased seed germination	118
Phyllanthus fraternus	Leaf and plant extracts	Peanut green mosaic virus	Root extracts reduced the infectivity of the virus by 98%	119
Polyalthia longifolia	Ethanol extracts	*A. flavus*	Reduced in vitro incidence of *A. flavus*	114
		A. niger	Reduced in vitro incidence of *A. niger*	114
		R. bataticola	Reduced in vitro incidence of *R. bataticola*	114

of peanut diseases using biocontrol fungi and bacteria. The results summarized in this chapter indicate that biological control of fungal diseases of peanut is possible using biocontrol agents as major constituents of IDM. Most of these studies were confined to greenhouse environments, and few have reached the point of commercial application, which reflects the necessity for more research and better understanding of the impact of field environment on biocontrol agents and their application technology. Although progress has been made in the experimental biological control of peanut diseases, very little is known about the mechanism of action of biocontrol agents, especially for the control of foliar diseases. Further research is needed into the basic biology of biocontrol agents and plant pathogens in order to understand their ecology, physiology, biochemistry, and genetics necessary to predict the behavior of the biocontrol agents under different environmental conditions and to develop reliable biocontrol systems.

The antagonistic activity of biocontrol agents against plant pathogens is highly specific for both a pathogen and different races of pathogen. Further large-scale screenings will be carried out in search of biocontrol agents with a broad spectrum of activity. Moreover, biocontrol agents that can elicit systemic resistance in plants can offer protection against a wide range of pathogens. In-depth studies of the mechanisms of action of biocontrol agents must be made before using them on a large scale. Development of new formulations and delivery systems suitable for field application will help the commercialization of biocontrol agents for peanut diseases.

Several plant extracts and their components proved to have antagonistic activity against major peanut pathogens, and several have been evaluated for disease control (Table 2). There is great potential for control of major peanut diseases and storage fungi using plant extracts, which enormously reduces the cost of cultivation of the crop and losses during storage. Intensive investigations need to be done in this regard to develop plant extracts as an alternative to chemical fungicides.

REFERENCES

1. FAO Year Book, Production. Rome: FAO, 1998, pp. 103–104.
2. P Subrahmanyam, DVR Reddy, SB Sharma, VK Mehan, D McDonald. A world list of groundnut diseases. Legumes Pathology, Progress Report 12, ICRISAT, Patancheru, India, 1990.
3. RA Taber, RE Pettit, RE McGee, DH Smith. Reduction of sporulation of *Cercosporidium personatum* by *Hansfordia* in Texas (abstr). Am Peanut Res Ed Assoc 11:45, 1979.
4. A Krishna, RA Singh. *Hansfordia pulvinata* mycoparasitic on *Cercospora* species causing "Tikka disease" of groundnut. Indian Phytopath 32:318–320, 1979.
5. FM Shokes, RA Taber. Occurrence of a hyperparasite of *Cercosporidium personatum* on peanut in Florida (abstr). Phytopathology 73:505–506, 1983.

6. JK Mitchell, RA Taber. Factors affecting the biological control of *Cercosporidium* leaf spot of peanuts by *Dicyma pulvinata*. Phytopathology 76:990–994, 1986.
7. JK Mitchell, RA Taber, RE Pettit. Establishment of *Dicyma pulvinata* in *Cercosporidium personatum* leaf spot of peanuts: effect of spray formulation, inoculation time, and hours of leaf wetness. Phytopathology 76:1168–1171, 1986.
8. JK Mitchell, DH Smith, RA Taber. Potential for biological control of *Cercosporidium personatum* leafspot of peanuts by *Dicyma pulvinata*. Can J Bot 65:2263–2269, 1987.
9. MP Ghewande. Biological control of late leaf spot (*Phaeoisariopsis personata*) of groundnut (*Arachis hypogaea*). Indian J Agric Sci 59:189–190, 1989.
10. P Subrahmanyam, PM Reddy, D McDonald. Parasitism of rust, early and late leaf spot pathogens of peanut by *Verticillium lecanii*. Peanut Sci 17:1–4, 1990.
11. MP Ghewande, S Desai, P Narayan, AP Ingle. Integrated management of foliar diseases of groundnut (*Arachis hypogea* L.) in India. Int J Pest Manage 39:375–378, 1993.
12. T Sommart, N Kitjoa. Feasibility of using phylloplane and rhizosphere microfloras as a biological control agent against late leaf spot of groundnut. Proceedings of the Seventh Thailand National Groundnut Meeting for 1987, Pattaya, Chonburi, Thailand, 1989, pp 170–174.
13. C Leben. Epiphytic microorganisms in relation to plant disease. Annu Rev Phytopathol 3:209–230, 1965.
14. N Kokalis-Burelle, PA Backman, R Rodriguez-kabana, LD Ploper. Potential for biological control of early leafspot of peanut using *Bacillus cereus* and chitin as foliar amendments. Biol Control 2:321–328, 1992.
15. GR Knudsen, HW Spurr Jr. Field persistence and efficacy of five bacterial preparations for control of peanut leaf spot. Plant Dis 71:442–445, 1987.
16. CVSS Rajesh. Evaluation of chitinase production and ACC utilization for the selection of groundnut rhizobacteria. M.Sc. thesis, University of Hyderabad, Hyderabad, India, 1998.
17. G Krishna Kishore, S Pande, AR Podile, J Narayana Rao. Potential for biological control of late leaf spot of groundnut using antagonistic rhizobacteria. Proceedings of seminar on biological control and plant growth promoting rhizobacteria (PGPR) for sustainable agriculture, Department of Plant Sciences, University of Hyderabad, Hyderabad, India. AB-05, 2000.
18. MP Ghewande. Biological control of groundnut (*Arachis hypogea* L.) rust (*Puccinia arachidis* Speg.) in India. Trop Pest Manage 36:17–20, 1990.
19. BJ Gowdu, R Balasubramanian. Biocontrol potential of rust of groundnut by *Acremonium obclavatum*. Can J Bot 71:639–643, 1993.
20. M Sathiyabama, R Balasubramanian. Effect of *Acremonium obclavatum* glucan on groundnut leaf rust disease caused by *Puccinia arachidis* Speg. World J Microbiol Biotechnol 14:783–784, 1998.
21. N Mathivanan, V Kabilan, K Murugesan. Production of chitinase by *Fusarium chlamydosporum*, a mycoparasite to groundnut rust, *Puccinia arachidis*. Indian J Exp Biol 35:890–893, 1997.
22. N Mathivanan, K Murugesan. *Fusarium chlamydosporum*, a potent biocontrol

agent to groundnut rust, *Puccinia arachidis*. Zietschr Pflanzenkr Pflanzenschutz 107:225–234, 2000.

23. N Mathivanan, V Kabilan, K Murugesan. Purification, characterization, and antifungal activity of chitinase from *Fusarium chlamydosporum*, a mycoparasite to groundnut rust, *Puccinia arachidis*. Can J Microbiol 44:646–651, 1998.

24. N Mathivanan, K Murugesan. Isolation and purification of an antifungal metabolite from *Fusarium chlamydosporum*, a mycoparasite to *Puccinia arachidis*, the rust pathogen of groundnut. Indian J Exp Biol 37:98–101, 1999.

25. V Govindasamy, R Balasubramanian. Biological control of groundnut rust, *Puccinia arachidis*, by *Trichoderma harzianum*. Zietschr Pflanzenkr Pflanzenschutz 96: 337–345, 1989.

26. KH Garren, CM Christensen, DM Porter. The mycotoxin potential of peanuts (groundnuts): the USA viewpoint. J Stored Product Res 5:265–273, 1969.

27. Y Harder, I Chet, Y Henis. Biological control of *Rhizoctonia solani* damping-off with wheat bran culture of *Trichoderma harzianum*. Phytopathology 69:64–68, 1979.

28. SM Lashin, HIS El-Nasr, MAA El-Nagar, MA Nofal. Biological control of *Aspergillus niger* the causal organism of peanut crown rot by *Trichoderma harzianum*. Ann Agric Sci 34:795–803, 1989.

29. A Karthikeyan. Effect of organic amendments, antagonist *Trichoderma viride* and fungicides on seed and collar rot of groundnut. Plant Dis Res 11:72–74, 1996.

30. C Dileep, BSD Kumar, HC Dube. Disease suppression and plant growth promotion of peanut by fluorescent pseudomonad strains, FPC 32 and FPO 4, in soil infested with collar-rot fungi, *Aspergillus niger*. Ind Phytopath 52:415–416, 1999.

31. AR Podile, AP Prakash. Lysis and biological control of *Aspergillus niger* by *Bacillus subtilis* AF1. Can J Microbiol 42:533–538, 1996.

32. PR Sailaja, AR Podile, P Reddanna. Biocontrol strain of *Bacillus subtilis* AF1 rapidly induces lipoxygenase in groundnut (*Arachis hypogaea* L.) compared to crown rot pathogen *Aspergillus niger*. Eur J Plant Pathol 104:125–132, 1998.

33. PR Sailaja, AR Podile. A phytoalexin is modified to less fungistatic substances by crown rot pathogen in groundnut (*Arachis hypogaea* L.). Indian J Exp Biol 36: 631–634, 1998.

34. K Manjula. Characterization of chitinolytic ability of *Bacillus subtilis* AF 1 for its use in improved formulations for plant growth promotion and disease control. Ph.D. thesis, University of Hyderabad, Hyderabad, India, 1999.

35. DJ Askew, MD Laing. The invitro screening of 118 *Trichoderma* isolates for antagonism to *Rhizoctonia solani* and an evaluation of different environmental sites of *Trichoderma* as sources of aggressive strains. Plant Soil 159:277–281, 1994.

36. EGC Gaminde. Biological control of *Rhizoctonia solani* Kuhn by soil fungal antagonist ANF-777. Philipp J Crop Sci 17S:46, 1995.

37. G Turhan. Further hyperparasites of *Rhizoctonia solani* Kuhn as promising candidates for biological control. Zietschrift Pflanzenkr Pflanzenschutz 97:208–215, 1990.

38. RAC Morris, JR Coley Smith, JM Whipps. The ability of the mycoparasite *Verticillium biguttatum* to infect *Rhizoctonia solani* and other plant pathogenic fungi. Mycol Res 99:997–1003, 1995.

39. BL Bertagnolli, FK Dal Soglio, JB Sinclair. Extracellular enzyme profiles of the fungal pathogen *Rhizoctonia solani* isolate 2B-12 and of two antagonists, *Bacillus megaterium* strain B153-2-2 and *Trichoderma harzianum* isolate Th008. I. Possible correlations with inhibition of growth and biocontrol. Physiol Mol Plant Pathol 48: 145–160, 1996.

40. Y Elad, I Chet, P Boyle, Y Henis. Parasitism of *Trichoderma* spp. on *Rhizoctonia solani* and *Sclerotium rolfsii*—scanning electron microscopy and fluorescence microscopy. Phytopathology 73:85–88, 1983.

41. N Benhamou, I Chet. Hyphal interactions between *Trichoderma harzianum* and *Rhizoctonia solani*: ultrastructure and gold cytochemistry of the mycoparasitic process. Phytopathology 83:1062–1071, 1993.

42. Y Elad, R Barak, I Chet. Possible role of lectins in mycoparasitism. J Bacteriol. 154:1431–1435, 1983.

43. RJ Tweddell, HSH Jabaji, PM Charest. Production of chitinases and β-1,3-glucanases by *Stachybotrys elegans*, a mycoparasite of *Rhizoctonia solani*. Appl Environ Microbiol 60:489–495, 1994.

44. JC Tu, O Vaartaja. The effect of the hyperparasite (*Gliocladium virens*) on *Rhizoctonia solani* and on *Rhizoctonia* root rot of white beans. Can J Bot 59:22–27, 1981.

45. CR Howell. Relevance of mycoparasitism in the biological control of *Rhizoctonia solani* by *Gliocladium virens*. Phytopathology 77:992–994, 1987.

46. Y Elad, Y Hadar, I Chet, Y Henis. Prevention with *Trichoderma harzianum* Rifai Aggr., of reinfestation by *Sclerotium rolfsii* Sacc. and *Rhizoctonia solani* Kuhn of soil fumigated with methyl bromide, and improvement of disease control in tomatoes and peanuts. Crop Prot 1:199–211, 1982.

47. MM El-Zayat, MM Satour, EM El-Sherif, SM Abd-El-Gafour, SN El-Sherbeiny. Studies of some factors influencing the potential of the biocontrol agent *Trichoderma harzianum* on *Rhizoctonia solani* affecting peanut. Assuit J Agri Sci. 24: 279–291, 1993.

48. DR Summer, DK Bell. Survival of *Rhizoctonia* spp. and root diseases in a rotation of corn, snap bean and peanut in microplots. Phytopathology 84:113–118, 1994.

49. KK Sabet, MA Mostafa, OHI El-Bana, EM El-Sherif. *Pseudomonas lindbergii* and *Coniothyrium minitans* as biocontrol agents effective against some soil fungi pathogenic to peanut. Egypt J Agric Res 70:403–414, 1992.

50. AR Podile, BSD Kumar, HC Dube. Antibiosis of rhizobacteria against some plant pathogens. Indian J Microbiol 28:108–111, 1988.

51. S Savithiry, SS Gnamanickam. Bacterization of peanut with *Pseudomonas fluorescens* for biological control of *Rhizoctonia solani* and for enhanced yield. Plant Soil 102:11–15, 1987.

52. JT Turner, PA Backman. Factors relating to peanut yield increases after seed treatment with *Bacillus subtilis*. Plant Dis 75:347–353, 1991.

53. J Badel, S Kelemu. In vitro inhibition of *Colletotrichum gloeosporioides* Penz and other phytopathogenic fungi by culture filtrates of *Bacillus subtilis*. Fitopatol Colomb 18:30–35, 1994.

54. M Chaudhary, LV Gangawane. Effect of dairy and distillery effuents on biological control of *Rhizoctonia bataticola*. Indian Bot Reptr 8:81–82, 1989.

55. Umamaheswari, G Ramakrishnan. Effect of seed treatment with *Trichoderma*

viride and moisture levels on root rot disease in groundnut. Madras Agric J 81: 553–555, 1994.

56. F Perdomo, R Echavez-Badel, M Alameda, EC Schroder. In vitro evaluation of bacteria for the biological control of *Macrophomina phaseolina*. World J Microbiol Biotechnol 11:183–185, 1995.

57. S Bullock, PB Adams, HJ Willetts, WA Ayers. Production of haustoria by *Sporidesmium sclerotivorum* in sclerotia of *Sclerotinia minor*. Phytopathology 76:101–103, 1986.

58. PB Adams, JAL Wong. The effect of chemical pesticides on the infection of sclerotia of *Sclerotia minor* by the biocontrol agent *Sporidesmium sclerotivorum*. Phytopathology 81:1340–1343, 1991.

59. PB Adams. Comparison of antagonists of *Sclerotinia* species. Phytopathology 79: 1345–1347, 1989.

60. AJL Phillips. Factors affecting the parasitic activity of *Gliocladium virens* on sclerotia of *Sclerotinia sclerotiorum* and a note on its host range. J Phytopathol 116: 212–220, 1986.

61. JC Tu. *Gliocladium virens*, a destructive mycoparasite of *Sclerotinia sclerotiorum*. Phytopathology 70:670–674, 1980.

62. D Jones, AH Gordon, JSD Bacon. Co-operative action by endo- and exo-β-(1,3) glucanases from parasitic fungi in the degradation of cell-wall glucans of sclerotinia sclerotiorum (Lib) de Bary. Biochem J 140:47–55, 1974.

63. PM Phipps, DM Porter. Evaluation of *Trichoderma harzianum* (TH-88) for biological control of sclerotinia blight of peanut. Biol Cult Tests Control Plant Dis 4:40, 1989.

64. HA Melouk, FA Chanakira, KE Conway. Inhibition of *Sclerotinia minor* by *Penicillium citrinum* (abstr). Phytopathology 75:502, 1985.

65. CN Akem, HA Melouk. Colonization of sclerotia of *Sclerotinia minor* by a potential biocontrol agent *Penicillium citrinum*. Peanut Sci 14:66–70, 1987.

66. K Jackson, HA Melouk. Biological control trials on sclerotinia blight of peanut. In: KE Jackson, JP Damicone, E Williams, HA Melouk, PW Pratt, CC Russell, Sholar, JR Stillwater, eds. Results of 1991 Plant Disease Control Field Studies. Research Report. Oklahoma, USA: Oklahoma Agricultural Experiment Station, Oklahoma State University, 1992, pp 83–88.

67. Y Henis, PB Adams, JA Lewis, GC Papavizas. Penetration of sclerotia of *Sclerotium rolfsii* by *Trichoderma* spp. Phytopathology 73:1043–1046, 1983.

68. N Benhamou, I Chet. Parasitism of sclerotia of *Sclerotium rolfsii* by *Trichoderma harzianum*: ultrastructural and cytochemical aspects of the interaction. Phytopathology 86:405–416, 1996.

69. S Desai, E Schlosser. Parasitism of *Sclerotium rolfsii* by *Trichoderma*. Ind Phytopath 52:47–50, 1999.

70. LT Kwee, TB Keng. Antagonism in vitro of *Trichoderma* species against several basidiomycetous soil-borne pathogens and *Sclerotium rolfsii*. Zietschr Pflanzenkr Pflanzenschutz 97:33–41, 1990.

71. L Madi, T Katan, J Katan, Y Henis. Biological control of *Sclerotium rolfsii* and *Verticillium dahliae* by *Talaromyces flavus* is mediated by different mechanisms. Phytopathology 87:1054–1060, 1997.

72. S Haran, H Schickler, A Oppenheim, I Chet. Differential expression of *Trichoderma harzianum* chitinases during mycoparasitism. Phytopathology 86:980–985, 1996.
73. R Barak, Y Elad, D Mirelman, I Chet. Lectins: A possible basis for specific recognition in the interaction of *Trichoderma* and *Sclerotium rolfsii*. Phytopathology 75: 458–462, 1985.
74. J Inbar, I Chet. Biomimics of fungal cell-cell recognition by use of lectin-coated nylon fibers. J Bacteriol 174:1055–1059, 1992.
75. J Inbar, I Chet. A newly isolated lectin from the plant pathogenic fungus *Sclerotium rolfsii*: purification, characterization and role in mycoparasitism. Microbiology 140: 651–657, 1994.
76. GC Papavizas, DJ Collins. Influence of *Gliocladium virens* on germination and infectivity of sclerotia of *Sclerotium rolfsii*. Phytopathology 80:627–630, 1990.
77. SA Kulkarni, S Kulkarni. Biological control of *Sclerotium rolfsii* Sacc.—a causal agent of stem rot of groundnut. Karnataka J Agric Sci 7:365–367, 1994.
78. S Sreenivasaprasad, K Manibhushanrao. Efficacy of *Gliocladium virens* and *Trichoderma longibrachiatum* as biological control agents of groundnut root and stem rot diseases. Int J Pest Manage 39:167–171, 1993.
79. K Rajappan, T Raguchander, K Manickam. Efficacy of UV-induced mutants of *Trichoderma viride* against *Sclerotium rolfsii*. Plant Dis Res 11:97–99, 1996.
80. Y Elad, I Chet, J Katan. *Trichoderma harzianum*: A biocontrol agent effective against *Sclerotium rolfsii* and *Rhizoctonia solani*. Phytopathology 70:119–121, 1980.
81. PA Backman, R Rodriguez-kabana. A system for the growth and delivery of biological control agents to the soil. Phytopathology 65:819–821, 1975.
82. D Matti, C Sen. Integrated biocontrol of *Sclerotium rolfsii* with nitrogenous fertilizers and *Trichoderma harzianum*. Ind J Agric Sci 55:464–468, 1985.
83. KS Jagadeesh, GS Geeta. Effect of *Trichoderma harzianum* grown on different food bases on the biological control of *Sclerotium rolfsii* Sacc. in groundnut. Environ Ecol 12:471–473, 1994.
84. M Muthamilan, R Jeyarajan. Integrated management of sclerotium root rot of groundnut involving *Trichoderma harzianum*, *Rhizobium* and carbendazim. Indian J Mycol Plant Pathol 26:204–209, 1996.
85. AS Csinos, DK Bell, NA Minton, HD Wells. Evaluation of *Trichoderma* spp., fungicides and chemical combinations for control of southern stem rot on peanuts. Peanut Sci 10:75–79, 1983.
86. P Bhattacharyya, N Mukherjee. *Rhizobium* challenges the root rot pathogen (*Sclerotium rolfsii*) on groundnut surfaces. Indian Agricu 34:63–71, 1990.
87. ZK Punja. The biology, ecology and control of *Sclerotium rolfsii*. Annu Rev Phytopathol 23:97–127, 1985.
88. A Ordentlich, Y Elad, I Chet. Rhizosphere colonization by *Serratia marcescens* for the control of *Sclerotium rolfsii*. Soil Biol Biochem 19:747–751, 1987.
89. A Ordentlich, Y Elad, I Chet. The Role of chitinase of *Serratia marcescens* in biocontrol of *Sclerotium rolfsii*. Phytopathology 78:84–88, 1988.
90. P Ganesan, SS Gnanamanickam. Biological control of *Sclerotium rolfsii* Sacc. in

peanut by inoculation with *Pseudomonas fluorescens*. Soil Biol Biochem 19:35–38, 1987.

91. JW Dorner, RJ Cole, PD Blankenship. Use of a biocompetitive agent to control preharvest aflatoxin in drought stressed peanuts. J Food Prot 55:888–892, 1992.

92. HK Chourasia, RK Sinha. Potential of the biological control of aflatoxin contamination in developing peanut (*Arachis hypogaea* L.) by atoxigenic strains of *Aspergillus flavus*. J Food Sci Technol 31:362–366, 1994.

93. JW Dorner, RJ Cole, PD Blankenship. Effect of inoculum rate of biological control agents on preharvest aflatoxin contamination of peanuts. Biol Control 12:171–176, 1998.

94. DJ Daigle, PJ Cotty. Formulating atoxigenic *Aspergillus flavus* for field release. Biocontrol Sci Technol 5:175–184, 1995.

95. ST Kim, YB Kim. The effect of some koji molds on production of aflatoxin by *Aspergillus flavus*. J Korean Agric Chem Soc 29:255–259, 1986.

96. AC Mixon, DK Bell, DM Wilson. Effect of chemical and biological agents on the incidence of *Aspergillus flavus* and aflatoxin contamination of peanut seed. Phytopathology 74:1440–1444, 1984.

97. JW Kloepper, JA McInroy, KL Bowen. Comparative identification by fatty acid analysis of soil, rhizosphere, and geocarposphere bacteria of peanut (*Arachis hypogaea* L.). Plant Soil 139:85–90, 1992.

98. CJ Mickler, KL Bowen, JW Kloepper. Evaluation of selected geocarposphere bacteria for biological control of *Aspergillus flavus* in peanut. Plant Soil 175:291–299, 1995.

99. HK Chourasia. Kernel infection and aflatoxin production in peanut (*Arachis hypogaea* L.) by *Aspergillus flavus* in presence of geocarposphere bacteria. J Food Sci Technol 32:459–464, 1995.

100. IJ Misaghi, PJ Cotty, DM Decianne. Bacterial antagonists of *Aspergillus flavus*. Biocontrol Sci Technol 5:387–392, 1995.

101. RG Cuero, G Osuju. Chitosanase bioinduction by two strains of *Bacillus* sp. and chitosan in peanut: an effective biocontrol of pathogenic and toxigenic fungi. Mededelingen Faculteit Landbouwwetenschappen, Rijksuniversiteit Gent 56:1415–1425, 1991.

102. P Jatala. Biological control of plant-parasitic nematodes. Annu Rev Phytopathol 24:453–489, 1986.

103. DJ Patel, RV Vyas, BA Patel, RS Patel. Bioefficacy of *Paecilomyces liliacinus* in controlling *Meloidogyne javanica*. Int Arachis Newslett 15:46, 1995.

104. RV Vyas, HR Patel, DJ Patel, NB Patel. Biological suppression of root-knot nematode-white grub pest complex attacking groundnut. Int Arachis Newslett 17:40–41, 1997.

105. TS Dahiya, DP Singh. Inhibitory effect of *Aspergillus niger* culture filtrate on mortality and hatching of larvae of *Meloidogyne* sps. Plant Soil 86:145–146, 1985.

106. RM Sayre, MP Starr. *Pasteuria penetrans* (ex Thorne, 1940) nom. rev. comb.n., sp.n., a mycelial and endospore-forming bacterium parasitic in plant parasitic nematodes. Proceedings of the Helminthological Society of Washington 1989, pp 149–165.

107. DW Dickson, M Oostendorp, RM Giblin-Davis, DJ Mitchell. Control of plant-parasitic nematodes by biological antagonists. In: D Rosen, FD Bennett, JL Capinera, eds. Pest Management in the Subtropics, Biological Control—A Florida Perspective. Hampshire, UK: Intercept, 1994, pp 575–601.

108. R Mankau. *Bacillus penetrans* n. Comb. J Invertebrat Pathol 26:333–339, 1975.

109. ZX Chen, DW Dickson, R McSorley, DJ Mitchell, TE Hewlett. Suppression of *Meloidogyne arenaria* race 1 by soil application of endospores of *Pasteuria penetrans*. J Nematol 28:159–168, 1996.

110. ZX Chen, DW Dickson, DJ Mitchell, R McSorley, TE Hewlett. Suppression mechanisms of *Meloidogyne arenaria* race1 by *Pasteuria penetrans*. J Nematol 29:1–8, 1997.

111. GR Stirling. Biological control of *Meloidogyne javanica* with *Bacillus penetrans*. Phytopathology 74:55–60, 1984.

112. RK Bansal, AK Sobti. An economic remedy for the control of two species of *Aspergillus* on groundnut. Indian Phytopath 43:451–452, 1990.

113. MP Ghewande. Management of foliar diseases of groundnut (*Arachis hypogaea*) using plant extracts. Indian J Agric Sci 59:133–134, 1989.

114. AK Sobti, OP Sharma, AK Bhargava. A comparative study of fungicidal compounds and plant extracts against three pathogens of *Arachis hypogaea*. Indian Phytopath 48:191–193, 1995.

115. T Ganapathy, P Narayanasamy. Effect of plant products on the incidence of major disease of groundnut. Int Arachis Newslett 7:20–21, 1990.

116. V Chandrasekar, R Narayanaswami, R Ramabadran. Effect of foliar spray of potash and neem seed extract on the tikka leaf spot of groundnut. Indian Phytopath 47: 188–189, 1994.

117. M Usman, R Jaganathan, D Dinakaran. Plant disease management on groundnut with naturally occurring plant products. Madras Agric J 78:152–153, 1991.

118. TK Mahapatra, SN Tewari. *Ocimum sanctum* L. leaf extract toxicity against collar rot (*Aspergillus niger*) and yellow root (*A. flavus*) diseases of groundnut. Allelopathy J 1:114–117, 1994.

119. DVR Saigopal, V Siva Prasad, P Sreenivasulu. Antiviral activity in extracts of *Phyllanthus fraternus* webst (*P. niruri*). Curr Sci 55:264–265, 1986.

8

Biological Control of Sugarcane Diseases

D. Mohanraj, P. Padmanaban, and R. Viswanathan
Sugarcane Breeding Institute, Indian Council of Agricultural Research, Coimbatore, Tamil Nadu, India

I. INTRODUCTION

The importance of sugarcane as one of the world's most important commercial crops needs no special emphasis. Worldwide, sugarcane is cultivated over an area of 18.32 million ha with an annual production of 700.99 million tons of sugarcane and 85.57 million tons of sugar. The crop is a major source of employment and industrial development, particularly in some developing countries.

A. The Role of Diseases

The primary factor deciding the production and productivity of the crop is the cultivar in relation to its yield and quality (sugar content). The next most important constraint in the production and productivity of the crop is the role of diseases. It has been estimated that diseases annually cause a loss of 10–25% of the sugarcane crop worldwide, amounting to losses of millions of dollars.

In addition to yield loss, many diseases result in severe deterioration in the quality of juice from the infected canes and considerable problems are encountered during processing of the juice resulting in further reduction in the production of quality sugar.

B. Disease Management in Sugarcane

By far the most important strategy to manage diseases of sugarcane is the use of disease-resistant cultivars and agronomic and cultural measures. However, there are often severe constraints in adopting the above measures of disease control due to practical factors such as value of the crop in terms of yield and quality and expenditure likely to be incurred by following the control methods.

C. Inadequacy of Chemical Control

Use of chemicals, particularly fungicides, in the management of sugarcane diseases is normally restricted to seed or sett (seed cutting) treatment with fungicides such as carbendazim for the control of sett rot disease. It was at one time recommended that the setts should be treated with organo-mercurial compounds such as agrason, cereson, agallal, etc. for the prevention of sett- and soilborne diseases such as sett rot (pineapple disease). However, with the ban on the use of mercurial fungicides in the 1970s, it became necessary to use eco-friendly methods to manage sugarcane diseases. Now it is imperative that alternative methods be developed to manage sugarcane diseases effectively on a sustainable basis.

D. Fungicide Resistance

Continuous use of chemicals for disease control is known to result in the development of resistance. It has been observed that treatment of seed setts with organomercurial fungicides or carbendazim over a long duration results in the build-up of resistance and tolerance to these chemicals by the sugarcane pathogens.

E. The Impact of Chemical Control

For the last three decades nonspecific and broad-spectrum antimicrobial chemicals were routinely recommended for the management of certain sugarcane diseases. The use of organo-mercurials (e.g., agrasan, cerasan) was widely recommended to treat the seed setts for protection against soilborne plant pathogens. However, in view of the nonspecific and general toxic effects of these compounds on soil microflora, it has become desirable to discontinue their use and seek more eco-friendly methods to manage the sugarcane diseases. Some recent antifungal compounds such as thiophanate methyl are relatively specific against the red rot pathogen of sugarcane and result in much less soil damage.

Considering the above facts, it is clear that nonchemical (or biological) measures are the ultimate long-term solution to the problems of disease management in sugarcane.

F. The Alternatives

With increased awareness of the adverse effects of plant protection chemicals on the environment, particularly in the sugarcane ecosystem, the need to manage these diseases using biological approaches has assumed significance. Worldwide, considerable information has been generated to suggest the possible control of sugarcane diseases using biocontrol measures. In spite of the limitations expressed by many workers, there are specific advantages in adopting biocontrol measures for the control of sugarcane diseases.

G. Sugarcane—An Ideal Crop for Biocontrol

From the point of biocontrol of diseases, sugarcane can be considered ideal for control of its diseases using biological methods based on the following criteria:

1. Because sugarcane is a commercial crop of high economic value, any investment in biological control would be suitably rewarded in terms of higher productivity.
2. Cultivation of the crop often on a plantation scale over large and continuous areas enables easy application of biocontrol techniques.
3. Continuous presence of the crop in the same fields over extended periods by way of ratoons or as a result of monocropping renders the biocontrol agents self-sustainable without interruption.
4. Sugarcane crop management is very effectively carried out by efficient agencies such as development organizations, sugar mill managements, etc., providing adequate infrastructure to mass-produce biocontrol agents, their application, and obtain feedback.
5. The industry is capable of effectively supporting R&D efforts in this area through well-organized laboratories, scientific and technical manpower, etc., of both governmental and private research institutions.

II. BIOLOGICAL CONTROL OF SUGARCANE DISEASES

A. Red Rot Disease

1. Biocontrol Using Fungal Antagonists

Among the fungal diseases of sugarcane, red rot, caused by the fungal pathogen *Colletotrichum falcatum* [*Glomerella tucumanensis* (Speg.) Arx & Muller] is known to cause severe reduction in crop yields and quality [1]. Although the disease is primarily settborne, considerable transmission takes place through soil and water that harbor and transport propagules of the pathogen. These soil- and waterborne propagules are amenable for control using suitable antagonistic organisms. Alfonso and Cruz [2] observed that mycelial growth of *C. falcatum* was inhibited by species of *Trichoderma* and *Aspergillus* using dual culture techniques. Culture filtrates of these two fungi were also inhibitory to *C. falcatum*, and the action was attributed to antibiotic production by *Aspergillus* and *Tricho-*

derma. In addition, *Trichoderma* also exhibited hyperparasitic activity. Iqbal et al. [3], investigating fungal isolates from different sugarcane soils, found that many isolates of *Trichoderma harzianum*, *Chaetomium* spp., and *Penicillium* spp. were inhibitory to red rot pathogen. Filtrates of *T. harzianum*, *Acremonium* spp., and *Penicillium* spp. were very effective in suppressing spore germination of the fungus. Singh [4] studied the antagonistic effects of *Chaetomium globosum* and *Trichoderma* spp. against *C. falcatum* in detail. He has reported successful control of red rot under endemic field conditions by application of these fungi after large-scale multiplication.

Gururaja [5] carried out extensive studies on the possible use of biocontrol agents in the management of sugarcane red rot. He concluded that *Trichoderma harzianum* was the most effective organism to suppress mycelial growth and sporulation of *C. falcatum*. Antibiotic production was attributed to the antagonistic action with coiling of hyphae and lysis. Soil application of the antagonist was found to protect germinating plants from infection. Treatment of healthy seed cane with a suspension of 5×10^8 spores/mL appreciably reduced red rot infection in pathogen-infested soil. However, treatment of internally red rot–infected setts was not found to be effective in preventing disease development. Spray application with the fungal suspension at a concentration of 5×10^8 spores/mL on preinfected plants also reduced infection. In addition to the antagonist, its metabolites (culture filtrate), when sprayed, also effectively reduced disease development. The purified metabolite with inhibitory action against the pathogen was determined to be a protein with a molecular weight of 14.2 kDa (thermolabile). *T. harzianum* antagonistic to *C. falcatum* was favored by an acidic pH of around 5 with a temperature range of 25–30°C. Commercial formulations of *T. harzianum* in talc powder were prepared, which produced viable colonies up to 105 days. However, *T. harzianum* was severely inhibited by the fungicide carbendazim, which limits its use, suggesting the need to identify compatible fungicides for use with the antagonist concomitantly.

Extensive studies on the possible use of biocontrol agents against red rot utilizing bacterial and fungal antagonists have been conducted at this institute during the past 5 years. Native strains of *Pseudomonas* spp. and *Trichoderma* spp. were isolated from soil samples of sugarcane from different red rot endemic locations. About 43 isolates of fluorescent pseudomonads and 23 isolates of *Trichoderma* were obtained. Among these antagonists, 15 bacterial isolates and 11 fungal isolates were found effective against *C. falcatum* under field conditions. In addition, 13 isolates of *Pseudomonas* spp. and 8 of *Trichoderma* spp. were effective against sett rot and root rot diseases [6]. All the bacterial isolates caused lysis of the mycelium except one, which was fungistatic. Among the *Trichoderma* isolates T5 was competitive, T33, T36, T52, T66, and T74 were mycoparasitic, and T53 and T62 were antibiotic producers. Red rot–infected setts of cv CoC

671 treated with *Trichoderma* isolates showed less than 50% germination and subsequent disease incidence was 20%, whereas treatment with bacterial antagonists recorded 80% germination with less than 5% disease in settlings under greenhouse conditions [6].

Studies conducted by Singh [4] revealed that sugarcane sett treatment with *Chaetomium* sp. and *T. harzianum* significantly improved the germination in the pathogen inoculated setts. In the field *Chaetomium* sp. treatment showed reduced disease build up and enhanced cane yield. Apart from sett treatment, foliar spray of antagonists *Chaetomium* sp., *T. harzianum*, and *Fusarium moniliformae* was found effective against red rot disease development in the field.

Singh [7] reported efficacy of *Chaetomium globusum* against red rot disease under field conditions in subtropical India. Initially the setts (cv. CO7717) were infected with *C. falcatum* by dipping in a spore suspension (10^5 conidia/m^L) prepared in 0.5% carboxymethylcellulose (CMC) for 10 minutes. The inoculated setts were incubated for 24 hours for the development of pathogen infection before treating with antagonists. The antagonists *C. globosum*, *T. harzianum*, and *T. viride* were multiplied on autoclaved sugarcane leaf bits (25 days). While treating the setts, 500 g leaf bits were blended and mixed in 20 L of water containing 0.5% CMC and the setts were dipped in this for 10 minutes before planting in the field. The results revealed that sett treatment with *C. globosum* and the fungicide bavistin significantly improved germination of setts. Similarly, *C. globosum* treatments showed high reduction in disease and enhanced cane yields (Table 1).

Table 1 Effect of Fungal Antagonists on Red Rot Disease Development in Sugarcane and Cane Yield

Treatment	Germination percent	Red rot incidence (%)		Cane yield at harvest (t/ha)
		Initial stage	At harvest	
Healthy setts	43.2	0.0	1.3	107.6
Pathogen alone	23.8	11.2	15.5	59.3
Pathogen + *C. globosum*	36.2	1.7	4.6	96.8
Pathogen + *T. harzianum*	26.9	4.9	9.5	80.4
Pathogen + Ecoderma	28.8	5.9	6.5	87.2
Pathogen + *T. viride*	31.3	4.9	8.0	81.0
Pathogen + bavistin	35.0	6.0	7.3	77.7
Pathogen + OvisG	29.5	5.9	8.8	77.4
CD (P = 0.05)	10.3	4.4	5.0	21.6

Ecoderma—*T. viride* commercial formulation from Margo; Ovis G—chemical 100 g/7.5 m^2; bavistin—0.2%.

2. Strains of *Pseudomonas* spp. Against Red Rot Disease

Recently detailed investigations were conducted on the effect of *Pseudomonas* spp. on the suppression of *Colletotrichum falcatum* by Viswanathan [8]. Native *Pseudomonas* strains isolated from sugarcane rhizosphere and sugarcane stalks were utilized for the biological suppression of red rot disease. Antagonistic activity of the *Pseudomonas* strains was studied on oatmeal agar, potato dextrose agar, and King's B medium. The results revealed that in vitro growth inhibition by the bacterial strains was more on King's B medium followed by oatmeal agar and potato dextrose agar medium. Enhanced suppression of mycelial growth by different strains, namely CHA0, Pf1, ARR1, and ARR2, was associated with higher production of secondary metabolities in the medium [8].

Since the bacterial cultures were not suitable to study their efficacy under field conditions, a talc formulation was prepared as per Vidhyasekaran and Muthamilan [9]. The bacterial cultures were multiplied in King's B broth for 48 hours to 9×10^8 cfu/mL. Talc formulations were prepared by mixing 400 mL of the broth with 1 kg of sterilized talc powder (pH adjusted to neutral by adding $CaCO_3$ 15 g/kg and carboxymethylcellulose 10 g/kg). After shade drying overnight, the mixure was packed in polypropylene bags and sealed. The formulation with $2.5-3.0 \times 10^8$ cfu/g was used in different experiments.

3. *Pseudomonas*-Induced Systemic Resistance Against C. falcatum

Induction of systemic resistance by the bacterial strains against red rot was established in a disease-susceptible cultivar CoC 671. The strains in the form of talc formulation were applied in the rhizosphere by various methods. Among the different combination of bacterial applications, sett treatment while planting followed by soil application 60 and 120 days after planting was found to induce higher resistance in sugarcane against the pathogen. In the treated cane stalks the pathogen was inoculated artificially by three different methods. The pathogen was challenged by these three methods to prove the efficacy of different bacterial strains in inducing systemic resistance against the pathogen. It was found that the different bacterial strains had induced systemic resistance against the pathogen when challenged by three different inoculation methods, namely plug method, nodal swabbing method, and controlled condition testing. In the first two methods the pathogen was inoculated on live canes in the field, whereas in the third method the treated canes were brought to a disease-testing chamber for inoculation. The results revealed that pathogen penetration and spread were significantly reduced in the bacteria-treated canes as compared to the untreated canes [10].

Assessment of pathogen colonization by ELISA at different nodal positions of the bacteria-treated canes revealed that the induced systemic resistance (ISR) effect had markedly reduced the pathogen colonization in the nodal region. In the *Pseudomonas*-treated canes the pathogen could colonize only few upper nodes from the point of pathogen inoculation, whereas in the untreated canes the pathogen traversed throughout the cane and caused drying in 30 days [11] (Table 2).

Of the different methods of application of bacterial formulations attempted, sett treatment while planting followed by soil application 60 and 120 days after planting showed higher ISR effects as compared to other methods [12]. Later studies of the same authors revealed that *Pseudomonas* spp.–induced resistance was maintained up to 90 days in the host [10].

Although ISR effect against *C. falcatum* was proved in sugarcane, efficacy of the bacterial strains against soilborne inoculum carried by infected crop debris was not assessed. Recent studies by Viswanathan and Samiyappan [13] revealed that *P. fluorescens* strain CHA0 treatment protected the crop up to 8 months in a sick soil containing viable pathogen inoculum. Treatment with other strains (KKM1 and VPT4) resulted in disease-free crops up to 6 and 4 months, respectively. At 12 months, the control treatment had red rot in 35.5% of sugarcane clumps, and in bacteria treatments the percent infection varied from 5.88 to 16.67%. This information suggests that the selected bacterial strains in addition to the ISR effect have direct antagonistic activity against the pathogen. In a pathogen sick soil, germination of sugarcane setts was adversely affected. However, *Pseudomonas* treatment was found to protect the setts from the time of germination onwards.

Table 2 Assessment of *C. falcatum* Colonization in *Pseudomonas fluorescens* Strain (VPT4)–Treated Sugarcane Stalks by ELISA

	OD value at 405 nm	
Nodal position	*Pseudomonas*-treated	Control
4[a]	1.644	1.412
5	0.943	1.709
6	1.143	1.606
7	0.992	1.504
8	0.717	1.383
9	0.668	0.955
10	0.577	0.924

[a] Pathogen inoculation was done just below the 4th node.
Source: Ref. 11.

4. Mechanism of *Pseudomonas*-Induced Resistance

It was found that *Pseudomonas* strain (KKM1) treatment increased the defense enzymes (chitinase, β-1,3-glucanase, peroxidase, and phenylalanine ammonia lyase) in sugarcane. The challenge inoculation of the pathogen triggered multifold increase in the enzyme activities in both bacteria-treated and untreated canes. However, the increase in enzyme activity was significantly higher in the bacteria-treated cane tissues as compared to the untreated control [10,12,14]. Western blot studies revealed that *Pseudomonas* strains induce certain pathogenesis-related (PR) proteins in sugarcane. Barley chitinase antiserum detected four new chitinase isoforms with molecular weights of 12.0, 34.5, 53.5, and 63 kDa in the KKM1-treated plants after *C. falcatum* challenge, whereas the control plants showed none of the chitinases after pathogen inoculation. In addition to chitinases, 40 and 43 kDa β-1,3-glucanases (PR-2) were found to be involved in ISR against *C. falcatum* in sugarcane. Similarly, a 42 kDa thaumatin-like protein (TLP), (PR-5), was also found associated with the systemic resistance [15]. Specific induction and identification of chitinase and other PR proteins for the first time in sugarcane gives new dimensions to the understanding of red rot resistance. These findings indicate that induction of antifungal PR proteins in a red rot–susceptible variety results in enhanced resistance to the pathogen as noticed in red rot–resistant varieties.

5. Bacterial Metabolites

Production of metabolites such as pyocyanine, pyrolnitrin, 2-4-diacetylphloroglucinol, and phenazine in the culture medium by the *P. fluorescens* has been proved. Existence of good correlation between metabolite production by the bacterial strains and *C. falcatum* mycelial growth inhibition under in vitro condition was found. Further, the strains have been found to produce siderophores by Pf1, VPT4, and VPT10 in the medium. Similarly the strains CHA0, ARR10, and Pf1 produce higher levels of salicylic acid in the medium [8]. The results suggest that enhanced production of these metabolites may have a positive role in plant growth promotion and induced systemic resistance against the pathogen.

B. Sett Rot

Sett rot disease caused by *Ceratocystis paradoxa* Moreau is an important soilborne disease that affects germination of setts at early stages of planting. The disease is also referred to as pineapple disease, since the split-open infected canes emit an odor reminiscent of pineapple fruits. Occasionally standing canes are also affected, depending on environmental conditions. This disease is widely distributed in warm temperate and tropical regions of the world and almost all the sugarcane–growing regions [16]. The pathogen (*C. paradoxa*) is a facultative parasite inhabiting sugarcane soils and infects the seed setts through the cut ends, particularly in ill-drained and waterlogged conditions. The fungus has a high saprophytic survival potential and is grossly affected by soil biotic and abiotic factors.

Species of *Trichoderma* are the most common antagonists reported to be effective against *C. paradoxa*. Contradicting findings have been reported on the prevalence of *Trichoderma* spp. in sugarcane soils in relation to sett rot control. Agnihotri and Singh [17] reported very high populations of *Trichoderma* spp. in the sugarcane rhizosphere compared to a crop-free soil.

Dipping of the setts in a fungicide solution before planting is recommended to control the disease. Fungicides such as carbendazim, bayleton, aretan, benomyl, thiophanate methyl, fluzolazole, and ethyltrianal have been found effective in reducing the disease build-up in the soil [18–20]. Studies on the possibilities of using biocontrol agents for the management of sett rot were started at this institute. Different strains of *Trichoderma harzianum* and *T. viride* showed high antagonism against *C. paradoxa*. Sett treatment with *T. harzianum* strain (5 × 10^8 cfu/m^L) and or in combination with vitavax 300 ppm was found highly effective in the control of sett rot. The pathogen was able to survive in the field for up to 60 days in the soil. However, in *T. harzianum*–treated soil the pathogen could not survive beyond 15 days. When different fungicides were tested for compatibility with the efficient *T. harzianum* strains, carbendazim, captan, and prochloraz were found to be inhibitory to the pathogen at 100 ppm itself. Vitavax and emisan were inhibitory to the pathogen at 750 ppm. A talc-based commercial formulation was prepared, and viable colonies of *T. harzianum* were obtained up to 120 days [21] (Table 3). The results of the study indicated that biocontrol measures in combination with fungicides could effectively manage the sett rot disease in sugarcane.

C. Biological Control of Wilt Disease of Sugarcane

Among the diseases that affect the sugarcane stalk, wilt disease is considered to be important in certain situations. It is primarily a disease of the stalk. The external

Table 3 Effect of Sett Treatment with *T. harzianum* on Sett Rot Incidence

Treatment	Percent germination on 35th day
T. harzianum–treated setts in pathogen-infested soil	80.00
T. harzianum + vitavax 300 ppm–treated setts in pathogen-infested soil	85.00
Carbendazim (500 ppm)–treated setts in pathogen-infested soil	60.00
Untreated setts in pathogen-infested soil	0.50
Untreated setts in pathogen-free soil	90.00

symptoms of the disease include yellowing and drying of leaves, desiccation and shrinking of stalks, and stunted growth. Internally, the stalk tissues become discolored brown, with drying and cavity formation. In the advanced stages, growth of the associated fungi can be observed in these cavities. The symptoms are usually expressed after the growth phase of the crop, i.e., about 6–7 months after planting. In addition to yield loss, the disease results in quality loss due to reduction in juice content, decreased sugar in the juice, and accumulation of non-sugar components in the juice [1].

Wilt has been associated with two fungi: *Fusarium moniliforme* and *Cephalosporium sacchari*. Although they are observed to be weak pathogens, colonization of the host by these fungi is known to precede disease development. Studies at the Sugarcane Breeding Institute, Coimbatore, India, have indicated the preferential colonization of the roots and lower parts of the stalks by *Fusarium* and the middle and upper parts by *C. sacchari*. Simultaneous inoculation with both fungi associated with stress results in better disease expression. Abiotic soil stress such as drought, waterlogging, soil reaction, or nutrient imbalances and biotic stress such as soil pests are found to be essential predisposing factors for disease expression. Both *F. moniliforme* and *C. sacchari* causing sugarcane wilt are soil inhabitants and infect the host through roots or subsoil parts of the stalk. Being soilborne organisms, they are suitable targets for inhibition by other microbes in the soil and in the rhizosphere. Since the inoculum is constantly available in the soil and the crop can be infected at all stages, depending on the availability of predisposing factors, fungicidal treatments of the seed cuttings are not effective in controlling the disease.

Although extensive observations and reviews are available on the biological control of *Fusarium* spp., reports on the specific control of sugarcane wilt fungi by biocontrol agents are not common. The potential for biological control of *F. moniliforme* with rhizobacteria was examined by Bacon and Williamson [22]. Application of soil amendments has been reported to reduce severity of sugarcane wilt. However, Singh et al. [23] attributed this effect to an increase in the activity of organisms antagonistic to the wilt fungi in the soil. They also observed that of the two organisms associated with wilt, *Fusarium* was more sensitive to inhibition by the soil organisms.

Bhatti and Chohan [24] found strains of *Streptomyces* and *Bacillus* to be highly antagonistic to *Cephalosporium sacchari* and indicated possibilities of suppressing it in soil through their use. Sugarcane wilt very much resembles the stalk rot and late wilt of maize associated with strains of *F. moniliforme* and *C. maydis*, respectively. Many reports are available on the control of *Fusarium* associated with maize, and attempts to follow similar approaches with sugarcane wilt may yield promising results. Selected isolates of fungi, bacteria, and actinomycetes were antagonistic to *Fusarium* depending on the soil source from which they were obtained.

Ramu [25] investigated in detail some aspects of biological control of sugarcane wilt. Isolates of *Trichoderma viride* and *Pseudomonas* were inhibitory to wilt fungi in vitro and also reduced their population in soil. In addition, the antagonists markedly reduced colonization of the host by the pathogens in infested soil. Evidence was obtained to suggest that both competition and antibiosis were involved in the inhibitory action of the antagonists against the wilt fungi.

Considering the fact that sugarcane wilt develops only in the presence of essential predisposing biotic and abiotic stress factors, the role of soil microbes is quite apparent. This suggests the potential to manage the disease by use of antagonistic organisms in soils along with suitable soil management practices that would ensure sustained multiplication and survival of the antagonists in the sugarcane rhizosphere.

D. *Pythium* Root Rot

Sugarcane is commonly propagated by seed cuttings (setts) of the stalk. However, in the varietal improvement programs true seed of sugarcane obtained through hybridization are used to generate genetic variability for selection of high-yielding and high-quality genotypes. These seeds are of very small size, and large numbers of them are sown in seed beds of specialized soil mixture for subsequent transplantation in the field for evaluation and selection.

These seedlings during the early phases of growth are very sensitive to microbial attack. One of the common problems encountered in these seed beds is root rot of the seedlings associated with species of *Pythium*. The symptoms are characterized by yellowing and drying of the shoots, dark necrotic lesions on the roots, followed by death of the seedlings along with rotting of roots. Severe large-scale incidence of the disease results in extensive patchy stand of the seedlings. Death of the seedlings causes loss of potentially promising genotypes, which could be developed into valuable commercial cultivars. Also, genotypes that could become genetic stocks as donors of desirable attributes are also lost. Thus, the disease is a serious constraint in the breeding and selection programs. High seedling density and excess soil moisture usually result in high levels of root rot incidence. Occasionally, standing canes are also affected by the disease, resulting in rotting of roots and yellowing and dying of the clumps. This occurs particularly under waterlogged and ill-drained conditions.

The common fungus associated with sugarcane root rot is *Pythium graminicolum* Subr. The traditional method of managing the problem was drenching the seed beds with fungicides. Padmanaban and Alexander [26] reported studies on the use of fungicides and biocontrol agents to control sugarcane seedling root rot. They identified an isolate of *Trichoderma viride* antagonistic to *P. graminicolum*. Among the fungicides used, demosan, difolatan, plantvax, ziram, and fenaminosulf were inhibitory to *P. graminicolum* at the concentrations tried. The antagonistic isolate of *Trichoderma* sp. was very sensitive to the fungicides ziram, agallol, fytolan, and difolaton. The fungicides fenaminosulf, plantvax, and demosan favored the growth of *T. viride*. Since fenaminosulf was also studied against *P. graminicolum*, combinations of *T. viride* and fenaminosulf were studied for root rot control. Supplementary *Trichoderma* application with fenaminosulf drenching significantly reduced incidence of root rot while increasing root and shoot growth of the seedlings. It also reduced the number of drenchings required for effective disease control. Subsequent studies showed that with an increase in the quantities of the antagonist applied, *T. viride* alone was very effective in reducing the incidence of sugarcane seedling root rot. Protection of seedlings in the seedling pans recorded greater seedling height and root length (Table 4).

While the mechanism of antagonism of *T. viride* against *P. graminicolum* was examined, it was observed that no coiling of hyphae or lysis of mycelium was involved. In dual culture tests *T. viride* completely suppressed the growth of *P. graminicolum* within 4 days. Thus, growth suppression appears to be one of the mechanisms of antagonism of *T. viride* against *P. graminicolum*. In addition to competition, production of toxic metabolites by *T. viride* inhibitory to *P. graminicolum* was also observed. Many workers have reported competition as one of the mechanisms of antagonism by *T. viride*. In addition to competition,

Table 4 Effect of *Trichoderma* sp. and Fenaminosulf on Control of Root Rot Disease

Treatment	Percent disease incidence	Seedling height (cm)	Root length (cm)
Sterile soil			
Control	12.6 (15.6)	5.7	9.9
Pathogen inoculated	49.0 (33.3)	4.1	7.6
Pathogen + *Trichoderma*	7.2 (10.9)	10.7	10.7
Pathogen + *Trichoderma* + fenaminosulf	4.3 (8.9)	12.7	15.2
Unsterile soil			
Control	27.9 (23.9)	4.9	9.9
Pathogen inoculated	57.6 (37.0)	3.8	6.4
Pathogen + *Trichoderma*	3.5 (7.9)	10.4	12.1
Pathogen + *Trichoderma* + fenaminosulf	1.6 (3.5)	11.5	16.9
CD at 5%	9.8	2.9	2.9

Figures in parentheses are transformed values.

species of *Trichoderma* are known to produce toxic metabolites inhibitory to plant pathogens [27]. Such a phenomenon was also observed with *T. viride* and *P. graminicolum* interaction, where toxic metabolites produced in vitro by *T. viride* effectively inhibited *P. graminicolum.*

Recently fluorescent pseudomonad strains native to sugarcane were tested for their efficacy against seedling root rot disease and seedling parameters. The talc formulation of five *Pseudomonas* strains tested showed significantly improved germination. In the treated pots seedling stand was better in the *Pseudomonas*-treated ones as compared to the untreated ones. The control treatments lost more than 60% of their original population, and in the bacteria-treated pots it varied from 36 to 48.0% by the 120th day. The strains CHA0 and VPT4 were found to be highly effective in improving seed germination and maintaining seedling stand as compared to the other strains (Table 5) [28].

Nallathambi et al. [29] studied the efficacy of certain *Pseudomonas* and *Trichoderma* sp. strains native to sugarcane against seedling rot caused by *P. graminicola*. *Pseudomonas* strains were found to be comparatively more effective in suppressing seedling rot and enhancing seedling growth as compared to *Trichoderma* strains. Colonization by bacterial strains in the root tissues was determined up to 7 weeks after treatment in the soil.

E. Mycorrhizae and Control of Sugarcane Diseases

Mycorhizal fungi such as *Glomus*, *Sclerocystis* spp., *Acaulospora* spp., and *Scutellespora* spp. have been known to be associated with sugarcane roots. Of these,

Table 5   Effect of *Pseudomonas* sp. Strains on Sugarcane Seed Germination and Seedling Growth

				Seedling growth parameters, 120th day			
Bacterial strain	Seedling population (days after sowing)			Shoot length (cm)	Shoot dry weight (g)	Root length (cm)	Root dry weight (g)
	30	60	120				
CHA0	52.33[a]	39.33[a]	37.00[a]	73.43[a]	8.70[a]	22.42[a]	3.70[ab]
Pf 1	29.33[bc]	18.33[bc]	17.37[bc]	55.48[abc]	6.20[b]	15.42[bc]	3.16[b]
VPT 1	25.00[cd]	14.67[cd]	13.33[c]	52.08[bc]	6.73[b]	17.30[abc]	3.41[b]
VPT 4	35.00[b]	23.00[bc]	21.00[b]	60.13[ab]	8.73[a]	20.75[ab]	4.60[a]
VPT 10	27.67[bc]	16.00[bc]	15.00[bc]	52.03[bc]	6.50[b]	17.25[abc]	3.57[b]
Control	18.30[d]	8.33[d]	5.33[d]	38.28[c]	4.88[c]	14.25[c]	2.78[b]

In a column, means followed by the same alphabets are not significantly different at the 5% by DMRT.
Source: Ref 28.

Glomus mosseae was found to be predominant. Association of sugarcane roots with *G. mosseae*, either alone or in combination with fungicides and *Trichoderma viride*, resulted in effective control of *Pythium* root rot of sugarcane [30]. Hence more detailed studies on the use of mycorhizae in the integrated management of soilborne transmission of sugarcane diseases is expected to yield useful results.

F. Leaf Scald Disease

Among the major bacterial diseases of sugarcane, not much information is available on their biological control except for leaf scald disease.

Leaf scald disease is widely distributed in many sugarcane growing areas of the world. Initially the disease appears as long brownish pencil line–like streaks parallel to the veins along the length of the leaf lamina. Subsequently these lesions coalesce and the leaf presents a scalded appearance. With the progress of the disease, all the leaves show scalding symptoms and dry up. There is marked stunting of the stalks with shortening of internodes. Internal tissues also show linear reddish-to-brown streaks along the vascular tissues with subsequent development of extensive lesions and necrosis of the tissues. There is extensive sprouting of the lateral buds. Severe incidence of the disease causes significant yield and quality loss.

The disease is caused by a bacterium, *Xanthomonas albilineans*, and is primarily transmitted through infected seed cuttings. Hence, use of disease-free seed cane is the major control measure against the disease. Biological control attempts against the disease have been mainly through a biotechnological approach. The bacterium *Pantoea dispersa* is antagonistic to *Xanthomonas albilineans*. The pathogen produces a toxin, albicidin, which is associated with pathogenicity and symptom development. Strains of *P. dispersa* (*Erwinia herbicola*) resistant to albicidin produced by *X. albilineans* were isolated from sugarcane tissues infected by the bacterium and screened using a simple assay to distinguish resistance mechanism. One strain with a strong capacity for enzymatic detoxification of albicidin was identified. The strain designated 531403 provided almost complete biocontrol against leaf scald disease when inoculated with a 10-fold excess of *X. albilineans* cells onto a highly susceptible sugarcane variety [31].

A gene (*albD*) from *P. dispersa* has been cloned and sequenced and has been shown to code for a peptide of 235 amino acids to detoxify the phytotoxin albicidin. The gene showed no significant homology at the DNA or protein level to any known sequence, but the gene product contained a GCG motif that was a serine hydrolase. The albD protein purified to homogenicity by means of a glutothione-S-transferase gene fusion system showed strong esterase activity (*p*-nitrophenylbutyrate) and released hydrolytic products during detoxification of albicidin. The *albD* hydrolysis of *p*-nitrophenylbutyrate and detoxification of albicidin required no complex cofactors. The data suggested that *albD* is an albicidin hydrolase. The enzyme detoxified albicidin efficiently over a pH range of 5.0–8.0 with a broad temperature optimum of 15–35°C. Expression of *albD* in transformed *X. albilineans* strains abolished the capacity to produce albicidin toxin and to incite disease symptoms in sugarcane [32]. Zhang et al. [33] further explained the mechanism of albicidin detoxification and predicted the potential to pyramid genes for different mechanisms in transgenic plants to protect plastid DNA replication from inhibition by albicidin.

As in the leaf scald disease, biotechnological approaches using transformed microbes may be attempted with other sugarcane diseases. The red rot pathogen *Colletotrichum falcatum* produces phytotoxins that produce most of the symptoms of the disease, including the induction of phytoalexin (anthocyanidin) pigments, in resistant varieties [34]. Recent studies of Viswanathan (unpublished) indicated that an antagonistic bacterial strain VPT4 could completely inactivate *C. falcatum* toxin activity. Further studies on identifying the toxin-inactivating principle are expected to yield more fruitful results.

III. SUMMARY AND CONCLUSION

The need for developing biocontrol methods to manage sugarcane diseases is well recognized in view of their advantages in relation to the uniquely suitable aspects of the crop. Although the effectiveness of many biocontrol agents against major sugarcane diseases such as red rot, wilt, sett rot, and root rot have been demonstrated, their wide spread and systematic field level adoption is not yet common. This would be further facilitated by identification of more effective antagonistic strains of biocontrol agents and development of efficient techniques for their mass multiplication, formulation, and field application. Altering the sugarcane environment to favor the rapid multiplication, survival, and sustenance of biocontrol agents would be an added advantage.

Few reports are available on the biological control of foliar diseases of sugarcane, such as leaf spot and rust, which are very important. The role of phyllosphere microflora and hyperparasitism reported to be effective against similar diseases in other crops could be deployed to manage foliar diseases of sugarcane.

With recent advances in the area of molecular biology, biotechnological approaches using biocontrol agents are bound to become more common. The results already obtained in this regard, such as those relating to PGPR, pathogenicity-related proteins, transgenic antagonists, and induced resistance by biocontrol agents, etc. appear to be promising developments in the biological control of sugarcane diseases.

REFERENCES

1. KC Alexander, R Viswanathan. Major diseases affecting sugarcane production in India and recent experiences in quarantine. In: BJ Croft, CM Piggin, ES Wallis, DM Hogarth, eds. Sugarcane Germplasm Conservation and Exchange. Proceedings of Australian Centre for International Agricultural Research No. 67, 1996, pp 46–48.

2. F Alfanso, B Cruz. Antagonistic activity in vitro of *Trichodema* spp. and *Aspergillus* spp. against *Colletotrichum falcatum*. Rev Prot Veg 2:119–124, 1987.

3. SM Iqbal, CA Rauf, S Rahat, CM Aktar. Antagonism to *Colletotrichum falcatum* Went the cause of sugarcane red rot. Sarhad J Agric 10:575–579, 1994.

4. N Singh. *Trichoderma harzianum* and *Chaetomium* sp. as potential biocontrol fungi in management of red rot disease of sugarcane. J Biol Cont 8:65–67, 1994.

5. N Gururaja. Studies on the possibility of control of red rot disease of sugarcane through chemical, physical and biological means. M.Sc. (Agri.) thesis, Tamil Nadu Agricultural University, Coimbatore, 1992 p 124.

6. P Nallathambi, D Mohanraj, P Padmanaban. Biocontrol of red rot caused by *Colletotrichum falcatum* Went in sugarcane. Proceedings of International Symposium on Integrated Disease Management for sustainable Agriculture, New Delhi, 1997.

7. N Singh. Biological control of red rot. Annual report, Sugarcane Breeding Institute, 1998–99, p 56.

8. R Viswanathan. Induction of systemic resistance against red rot disease in sugarcane by plant growth promoting rhizobacteria. Coimbatore, India, Tamil Nadu Agricultural University, PhD thesis, 1999 p 167.

9. P Vidhyasekaran, M Muthamilan. Development of formulation of *Pseudomonas fluorescens* for control of chick pea wilt. Plant Dis 79:782–786, 1995.

10. R Viswanathasn, R Samiyappan. Induction of systemic resistance by plant growth promoting rhizobacteria against red rot disease in sugarcane. Sugar Tech 1:67–76, 1999.

11. R Viswanathan, R Samiyappan, P Padmanaban. Specific detection of *Colletotrichum falcatum* in sugarcane by serological techniques. Sugar Cane 3:18–23, 1998.

12. R Viswanathan, R Samiyappan. Plant growth promoting rhizobacteria for the management of red rot disease in sugarcane. Proceedings of International Symposium on Integrated Disease Management for sustainable Agriculture, New Delhi, 1997.

13. R Viswanathan, R Samiyappan. Efficacy of *Pseudomonas* spp. strains against soil borne and sett borne inoculum of *Colletotrichum falcatum* causing red rot disease in sugarcane. Sugar Tech 2(3):26–29, 2000.

14. R Viswanathan, R Samiyappan. Identification of antifungal chitinases from sugarcane. ICAR News 5(4):1–2, 1999.

15. R Viswanathan. Induction of antifungal pathogenesis related proteins in sugarcane. Sugarcane Breeding Institute Newslett 18(4):1–3, 1999.

16. CA Wismer, RA Bailey. Pineapple disease In: C Ricaud, BT Egan, AG Gillaspie Jr, CG Hughes, eds. Diseases of Sugarcane, Major Diseases. Amsterdam: Elsevier 1989, pp 145–155.

17. VP Agnihotri, N Singh. Microbial population in rhizosphere of leaf scald affected and healthy plants of sugarcane. Indian J Plant Pathol 4:1–3, 1986.

18. S Natarajan, S Muthuswamy. Effect of systemic fungicides in the control of sett rot of sugarcane. Pesticides 16:19–20, 1982.
19. RN Raid. Fungicidal control of pineapple disease of sugarcane. J Am Soc Sugarcane Technol 10:45–50, 1990.
20. CC Ryan. Bayleton controls pineapple disease. Bureau Sugar Exp Stations Bull 1: 14, 1983.
21. R Thennarasu. Biological control of sett rot (*Ceratocyctis paradoxa*) of sugarcane. M.Sc. (Agri.) thesis, Tamil Nadu Agricultural University, Coimbatore, 1997, p 93.
22. CW Bacon, JW Williamson. Interaction of *Fusarium moniliformae*, its metabolites and bacteria with corn. Mycopathologia 117:65–71, 1992.
23. N Singh, RP Singh, S Lal. Effect of soil amendment on sugarcane wilt and its pathogens. Acta Botan Indica 13:212–217, 1985.
24. DS Bhatti, KL Chohan. Antagonism of certain microorganisms to *Cephalosporium sacchari* Butl. J Res (Punjab Agricultural Univ) 7:631–635, 1990.
25. E Ramu. Effect of certain soil factors on the population of sugarcane wilt fungi. M.Sc. (Agri.) thesis, Tamil Nadu Agricultural University, Coimbatore, 1994, p 159.
26. P Padmanaban, KC Alexander. Studies on sugarcane seedling root rot in seed bed nurseries. Pestology 6(9):9–12, 1982.
27. GC Papavizas, RD Lumsden. Biological control of soil borne fungal propagules. Annu Rev Phytopathol 187:389–413, 1980.
28. R Viswanathan, R Samiyappan. Management of damping off disease in sugarcane using plant growth promoting rhizobacteria. Madras Agric J 86:647–649, 1999.
29. P Nallathambi, D Mohanraj, NR Prasad. Biological control of seedling rot of sugarcane using *Trichoderma* and *Pseudomonas fluorescens*. Proceedings of International Symposium on Integrated Disease Management for Sustainable Agriculture, New Delhi, 1997.
30. N Prakasam, Vesicular arbuscular mycorrhizal (VAM) fungal association in sugarcane. Annual Report 1995–96, Sugarcane Breeding Institute, Coimbatore, 1996, pp 58–59.
31. L Zhang, RG Birch. Biocontrol of sugarcane leaf scald disease by an isolate of *Pantoea dispersa* which detoxifies albicidin phytotoxin. Lett Appl Microbiol 22: 132–136, 1996.
32. L Zhang, RG Birch. Mechanism of biocontrol by *Pantoea dispersa* of sugarcane leaf scald disease caused by *Xanthomonas albilineans*. J Appl Microbiol 82:448–452, 1997.
33. L Zhang, J Xu, RG Birch. Evaluation of two albicidin resistance genes against sugarcane leaf scald disease. Proceedings of International Society of Sugarcane Technologists: Pathology and Molecular Biology Workshop, Kwazulu-Natal, 1997.
34. R Viswanathan, D Mohanraj, P Padmanaban, KC Alexander. Synthesis of phytoalexins in sugarcane in response to infection by *Colletotrichum falcatum* Went. Acta Phytopathol Entomol Hungar 31:229–237, 1996.

9
Biological Control of Potato Pathogens

Barry Jacobsen
Montana State University, Bozeman, Montana

The term *biological control* has many different definitions [1]. In this chapter I will use the definition proposed by Baker and Cook [2]: "Biological control is the reduction of inoculum or disease producing activity of a pathogen accomplished by one or more organisms other than man." This definition is broad enough to encompass classical approaches to biological control that directly influence pathogen populations via antibiosis, parasitism, or predation as well as approaches that afford disease reduction through competition for nutrients or infection niches, induced systemic resistance, altered plant physiology, or use of transgenic strategies. While specific biological controls will be discussed in detail, it is important to understand that resident soil microflora and fauna are responsible for much of the disease reduction attributed to crop rotation or incorporation of organic amendments. The specific role(s) of individuals or communities of these organisms in affecting either pathogen populations or disease severity is largely unknown.

Review of the literature reveals significant research efforts on biological control of potato diseases, with the majority of effort focused on bacterial soft rot, blackleg, Rhizoctonia black scurf and canker, Verticillium wilt, Fusarium dry rot, silver scurf, and nematodes. While efficacy has been field demonstrated for the majority of diseases or pathogens above, there are few products for potato producers to use at this time.

I. DISEASES CAUSED BY BACTERIAL PATHOGENS

A. Bacterial Soft Rot

This disease, caused by *Erwinia carotovora* subsp. *carotovora* (*Ecc*), is involved in seed piece decay, aerial stem rot, and postharvest soft rot of tubers [3]. This pathogen can survive in soil and is commonly spread by infected potato seed and contaminated irrigation water [4]. Biological control of seed piece decay has been achieved by inoculation of seed pieces with various bacteria. This began with the work by Kloepper et al. [5–7], who described control of *Ecc* by fluorescent psuedomonads and the growth promotion effects of these rhizobacteria, coined as plant growth–promoting rhizobacteria (PGPR). Disease control was attributed to production of siderophores [5] that complex iron such that it is unavailable to *Ecc* and a group of rhizobacteria termed deleterious rhizobacteria (DRB) [8]. Growth promotion and yield increases were noted in these and many other studies. Since this work, others have demonstrated control of seed piece decay in the field with fluorescent psuedomonads [9,10]. While isolates of *Pseudomonas fluorescens* and *P. putida* have shown promise in the field, no commercial products are available, most likely owing to the difficulty of making commercially stable formulations. More recently, Sharga and Lyon [11] identified a *Bacillus subtilis* isolate BS 107 that controlled *Ecc* and the closely related *E. carotovora* subsp. *atroseptica* (*Eca*), the causal agent of blackleg. Control was attributed to antibiosis. The potential for commercial product development is greater with bacilli owing to the presence of endospores. These provide considerable resistance to mortality caused by environmental fluctuations. *Erwinia carotovora* subsp. *betavasculorum* isolates have been demonstrated to suppress *Eca* by both antibiosis [12] and competition [13].

Control of postharvest soft rot is achieved by reduction of injuries associated with harvesting and handling (wounds serve as infection sites) and by initial storage conditions that allow the formation of infection-resistant cork layers in wounds. *P. putida* strain M17 was shown to control postharvest soft rot when applied as a seed piece treatment or as a postharvest treatment [14].

B. Blackleg

While blackleg can be caused by both *Ecc* and *Eca*, *Eca* is the more common pathogen in temperate production areas. *Eca* does not survive in soils for long periods and is spread almost exclusively by contaminated tubers. Two of the fluorescent pseudomonad strains shown to provide control of *Ecc* also demonstrated control of *Eca* [15] as did *B. subtilis* strain BS 107 [11]. While most biological control efforts have focused on rhizosphere colonists that suppress *Ecc* or *Eca*, endophytes from potato tubers belonging to the genera *Curtobacterium* and *Pantoea* have been shown to suppress *Eca* in vivo and in vitro [16]. These authors hypothesize that induced systemic resistance may be involved, although they presented no evidence. They did demonstrate that more resistant cultivars had higher populations of *Eca*-inhibitory bacterial isolates and had higher total numbers of endophytic bacteria.

C. Ring Rot

The fluorescent pseudomonads *P. aurefaciens* and *P. fluorescens* biovar III were shown to be inhibitory to the causal agent of ring rot, *Clavibacter michiganensis* subsp. *sepedonicus* in vitro and in vivo [17]. In glasshouse trials these authors showed both reduced infection and reduced populations of the pathogen when roots of potato seedlings were dipped in suspensions of the antagonistic bacteria. This disease is controlled primarily by pathogen-free certification programs.

D. Bacterial Wilt (Brown Rot)

Bacterial wilt is caused by *Ralstonia solanacearum* (formerly *Pseudomonas solanacearum*) and is controlled by use of pathogen-free certification programs, crop rotation, and resistant varieties. Unfortunately, resistant varieties are resistant to only a few of many strains of the pathogen. Kempe and Sequiera [18] demonstrated reduced levels of infection and disease severity when seed tubers were treated with either avirulent or incompatible strains of *R. solanacearum* or *P. fluorescens*. Induced resistance was hypothesized as the mechanism of action. Preemptive colonization was hypothesized as the mechanism of biological control by *Bacillus polymyxa* strain FU-6 and by *P. fluorescens* [19].

E. Scab

Scab is caused by several species of *Streptomyces* commonly found in soils worldwide. *S. scabies* is considered to be the most common species causing scab. While resistant cultivars are available, many susceptible cultivars are used because of their specific market characteristics. Management of this disease can be achieved in part by management of soil moisture during early tuber formation, maintenance of low soil pH, use of green or animal manures, and the use of seed treatment fungicides to reduce seedborne inoculum [20,21]. Maintenance of high soil moisture during early tuber development allows for higher populations of antagonistic bacteria and lower populations of actinomycetes on the tuber surface as compared to dry conditions [22]. They demonstrated that the effect of moisture was not on *S. scabies* since it infected equally well in inoculated dry or wet sterile soils. It is likely that the control achieved through the use of green or animal manures is due to increased microbial activity. Biological control of scab associated with increased populations of fluorescent pseudomonads was found when a swine feces–based nonantibiotic "actinomycete biofertilizer" was incorporated into scab-infested soil in Japan [23].

Menzies [24] described scab decline in several soils and based on steaming and soil transfer experiments concluded that a biological factor was involved.

This same situation was observed in Minnesota [25]. These researchers identified scab-inducing and -suppressing species and attributed biological control to antibiosis. Liu et al. [26] indicated that both competition and antibiosis were involved with suppressive streptomycete strains. The best strains were both antibiotic producers and were nonpathogenic. Pathogenicity of *S. scabies* is characterized by the production of the pathotoxin thaxtomin [27]. Using *S. scabies* strains that do not produce thaxtomin to compete for infection sites on young tubers has been proposed by Loria [20]. It is likely that biological control products will be developed for scab control.

II. DISEASES CAUSED BY FUNGAL PATHOGENS

A. Rhizoctonia Black Scurf

Rhizoctonia black scurf and canker is caused by *Rhizoctonia solani*, a common soil-inhabiting fungus with worldwide distribution. Losses are from cankers on stems and stolons that directly reduce yield and from the presence of the black sclerotia on tubers that can create marketing problems. Biological control by antagonistic organisms has been reported with *Verticillium biguttatum, Trichoderma harzianum, T. viride, T. hamatum, Gliocladium virens,* binucleate *Rhizoctonia,* fluorescent pseudomonads, and *Bacillus* sp. applied to the soil or to seed tubers [28–32]. Proposed mechanisms for biological control by these organisms include mycoparasitism, antibiosis, and competition. Results of these biological control efforts have been highly variable, and to date no commercial products have been developed.

B. Fusarium Dry Rot

Fusarium sambuccinum (Gibberella pulicaris) is the primary cause of Fusarium dry rot [33]. This fungus is commonly found in most soils where potatoes are grown. Infection takes place through injuries to the periderm. Losses are both from seed piece decay and decay in storage. Control is based on prevention of injuries, adjusting storage temperatures and humidity to allow wound healing in the 10–14 days after binning, and the use of postharvest applications of the fungicide thiabendazole [22]. Unfortunately, widespread resistance to thiabendazole has limited the effectiveness of the fungicide. Because infections only take place through injuries and tissues are susceptible to infection for a relatively short time, it is logical to assume that a biological control agent placed on fresh wounds would provide control. Bacteria including *P. syringae, P. corrugata, P. fluorescens, Enterobacter cloacae, Pantoea agglomerans, and Bacillus* sp. have shown excellent control when used as postharvest treatments [34–36]. Based on work by Kiewnick and Jacobsen, EcoScience Inc. labeled two products BioSave

110 and BioSave 1000 for Fusarium dry rot control and silver scurf control. The products are based on *P. syringae* strain ESC-11 and are also used for control of postharvest decay on fruits. These products are formulated as frozen pellets to preserve viability of bacteria, which makes their use somewhat inconvenient. The *Bacillus* formulations used by Kiewnick and Jacobsen are dried powder formulations which have good storability and ease of handling. These *Bacillus* isolates are compatible with thiabendazole and have shown control of thiabendazole-resistant isolates.

C. Silver Scurf

Silver scurf is another important postharvest disease caused by the fungus *Helminthosporium solani*. Losses are the result of market defects caused by skin discoloration and sloughing and by increased moisture losses by infected tubers in storage [37]. This disease is spread primarily by infected tubers, although infested soils may play a role [38]. The disease is controlled by selection of resistant cultivars, seed treatment fungicides, and postharvest fungicide treatments [38]. However, resistance to the postharvest applied fungicide thiabendazole is widespread [39].

The potential for development of biological controls for silver scurf was shown by Adams et al. [40], who reported that soils high in bacterial counts reduced incidence of silver scurf. The first paper to deal with a specific biological control was published by Chun and Shetty [41], who demonstrated that a strain of *Pseudomonas corrugata* could reduce silver scurf severity when applied as a postharvest treatment and in addition reduced secondary transfer to daughter tubers from 18.6% to 2.7% under glasshouse conditions. Elson et al. [37] published a study of 430 bacteria, yeasts, and actinomycetes from 47 agricultural soils and 7 tuber samples relative to biological control of *H. solani*. These authors identified 12 soils as suppressive to development of the silver scurf disease and identified a *P. putida* isolate, a *Nocardia globerula* isolate, and a *Xanthomonas campestris* isolate as providing biological control in their laboratory test system. These antagonists were not good periderm colonists, and antibiosis was hypothesized as the mechanism of action.

D. Verticillium Wilt

The causal agents of Verticillium wilt are *Verticillium dahliae* and *Verticillium albo-atrum*. These fungi are common soil inhabitants wherever potato or other susceptible crops are grown. The host range is extremely large, with representation in more than 50 species of plants in 23 families. They survive in the soil as microsclerotia (*V. dahliae*) or as dark thick-walled mycelia (*V. albo-atrum*). Interactions with root lesion nematodes, *Pratylenchus* sp., create the disease potato early dying. Verticillium wilt and the early dying diseases are controlled by planting pathogen-free seed, crop rotation, the use of soil fumigants such as metham sodium, irrigation management, and planting resistant varieties [21]. Davis et al. [42] demonstrated that Sudan grass, oat, corn, and rape green manures reduced root infection by *V. dahliae* and that Sudan grass green manures actually reduced soil populations of *V. dahliae.* Suppressiveness was associated with increased microbial activity in soil. *Fusarium equiseti* populations and root colonization were increased by the green manures and were associated with disease suppression.

Biological control research has focused on the use of mycoparasites. Fravel et al. [43] demonstrated control of early dying using the fungus *Talaromyces flavus.* This fungus was shown to colonize potato roots from treated seed pieces and to colonize *V. dahliae* microsclerotia [44]. Results with this fungus have been inconsistent [45]. *Gliocladium roseum* was shown to reduce microsclerotial viability in a range of soils and water matrix potentials in laboratory assays [46]. The mechanism of action is thought to involve both mycoparasitism and production of antibiotic compounds. Based on the research done to date, it seems that broadcast preplant treatments with these mycoparasites will be more effective than planting time or seed tuber treatments.

III. DISEASES INDUCED BY NEMATODES

A wide range of nematodes attack potato, and management involves crop rotation, use of nematicides, resistant varieties, and green manures [47,48]. The majority of research has focused on root knot (*Meloidogyne chitwoodi, M. hapla, M. incognita, M. javanica,* and *M. arenaria*) and cyst nematodes (*Globodera pallida* and *G. rostochiensis*). Historically, the addition of animal or organic manures has been used to manage plant parasitic nematodes. The addition of manures increases microbial activity, and when amendments contain high amounts of ammonical nitrogen they release high amounts of NH_3, which is directly toxic to nematodes [48]. Increases in microbial populations include those that are antagonistic to nematodes. In related work, Rodriquez-Kabana et al. [39,49] demonstrated control of root knot nematodes by amending soils with chitin and chitin-urea mixtures. Chitin directly affects microbial communities [50,51]. Incorporation of large amounts of neem, castor, or groundnut oil cake have also been shown to provide control [52].

Biological control of nematodes that can infect potatoes has been reviewed by Jatala [48,53]. Numerous experiments have shown control by fungi (*Arthobotrys irregularis, Paecilomyces lilacinus, Verticillium chlamydosporium,* and many others) and by bacteria including *Bacillus penetrans* and several rhizobacteria [54]. The fungi are generally direct parasites on eggs or adult females. Two fungal products have been developed. These are *A. irregularis*-Royal-350 and *P. lilacinus*-Biocon. These products have given variable control and are most effective when used in conjunction with other control measures. Bacteria such as *B. penetrans* may be direct parasites or act by antibiosis, competition, or as inducers of induced systemic resistance (ISR) [55]. In this later paper, Rietz et al. demonstrated that *Rhizobium etli* strain G12 provides control of *G. pallida* by ISR. Reduced infection and ISR was associated with lipopolysaccharides produced by the bacteria, and the ISR was not associated with typical pathogenesis-related proteins.

IV. CONCLUSION

Biological control of potato pathogens has shown promise for a number of diseases. Research in this area has expanded greatly since 1974 at which time the Agricola database shows 278 publications. This same database shows 829 publications addressing biological control of plant pathogens between 1995 and 1999 [1]. However, the direct use of specific antagonists is limited by the paucity of products available. Product availability is limited by development of stable formulations, production of the antagonist, and the cost of introducing effective quantities of an antagonist into the soil. Research on mode of action and integration of biological controls with other control measures will certainly provide new products in the future.

Another approach is the use of transgenic disease resistance. Monsanto Inc. has pioneered this approach with the introduction of their trademarked New Leaf line of potatoes. These are primarily established varieties carrying transgenes for virus coat proteins (potato leaf roll virus and potato virus Y) or transgenes for resistance to Verticillium wilt or late blight. Unfortunately, due to problems associated with consumer acceptance of genetically modified crops, these cultivars are not available, and the New Leaf business unit has been dissolved until consumer acceptance issues are resolved. In 1998 Lorito et al. [56] reported the transformation of potato with the *ThEn-42* gene from *Trichoderma harzianum*. This gene encodes for an endochitinase, and transformed plants exhibit resistance to early blight (*Alternaria solani*), *Botrytis cinerea*, and *Rhizoctonia solani*. The concept of using genes from anatagonistic organisms may be useful against many other pathogens.

As pesticide regulation and cost reduces pesticide availability, biological controls will become more important. At that time products will be developed that provide control by antibiosis, competition, parasitism, or ISR or a combination of these mechanisms.

REFERENCES

1. CT Bull. Biological control. In: OC Malloy, TD Murray, eds. Encyclopedia of Plant Pathology, Vol. 1. New York: J. Wiley and Sons 2001, pp. 128–135.
2. RJ Cook, KF Baker. The Nature and Practice of Biological Control of Plant Pathogens. St. Paul, MN: APS Press, 1983.
3. MCM Perombelon, A Kelman. Blackleg and other potato diseases caused by soft rot *Erwinias*: proposal for revision of terminology. Plant Dis 71:283–285, 1987.
4. NJ Mccarter-Zorner, GD Franc, MD Harrison, MDJE Michaud, CE Quinn, DC Graham. Soft rot *Erwinia* bacteria in surface and underground waters in southern Scotland and in Colorado, United States. J Appl Bacteriol 57:95–105, 1984.
5. JW Kloepper. Effect of seed piece inoculation with plant growth-promoting rhizobacteria on populations of *Erwinia carotovora* on potato roots and daughter tubers. Phytopathology 73:217–219, 1983.
6. JW Kloepper, MN Schroth. Development of a powder formulation of rhizobacteria for inoculation of potato seed pieces. Phytopathology 71:590–592, 1981.
7. JW Kloepper, J Leong, M Teintze, MN Schroth. Enhanced plant growth by siderophores produced by plant growth-promoting rhizobacteria. Nature 286:885–886, 1980.
8. TV Suslow, MN Schroth. Role of deleterious rhizobacteria as minor pathogens in reducing crop growth. Phytopathology 72:111–115, 1982.
9. WJ Howie, E Echandi. Rhizobacteria: influence of cultivar and soil type on plant growth and yield of potato. Soil Biol Biochem 15:127–132, 1983.
10. GW Xu, DC Gross. Selection of fluorescent pseudomonads antagonistic to *Erwinia carotovora* and suppressive of potato seed piece decay. Phytopathology 76:414–422, 1986.
11. BM Sharga, GD Lyon. *Bacillus subtilis* BS 107 as an antagonist of potato blackleg and soft rot bacteria. Can J Microbiol 44:777–783, 1998.
12. PE Axelrood, M Rella, MN Schroth. Role of antibiosis in competition of *Erwinia* strains in potato infection courts. Appl Environ Microbiol 54:1222–1229, 1988.
13. JM Costa, J Loper. Derivation of mutants of *Erwinia carotovora betavasculorum* deficient in export of pectolytic enzymes with potential for biological control of potato soft rot. Appl Environ Microbiol 60:2278–2285, 1994.
14. PD Colyer, MS Mount. Bacterization of potatoes with *Pseudomonas putida* and its influence on postharvest soft rot diseases. Plant Dis 68:703–706, 1984.
15. DJ Rhodes, C Logan. Effects of fluorescent pseudomonads on the potato blackleg syndrome. Ann Appl Biol 108:511–518, 1986.
16. AV Sturz, BG Matheson. Populations of endophytic bacteria which influence host-resistance to *Erwinia*-induced bacterial soft rot in potato tubers. Plant Soil 184:265–271, 1996.
17. AR De La Cruz, AR Poplawsky, MV Wiese. Biological suppression of potato ring rot by fluorescent pseudomonads. Appl Environ Microbiol 58:1986–1991, 1992.
18. J Kempe, L Sequiera. Biological control of bacterial wilt of potatoes: attempts to induce resistance by treating tubers with bacteria. Plant Dis 67:499–503, 1983.
19. RB Aspiras, de la Cruz. Biocontrol of bacterial wilt in tomato and potato through pre-

emptive colonization using *Bacillus polymyxa* FU-^ and *Pseudomonas fluorescens.* Philipp J Crop Sci 11:1–4, 1986.

20. R Loria. Common scab and acid scab of potato: management, etiology and potential uses of phytotoxins produced by *Streptomyces* species. In: GW Zehnder, MK Powelson, RK Jansson, KV Raman, eds. Advances in Potato Pest Biology and Pest Management. St. Paul, MN: APS Press, 1994, pp. 149–154.

21. ML Powelson, KB Johnson, RC Rowe. Management of diseases caused by soilborne pathogens. In: RC Rowe, ed. Potato Health Management. St. Paul, MN: APS Press, 1993, pp. 149–158.

22. MJ Adams, DH Lapwood. Studies on lenticle development, surface microflora and infection by common scab (*Streptomyces scabies*) on potatoes. Ann Appl Biol 90: 335–343, 1978.

22. MJ Adams, DH Lapwood. Studies on lenticle development, surface microflora and infection by common scab (*Streptomyces scabies*) on potatoes. Ann Appl Biol 90: 335–343, 1978.

23. N Nanri, Y Gohda, M Ohno, K Miyabe, K Furukawa, S Hayashida. Growth promotion of fluorescent pseudomonads and control of potato common scab in field soil with non-antibiotic actinomycete-biofertilizer. Biosci Biotech Biochem 56:1289–1292, 1992.

24. JD Menzies. Occurrence and transfer of a biological factor in soil that suppresses potato scab. Phytopathology 49:648–652, 1959.

25. JM Lorang, D Liu, NA Anderson, JL Schottel. Identification of potato scab inducing and suppressive species of *Streptomyces*. Phytopathology 85:261–268, 1995.

26. D Liu, NA Anderson, LL Kinkel. Selection and characterization of strains of *Streptomyces* suppressive to the potato scab pathogen. Can J Microbiol 42:487–502, 1996.

27. RR King, CH Lawrence. Characterization of new thaxtomin A analogs generated in vitro by *Streptomyces scabies*. J Agric Food Chem 44:1108–1110, 1996.

28. G Jager, H Velvis, LG Lamers, A Mulder, J Roosjen. Control of *Rhizoctonia solani* in potato by biological, chemical and integrated measures. Potato Res 68:269–284, 1991.

29. JE Beagle-Ristaino, GC Papavizas. Biological control of Rhizoctonia stem canker and black scurf of potato. Phytopathology 75:560–564, 1985.

30. AR Escande, E Echandi. Protection of potato from Rhizoctonia canker with binucleate *Rhizoctonia* fungi. Plant Pathol 40:197–202, 1991.

31. S Kiewnick, BJ Jacobsen. Control of Rhizoctonia black scurf and Fusarium dry rot in potatoes with fungicides and antagonistic bacteria (abstr). Phytopathology 87: S51#P-1997-0360-AMA, 1997.

32. BJ Jacobsen, S Kiewnick. Novel fungicides and biocontrol agents for control of Rhizoctonia black scurf on potato (abstr). Am J Potato Res 75:280, 1998.

33. AEW Boyd. Potato storage diseases. Rev Plant Pathol 51:297–321, 1972.

34. DA Schisler, PJ Slininger. Selection and performance of bacterial strains for biologically controlling Fusarium dry rot of potato incited by *Gibberella pulicaris*. Plant Dis 78:251–255, 1994.

35. DA Schisler, PJ Slininger, RJ Bothast. Effects of antagonistic cell concentration and two-strain mixtures on biological control of Fusarium dry rot of potatoes. Phytopathology 86:177–183, 1996.

36a. S Kiewnick, BJ Jacobsen. Control of Fusarium dry rot on potato with antagonistic bacteria (abstr). Am J Potato Res 75:284, 1998.

36b. S Kiewnick, BJ Jacobsen. Biological control of Fusarium dry rot of potato with antagonistic bacteria in commercial formulation (abstr). Phytopathology 88:S 47, 1998.

37. MK Elson, DA Schisler, RJ Bothast. Selection of microorganisms for biological control of silver scurf (*Helminthosporium solani*) of potato tubers. Plant Dis 81: 647–652, 1997.

38. GA Secor. Management strategies for fungal diseases of tubers. In: GW Zehnder, MK Powelson, RK Jansson, KV Raman, eds. Advances in Potato Pest Biology and Pest Management, St. Paul, MN: APS Press, 1994, pp. 155–165.

39. R Rodriquez-Kabana, D Boube, RW Young. Chitinous materials from blue crab for control of root-knot nematode. II. Effect of soybean meal. Nematropica 20:153–168, 1990.

40. A Adams, N Sandar, DC Nelson. Some properties of soils affecting russet scab and silver scurf of potatoes. Am Potato J 47:49–57, 1970.

41. WWC Chun, KK Shetty. Control of the silverscurf disease of potatoes caused by *Helminthosporium solani* Dur. and Mont. with *Pseudomonas corrugata* (abstr). Phytopathology 84:1090, 1994.

42. JR Davis, OC Huisman, DT Westerman, LH Sorensen, AT Schneider, JC Stark. The influence of cover crops on the suppression of Verticillium wilt of potato. In: GW Zehnder, MK Powelson, RK Jansson, and KV Raman, eds. Advances in Potato Pest Biology and Pest Management. St. Paul, MN: APS Press, 1994, pp. 332–341.

43. DR Fravel, JR Davis, LH Sorenson. Effect of *Talaromyces flavus* and metham on Verticillium wilt incidence and potato yield, 1984–1985. Biol Cult Cont Plant Dis 1:17, 1986.

44. NPM Nagtzaam, GJ Bollen. Colonization of roots of eggplant and potato by *Talaromyces flavus* from coated seed. Soil Biol Biochem 29:1499–1507, 1997.

45. DS Spink, RC Rowe. Evaluation of *Talaromyces flavus* as a biological control agent against *Verticillium dahliae* in potato. Plant Dis 73:230–236, 1989.

46. AP Keinath, DR Fravel, GC Papavizas. Potential of *Gliocladium roseum* for biocontrol of *Verticillium dahliae*. Phytopathology 81:644–648, 1991.

47. WJ Hooker, ed. Compendium of Potato Diseases. St. Paul, MN: American Phytopathological Soc, 1983.

48. P Jatala. Biology and management of nematode parasites of potato in developing countries. In: GW Zehnder et al., eds. Advances in Potato Pest Biology and Pest Management. St Paul, MN: APS Press, 1994, pp. 214–233.

49. R Rodriquez-Kabana, D Boube, RW Young. Chitinous materials from blue crab for control of root-knot nematode. I. Effects of urea and enzymatic studies. Nematropica 19:53–74, 1989.

50. GR Godoy, R Rodriquez-Kabana, RA Shelby, G Morgan-Jones. Chitin amendments for control of *Meloidogyne arenaria* in infested soil. II. Effects on microbial population. Nematropica 13:63–74, 1983.

51. JR Hallman, R Rodriquez-Kabana, JW Kloepper. Chitin-mediated changes in bacterial communities of the soil, rhizosphere and within roots of cotton in relation to nematode control. Soil Biol Biochem 31:551–560, 1999.

52. RS Singh, K Sitaramaiah. Incidence of root-knot of okra and tomato in oil cake amended soil. Plant Dis Rptr 50:668–672, 1966.
53. P Jatala. Biological control of plant-parasitic nematodes. Annu Rev Phytopathol 24: 453–489, 1986.
54. J Racke, RA Sikora. Isolation, formulation and antagonistic activity of rhizobacteria towards potato cyst nematode *Globodera pallida*. Soil Biol Biochem 24:521–526, 1992.
55. M Reitz, K Rudolph, I Schroder, S Hoffmann-Hergarten, J Hallmann, RA Sikora. Lipopolysaccharides of *Rhizobium etli* strain G12 act in potato roots as an inducing agent of systemic resistance to infection by the cyst nematode *Globodera pallida*. Appl Environ Microbiol 66:3515–3518, 2000.
56. M Lorito, SL Woo, IG Fernandez, G Colucci, GE Harmon, JA Pintor-Toro, E Filippone, S Muccifora, CB Lawrence, S Tuzun, F Scala. Genes from mycoparasitic fungi as a source for improving plant resistance to fungal pathogens. Proc Natl Acad Sci 95:7860–7865, 1998.

10
Biological Control of Soybean Diseases

Terry R. Anderson
Agriculture and Agri-Food Canada, Harrow, Ontario, Canada

I. INTRODUCTION

World soybean production has increased dramatically during the past 50 years. The increase has been facilitated by several factors, including market demand for soybean byproducts, adaptation and development of high-yielding varieties, and improvement in production practices and equipment. Soybean is highly adaptable, and the crop is grown in all areas from the tropics to very short-season areas in northern China and Canada. The expansion in production is the result of new areas of production, e.g., Argentina and Brazil, and expanded production and more frequent cropping of traditional areas, e.g., the United States and China. During the past 25 years there has been less crop rotation and monoculture has increased in many areas. Decreased rotation has resulted in increased disease problems, and even in new soybean production areas in South America, there have been enough crop cycles to promote an increase in soybean diseases.

More than 100 pathogens of soybean caused by viruses, bacteria, fungi, and nematodes have been identified [1]. The importance of individual diseases varies with geographic location; however, most diseases are found to some extent in all soybean-producing areas. There are a few exceptions; soybean rust, which is primarily a problem in Asia, has not been found in the north central U.S. production belt, and stem canker, which is important in the Western Hemisphere, is not considered important in Asia.

For the purposes of this chapter, biological control is nonchemical control that includes variety resistance, cultural practices, and exclusion. Soybean production has not relied on classic biological control involving the direct application of a biological agent to control disease, although there are a few examples and opportunities to incorporate classical methods with numerous soybean diseases.

Because of the large number of diseases of soybean, it is not possible to discuss controls for each one; therefore, the most important diseases will be considered in this chapter. Disease loss estimates of the 10 top soybean-producing countries [2] and additional information on diseases worldwide in 1998 provide the best available assessment of the importance of diseases in each area of the world.

The 10 most important diseases affecting world production in descending order are:

1. Soybean cyst nematode (SCN) (*Heterodera glycines* Ichinohe)
2. Brown spot (*Septoria glycines* Hemmi.)
3. Charcoal root rot (*Macrophomina phaseolina* [Tassi] Goid)
4. Sclerotinia stem rot (*Sclerotinia sclerotiorum* [Lib.] de Bary)
5. Purple seed stain (*Cercospora kikuchii* [T. Matsu. and Tomoyosu] Gardener)
6. Stem canker (Diaporthe spp.)
7. Viruses
8. Phytophthora root rot (*Phytophthora sojae* [Kaufmann and Gerdeman]
9. Pod and stem blight (*Phomopsis longicolla* [Hobbs] and *Diaporthe phaseolorum* [Cke. and Ell.] Sacc. var. sojae [Lehman] Wehm)
10. Sudden death syndrome (*Fusarium solani* f. sp. *glycines* [Roy])

Losses caused by SCN are more than double the next most important pathogen. Certain diseases are important locally, and some are widely spread and increasing in severity.

II. SOYBEAN CYST NEMATODE (*HETERODERA GLYCINES* ICHINOHE)

A. Resistance

Resistant cultivars are used in the United States, China, Argentina, and Brazil and are the main component of any management strategy. Resistance is controlled by several single genes, some of which are closely linked. In addition to single dominant and recessive genes, soybean varieties exhibit varying degrees of tolerance [3]. Linked single genes, tolerance, and variation in SCN screening populations add to the complexity of developing resistant cultivars; however, numerous high-yielding resistant lines are now available. The main sources of resistance in the United States are Peking, PI90763, PI89772, Cloud, PI209332, PI88788, 87631-1, and PI 437654. Resistance inhibits development and maturation of SCN by preventing the establishment of a feeding site but does not prevent infection. Both conventional and molecular methods are used to identify new genes for resistance [5–8]. Plant introductions (PIs) and cultivars have been screened for novel sources of resistance [4,8–10]. A cultivar is considered resistant if it has 90% fewer cysts than a susceptible control such as cv. Lee. Although resistance is not complete, it is effective in reducing yield losses and reducing soil populations of SCN. Soybean cultivars differ in tolerance to SCN. Reproduction of SCN occurs on roots of tolerant cultivars, but yield is not suppressed [11,12].

B. Management

Crop rotation with nonhosts is frequently used to control nematodes and other crop pests. An excellent summary of current and potential management practices for SCN has recently been published [13]. Rotation studies were initiated in North Carolina shortly after SCN was identified there in 1954. Preliminary studies suggested that nonhosts such as cowpea were effective in reducing damage [14]. In Minnesota, moderate development of SCN occurred on adzuki bean and low levels developed on peas [15]. Numerous additional experiments have been conducted since that time, which included resistant varieties once they became available. Certain resistant cultivars such as Bedford have been grown continuously without increasing the soil populations of SCN; however, this practice was not recommended [16]. More recently rotation of both resistant and susceptible cultivars with a nonhost crop was found to produce higher yields. Rotation with corn improved the yields of susceptible cultivars on SCN-infested soil [17]. This was confirmed in a later study in a field infested with both *Meloidogyne* spp. and SCN [18]. Crop rotation with corn, cotton, and rice is used in Brazil, and no-tillage with a fall-spring cover crop is also recommended in Brazil [19], Argentina [20], and the United States [21]. A 3- to 5-year rotation is recommended in China [22].

Balanced fertility levels will reduce crop stress and consequently the effects of SCN. Application of fertilizer has not been adequately studied; however, foliar deficiencies occur when plant roots are invaded by SCN. Application of potassium chloride did not increase the yield of soybeans in SCN-infested soil but did increase the foliar potassium content of the SCN-susceptible cultivar Essex [23]. Application of complete fertilizer, ammonium nitrate, superphosphate or muriate of potash did not affect SCN numbers or yield in Alabama [24].

Organic amendments in the form of swine manure compared to NPK fertilizer with similar analysis were compared for effects on SCN populations and soybean yields. The treatments were applied in-furrow, between rows, and broadcast prior to planting. SCN eggs and second stage juveniles were sampled before planting, at midseason, and after harvest. Swine manure in-furrow reduced egg hatch compared to the controls as indicated by midseason and final egg counts. End-of-season egg densities were greater in soybean plots treated with manure in-furrow, and yields were increased. Manure may have increased root growth and subsequently increased reproduction of SCN later in the season [25].

Zinc ions have been shown to stimulate hatching of SCN [26]; however, attempts to influence hatching in soil by addition of zinc have been unsuccessful. Zinc was applied to corn at incremental rates in Iowa prior to planting soybeans. Although commercial zinc sulfate fertilizer stimulated hatch in vitro, no response was noted under field conditions possibly because the zinc ions were adsorbed to clay particles [27].

The effect of planting date on yield loss caused by SCN may vary with geographical location and soil temperature, which affect SCN activity. In Missouri, yields of soybeans planted in late May were higher in 1985 than yields of soybeans planted earlier or later. In 1986 there was no significant difference [28]. Early season cultivars planted late tend to reduce SCN soil populations compared to long-season cultivars planted early [29].

Tillage practices have shifted from conventional or complete tillage to minimum or no-till. This shift has occurred for conservation and economic reasons, and the effects of reduced tillage on plant diseases are still under investigation. The effect of tillage on survival of SCN has not been resolved because there are many factors involved such as the nematode population, soil type, crop sequence, location, and varieties. Results from an 3-year tillage-rotation study suggest no-till was positive or neutral to soybean yield early in the study but detrimental in the final years of the study mainly because of weed problems. Numbers of SCN fluctuated in an unpredictable manner from year to year but declined when a nonhost was planted [30]. In another study, the effects of wheat, minimum tillage, and no-tillage on SCN populations and yield of a susceptible soybean cultivar were assessed. Tillage did not affect SCN populations as much as wheat residue, which reduced SCN cyst populations at harvest. It was concluded that no-till double crop soybean following wheat can reduce SCN numbers [31]. In a major 2-year study involving 18 sites in the north central U.S. states, it was concluded that resistant soybean varieties increased yield at 88% of the sites, no-tillage plots had higher yield than conventional tillage plots at 55% of sites in 1998, and SCN populations did not increase with increasing row spacing [32].

C. Biological Control

SCN is one major pest of soybean that has generated significant research surrounding biological control. A number of potential biocontrol organisms for cyst nematodes have been identified, including fungi [33–36] and the bacteria *Pasteuria penetrans* [37]. A summary of research and overview of the requirements for a successful SCN biocontrol agent have been suggested [38]. Most researchers recognize the need to alter field conditions that could promote activity of those agents already present in soybean fields rather than direct applications of a new biocontrol agent.

Verticillium lecanii parasitizes eggs of SCN. In greenhouse experiments, benomyl-resistant mutants significantly reduced SCN cyst numbers when applied at a high rate in alginate prills [39]. Alginate prills may contribute to a suitable delivery system because the fungus remains viable for extended periods [40]. *V. lecanii*, an SCN sex pheromone (vanillic acid), and chemical analogs reduced mid-season cyst counts compared to controls. In microplot tests at harvest, cyst numbers were lowest with *V. lecanii* and with vanillic acid treatments. The treatments were shown to have potential for inclusion in SCN management schemes [41].

A nonsporulating fungus (ARF18) isolated from soils in Arkansas and the mid-central United States has received attention as an active parasite of SCN cysts. The distribution in field soils and characterization of isolates of this fungus were studied recently. Sterile ARF 18 was isolated from eggs and cysts in Kentucky, Louisiana, Mississippi, and Tennessee. Isolates could be divided into two groups by colony morphology and further characterized by restriction fragment length polymorphisms (RFLPs) of mitochondrial DNA. One morphological group, ARF18-C, was found to be more aggressive in parasitizing SCN eggs [42].

A number of major studies have been made to identify fungi associated with SCN eggs, cysts, and juveniles. The fungi were isolated from SCN in field soil in different areas of the United States [34,36,43,44], Columbia [45], and Brazil [46].

Sixty-one species of fungi and 20 nonsporulating fungi were identified from infective juveniles, eggs, or cysts in Tennessee under six cropping treatments involving wheat double cropping and reduced or no tillage. *Fusarium solani* was the most frequently isolated of 47 species infecting cysts. *F. solani* and *F. oxysporum* were the most frequent of 20 species isolated from field females, and *F. oxysporum* was more frequent isolated from greenhouse-grown females. *Paecilomyces lilacinus* was most common from eggs. The percentage of parasitized eggs and females was similar for all treatments; however, incidence of *P. lilacinus* increased with tillage (discing) and *V. chlamydosporium* increased with ploughing. A comparison of the fungal species isolated in this and previous studies indicated that sites could differ in the kinds of fungi associated with cysts and eggs, but many of the fungi overlapped in distribution. The authors point out several factors to be considered, and the presence of the fungus does not ensure it will be a successful biocontrol agent [44].

In Florida, 18 fungal species were evaluated for pathogenicity to eggs of SCN on water agar. The highest potential for biocontrol activity was shown by *Verticillium chlamydosporium, Pyrenochaeta terrestris*, and two sterile fungi. Several other soil fungi were moderately pathogenic to eggs. In greenhouse tests, plant weights and heights generally increased in soil treated with the fungi [36]. It was recently demonstrated that in vitro tests of biocontrol agents may not be as suitable as heat-treated soil to evaluate aggressiveness [47].

In Brazil, *F. solani, F. oxysporum, Stagnospora* sp., and *Gliocladium* sp. were found infecting eggs of SCN [46].

Pasteuria nishizawa is a very promising biocontrol agent, which was first identified in Japan parasitizing *Heterodera* spp. [48], and cultures of *Pasteuria* sp. were subsequently recovered from SCN field plots in Illinois [37]. Endospores of the bacteria were only attached to SCN and not other nematodes in the samples, indicating the bacteria may be species specific. Although *Pasteuria* spp. are obligate parasites, new techniques for extracting and quantifying the organism will facilitate future research [49].

The effect of delta-endotoxins from *Bacillus sphaericus, B. thuringiensis* var. *israelensis*, and *B. thuringiensis* var. *kurstaki* on oviposition and juvenile hatching was studied in Brazil. The egg production of SCN varied with the amount of toxin of *B. sphaericus* applied to soil. This toxin also reduced hatching of SCN in the absence of the host plant [50].

III. BROWN SPOT

A. Resistance

There are no known sources of resistance, but varieties differ in susceptibility to *Septoria glycines* and efforts have been made to select better lines [51,52]. Regenerated soybean plants exposed to the host specific toxin from culture filtrates varied in response to the disease in the field. Disease development was delayed in some lines for up to 5 weeks. In the next generation, only seed from BSR201 continued to express resistance [53]. In another study, regenerated lines from soybean calli had intermediate and susceptible responses. The heritability of resistance (23%) was low. The progress in developing resistant lines using these techniques is difficult but promising [54].

B. Management

In a maximum yield trial with surface irrigation, it was noted that disease severity varied with genotype but not specifically plant type (determinant vs. indeterminant). Yield increases were greater in 17 cm rows than 75 cm rows. With surface irrigation, brown spot was considered an important component of yield. Yield losses of between 7 and 21% were noted in a high-yield environment [55] and 14.1% when inoculated with *Pseudomonas glycinea* [56]. In inoculated plots losses of 13–30% for the variety Wells and 16–22% for cv. Williams were observed, while noninoculated plots had 8% yield loss [57].

IV. CHARCOAL ROT

A. Resistance

In the United States the soybean varieties Delta-Pineland 3478, Hamilton, and Jackson II are considered moderately resistant and Davis and Asgrow 3715 are considered tolerant to charcoal rot in field trials [61]. From studies involving maturity groups II, III, IV, and V, SCN resistance, and planting date, it was concluded that SCN-resistant varieties did not have higher yields than SCN-susceptible varieties when charcoal rot was present and later maturing cultivars that were planted late had less disease and greater yields. The increased yield in later cultivars was attributed to less heat and drought stress during the period of pod fill. In Kansas, *M. phaseolina* colonization of maturity group V varieties was less than in earlier varieties [62]. These results agree with a previous study that resulted in reduced charcoal rot by planting full season varieties and found that there was no advantage to planting blended soybean lines [63].

B. Management

Charcoal rot occurs in seedlings [58] and in dry soil [59]. The effect of tillage on the incidence and severity of charcoal rot is unclear. Minimum or no-tillage with a cover crop is recommended for control of charcoal rot in Brazil [19], but the effect of tillage in the United States is unknown or inconclusive [21]. In Tennessee, disc-tillage, no-tillage, and moldboard plough treatments were compared for their effect on populations of *M. phaseolina* in soil and soybean roots. At the conclusion of the experiment, there were higher populations of *M. phaseolina* in the 0–7.5 cm layer of soil in minimum tillage plots than either the disc or moldboard plough treatments. Root segments were equally infected in all treatments, and there was no correlation between *M. phaseolina* population and yield [64]. Tillage practices that conserve moisture in areas with limited rainfall may reduce the incidence of disease by reducing stress.

The benefits of mycorrhizal associations with plants is well known. The effect of arbuscular mycorrhizal fungi on *H. glycines* and *M. phaseolina* was studied in SCN-infested and noninfested fields. In the presence of SCN, SCN-susceptible varieties had an eightfold increase in colonization by *M. phaseolina* compared to SCN-resistant varieties. There were no differences in charcoal rot between resistant and susceptible varieties in the absence of SCN. SCN infection and development contributed significantly to the severity of charcoal rot [65].

In general, charcoal rot will be reduced by any cultural practices that reduces stress on the soybean crop, especially during pod fill. Improperly applied herbicides can contribute to stress and increase the severity of charcoal rot [66]. Rotation is recommended to reduce inoculum, but this practice will not eliminate the disease because of the longevity of microsclerotia [60] and the potential for reproduction on numerous hosts.

V. SCLEROTINIA STEM DECAY

A. Resistance

Soybean varieties differ in response to inoculation with *S. sclerotiorum* [68] and in the incidence and severity of disease in the field; however, genes for resistance have not been identified. At least six tolerant varieties are available in Argentina [20]. In the north central soybean production area of the United States, S19-90 and Corsoy 79 are considered the most resistant varieties [68] and are used where possible as resistant controls in most screening trials. NKD 19-90 (S19-90), A2506, Colfax, and Corsoy 79 were rated most resistant in recent Illinois field trials [69]. In the short-season area of Canada, Maple Arrow, Ace, Maple Presto, and McCall were found most resistant [70]. Maple Presto and McCall were found resistant in the northern United States [71]. Evaluation of resistance or tolerance has been difficult because early-maturing cultivars frequently escape the disease [72], so it is important to compare varieties with the same maturity requirements. Varieties differ in the number of sclerotia produced per diseased plant and per square meter, which could greatly affect inoculum levels in future crops [72]. Recently, a major project was initiated to screen plant introductions (PIs) and varieties in several northern U.S. states for resistance to Sclerotinia stem decay (C. Grau, University of Wisconsin, personal communication). Transgenic soybeans containing the oxalate oxidase (OXO) gene from wheat have been developed which have higher resistance than S19–90 [73].

It has been suggested that widespread use of susceptible cultivars such as Williams and A3127 in breeding programs may have contributed to disease susceptibility in contemporary cultivars [69].

B. Management

Weed control in soybean fields is important because *S. sclerotiorum* has a very broad host range [67]. Minimum or no-tillage can reduce losses from Sclerotinia stem decay compared to complete tillage in soybean fields [21] perhaps by exposing sclerotial inoculum to weather. Minimum tillage and a thick layer of organic residue prevents emergence and elongation of the apothecial stipe in Brazil [74,75]. Irrigation can influence the incidence of several soybean diseases. Overhead irrigation during the flowering period increases the incidence of white mold in soybean field trials [76]. In Brazil, overhead irrigation in seed production fields has resulted in increased white mold and the spread of the disease with seed

(L. C. B. Nasser, Cerrados Agricultural Research Centre, EMBRAPA, Planaltina, Brazil, personal communication).

The effect of row spacing on the incidence of Sclerotina stem rot has never been completely defined. Mold increased slightly but not significantly as row spacing decreased from 69 to 45 to 23 cm and yield increased [72]. In Wisconsin, disease severity was greater at 25 and 38 cm than at 76 cm in 2 of 3 years and yield was greater at 76 cm [76]. It appears that other factors affect the incidence of mold more than row spacing.

Early-maturing cultivars tend to have less mold, possibly because there is less canopy closure or a shorter flowering period. Planting high-yielding varieties that mature early is a feasible method to avoid disease and the build-up of inoculum in the field [72]. This strategy is most useful when there are a range of cultivars available that differ in maturity group and growing conditions allow selection in planting date.

Crop rotation with nonhosts such as corn or small grain is an important strategy to help reduce soilborne inoculum, although in short rotations this management strategy seldom eliminates the disease. Rotation with crops that are more susceptible to white mold than soybean, such as dry edible bean or sunflowers, should be avoided.

Although soybean seed infection by *S. sclerotiorum* has been known to occur for some time [77,78], recent recognition of the extent to which seedborne infection can occur has put new emphasis on this means of dispersal. Seed from diseased fields may be accompanied by sclerotia or may have internal infection, which is capable of establishing the disease in new areas. Seed infection was found to range from 0.3 to 0.7% in Illinois depending on variety [79]. Sclerotia were recovered from 71 of 81 seed lots suspected of being contaminated, and internal seed infection ranged from 0.07 to 0.1% in another Illinois study [80]. Most infected seed fail to germinate, but sclerotia develop on the seed after planting. In field experiments, many sclerotia and some apothecia were produced after infected seed was covered with soil. Treatment with a mixture of fungicide seed treatments applied prior to planting can reduce the incidence of sclerotia by 98% [81].

C. Biological Control

Sclerotinia stem rot is one of the few diseases of soybean that has good potential for control involving application of a biological control agent. Because of a broad host range and worldwide distribution, extensive research has been conducted with numerous crops and *Sclerotinia* spp. Most results that focus on control of the overwintering sclerotia are applicable to all crops. Numerous fungi have been identified as potential biocontrol agents to reduce sclerotial inoculum in soil, including *Trichoderma harzianum* [82–85]; *Trichoderma* spp. [86–88], *Gli-*

ocladium virens [88], *Sporidesmium sclerotivorum* (*Teratosperma sclerotivora*) [89–93], *Coniothyrium minitans* [94–96], *Talaromyces flavus* [94], *Epicoccum* sp. [97,98], *Trichothecium roseum* [85,99], and *Erwinia herbicola* [100].

The majority of biocontrol experiments have emphasized fungal agents applied to sclerotia and/or soil. In field plots, *T. harzianum* reduced germination of sclerotia by 62% and increased soybean plant survival by 40% [82]. The researchers detected chitinase and 1,3-glucanase produced by *T. harzianum* in a medium with *S. sclerotiorum* as a sole carbon source. These results indicate the fungus attacks sclerotia directly. In another study, involving soil and seed treatments of cucumber and lettuce and soil applications of *T. harzianum*, it was concluded that hyphal mycoparasitism of *S. sclerotiorum* occurred rather than sclerotial parasitism [84]. *T. harzianum* may attack both sclerotia and mycelium of *S. sclerotiorum*. Of seven fungi isolates from soil in India that inhibited growth of *S. sclerotiorum* on agar plates, *T. harzianum* was the most promising in vitro. Field application of the fungus in wheat bran culture reduced disease in peas and increased yield. A mycelial preparation was more effective than spores [85]. *T. harzianum* in alginate pellets readily colonized sclerotia in steamed soil rather than untreated soil, at 25°C rather than at 15°C, and under more humid conditions. Field inoculation was less effective in reducing sclerotia, possibly because of dry weather [83].

In a recent study, *Coniothyrium minitans* was found to be worldwide in distribution to 29 countries on all continents except South America. Seven colony types were observed on PDA [101]. *C. minitans* has been successfully used as a biocontrol agent applied to the field crops, sunflower [94], and beans, wheat, and barley [96]. *C. minitans* was more effective in reducing the apothecial production of *S. sclerotiorum* when applied directly to buried sclerotia than in fields with a high population of those naturally distributed; however, numbers were greatly reduced. The parasite also spread to sclerotia produced on the bean crop, which would reduce the viability of future inoculum [96]. Application of biocontrol agent in a rotation year with the nonhosts wheat and barley could also be effective. Once applied, *C. minitans* can establish control for at least 2 years [94]. *Talaromyces flavus* did not prove effective under irrigated conditions [96]. *C. minitans* and/or *Trichoderma* spp. were implicated in the decline of disease for 6 and 5 years, respectively, in two fields monocropped with sunflower. Even the annual addition of new sclerotia failed to increase disease [87]. A reduction in sclerotia carpogenic germination and production of apothecia was noted in oilseed rape fields following application of *C. minitans*, but no disease control was obtained [95].

Sporidesmium sclerotivorum is an obligate mycoparasite of sclerotia with a broad distribution worldwide [89,102]. The fungus has reduced inoculum density of *Sclerotinia* spp. and appears to be effective at low application rates [89]. Comparison with other sclerotial mycoparasites indicated that *S. sclerotivorum* and *Teratosperma oligocladium* were more effective than *Dictyosporium elegans* and *Coniothyrium minitans*, which were more effective than *Penicillium citrinum, Talaromyces flavus, Trichoderma* sp., and *Gliocladium virens* in reducing numbers of sclerotia in soil [88]. *S. sclerotivorum* is stimulated to germinate in the presence of sclerotia by a compound present in the melanized layer termed sporigermin [103]. Sclerotia trigger germination in vitro as well as in soil [104]. In field trials with soybean, it was found that fall application of 20 spores/cm² of soil had the potential to become a successful biological control for Sclerotinia stem rot. Disease did not develop in the study area, but 15% plant infection occurred outside the area. Autumn application of the agent was more effective than spring application [93].

T. *roseum* has excellent potential as a biocontrol agent. Application of *T. roseum* to sclerotia in soil completely inhibited germination after 30 days [85]. In another study *T. roseum* infected and destroyed sclerotia in dual cultures on PDA, and 54 and 43% of sclerotia were destroyed in moist sand. Differences in isolates were noted [99]. Considerable information and research has been conducted around the microflora of petals and their potential use to reduce infection by Sclerotinia ascospores. Filamentous fungi from bean and rapeseed petals were applied to flowers up to 24 hours after inoculation with ascospores of *S. sclerotiorum*. The most suppressive fungi included *Alternaria alternata, Dreshslera* sp., *Epicoccum nigrum, Fusarium graminearum, F. heterosporum,* and *Myrothecium verrucaria,* but they did not provide consistent control in four field trials. *Drechslera* sp. and *E. nigrum* significantly reduced white mold in one and two trials, respectively [98]. In a similar manner, sterile culture filtrates from *Epicoccum purpurascens* decreased severity of white mold on bean and increased pod yield [97].

Erwinia herbicola inhibited ascospore germination and development of white mold lesions in a bioassay but failed to reduce disease severity in the field in western Nebraska. Low temperatures may have limited the ability of *E. herbicola* to multiply to protect expanding blossoms of dry edible beans [100]. Most biocontrol agents are greatly affected by environment. The effectiveness of *E. nigrum* was the only one of several fungi studied that was independent of environment. Suppression of disease was most effective under conditions that were least conducive for disease [105].

VI. CERCOSPORA BLIGHT AND PURPLE SEED STAIN

A. Resistance

Resistance to *C. kikuchii* has proven difficult to identify in greenhouse and field trials. Differences among cultivars could be detected in the greenhouse and in the field, but ranking of the cultivars was not similar [107]. In this study, latent infection, foliar disease, and seed stain were evaluated in 17 cultivars in MG I-IV. Leaves and pods were inoculated in growth room experiments. Field results were based on natural inoculum. Foliar disease, latent infection, and seed infection were obtained in the greenhouse, but differences among cultivars were not significant. Significant differences were noted in the field, but results varied by year. Four susceptible cultivars were BSR 101, Amsoy 71, Hack, and Miami. Resnick was resistant to foliar infection in both greenhouse and field trials. Incidence of stained seed was not related to foliar rating in any cultivar [106,107]. In field trials conducted in 1980, cv. Tracy was the most resistant to foliar blight. Lee 74 and Davis were almost as resistant. Bragg, Hood 75, and Forrest were very susceptible [108]. Resistance to seed infection has been directly associated with the length of interval between growth stage R7 and R8. A short interval and rapid seed moisture loss results in less seed infection by *C. kikuchii* and *Phomopsis* sp. [109,110].

B. Management

Since the fungus survives between crops on soybean debris and is a primary source of inoculum [111], ploughing should be practiced where feasible. Recently, weed hosts of the disease have been identified in Mississippi. Cocklebur, morning glory, and sicklepod had latent infections, and spores from the infected weeds caused infection on soybean seedlings [112]. Weed control is an important mechanism to reduce inoculum. Use of disease-free seed can prevent introduction of the disease to new fields and reduce seedling infections.

Since high relative humidity [113] and leaf wetness periods of 18 hours or longer are required to establish foliar disease [114] and pod infection required 24 hours [115], overhead irrigation should be used with discretion to minimize wet periods.

VII. DIAPORTHE-PHOMOPSIS COMPLEX

Diseases caused by Diaporthe-Phomopsis Complex cause severe losses in many areas of the world. The complex can be divided into individual diseases as follows: seed decay caused by *Phomopsis longicolla*, pod and stem blight caused by *Diaporthe phaseolorum* var. *sojae*, and stem canker caused by *D. phaseolorum* var. *caulivora* and *D. phaseolorum* var. *meridionalis*. All of the fungi can cause seed decay and poor germination. The pathogens are generally distinguished by cultural characteristics on acidified PDA [1], but recently methods of molecular identification and phylogenic grouping have been refined to separate members of the complex [116].

A. Pod and Stem Blight (*Phomopsis* sp. or *Diaporthe* sp.)

1. Resistance

Some hard-seeded lines, e.g., D67-5677-1, are resistant to infection by *Phomopsis* sp., but this may not be true for all hard-seeded lines. Part of the screening process utilized a pod injection technique that may be utilized in future screening procedures [117]. Some sources of resistance in plant introductions have been reported for Phomopsis seed decay and pod and stem blight [1]. Most PIs that are reported resistant are resistant to both diseases. Resistance to both diseases is present in PI 417479 [118]. Using classical backcrosses to susceptible parents, Agripro 350 and PI 91113, and progeny screening it was determined that resistance was due to two complementary dominant genes. Gene expression is strongly influenced by environment [118]. A restriction fragment length polymorphism (RFLP) marker associated with the resistance in PI 417479 was recently identified, which should assist selection for Phomopsis seed mold resistance [119].

Near-isogenic lines of cv. Clark that differed in stem termination, time of flowering, and maturity were evaluated for susceptibility to *Phomopsis longicolla*. Determinate lines had 20% more infection and a 16% reduction in germination compared to indeterminate lines. Late flowering and maturity reduced infection by 48% and improved germination 97% compared to earlier lines. It was concluded that early-flowering determinate cultivars could increase the risk of seed mold in the mid-Atlantic area of the United States [120].

Observations of seeds with high amounts of infection by *Phomopsis phaseoli* with a scanning electron microscope revealed that seeds with low infection (0–5%) had fewer pores and those with high infection (65–83%) had multiple pores. In addition, seed with low infection had closed micropyles, and those with high infection had open micropyles. Resistance to *Phomopsis* may be due to mechanical mechanisms [121]. A comparison of two lines that differed in susceptibility to Phomopsis seed decay indicated that pods and seeds of the susceptible variety were infected approximately one week before the resistant variety. The susceptible variety also had more stomata per mm^2 and more trichomes per mm^2 than the resistant variety [122].

Resistance to seedborne disease has also been shown to relate to the length of time required for the soybean variety to move from the R7 to R8 stages of growth. Resistant genotypes take less time to complete the R7 stage and moisture loss was greater than for susceptible genotypes. Infection of seed occurred as moisture in pods and seeds decreased from 30–35% to 15–18% [123].

In North America, the incidence of Phomopsis seed mold is influenced by planting date and maturity requirements of the variety. Six cultivars ranging from MG 00 to IV were planted in April, mid-May, and early and late June in Lexington, Kentucky. *P. longicolla*–infected seed was greater than 30% when planting occurred in April and May. Seed quality was improved with later planting of early varieties or by planting full season varieties [124]. In another study in Portageville, Missouri, early planting with early varieties resulted in greater Phomopsis seed infection. A full season variety had less seed mold [125]. Similar results were obtained in Arkansas, but it was noted that the seed mold–resistant PI 417479 had low infection in all environments [126].

2. Management

Soil fertility is important to the health of soybean seed, and potassium appears to be an essential component. Increasing K fertilizer significantly increased soybean yield and quality of delayed harvest soybeans in Texas. Increasing K reduced Phomopsis seed infection from 35% to about 1% in beans harvested at maturity and in 2-week-delayed harvest seed infection was reduced from 52% to 28% in plots receiving no K fertilizer and 450 kg/ha, respectively [127]. In another study, K fertilization decreased Phomopsis infection in lower seeds by 23% and in upper seeds by 14%, but in some instances germination and seed yield increases were not consistently evident [128]. The levels of Ca in pod walls may also be important in preventing Phomopsis seed mold [129].

Rotation with nonhosts is recommended to reduce pod and stem blight and seed mold [1]. A number of weed hosts have been identified including *Ambrosia trifida*, *Xanthium strumarium*, *Euphorbia maculata*, and *Rumex crispus* in the United States [130], *Amaranthus spinosus*, *Leonotis nepetaefolia*, and *Leonotis sibiricus* in Brazil [131], and others [132]. Rotation corps should be maintained as weed-free as possible to reduce the carryover of inoculum.

Although no-till and minimum tillage has proven advantageous for soil and moisture conservation, soybean debris left on the soil surface is a prime source of inoculum for seed mold and pod and stem blight. In studies involving the survival of fungi on crop debris over the winter in no-till with and without a cover crop, the incidence of *D. phaseolorum* var. *sojae* was higher in soybean debris in May than in preceding months [133]. Monoculture of soybeans and minimum tillage results in the accumulation of inoculum.

Because Phomopsis-Diaporthe Complex is seedborne and can affect germination and stand, it is beneficial to have prior knowledge of seed infection prior to planting. Recent advances in detecting asymptomatic infection have been developed with ultrasound analysis [134] and color classification using image analysis [135]. Knowledge of the incidence of mold allows planning a strategy for planting, e.g., overseeding, seed treatment, or different seed source. Detection of pathogens and degree of infection by ELISA [136] in seed lots will facilitate decision making by growers and seed dealers.

B. Northern and Southern Stem Canker

1. Resistance

D. p. var. *caulivora* produces a phytotoxin that is directly involved in symptom production and pathogenicity [137]. In the United States, northern soybean cultivars are resistant to southern U.S. isolates of *D. p.* var. *meridionalis*. Resistance in southern soybean germplasm is found in cv. Crockett and cv. Dowling, which have the dominant genes for resistance Rdc3 and Rdc4, respectively. These genes are distinct from the two dominant genes Rdc1 and Rdc2 found in cv. Tracy-M [138].

2. Management

Adequate soil fertility is important in the control of stem canker. In a fertility trial involving complete NPK, NH3NO4, PO4, and KO, application of any of the four fertilizers reduced stem canker [24]. The incidence of stem canker increases when plants are under stress; therefore, reduced plant populations to alleviate stress are recommended in Argentina to control stem canker [20].

Inoculum of stem canker overwinters on soybean debris; therefore, crop rotation and full tillage will help to reduce sources of inoculum and disease. *D. phaseolorum* var. *caulivora* has a broad host range on common weeds of soybean fields. Numerous weeds were identified as hosts, although many were asymptomatic [139]. Hosts for southern stem canker *D. phaseolorum* var. *meridionalis* in Brazil have been documented [140].

Few biological controls have been described for stem canker, but isolates of *Chaetomium globosum* were evaluated for their effect on survival of *D. phaseolorum* var. *meridionalis* colonized soybean stems. The antibiotic-producing strain suppressed sporulation more than the nonantibiotic strain under moist conditions, but both isolates were effective under drier conditions. The antibiotic strain reduced stem disease when applied as a seed treatment of soybean planted in stem canker–infested soil [141]. This type of research should be expanded, especially with the increases in minimum tillage and short rotations.

VIII. VIRAL DISEASES

Numerous viruses have been reported to cause disease in soybean under field conditions; however, the severity and distribution of each virus varies considerably [1]. Viruses such as soybean mosaic, tobacco ring spot, and bean yellow mosaic are worldwide in distribution, but others are more local, such as black gram mottle (Thailand), cowpea severe mosaic (Brazil, Puerto Rico, Trinidad, and part of the United States), soybean yellow vein (Thailand), Indonesian soybean dwarf (Indonesia, Thailand), African soybean dwarf (Nigeria), soybean chlorotic mottle (Japan), and soybean severe stunt (Delaware) [1]. The estimated yield losses caused by virus diseases doubled between 1994 and 1998. A significant proportion of the increased loss has occurred in India—105,000 metric tons in 1994 [2] to 438,500 metric tons in 1998 (J.A. Wrather, personal communication)—and the United States, where losses increased from 70,600 metric tons in 1994 to 250,600 metric tons in 1998 (J.A. Wrather, personal communication). An increase in total production may be responsible for some of the apparent loss, but it is evident that viral diseases are increasing in importance. Recent reports indicate that five viral diseases are important in Argentina [20] and eight viruses are important in China [22]. In both countries soybean mosaic virus (SMV) was considered very important. Management of viral diseases depends on the characteristics of each virus. The following examples will provide some insight into the management of viral diseases.

A. Soybean Mosaic Virus

1. Resistance

Resistance to SMV is available in a number of lines and cultivars such as PI96983, Ogden, York, Marshall, and Kwanggyo used to differentiate strains of SMV. Recently, it was demonstrated that the individual genes in these lines were alleles at the Rsv locus for SMV resistance [143]. The lines Buffalo and HLS also contain single genes for resistance located on separate loci, but it is not known if one of the loci is Rsv1 [144]. PI486355 contains two independent genes for resistance, one of which is at the Rsv1 locus [145]. PI lines from China contain genes at the Rsv1 locus and PI556950 may possess a nonallelic gene for resistance [146]. Molecular marker techniques suggest that the lines Columbia, Holladay, Peking, Virginia, FF4-471, PI507403, and PI556949 may have novel genes independent of Rsv1 [147]. Although these and other sources of resistance are available [1], the existence of SMV strains prevents complete control of the virus by resistance alone.

2. Management

Although seed transmission of SMV is generally low, plants infected early suffer the greatest yield loss [142]. Seed coat mottling cannot be used as an indicator of seed transmission, but varieties differ significantly in the rate of seed transmission [148]. Use of varieties with low transmission rates will reduce the infection foci for secondary spread of the virus. Using strain-specific monoclonal antibodies it was demonstrated that field spread from a source was random in 1991 and aggregated in 1992 and 1993 [149]. Field spread may be dependent on the activity and numbers of aphid vectors, which can vary from location to location and year to year. Studies have shown that aphids do not prefer mosaic-infected plants based on color or odor over healthy plants, but they do not feed as long on SMV-infected plants and therefore move more quickly to new plants, creating a greater opportunity to transmit the virus, which is nonpersistent [150].

Since aphids are the primary vectors of SMV, intercropping to interfere with movement of the vector has been an effective management strategy. SMV-induced mottling was reduced from 4.01% in monocropped soybeans and 2.02 and 2.07% in soybeans intercropped with dwarf and tall sorghum, respectively. Although yields were depressed with intercropping, this type of study should be investigated more extensively [151]. Silver plastic strips are used in soybean fields in China to confuse aphid landing [22]. Double-cropped soybeans are planted late in the growing season, which increases SMV infection. Double-cropping with SMV-resistant varieties in Kentucky resulted in no seed transmission and low seed coat mottling [152]. Fertility may also affect SMV infection. In recent trials complete fertilizer or superphosphate up to 100 kg/ha decreased the incidence of SMV in Alabama. Increasing the fertilizer rates to 150 and 200 kg/ha increased the incidence of SMV, as did lower rates of muriate of potash [24]. Since SMV has a broad host range [1], weed control of potential hosts is important, especially in the vicinity of seed production fields.

B. Tobacco Ringspot

1. Resistance

Plant introductions PI92713 and PI154194 are resistant to tobacco ringspot virus (TRSV) [1]. Strains of TRSV affect cultivars differently [153]. In an initial screening of 630 plant introductions, PI407287 was resistant when inoculated at the seedling stage and 13 other introductions were somewhat resistant [154].

2. Management

Since seed transmission is very important in the spread of this disease, avoiding diseased seed for planting is essential. In China, movement of seed from areas with TRSV is prohibited by quarantine regulations [22].

C. Phytophthora Rot

Phytophthora rot caused by *Phytophthora sojae* was first observed in Indiana in 1948 and Ohio in 1951. The disease quickly became a major problem in the Great Lakes Basin of North America, and the disease was a limiting factor in soybean production for two decades. The disease is now found in North and South America, Europe, Russia, China, Japan, and Australia [1]. It was found in

Argentina in 1978 but did not cause problems until the 1990s [20]. Phytophthora rot was found in China in 1989 and has since been reported in several areas [22]. The disease is most severe on poorly drained, wet soils, but it can also cause losses on lighter soils that are flooded.

1. Resistance

At present, 13 single dominant genes for resistance located at seven loci have been identified to provide vertical resistance to *P. sojae* [1]. Rps 1-a was the first resistance gene identified and used shortly after Phytophthora rot was identified in the United States. In countries where Phytophthora rot has been recently diagnosed, Race 1, virulent on Rps 7, is commonly isolated [22,156,157]. Cultivars with the Rps 1a gene for resistance reduced losses for approximately 10 years in the United States and Canada before new races of the pathogen developed [158]. The most commonly used resistance genes in the United States and Canada are Rps 1-a, Rps1-c, and Rps1-k [159]. Many races of *P. sojae* recently identified are virulent on Rps1-a and Rps1-c [155,160]. In addition, many isolates of *P. sojae* have virulence against genes for resistance that have not been available in local soybean varieties [155,159]. New genes for resistance will be required in the future. New techniques using pod inoculation in the field will facilitate screening [161]. Recently, putative new genes or gene combinations have been recently identified in soybean germplasm from China [162,163] and Japan [164]. The southern Chinese provinces of Hubei, Jiangsu, and Sichuan appear to be valuable sources of resistance [163]. Molecular techniques such as RFLP markers are utilized more frequently to characterize resistance alleles to reduce the reliance on traditional breeding techniques [165]. Markers are also being used to locate specific resistance genes [166,167]. Novel genes for resistance may be present in the related *Glycine sojae*. Resistance was detected in *G. sojae* accessions using new screening techniques [168].

The use of multirace resistance has been a valuable strategy in reducing losses from *P. sojae*; however, this type of resistance is best utilized if combined with disease tolerance or field resistance. Tolerance is the ability of the cultivar to sustain infection but still produce adequate yields. It is not race specific. Tolerance was first used in the late 1970s in North America, but it was found to be unstable under very wet conditions that favored the disease. Tolerance is still a valuable tool and has been used successfully in Australia [151]. Tolerance is best utilized if combined with race-specific resistance. Tolerance prevents major plant loss if isolates of the pathogen shift to become virulent on the race-specific resistance gene. Tolerance is best detected and evaluated in field trials [169,170]. Numerous tests for tolerance have been devised [171–173] with severe disease pressure, but it is difficult to assess in laboratory and greenhouse trials [174,175].

In addition to tolerance and race-specific resistance that is detected by hypocotyl inoculation, there is growing evidence supporting the concept of a set of race-specific genes that function only in the roots of soybean cultivars. During screening trials to detect tolerance with different races of *P. sojae*, it was determined that soybean reactions indicated some isolates differed in virulence and pathogenicity and that tolerance was not universal [176]. Differences in root resistance were more evident when a series of cultivars were screened with Race 1 and Race 10 of *P. sojae*. Based on plant survival, some varieties reacted similarly to both races, some were intermediate, and a few had a clear differential response to root inoculation [177]. Recent race surveys in Canada have produced isolates that are virulent on hypocotyls of Rps1-k varieties but are avirulent on roots, whereas other isolates are virulent on hypocotyls and roots of Rps1-k varieties (T. Anderson, unpublished). This suggests that *Phytophthora* isolates have genes for virulence and avirulence on roots of certain soybean cultivars.

2. Management

Minimum and no-tillage systems are conducive to Phytophthora rot because of increased soil moisture and cooler conditions [1]. In a survey of *P. sojae* under different tillage systems in Illinois, Indiana, Iowa, and Minnesota, *P. sojae* inoculum was more prevalent in surface samples under minimum tillage conditions than under conventional tillage. It was concluded that the possibility of damping-off by *P. sojae* was greater under minimum tillage [178]. Phytophthora rot has also been a problem in Argentina under minimum tillage [20].

Phytophthora rot is prevalent on fine-textured soils with poor drainage [1]. Soil compaction increased the prevalence of disease in a 2-year study in Illinois. Emergence was less and disease incidence was significantly greater under compacted conditions [179].

Application of muriate of potash increased the incidence of Phytophthora rot, but application of complete fertilizer reduced the incidence of the disease in northern Alabama [24]. In an Ontario, Canada, study, fertility increased root rot. The incidence of Phytophthora rot averaged 3.2. 23.0, 32.3, and 41.3% diseased plants at 0, 224, 448, and 672 kg/ha of 8-32-16 fertilizer, respectively [180]. In the same study it was noted that the incidence of Phytophthora rot increased with increasing distance from drainage tile, which agrees with an Australian study that found field-resistant cultivars were not affected by saturated soil culture and furrow irrigation. Susceptible varieties were killed in either system [181].

Seed treatments containing *Actinoplanes missouriensis*, *A. utahensis*, *Amorphosporangium auranticolor*, *Micromonospora* sp., and *Hyphochytrium catenoides* were evaluated in greenhouse experiments for their effect on Phytophthora rot in naturally infested field soil. *A. missouriensis*, *A. utahensis*, and *Micromonospora* sp. improved stands of soybean cv. Corsoy, demonstrating that application of biocontrol organisms to soybean seed could be effective [182]. *Bacillus cereus* (UW85)–treated seed improved the yields of a *Phytophthora*-susceptible cultivar in five of five growing seasons. All cultivars regardless of resistance had increased yield in one year with seed treatments when disease pressure was high. Efficiency was influenced by the formulation of the treatment [183]. Biological seed treatments such as Kodiak are commercially available.

IX. SUDDEN DEATH SYNDROME

Sudden death syndrome (SDS) has been reported from most states in the central U.S. soybean production area, especially in the Mississippi river basin [1,184] Argentina [20], Brazil [19], and Canada [185]. Losses in metric tons in 1998 were estimated as follows: Argentina, 147,200; Brazil 200,000; Canada, 1,800; Paraguay 5,000; and the United States 900,600 (Wrather, personal communication). The total world losses doubled from 746,650 in 1994 (2) to 1,493,300 metric tons in 1998 (Wrather, personal communication), which suggests that this recently reported disease is increasing in severity in the Western Hemisphere. Yield losses vary considerably depending on the severity of infection, but yield of soybean was 46.2% less in nonreplicated field samples from areas with high and low incidence of SDS in Illinois [186].

The disease is caused by *Fusarium solani* f. sp. *glycines*, although confusion with other *F. solani* isolates existed for some time [187].

A. Resistance

As with many soybean diseases, resistant cultivars offer the most economic means of control; however, it has been difficult to assess resistance because of high environmental interaction with disease development and the absence of consistent uniform symptoms. Field resistance to SDS was identified in cvs. Forrest, Jack, and Ripley [189]. Field resistance or rate-reducing resistance is expressed over the growing season [188,189], so it is difficult or impossible to identify with a one-time inoculation or rating. In addition, strains of the fungus may differ with locality because cv. Ripley developed foliar and root symptoms following inoculation with isolates from Ontario, Canada [185]. Some cultivars may have tolerance to SDS because roots of Essex and Asgrow 5403 can be infected without foliar symptoms [189]. Using DNA molecular markers, field resistance to SDS was found to be associated with four quantitative trait loci [190,191]. Identification of resistance may be simplified by exposing calli of soybean cultivars to culture filtrates of the pathogen. Discoloration of calli corresponded to field response of the cultivars evaluated [192]. Novel genes for resistance may have been found recently in plant introductions from China [193]. Determination of the extent of root colonization was an effective way to identify resistance and can be used in the absence of foliar symptoms [194].

B. Management

The disease tends to be more severe under cool wet conditions, and wet conditions are necessary after planting to initiate infection, which may not be expressed as foliar symptoms until later in the season. Improving field drainage and avoiding excessive irrigation helps to reduce the incidence of SDS. Minimum tillage may also contribute to disease severity because the practice leads to cooler soils with greater moisture content than ploughed fields. Late planting combined with increased tillage reduced foliar symptoms caused by SDS [195]. Delayed planting reduced SDS considerably in Kentucky [196]. Complete tillage and delayed planting are a means of avoiding cool wet spring conditions.

Because SDS may be more severe in the presence of SCN [187,197], management practices to reduce SCN should help to control SDS. SCN-resistant varieties and rotation with nonhosts should reduce SDS.

Since mung bean, green bean, lima bean, and cowpea can be hosts for *F. solani* f. sp. *glycines* [198], they should be avoided as rotation crops; however, there is evidence that the pathogen can survive in soil as chlamydospores regardless of cropping practices.

Some biological control experiments have been conducted with SDS. Fungi isolated from the rhizosphere of healthy or mildly affected soybean plants in areas with severe SDS were evaluated as potential biological control agents. In greenhouse trials, 46 of 151 isolates had some control activity. The predominant fungi that demonstrated control were *F. solani* and *F. oxysporum*. Field tests did not result in significant control [199], but continued research in this area may eventually prove rewarding.

X. SUMMARY

SCN is the most economically significant disease that affects soybean production on a global basis. Reducing the losses caused by SCN requires an integrated approach involving resistance and crop management. Long-term management of SCN will be directly dependant on biological control by manipulation of the field environment. More research is required on antagonists, nonhost crops, and soil physical factors that suppress populations of SCN. It is also important to monitor soybean production areas for SCN in order to initiate management practices before nematode populations become difficult to control. Among biological control agents, antagonistic fungi and bacteria such as *Pasteuria* spp. appear very promising. Reviews of research on SCN in the United States [3] and Brazil [200] have been published recently.

Septoria brown spot is considered the second most important disease of soybean, but very little progress has been made in developing control programs.

Effective controls involving resistance from exotic sources may be necessary in addition to agronomic practices to suppress disease development.

There are extensive opportunities to implement biological controls for Sclerotinia stem decay and for seedling and root diseases. A number of products have been commercialized and many more organisms have potential as soil amendments and seed treatments.

It is also noteworthy that significant information on soybean disease incidence, epidemiology, and management worldwide has been published [1,201]. The series of World Soybean Research Conferences has provided a valuable forum for exchange of information among soybean pathologists.

REFERENCES

1. GL Hartman, JB Sinclair, JC Rupe. Compendium of Soybean Disease. St. Paul, MN: APS Press, 1999.
2. JA Wrather, TR Anderson, DM Arsyad, J Gai, LD Ploper, A Porta-Puglia, HH Ram, JT Yorinori. Soybean disease loss estimates for the top 10 soybean producing countries in 1994. Plant Dis 81:107–110, 1997.
3. RD Riggs, JA Wrather. Biology and Management of Soybean Cyst Nematode. The American Phytopathological Society, St Paul, MN, 1992.
4. RD Riggs, L Rakes, D Dombek. Responses of soybean cultivars and breeding lines to races of *Heterodera glycines*. J Nematol 27(suppl):592–601, 1995.
5. J Faghihi, RA Vierling, JM Halbrendt, VR Ferris, JM Ferris. Resistance genes in a 'Williams 82' × 'Hartwig' soybean cross to an inbred line of *Heterodera glycines*. J Nematol 27:418–421, 1995.
6. RA Vierling, J Faghihi, VR Ferris, JM Ferris. Association of RFLP markers with loci conferring broad-based resistance to the soybean cyst nematode (*Heterodera glycines*). Theor Appl Genet 92:83–86, 1996.
7. VC Concibido, DA Lange, RL Denny, JH Orf, ND Young. Genome mapping of soybean cyst nematode resistance genes in 'Peking', PI 90763 and PI 88788 using DNA markers. Crop Sci 37:258–264, 1997.
8. BW Diers, PR Arelli. Management of parasitic nematodes of soybean through genetic resistance. In: HE Kauffman, ed. Champaign, IL: World Soybean Research Conference VI, Superior Printing, 1999, pp 300–306.
9. RD Riggs, ML Hamblen, L Rakes. Resistance in commercial soybean cultivars to six races of *Heterodera glycines* and to *Meloidogyne incognita*. Ann Appl Nematol 2:70–76, 1988.
10. TC Todd, WT Schapaugh Jr, JH Long, B Holmes. Field response of soybean in maturity groups III-V to *Heterodera glycines* in Kansas. J Nematol 27(suppl):628–633, 1995.
11. RS Hussey, HR Boerma. Tolerance in maturity groups V-VIII soybean cultivars to *Heterodera glycines*. J Nematol 21(suppl):686–692, 1989.
12. SR Koenning, SC Anand, GO Myers. An alternative method for evaluating soy-

bean tolerance to *Heterodera glycines* in field plots. J Nematol 24:177–182, 1992.

13. TL Niblack. Long-term solutions to management of soybean cyst nematode. In: HE Kauffman, ed. Champaign, IL: World Soybean Research Conference VI. Superior Printing, 1999, pp 295–299.

14. JP Ross. Crop rotation effects on the soybean cyst nematode population and soybean yields. Phytopathology 52:815–818, 1962.

15. ME Sortland, DH MacDonald. Effect of crop and weed species on development of a Minnesota population of *Heterodera glycines* race 5 after one to three growing periods. Plant Dis 71:23–27, 1987.

16. LD Young, EE Hartwig. Selection pressure on soybean cyst nematode from soybean cropping sequences. Crop Sci 28:845–847, 1988.

17. LD Young. Influence of soybean cropping sequences on seed yield and female index of the soybean cyst nematode. Plant Dis 82:615–619, 1998.

18. DB Weaver, R Rodriguez-Kabana, EL Carden. Comparison of crop rotation and fallow for management of *Heterodera glycines* and *Meloidogyne* spp. in soybean. J Nematol 27(suppl):585–591, 1995.

19. JT Yorinori. Management of economically important diseases in Brazil. In: HE Kauffman, ed. Champaign, IL: World Soybean Research Conference VI, Superior Printing, 1999, p 290.

20. LD Ploper. Management of economically important diseases of soybean in Argentina. In: HE Kauffman, ed. Champaign, IL: World Soybean Research Conference VI, Superior Printing, 1999, pp 269–280.

21. CR Grau. Management of economically important diseases of soybean in the United States. In: HE Kauffman, ed. Champaign, IL: World Soybean Research Conference VI, Superior Printing, 1999, pp 264–268.

22. Yujun Tan, Zilin Yu and Yuhua Peng. Management of economically important soybean diseases in China. In: HE Kauffman, ed. Champaign, IL: World Soybean Research Conference VI, Superior Printing, 1999, pp 281–289.

23. RG Hanson, JH Muir, PM Sims, JK Boon. Response of three soybean cultivars to cyst nematode and KCl fertilization. J Prod Agric 1:327–331, 1988.

24. RP Pacumbaba, GF Brown, RO Pacumbaba, Jr. Effect of fertilizers and rates of application on incidence of soybean diseases in northern Alabama. Plant Dis 81: 1459–1460, 1997.

25. DA Reynolds, GL Tylka, CA Martinson. Swine manure affects *Heterodera glycines* soil population densities. ASP/SON Annual Meeting, Monterey, CA, 1999, p 102.

26. PM Tefft and LW Bone. Zinc-medicated hatching of eggs of soybean cyst nematode, *Heterodera glycines*. J Chem Ecol 10:361–372, 1984.

27. JE Behm, GL Tylka, TL Niblack, WJ Wiebold, PA Donald. Effects of zinc fertilization of corn on hatching of *Heterodera glycines* in soil. J Nematol 27:164–171, 1995.

28. SR Koening, SC Anand. Effects of wheat and soybean planting date on *Heterodera glycines* population dynamics and soybean yield with conventional tillage. Plant Dis 75:301–304, 1991.

29. DP Schmitt. Management of *Heterodera glycines* by cropping and cultural practices. J Nematol 23:348–352, 1991.

30. SR Koenning, DP Schmitt, KR Barker, ML Gumpertz. Impact of crop rotation and tillage system on *Heterodera glycines* population density and soybean yield. Plant Dis 79:282–286, 1995.
31. DE Hershman, PR Bachi. Effect of wheat residue and tillage on *Heterodera glycines* and yield of doublecrop soybean in Kentucky. Plant Dis 79:631–633, 1995.
32. J Wang, PA Donald, TL Niblack, GW Bird, J Faghihi, JM Ferris, C Grau, DJ Jardine, PE Lipps, AE MacGuidwin, H Melakeberhan, GR Noel, P Pierson, RM Riedel, PR Sellers, WC Stienstra, TC Todd, GL Tylka, TA Wheeler, DS Wysong. Soybean cyst nematode reproduction in the north central United States. Plant Dis 84:77–82, 2000.
33. BR Kerry. Fungal parasites of cyst nematodes. Agric Ecosystems Environ 24:293–305, 1988.
34. LM Carris, DA Glowe, CA Smyth, DI Edwards. Fungi associated with populations of *Heterodera glycines* in two Illinois soybean fields. Mycologia 81:66–75, 1989.
35. DG Kim, RD Riggs. Characteristics and efficiency of a sterile hyphomycete (ARF18). A new biocontrol for *Heterodera glycines* and other nematodes. J Nematol 23:275–282, 1991.
36. SY Chen, DW Dickson, DJ Mitchell. Pathogenicity of fungi to eggs of *Heterodera glycines*. J Nematol 28:148–158, 1996.
37. GR Noel, BA Stanger. First report of *Pasteuria* sp. attacking *Heterodera glycines* in North America. J Nematol 26(suppl):612–615, 1994.
38. DG Kim, RD Riggs. Biological Control. In: RD Riggs, JA Wrather, eds. Biology and Management of the Soybean Cyst Nematode. St. Paul, MN: APS Press, 1992, pp 133–142.
39. SLF Meyer, RJ Meyer. Greenhouse studies comparing strains of the fungus *Verticillium lecanii* for activity against the nematode *Heterodera glycines*. Fund Appl Nematol 19:305–308, 1996.
40. SLF Meyer, RJ Meyer. Survival of the nematode-antagonistic fungus *Verticillium lecanii* in alginate prills. Nematologica 42:114–123, 1996.
41. SLF Meyer, G Johnson, M Dimock, JW Fahey, RN Huettel. Field efficacy of *Verticillium lecanii*, sex pheromone, and pheromone analogs as potential management agents for soybean cyst nematode. J Nematol 29:282–288, 1997.
42. DG Kim, RD Riggs, JC Correll. Isolation, characterization, and distribution of a biocontrol fungus from cysts of *Heterodera glycines*. Phytopathology 88:465–471, 1998.
43. B Gintis, G Morgan-Jones, R Rodriguez-Kabana. Fungi associated with several developmental stages of *Heterodera glycines* from an Alabama field soil. Nematropica 13:181–200, 1983.
44. EC Bernard, LH Self, DD Tyler. Fungal parasitism of soybean cyst nematode, *Heterodera glycines* (Nemata: Heteroderidae), in differing cropping-tillage regimes. Appl Soil Ecol 5:57–70, 1996.
45. G Morgan-Jones, R Rodriquez-Kabana and J Gomes Tovar. Fungi associated with cysts of *Heterodera glycines* in the Cauca Valley, Colombia. Nematropica 14:173–177, 1984.
46. JFV Silva, SMT Piza, RG Carneira. Fungos associados a ovos de *Heterodera glycines* no Brasil. Nematol Brasil 18:73–78, 1994.

47. P Timper, RD Riggs. Variation in efficacy of isolates of the fungus ARF against the soybean cyst nematode *Heterodera glycines*. J Nematol 30:461–467, 1998.

48. T Nishizawa. A decline phenomenon in a population of the upland cyst nematode, *Heterodera elachista*, caused by a bacterial parasite, *Pasteuria penetrans*. J Nematol 19:546, 1987.

49. SY Chen, J Charnecki, JF Preston, DW Dickson. Extraction and purification of *Pasteuria* spp. endospores. J Nematol 32:78–84, 2000.

50. RD Sharma, AC Gomes. Effects of Bacillus spp. toxins on oviposition and juvenile hatching of *Heterodera glycines*. Nematol Brasil 20:53–62, 1996.

51. HS Song, SM Lim, JM Clark Jr. Purification and partial characterization of a host-specific pathotoxin from culture filtrates of *Septoria glycines*. Phytopathology 83: 659–661, 1993.

52. GB Lee, GL Hartman. Reactions of glycines species and other legumes to *Septoria glycines*. Plant Dis 80:90–94, 1996.

53. HS Song, SM Lim, JM Widholm. Selection and regeneration of soybeans resistant to the pathogenic culture filtrates of *Septoria glycines*. Phytopathology 84:948–951, 1994.

54. GB Lee, GL Hartman, SM Lim. Brown spot severity and yield of soybeans regenerated from calli resistant to a host-specific pathotoxin produced by *Septoria glycines*. Plant Dis 80:408–413, 1996.

55. RL Cooper. Soybean yield response to benomyl fungicide application under maximum yield conditions. Agron J 81:847–849, 1989.

56. DJ Williams, RF Nyvall. Leaf infection and yield losses caused by brown spot and bacterial blight diseases of soybean. Phytopathology 70:900–902, 1980.

57. SM Lim. Brown spot severity and yield reduction in soybean. Phytopathology 70: 974–977, 1980.

58. PR Bristow, TD Wyllie. *Macrophomina phaseolina*, another cause of the twin-stem abnormality disease of soybean. Plant Dis 70:1152–1153, 1986.

59. G Olaya, GS Abawi. Effect of water potential on mycelial growth and on production and germination of sclerotia of *Macrophomina phaseolina*. Plant Dis 80:1347–1350, 1996.

60. GL Cloud, JC Rupe. Comparison of three media for enumeration of sclerotia of *Macrophomina phaseolina*. Plant Dis 75:771–772, 1991.

61. GS Smith, ON Carvil. Field screening of commercial and experimental soybean cultivars for their reaction to *Macrophomina phaseolina*. Plant Dis 81:363–368, 1997.

62. TC Todd. Soybean planting date and maturity effects on *Heterodera glycines* and *Macrophomina phaseolina* in southeastern Kansas. J Nematol 25(suppl):731–737, 1993.

63. CR Bowen, WT Schapaugh Jr. Relationships among charcoal rot infection, yield, and stability estimates in soybean blends. Crop Sci 29:42–45, 1989.

64. JA Wrather, SR Kendig. Tillage effects on *Macrophomina phaseolina* population density and soybean yield. Plant Dis 82:247–250, 1998.

65. HE Winkler, BAD Hetrick, TC Todd. Interactions of *Heterodera glycines*, *Macrophomina phaseolina*, and mycorrhizal fungi on soybean in Kansas. J Nematol 23: 675–682, 1994.

66. CH Canaday, DG Helsel, TD Willie. Effects of herbicide-induced stress on root colonization of soybeans by *Macrophomina phaseolina*. Plant Dis 70:863–866, 1986.

67. GJ Boland, R Hall. Index of plant hosts of *Sclerotinia sclerotiorum*. Can J Plant Pathol 16:93–108, 1994.

68. SN Wegulo, XB Yang, CA Martinson. Soybean cultivar responses to *Sclerotinia sclerotiorum* in field and controlled environmental studies. Plant Dis 82:1264–1270, 1998.

69. HS Kim, CH Sneller, BW Diers. Evaluation of soybean cultivars for resistance to Sclerotinia stem rot in field environments. Crop Sci 39:64–68, 1999.

70. GJ Boland, R Hall. Evaluating soybean cultivars for resistance to *Sclerotinia sclerotiorum* under field conditions. Plant Dis 71:94–936, 1987.

71. BD Nelson, TC Helms, MA Olson. Comparison of laboratory and field evaluations of resistance in soybean to *Sclerotinia sclerotiorum*. Plant Dis 75:662–665, 1991.

72. RI Buzzell, TW Welacky, TR Anderson. Soybean cultivar reaction and row width effect on Sclerotinia stem rot. Can J Plant Sci 73:1169–1175, 1993.

73. P Donaldson, T Anderson, D Simmonds. Development of white mould resistant soybean. 7th Biennial Conference of the Molecular and Cellular Biology of Soybean. Knoxville, TN, July 26–29, 1998, p F4.

74. R Hall, LCB Nasser. Practice and precept in cultural management of bean diseases. Can J Plant Pathol 18:176–185, 1996.

75. LCB Nasser, JC Sutton. Palhada de arroz pode controlar importante doenço do feizoeiro. Cerrados Pesquisa Tecnol 3:6, 1993.

76. CR Grau, VL Radke. Effects of cultivars and cultural practices on Sclerotinia stem rot of soybean. Plant Dis 68:56–58, 1984.

77. JF Nicholson, OD Dhingra, JB Sinclair. Internal seed-borne nature of *Sclerotinia sclerotiorum* and *Phomopsis* sp. and their effect on soybean seed quality. Phytopathology 62:1261–1263, 1972.

78. TR Anderson. Seed molds of soybean in Ontario and the influence of production area on the incidence of *Diaporthe phaseolorum* var. *caulivora* and *Phomopsis* sp. Can J Plant Pathol 7:74–78, 1985.

79. DD Hoffman, GL Hartman, DS Mueller, RA Leits, CD Nickell, WL Pedersen. Yield and seed quality of soybean cultivars infected with *Sclerotinia sclerotiorum*. Plant Dis 82:826–829, 1998.

80. GL Hartman, L Kull, YH Huang. Occurrence of *Sclerotinia sclerotiorum* in soybean fields in east-central Illinois and enumeration of inocula in soybean seed lots. Plant Dis 82:560–565, 1998.

81. DS Mueller, GL Hartman, WL Pedersen. Development of sclerotia and apothecia of *Sclerotinia sclerotiorum* from infected soybean seed and its control of fungicide seed treatments. Plant Dis 83:1113–1115, 1999.

82. AB Menendez, A Godeas. Biological control of *Sclerotinia sclerotiorum* attacking soybean plants. Degradation of the cell walls of this pathogen by *Trichoderma harzianum* (BAFC 742). Mycopathologia 142:153–160, 1998.

83. GR Knudsen, DJ Eschen, LM Dandurand, L Bin. Potential for biocontrol of *Sclerotinia sclerotiorum* through colonization of sclerotia by *Trichoderma harzianum*. Plant Dis 75:466–470, 1991.

84. J Inbar, A Menendez, I Chet. Hyphal interaction between *Trichoderma harzianum* and *Sclerotinia sclerotiorum* and its role in biological control. Soil Biol Biochem 28:757–763, 1996.

85. D Singh. Biocontrol of *Sclerotinia sclerotiorum* (Lib.) de Bary by *Trichoderma harzianum*. Trop Pest Manage 37:374–378, 1991.

86. HC Huang, JW Huang, G Saindon, RS Erickson. Effect of allyl alcohol and fermented agricultural wastes on carpogenic germination of sclerotia of *Sclerotinia sclerotiorum* and colonization by *Trichoderma* spp. Can J Plant Pathol 19:43–46, 1997.

87. HC Huang, GC Kozub. Monocropping to sunflower and decline of Sclerotinia wilt. Bot Bull Acad Sin 32:163–170, 1991.

88. PB Adams. Comparison of antagonists of *Sclerotinia* species. Phytopathology 79: 1345–1347, 1989.

89. PB Adams, WA Ayers. *Sporidesmium sclerotivorum*: distribution and function in natural biological control of sclerotial fungi. Phytopathology 71:90–93, 1981.

90. PB Adams, DR Fravel. Economical biological control of Sclerotinia lettuce drop by *Sporidesmium sclerotivorum*. Phytopathology 80:1120–1124, 1990.

91. VS Chaban, IV Yakubova, MP Sokolovskaya. A mycophylic fungus *Sporidesmium sclerotivorum* Uecker, Ayers et Adams new for the Ukraine on sclerotia of *Sclerotinia sclerotiorum* (Lib.) de Bary. Mikologiya I Fitopatologiya 27:61–63, 1993.

92. DR Fravel. Use of *Sporidesmium sclerotivorum* for biocontrol of sclerotial plant pathogens. In: GJ Boland, LD Kuykendall, eds. Plant-Microbe Interactions and Biological Control. New York: Marcel Dekker, 1998, pp 37–47.

93. LD Rio, CA Martinson, XB Yang. Control of Sclerotinia stem rot of soybeans with *Sporidesmium sclerotivorum*. Proceedings of the 1998 International Sclerotinia Workshop, Fargo, ND, 1998, pp 64–65.

94. DL McLaren, HC Huang, C Kozub, SR Rimmer. Biological control of Sclerotinia wilt of sunflower with *Talaromyces flavus* and *Coniothyrium minitans*. Plant Dis 78:231–235, 1994.

95. MP McQuilken, SJ Mitchell, SP Budge, JM Whipps, JS Fenlon, SA Archer. Effect of *Coniothyrium minitans* on sclerotial survival and apothecial production of *Sclerotinia sclerotiorum* in field-grown oilseed rape. Plant Pathol 44:883–896, 1995.

96. DL McLaren, HC Huang, SR Rimmer. Control of apothecial production of *Sclerotinia sclerotiorum* by *Coniothyrium minitans* and *Talaromyces flavus*. Plant Dis 80: 1373–1378, 1996.

97. T Zhou, RD Reeleeder, SA Sparace. Interactions between *Sclerotinia sclerotiorum* and *Epicoccum purpurascens*. Can J Bot 69:2503–2510, 1991.

98. GD Inglis, GJ Boland. Evaluation of filamentous fungi isolated from petals of bean and rapeseed for suppression of white mold. Can J Microbiol 38:124–129, 1992.

99. HC Huang, EG Kokko. *Trichothecium roseum*, a mycoparasite of *Sclerotinia sclerotiorum*. Can J Bot 71(12):1631–1638, 1993.

100. GY Yuen, ML Craig, ED Kerr, JR Steadman. Influences of antagonist population levels, blossom development stage, and canopy temperature on the inhibition of *Sclerotinia sclerotiorum* on dry edible bean by *Erwinia herbicola*. Phytopathology 84:495–501, 1994.

101. C SandysWinsch, JM Whipps, M Gerlagh, M Kruse. World distribution of the sclerotial mycoparasite *Coniothyrium minitans*. Mycol Res 97:1175–1178, 1993.

102. CE Sanford, JL Coley-Smith, D Parfitt. *Sporidesmium sclerotivorum* on sclerotia of *Sclerotinia*. Plant Pathol 36:411–412, 1987.

103. S Mischke, CF Mischke, PB Adams. A rind-associated factor from sclerotia of *Sclerotinia minor* stimulates germination of a mycoparasite. Mycol Res 99:1063–1070, 1995.

104. S Mischke, PB Adams. Temporal and spatial factors affecting germination of macroconidia of *Sporidesmium sclerotivorum*. Mycologia 88:271–277, 1996.

105. DJ Hannusch, GJ Boland. Influence of air temperature and relative humidity on biological control of white mold of bean (*Sclerotinia sclerotiorum*). Phytopathology 86:156–162, 1996.

106. RK Velichetti, JB Sinclair. Production of cercosporin and colonization of soybean seed coats by *Cercospora kikuchii*. Plant Dis 78:342–346, 1994.

107. CE Orth, W Schuh. Resistance of 17 soybean cultivars to foliar, latent, and seed infection by *Cercospora kikuchii*. Plant Dis 78:661–664, 1994.

108. HJ Walters. Soybean leaf blight caused by *Cercospora kikuchii*. Plant Dis 64:961–962, 1980.

109. TS Abney, LD Ploper. Growth regulator effects on soybean seed maturation and seedborne fungi. Plant Dis 75:585–589, 1991.

110. LD Ploper, TS Abney, KW Roy. Influence of soybean genotype on rate of seed maturation and its impact on seedborne fungi. Plant Dis 76:287–292, 1992.

111. KL Athow. Fungal diseases. In: BE Caldwell, ed. Soybeans Improvement, Production and Uses. Madison, WI: Amer Soc Agron, 1973, pp 459–489.

112. KS McLean, KW Roy. Purple seed stain of soybean caused by isolates of *Cercospora kikuchii* from weeds. Can J Plant Pathol 10:166–171, 1988.

113. W Schuh. Influence of interrupted dew periods, relative humidity, and light on disease severity and latent infections caused by *Cercospora kikuchii* on soybean. Phytopathology 83:109–113, 1993.

114. W. Schuh. Influence of temperature and leaf wetness period on conidial germination in vitro and infection of *Cercospora kikuchii* on soybean. Phytopathology 81:1315–1318, 1991.

115. W Schuh. Effect of pod development stage, temperature, and pod wetness duration on the incidence of purple seed stain of soybeans. Phytopathology 82:446–451, 1992.

116. AW Zhang, L Riccioni, WL Pedersen, KP Kollipara, GL Hartman. Molecular identification and phylogenetic grouping of *Diaporthe phaseolorum* and *Phomopsis longicolla* isolates from soybean. Phytopathology 88:1306–1314, 1998.

117. KW Roy, BC Keith, CH Andrews. Resistance of hardseeded soybean lines to seed infection by *Phomopsis*, other fungi, and soybean mosaic virus. Can J Plant Pathol 16:122–128, 1994.

118. MS Zimmerman, HC Minor. Inheritance of Phomopsis seed decay resistance in soybean PI 417479. Crop Sci 33:96–100, 1993.

119. GU Berger, HC Minor. An RFLP marker associated with resistance to Phomopsis seed decay in soybean PI 417479. Crop Sci 39:800–805, 1999.

120. PR Thomison, WJ Kenworthy, MS McIntosh. Phomopsis seed decay in soybean

isolines differing in stem termination, time of flowering, and maturity. Crop Sci 30:183–188, 1990.

121. MM Kulik, RW Yaklich. Soybean seed coat structures: relationship to weathering resistance and infection by the fungus *Phomopsis phaseoli*. Crop Sci 31:108–113, 1991.

122. TR Anderson, RI Buzzell, BR Buttery, VA Dirks. Incidence of pod and seed infection in two soybean lines differing in resistance to Phomopsis seed decay. Can J Plant Sci 75:543–545, 1995.

123. LD Ploper, TS Abney, KW Roy. Influence of soybean genotype on rate of seed maturation and its impact on seedborne fungi. Plant Disease 76:287–292, 1992.

124. DM TeKrony, LJ Grabau, M DeLacy, M Kane. Early planting of early-maturing soybean: effects on seed germination and *Phomopsis* infection. Agron J 88:428–433, 1996.

125. JA Wrather, SR Kendig, WJ Wiebold, RD Riggs. Cultivar and planting date effects on soybean stand, yield, and *Phomopsis* sp. seed infection. Plant Dis 80:622–624, 1996.

126. WL Mayhew, CE Caviness. Seed quality and yield of early-planted, short-season soybean genotypes. Agron J 86:16–19, 1994.

127. JW Sij, FT Turner, NG Whitney. Suppression of Anthracnose and Phomopsis seed rot on soybean with potassium fertilizer and benomyl. Agron J 77:639–642, 1985.

128. DL Jeffers, AF Schmitthenner, ME Kroetz. Potassium fertilization effects on *Phomopsis* seed infection, seed quality, and yield of soybeans. Agron J 74:886–890, 1982.

129. PR Thomison, DL Jeffers, AF Schmitthenner. Phomopsis seed decay and nutrient accumulation in soybean under two soil moisture levels. Agron J 79:913–918, 1987.

130. KW Roy, S Ratnayake, K McLean. Colonization of weeds by *Phomopsis longicolla*. Can J Plant Pathol 19:193–196, 1997.

131. RF Cerkauskas, OD Dhingra, JB Sinclair, G. Asmus. *Amaranthus spinosus, Leonotis nepetaefolia*, and *Leonurus sibiricus*: new hosts of *Phomopsis* spp. in Brazil. Plant Dis 67:821–824, 1983.

132. KW Roy. Host range of the *Diaporthe/Phomopsis* complex from soybean. In: AJ Pascale, ed. Proc. World Soybean Research Conference IV. Boulder, CO: Westview Press, 1989, pp 1707–1711.

133. RE Baird, BG Mullinix, AB Peery, ML Lang. Diversity and longevity of the soybean debris mycobiota in a no-tillage system. Plant Dis 81:530–534, 1997.

134. RR Walcott, DC McGee, MK Misra. Detection of asymptomatic fungal infections of soybean seeds by ultrasound analysis. Plant Dis 82:584–589, 1998.

135. IS Ahmad, JF Reid, MR Paulsen, JB Sinclair. Colour classifier for symptomatic soybean seeds using image processing. Plant Dis 4:320–327, 1999.

136. LM Brill, RD McClary, JB Sinclair. Analysis of two ELISA formats and antigen preparations using polyclonal antibodies against *Phomopsis longicolla*. Phytopathology 84:173–179, 1994.

137. B Lalitha, JP Snow, GT Berggren. Phytotoxin production by *Diaporthe phaseolorum* var. *caulivora*, the causal organism of stem canker of soybean. Phytopathology 79:494–504, 1989.

138. GR Bowers, K Ngeleka, OD Smith. Inheritance of stem canker resistance in soybean cultivars Crockett and Dowling. Crop Sci 33:67–70, 1993.

139. BD Black, GB Padgett, JS Russin, JL Griffin, JP Snow, GT Berggren Jr. Potential weed hosts for *Diaporthe phaseolorum* var. *caulivora*, causal agent for soybean stem canker. Plant Dis 80:763–765, 1966.

140. J Pereira, LAC Valle. Hosts of *Phomopsis phaseoli* f. sp. *meridionalis*, causal agent of soybean stem canker. Fitopatologia-Brasileira 22:553–554, 1997.

141. J Pereira, OD Dhingra. Suppression of *Diaporthe phaseolorum* f. sp. *meridionalis* in soybean stems by *Chaetomium globosum*. Plant Pathol 46:216–223, 1997.

142. Q Ren, TW Pfeiffer, SA Ghabrial. Soybean mosaic virus incidence level and infection time: interaction effects on soybean. Crop Sci 37:1706–1711, 1997.

143. P Chen, GR Buss, CW Roane, SA Tolin. Allelism among genes for resistance to soybean mosaic virus in strain-differential soybean cultivars. Crop Sci 31:305–309, 1991.

144. GR Bowers Jr., EH Paschall, RL Bernard, RM Goodman. Inheritance of resistance to soybean mosaic virus in "Buffalo" and HLS soybean. Crop Sci 32:67–72, 1992.

145. P Chen, GR Buss, SA Tolin. Resistance to soybean mosaic virus conferred by two independent dominant genes in PI486355. J Hered 84:25–28, 1993.

146. Y Wang, RL Nelson, Y Hu. Genetic analysis of resistance to soybean mosaic virus in four soybean cultivars from China. Crop Sci 38:922–925, 1998.

147. YG Yu, MA Saghai-Maroof, GR Buss. Divergence and allelomorphic relationship of a soybean virus resistance gene based on tightly linked DNA microsatellite and RFLP markers. Theor Appl Genet 92:64–69, 1996.

148. RP Pacumbaba. Seed transmission of soybean mosaic virus in mottled and nonmottled soybean seeds. Plant Dis 79:193–195, 1995.

149. FW Nutter, PM Schultz, JH Hill. Quantification of within-field spread of soybean mosaic virus in soybean using strain-specific monoclonal antibodies. Phytopathology 88:895–901, 1998.

150. A Fereres, GE Kampmeier, ME Irwin. Aphid attraction and preference for soybean and pepper plants infected with potyviridae. Ann Entomol Soc Am 92:542–548, 1999.

151. H Bottenberg, ME Irwin. Using mixed cropping to limit seed mottling induced by soybean mosaic virus. Plant Dis 76:304–306, 1992.

152. Q Ren, TW Pfeiffer, SA Ghabrial. Soybean mosaic virus resistance improves productivity of double-cropped soybean. Crop Sci 37:1712–1718, 1997.

153. JC Tu. Strains of tobacco ringspot virus isolated from soybean in southwestern Ontario. Can J Plant Sci 66:491–498, 1986.

154. RG Orellana. Resistance to bud blight in introductions from the germ plasm of wild soybean. Plant Dis 65:594–595, 1981.

155. TS Abney, JC Melgar, TL Richards, DH Scott, J Grogan, J Young. New races of *Phytophthora sojae* with Rps1-d virulence. Plant Dis 81:653–655, 1997.

156. D Barreto, P Grijalba, M Gally, S Vallone, D Ploper. Prevalencia de *Phytophthora sojae* en la region pampeana norte (Argentina), caracterizacion de razas y reaccion de cultivares. Fitopatol Brasil 23:54–57, 1998.

157. AD Heritage, CJ Castles, JA Pierpoint. Surveys of Phytophthora root and stem rot

in early maturing soybean crops in southern temperate regions of Australia from 1985 to 1988. Australasian Plant Pathol 22:131–136, 1993.

158. AF Schmitthenner. Evidence for a new race of *Phytophthora megasperma* var. *sojae* pathogenic to soybean. Plant Dis Rep 56:536–539, 1972.

159. AF Schmitthenner, M Hobe, RG Bhat. *Phytophthora sojae* races in Ohio over a 10-year interval. Plant Dis 78:269–276, 1994.

160. TR Anderson, RI Buzzell. Diversity and frequency of races of *Phytophthora megasperma* f. sp. *glycinea* in soybean fields in Essex County, Ontario, 1980–1989. Plant Dis 76:587–589, 1992.

161. JM Hegstad, DE Kyle, CD Nickell. Pod inoculation technique with *Phytophthora sojae* to evaluate soybean populations for RPS alleles in field plantings. Crop Sci 36:1706–1708, 1996.

162. DG Lohnes, CD Nickell, AF Schmitthenner. Origin of soybean alleles for *Phytophthora* resistance in China. Crop Sci 36:1689–1692, 1996.

163. DE Kyle, CD Nickell, RL Nelson, WL Pedersen. Response of soybean accessions from provinces in southern China to *Phytophthora sojae*. Plant Dis 82:555–559, 1998.

164. BD Rennie, RI Buzzell, TR Anderson, WD Beversdorf. Evaluation of four Japanese soybean cultivars for Rps alleles conferring resistance to *Phytophthora megasperma* f. sp. *glycinea*. Can J Plant Sci 72:217–222, 1992.

165. JM Hegstad, CD Nickell, LO Vodkin. Identifying resistance to *Phytophthora sojae* in selected soybean accessions using RFLP techniques. Crop Sci 38:50–55, 1998.

166. T Kasuga, SS Salimath, J Shi, M Gijzen, RI Buzzell, MK Bhattacharyya. High resolution genetic and physical mapping of molecular markers linked to the *Phytophthora* resistance gene Rps1-k in soybean. Mol Plant-Microbe Interact 10:1035–1044, 1997.

167. DG Lohnes, AF Schmitthenner. Position of the *Phytophthora* resistance gene Rps7 on the soybean molecular map. Crop Sci 37:555–556, 1997.

168. DL Pazdernik, GL Hartman, YH Huang, T. Hymowitz. A greenhouse technique for assessing Phytophthora root rot resistance in *Glycine max* and *G. soja*. Plant Dis 81:1112–1114, 1997.

169. AF Schmitthenner. Problems and progress in control of *Phytophthora* root rot of soybean. Plant Dis 69:362–368, 1985.

170. RI Buzzell, TR Anderson. Plant loss response of soybean cultivars to *Phytophthora megasperma* f. sp. *glycinea* under field conditions. Plant Dis 66:1146–1148, 1982.

171. BA McBlain, JK Hacker, MM Zimmerly, AF Schmitthenner. Tolerance to Phytophthora rot in soybean. II. Evaluation of three tolerance screening methods. Crop Sci 31:1412–1417, 1991.

172. RE Wagner, SG Cramer, HT Wilkinson. Evaluation and modelling of rate-reducing resistance of soybean seedling to *Phytophthora sojae*. Phytopathology 83:187–192, 1993.

173. JAG Irwin, PW Langdon. A laboratory procedure for determining relative levels of field resistance in soybean to *Phytophthora megasperma* f. sp. *glycinea*. Aust J Agric Res 33:33–39, 1982.

174. BA McBlain, MM Zimmerly, AF Schmitthenner, JK Hacker. Tolerance to Phy-

tophthora rot in soybean. I. Studies of the cross 'Ripley' X 'Harper'. Crop Sci 31: 1406–1411, 1991.

175. KD Glover, RA Scott. Heritability and phenotypic variation of tolerance to Phytophthora root rot of soybean. Crop Sci 38:1495–1500, 1998.

176. PR Thomison, CA Thomas, WJ Kenworthy, MS McIntosh. Evidence of pathogen specificity in tolerance of soybean cultivars to Phytophthora rot. Crop Sci 28:714–715, 1988.

177. PR Thomison, CA Thomas, WJ Kenworthy. Tolerant and root-resistant soybean cultivars: reactions to Phytophthora rot in inoculum-layer tests. Crop Sci 31:73–75, 1991.

178. F Workneh, XB Yang, GL Tylka. Effect of tillage practices on vertical distribution of *Phytophthora sojae*. Plant Dis 83:1258–1263, 1998.

179. CK Moots, CD Nickell, LE Gray. Effects of soil compaction on the incidence of *Phytophthora megasperma* f. sp. *glycinea* in soybean. Plant Dis 72:896–900, 1988.

180. VA Dirks, TR Anderson, EF Bolton. Effect of fertilizer and drain location on incidence of Phytophthora root rot of soybeans. Can J Plant Pathol 2:179–183, 1980.

181. RJ Troedson, DE Byth, MJ Ryley, JAG Irwin. Effect of Phytophthora root and stem rot on the response of field-grown soybean to saturated soil culture. Aust J Agric Res 42:791–799, 1991.

182. AB Filonow, JL Lockwood. Evaluation of several actinomycetes and the fungus *Hyphochytrium catenoides* as biocontrol agents for Phytophthora root rot of soybean. Plant Dis 69:1033–1036, 1985.

183. RM Osburn, JL Milner, ES Oplinger, RS Smith, J Handelsman. Effect of *Bacillus cereus* UW85 on the yield of soybean at two field sites in Wisconsin. Plant Dis 79:551–556, 1995.

184. KW Roy, DE Hersman, JC Rupe, TS Abney. Sudden death syndrome of soybean. Plant Dis 81:1100–1110, 1997.

185. TR Anderson, AU Tenuta. First report of *Fusarium solani* f. sp. *glycines* causing sudden death syndrome of soybean in Canada. Plant Dis 82:448, 1998.

186. GL Hartman, GR Noel, LE Gray. Occurrence of soybean sudden death syndrome in east-central Illinois and associated yield losses. Plant Dis 79:314–318, 1995.

187. KW Roy. *Fusarium solani* on soybean roots: nomenclature of the causal agent of sudden death syndrome and identity and relevance of *F. solani* form B. Plant Dis 81:259–266, 1997.

188. H Scherm, XB Yang, P Lundeen. Soil variables associated with sudden death syndrome in soybean fields in Iowa. Plant Dis 82:1152–1157, 1998.

189. VN Njiti, RJ Suttner, LE Gray, PT Gibson, DA Lightfoot. Rate-reducing resistance to *Fusarium solani* f. sp. *phaseoli* underlies field resistance to soybean sudden death syndrome. Crop Sci 37:132–138, 1997.

190. SJC Chang, TW Doubler, V Kilo, R Suttner, J Klein, ME Schmidt, PT Gibson, DA Lightfoot. Two additional loci underlying durable field resistance to soybean sudden death syndrome (SDS). Crop Sci 36:1684–1688, 1996.

191. N Hnetkovsky, SJC Chang, TW Doubler, PT Gibson, DA Lightfoot. Genetic mapping of loci underlying field resistance to soybean sudden death syndrome (SDS). Crop Sci 36:393–400, 1996.

192. J Jin, GL Hartman, CD Nickell, JM Widholm. Phytotoxicity of culture filtrate from

Fusarium solani, the causal agent of sudden death syndrome of soybean. Plant Dis 80:922–927, 1996.

193. GL Hartman, YH Huang, RL Nelson, GR Noel. Germplasm evaluation of *Glycine max* for resistance to *Fusarium solani*, the causal organism of sudden death syndrome. Plant Dis 81:515–518, 1997.

194. Y Luo, O Myers, DA Lightfood, ME Schmidt. Root colonization of soybean cultivars in the field by *Fusarium solani* f. sp. *glycines*. Plant Dis 83:1155–1159, 1999.

195. JA Wrather, SR Kendig, SC Anand, TL Niblack, GS Smith. Effects of tillage cultivar, and planting date on percentage of soybean leaves with symptoms of sudden death syndrome. Plant Dis 79:560–562, 1995.

196. DE Hershman, JW Hendrix, RE Stuckey, PR Bachi, G Henson. Influence of planting date and cultivar on soybean sudden death syndrome in Kentucky. Plant Dis 74:761–766, 1990.

197. KS McLean, GW Lawrence. Interrelationship of *Heterodera glycines* and *Fusarium solani* in sudden death syndrome of soybean. J Nematol 25:434–439, 1993.

198. J Melgar, KW Roy. Soybean sudden death syndrome: cultivar reactions to inoculation in a controlled environment and host range and virulence of causal agent. Plant Dis 78:265–268, 1994.

199. JC Rupe, CM Becton, KJ Williams, P Yount. Isolation, identification and evaluation of fungi for the control of sudden death syndrome of soybean. Can J Plant Path 18:1–6, 1996.

200. L Azevedo. O nematóide de cisto da soja: A experiência Brasileira. Sociedade Brasileira de Nematologia, Jaboticabal, Artsigner ditores, 1999.

201. HE Kauffman. World Soybean Research Conference. VI. Superior Printing, Champaign, IL, 1999.

11
Biological Control of Tomato Diseases

Nancy Kokalis-Burelle
Agricultural Research Service, U.S. Department of Agriculture, Fort Pierce, Florida

I. INTRODUCTION

Pressure to reduce dependence on chemicals to protect crops from pests combined with increased regulation of pesticide registration has necessitated the development of biological methods for agricultural pest management. It is particularly important to employ integrated pest management (IPM) programs in high-value cropping systems such as tomato, where reduced efficacy of chemicals and increased regulations have reduced pest control options. This chapter will provide a review of plant disease management tactics for tomato, which are not reliant on synthetic chemicals, as well as information on specific biological control strategies or agents when available.

Tomato (*Lycopersicon esculentum* Mill.) is in the family Solanaceae, which is native to Peru, Ecuador, and Chile on the west coast of South America. Other important plants in the family Solanaceae are pepper, potato, eggplant, and tobacco. Domestication and cultivation of tomato first occurred in early Mexican civilizations. The tomato was introduced in Europe in 1544 but was thought to be poisonous because of its relationship to nightshade, belladonna, and mandrake [1].

Worldwide production of tomato has increased substantially during the last 30 years to more than 60 million metric tons [2]. In the United States, California, Florida, Ohio, Indiana, and Michigan produce the most acreage of tomatoes, with the entire U.S. crop being produced in Florida from November through May. Consumption of tomatoes in the United States exceeds all vegetables except potatoes [1].

Tomato is classified as a diploid, self-pollinating, tender, herbaceous, perennial vegetable with an optimum mean growth temperature of 21–23°C (70–

75°F). The flower of tomato is perfect, possessing both male and female functional parts. Fruit maturation from pollination to ripening varies from around 6–
10 weeks depending on environmental conditions and variety. Environmental
conditions can greatly influence growth rate, fruit set, yield, and quality of fruit
[2].

II. DISEASE MANAGEMENT PRACTICES

There are approximately 200 known diseases of tomato. Integrated pest management programs for tomato include use of host resistance, pathogen exclusion,
eradication, and protection. Use of biological control agents for specific pathogens also has potential for success, provided that informed decisions are made
regarding optimization of growing conditions and cultural practices are utilized
to reduce disease incidence.

A. Resistant Varieties

Significant effort has gone toward the development of disease-resistant tomato
cultivars, resulting in cultivars suitable for a variety of environments, production
practices, and uses. Demand for high yields, fruit quality, and disease resistance
has resulted in hybrid cultivars that account for approximately 85% or more of
North American fresh market production [3]. The selection of cultivars that are
adapted to local conditions, or that are resistant to common pathogens, is important in reducing disease. In this chapter, availability of resistant cultivars or the
status of research in developing resistance will be covered under sections for
specific diseases.

B. Fertility

The promotion of balanced plant growth and vigor will reduce disease incidence.
Optimal soil pH for tomato production is 6.0–6.5. Fertility management is important to reduce damage to roots, which are primary infection sites for less aggressive pathogens. Plants that are stressed due to low levels of potassium and calcium
can be more susceptible to infection by pathogens such as bacterial wilt [1].
Physiological abnormalities are often the result of nutrient deficiencies. Poor fruit
skin condition and soft fruit are the result of potassium deficiency. Poor fruit
development and root growth are associated with low phosphorus. Blossom end
rot is caused by a calcium deficiency and low levels of boron can result in tip
dieback, fruit russeting, and brittle stems [1]. Some pathogens like Fusarium wilt
can be managed by increasing the ratio of nitrate to ammonium nitrogen fertilizers
(see Sec. III.A.2).

C. Moisture

It is important that tomatoes be grown on well-drained soils. Periods of excessive moisture in the soil can cause severe crop loss due to oxygen depletion and increased disease incidence [4]. There are several irrigation systems commonly used in tomato production, including overhead sprinkler, micro or drip, furrow, level-basin, and subirrigation. Drawbacks of the different systems include increased potential for foliar diseases with overhead irrigation and inadequate drainage in subirrigation [4].

In order to reduce foliar pathogens, wider rows can be used to increase airflow between the rows. Staking or trellising plants can improve overall production and fruit quality by allowing more air movement through the canopy and reducing incidence of foliar diseases. If overhead irrigation is used, it should be applied early in the day to allow the foliage to dry before evening. Working in the fields when leaves are dry will decrease the spread of waterborne pathogens.

D. Seed/Transplant Treatment

Reduction of pathogen inoculum in the field can be done by using certified disease-free seed or seedlings that have been inspected at all stages of production. Physiological seed treatments such as seed priming have been used to quicken seed germination and improve seedling survival [5]. During seed priming, the seeds are placed under controlled environmental conditions in an aerated osmotic solution of known water potential. This results in the imbibition of water and the completion of early metabolic processes of germination short of the emergence of the radicle. Osmotically priming seed requires specialized equipment and consequently researchers have explored alternative methodology including solid matrix priming [6]. With tomato seeds, combining matrix priming with *Trichoderma harzianum* or *T. koningii* resulted in improved seedling emergence and reduced incidence of damping-off [6].

Seed treatment is a practical delivery system for both fungal and bacterial biocontrol agents. Biological control agents applied to seed have been shown to protect seed in a variety of crops, as well as increase plant growth and vigor [7–9]. Research indicates that the use of biological agents as seed treatments is valuable but provides more variable and less effective protection than chemical seed treatments [10].

Most vegetables intended for transplanting, including tomato, are produced in small-celled flats in commercial potting media that consist of sphagnum peat, nutrients, and lime. For transplanted crops such as tomato, there exists an opportunity to introduce biocontrol agents into the transplant mix during the greenhouse production phase. This practice gives the biocontrol agents a relatively unchallenged time frame in which to colonize the rhizosphere and become established before transplanting into the field. Recent work has shown that transplant mixes amended with a formulated mixture of plant growth–promoting rhizobacteria (PGPR) increased transplant survival and yield of tomato in Florida under heavy pressure from root-knot nematode and Fusarium root and crown rot caused by *Fusarium oxysporum* f. sp. *radicis-lycopersici* (FORL) [11]. Field experiments with naturally occurring populations of FORL were conducted to evaluate *Trichoderma harzianum* as a seed coating or as a wheat bran/peat preparation [12]. *Trichoderma*-treated plots had less disease and 26% higher yields than control plots. *T. harzianum* populations were isolated from field-grown treated plants up to 20 weeks after planting. The antagonist was detected at highest concentrations on the root tips, resulting in elimination of *Fusarium* spp. from those segments of the root. Nemec et al. [13] found that *T. harzianum* delivered in the transplant plug system effectively controlled Fusarium crown rot, and that *Bacillus subtilis* delivered in this manner provided effective control of *Phytophthora* in the field. Nemec [14] also found that isolates of *Bacillus* and *Trichoderma* survived better in planting mixes than several other potential biocontrol agents including species of *Pseudomonas* and *Serratia*. The survival of the *Pseudomonas* isolates tested was greater when mixed with propagules of the fungus *Glomus intraradices*.

E. Crop Rotation

Sustainable crop management practices include the use of crop rotation. The management of many types of plant pathogens is often based on rotation to less susceptible, nonhost, or pathogen-suppressive crops. Rotations including antagonistic plants have been reported to alter the soil microflora, enhance populations of microorganisms known to be antagonistic toward pathogens, increase plant growth, or induce systemic resistance in the host [15]. Yield reductions in continuous cropping systems have been correlated with increased populations of pathogens. Adjustment of the root-soil environment and balance of microorganisms present in the rhizosphere can positively influence plant growth [16]. Crop rotation is often most effective against pathogens that attack only one crop. Many pathogens of tomato such as the wilts and root rots have wide host ranges, which makes crop rotation less likely to reduce disease. Recommendations for rotation practices will be made, when available, in sections on individual pathogens.

F. Organic Amendments

Increasing soil organic matter can help reduce plant stress by increasing the water holding capacity of the soil. There have been many reviews of the effects of organic amendments on soilborne pathogens, plant parasitic nematodes, and soil microbial ecology [17–19]. Ammonia released from animal manures has been widely reported to reduce survival and germination of some soilborne fungal pathogens [20] and viability of plant parasitic nematodes [19]. Many plant metabolites released from debris after incorporation into soil are known to be pesticidal, such as glucosinolates, which are hydrolyzed to release antimicrobial sulfur compounds. Members of the Cruciferae plant family contain high levels of glucosinolates in their tissues, and effects of incorporation of crop residues on pathogenic organisms have been investigated [21]. Plants in the family Compositae produce insecticidal pyrethrin compounds as metabolites, and members of the genus *Artemisia* produce terpenoid compounds [21]. Antifungal volatile compounds including allylisothiocyanate are found in many *Brassica* spp. [22]. Plant pathogens such as *Pythium* spp. and *Sclerotium rolfsii* are more vulnerable to toxic effects of volatile compounds than saprophytic soil microorganisms, which are relatively insensitive [23,24].

Increasing soil organic matter also increases the populations of beneficial microorganisms in the soil and reduces the populations of some tomato root pathogens. Volatile compounds, such as alcohols and aldehydes, released from organic matter can stimulate germination of fungal propagules and increase microbial activity in soil [25]. Increases in microbial activity in organic soils and composts are often correlated with suppression of soilborne pathogens [23,24]. Reduction of diseases caused by *S. rolfsii* and *Rhizoctonia solani* have been correlated with increases in soil enzyme activity in response to the addition of pine bark to soil [26,27]. Species of *Paecilomyces* and *Penicillium* that compete with pathogens by mycoparasitism and antibiotic production were the predominant colonizers of amended soil in these studies.

G. Soil Solarization

Soil solarization has proven effective against some pathogens under certain environmental conditions. Solarization is accomplished by covering the surface of the soil with clear plastic film to trap solar radiation and accumulate heat levels lethal to many plant pathogens, weeds, and nematodes. Soil is typically solarized for 4 or more weeks in order to raise temperatures effectively to a depth of 45–60 cm.

Solarization reduces the amount of pathogen inoculum in several ways, including direct thermal destruction of propagules; shifts in populations and activity of soil microorganisms; changes in physical and chemical properties of soil; and accumulation of volatile compounds produced by physical or microbial decomposition of organic matter [28]. An induction of soil suppressiveness, which prevents reestablishment of pathogens, has been reported following solarization [29]. Increased soil populations of fluorescent pseudomonads and *Bacillus* spp., known for antibiotic production, may be an important factor in suppression of pathogens in solarized soil [29]. Solarization has also been used successfully to control Verticillium wilt of tomato (see below).

The use of organic amendments and fertilizers may be a way to improve the effectiveness of solarization. Heating soil to a temperature of 45°C alone is effective in reducing viability of some fungal pathogens, including *S. rolfsii* and *Pythium ultimum*. However, heating soil amended with cabbage residue or composted chicken manure has resulted in better control at lower soil temperatures (38°C) [30,31]. Compost-amended soil reaches higher temperatures (2–3°C) under the same conditions than nonamended soil, which may be an important factor in improving control of organisms such as root-knot nematodes that are more resistant to heat. Thermal conductivity, exothermic microbial activity, and evolution of volatile compounds in solarized amended soil may also contribute to increased control of *Meloidogyne incognita* [32].

Some of the limits of solarization include the length of time that land must remain out of production (4–6 weeks) and its dependency on the weather. Also, solarization does not control many important pathogens such as root-knot nematodes in deep sandy soils and bacterial wilt caused by *Ralstonia solanacearum* [33]. Solarization can contribute significantly to disease reduction by controlling weeds that act as hosts to many tomato pathogens [34].

H. Sanitation

Sanitation practices such as the removal of diseased plant material from greenhouses and fields will lessen the amount of pathogen inoculum and reduce the spread of disease. Knives, tires, and implements should be cleaned in a 0.5–1.0% solution of quaternary ammonium chloride or other recommended disinfectants after working in an infested field. Tomato stakes should be rinsed free of soil and treated with a 10% bleach solution or can be disinfested by solarization.

III. BIOLOGICAL CONTROL

The majority of research on biological control of tomato pathogens has been on soilborne fungal pathogens and nematodes. This is due, in part, to the extensive host range of some of these organisms but mostly to the destructive potential of these pathogens and their economic importance. It is also more difficult to manipulate microbial populations in the phylloplane using inundative application of biocontrol agents or practices such as organic amendments or crop rotation. In many cases, foliar and fruit pathogens can be effectively controlled or reduced by adjusting crop management practices such as plant spacing, sanitation, and irrigation.

A. Epiphytic Fungal and Bacterial Pathogens

1. Anthracnose (*Colletotrichum* spp.)

Anthracnose is primarily a disease of ripe to overripe fruit and is most prevalent on processing and garden tomatoes, due to the extended period of time the fruit remains in the field. Anthracnose occurs in Asia, Europe, Africa, the East Indies, and North America [35]. Initial symptoms may occur during green stages and include small, circular, sunken, water-soaked lesions that spread to form a soft internal decay. Lesions may be salmon colored with black microsclerotia visible.

Reducing stress, reducing insect damage, and avoidance of overhead irrigation will also reduce losses due to anthracnose [36]. Due to the wide host range of this pathogen, weed control is extremely important and will reduce disease incidence. Avoidance of excessive overhead irrigation and 2-year rotations to nonsolanaceous crops are also recommended to reduce disease incidence [35,36].

2. Early Blight (*Alternaria solani*)

Early blight occurs in all tomato-producing regions. It is more destructive under humid conditions, such as in the Southeast and middle Atlantic regions of the United States, than under dryer conditions. Initial symptoms of early blight are small brownish-black lesions on the older foliage that enlarge rapidly and produce concentric rings. The tissue surrounding the spots may become chlorotic, and the entire leaf may become chlorotic as the infection spreads [36]. Stem-infected seedlings will usually die due to stem girdling. Fruit lesions may become extensive, often covering the entire fruit. Infected fruit frequently drop and may result in losses of up to 50% [36].

Control of early blight can be increased with the use of resistant cultivars, long rotations, weed control, and proper fertilization to keep plants vigorous [37]. Resistance to early blight has been identified in *Lycopersicon hirsutum*, and attempts are being made to breed this into commercial varieties [38].

3. Gray Mold (*Botrytis cinerea*)

Gray mold occurs wherever tomato is grown and consistently causes minor losses with occasional major outbreaks in the field. Conditions for gray mold development include cool temperatures and moisture. Gray mold occurs on all aboveground plant parts and appears as fuzzy gray fungal growth from necrotic tissue. In greenhouse-grown tomatoes, gray mold can be a serious problem due to high relative humidity and free moisture present on the plant surfaces, which is required for conidial germination [39]. Poor air circulation will contribute to disease severity. Acid soils should be limed to increase calcium in the plants and reduce susceptibility to the pathogen.

Several saprophytic bacteria and fungi are reportedly effective biocontrol agents against *B. cinerea* [40]. *Trichoderma harzianum* provided good control of *B. cinerea* when the organisms were applied simultaneously, rather than being applied once the infection was established [41]. Although disease was not reduced once the infection was established, *T. harzianum* did reduce the amount of sporulation from resulting lesions. Dik and Elad [42] tested *Trichoderma harzianum* and *Aureobasidium pullulans* for efficacy against *B. cinerea* in greenhouse experiments under different climatic conditions. Stem lesions and plant death were reduced from 40 to 100%, and in some cases control was better than that provided with the fungicides tolyfluanid and iprodione. Control of stem lesions and wilting was better than control of symptoms on fruit.

Shtienberg and Elad [43] developed an integrated strategy for control of *Botrytis cinerea* that incorporates weather forecasting to decide between spraying a chemical fungicide or the biological control agent *Trichoderma harzianum*. This integrated control strategy (BOTMAN, short for Botrytis manager) makes recommendations for application of the biocontrol agents during environmental conditions that are most favorable for survival and efficacy. The integrated strategy was compared with weekly applications of fungicides. Results showed similar disease control in both systems. Implementation of the integrated system would therefore result in similar disease control, a reduction of fungicide use, and, consequently, a reduction in the probability of the pathogen developing fungicide resistance.

The greatest damage by *B. cinerea* is fruit rot in the greenhouse, field, or in shipment [44]. Treatment of fruit with heated water or air is effective in controlling postharvest infection of *B. cinerea* [45]. Mari et al. [46] found that *Bacillus amyloliquefaciens* had a fungistatic effect on *B. cinerea* on mature green tomatoes stored at low temperature (10°C), and significantly reduced pathogen growth during the first 7 days of storage.

4.   Bacterial Spot (*Xanthomonas campestris* pv. *vesicatoria*)

Bacterial spot is present in all tomato-producing areas but is most serious in tropical and subtropical regions. This disease is of minor importance in the United States, with the exception of Florida, where it causes major yield and quality reductions. All above-ground plant parts are affected. Foliar symptoms of bacterial spot appear as brown, water-soaked, circular lesions usually less than 3 mm in diameter. The lesion may appear elongated on the leaf margins. Sometimes after a heavy rain, entire interveinal areas will become infected [36].

Early symptoms on the fruit appear as tiny black specks surrounded by a slightly lighter area. As the lesion enlarges, it becomes brownish, scab-like, slightly raised on the edges, and sunken in the middle. The epidermis eventually ruptures and curls back. Bacterial spot lesions usually extend only as far as halfway through the outer fleshy layer of the fruit [36].

Dissemination of the pathogen within the field is by rain, plant pruning, and aerosols [47]. Only disease-free transplants should be used. Once this pathogen is established in the field, it is difficult to eradicate. Fields should be rotated to avoid pathogen carry-over on volunteers and crop residue. Bacteria may also be perpetuated on contaminated seed [47].

5. Bacterial Speck (*Pseudomonas syringae* pv. *tomato*)

Bacterial speck occurs in tomato-growing regions worldwide but is only of importance under high-moisture, low-temperature conditions. Foliar infection by bacterial speck closely resembles bacterial spot. Fruit infection by bacterial speck appears as numerous, tiny, brown lesions less than 1.5 mm in diameter that do not extend below the epidermis of the fruit. Compared to a bacterial spot lesion, lesions of bacterial speck on the fruit are more restricted in size, are not raised, and do not cause the epidermis to rupture.

Bacterial speck is seedborne, and outbreaks are more severe during wet growing seasons [36]. Bacterial speck can reduce yield, but its primary effect is on fruit quality. Modification of irrigation practices including a decrease in overhead irrigation can reduce losses. Use of clean, disease-free seed and transplants will lessen disease incidence, which is difficult to control once established. Planting in the same field in consecutive seasons should be avoided and fields kept free of weeds and volunteers [48]. *Pseudomonas syringae* pv. *tomato* can colonize the surface of tomato plants and survive as an epiphyte for extended periods [49–51]. This trait makes it possible to use nonpathogenic bacteria to compete with the pathogen for colonization and infection sites. Reduction of bacterial speck occurred in greenhouse tests in response to application of a nonpathogenic transposon (Tn5) mutant of *P. syringae* pv. *tomato*. Applications of Kocide, preceding application of the nonpathogenic, copper-resistant mutant, resulted in greater reduction in disease than either treatment alone [52] (Table 1). Fluorescent pseudomonads have also been used to reduce bacterial speck in the field on both young and mature plants. Fluorescent *Pseudomonas* spp. partially controlled bacterial speck with slightly better residual activity than copper compounds [53].

6. Bacterial Stem Rot (*Erwinia caratovora* subsp. *carotovora*)

Bacterial stem rot is considered of minor importance, but occasionally substantial losses can occur. This disease occurs in the greenhouse and field, primarily on pruned, staked, or trellised tomatoes. Initial symptoms of bacterial stem rot occur at first fruit harvest and appear as a wilt. Eventually the pith disintegrates, causing a hollow stem.

Erwinia caratovora subsp. *carotovora* also causes soft rot in many vegetables including tomato fruit. This bacterium is ubiquitous and requires a fresh wound for infection, such as the removal of suckers or leaves. High relative

Table 1 Incidence of Bacterial Speck of Tomato Caused by Copper-Sensitive Strain PT12 of *Pseudomonas syringae* pv. *tomato* After Treatment with Kocide 101 and Coinoculation with Nonpathogenic Copper-Resistant Strains PT23.200 and PT23.201

Inoculum[a]	Preinoculation treatment with Kocide 101[b]	Lesions per leaflet[c]		
		Trial 1	Trial 2	Trial 3
PT12	−	115.0 a	23.6 a	5.2 a
	+	10.0 c	5.4 c	2.4 b
PT12 + PT232.200	−	36.0 b	10.7 b	2.2 b
	+	11.6 c	4.3 c	1.0 c
PT12 + PT23.201	−	19.6 c	3.6 c	0.6 c
	+	4.8 d	1.4 d	0.3 d

[a] Inoculum concentrations were approximately 2×10^7 cfu/mL for PT12 and 5×10^8 cfu/mL for PT23.200 and PT23.201.
[b] Kocide 101 was applied at the label rate of 2.4 g/L 1 day before bacterial inoculations.
[c] Mean of four leaflets from 12, 14, and 6 plants in trials 1, 2, and 3, respectively. Data were log-transformed before statistical analysis. The values presented are antilogs of the transformed means. Values followed by the same letter within a column do not differ significantly ($p = 0.05$) according to the Student-Newman-Keuls' test.
Source: Ref. 52.

humidity is required for the disease to develop. Sanitation is the most effective way of controlling this pathogen [54].

B. Soilborne Fungal and Bacterial Pathogens

1. Fusarium Crown and Root Rot (*Fusarium oxysporum* f. sp. *radicis-lycopersici*)

Symptoms of Fusarium crown rot (FORL) appear during cool season periods and include marginal yellowing on the lower leaves and an initially slow to increasingly rapid wilt that kills the plant. The lower stem exhibits vascular discoloration and pith necrosis [36]. Resistant cultivars are not yet available.

Manipulation of mineral fertilizers can be effective in controlling FORL by reducing pathogen growth [55], improving host defenses [56], or favoring indigenous populations of disease-suppressive bacteria [57]. Many fluorescent pseudomonad rhizosphere bacteria produce 2,4-diacetylphloroglucinol (Phl) [58,59], a secondary metabolite that has been shown to be toxic to bacteria [60], fungi [58,60], and nematodes [61]. Sharifi-Tehrani et al. [62] found a significant correlation between the amount of Phl produced on plates and the amount of protection those strains provided against Fusarium crown and root rot. A seedling assay was developed to evaluate pseudomonads for suppression of Fusarium crown and root rot in vitro, which provided results correlated with those from in vivo trials. Duffy and Défago [63] found that zinc amendments improved biocontrol of Fusarium crown and root rot by *Pseudomonas fluorescens* by 25%. Addition of minerals is an inexpensive way to improve biocontrol by creating a more favorable environment for disease suppression by reducing pathogen activity in the soil.

An endophytic strain of the bacterium *P. fluorescens* was evaluated for induction of resistance to FORL in tomato and was found to reduce colonization and restrict pathogen growth to the outer root tissues and intercellular spaces [64]. Typical host reactions included accumulation of electron-dense material in epidermal and outer cortical cells and most intercellular spaces. *Pseudomonas chlororaphis* strain PCL1391 was selected from over 70 bacterial isolates from the rhizosphere of tomato screened for activity against FORL [65] (Fig. 1). The biocontrol activity of this isolate was characterized at the molecular level and found to be mediated through the production of phenazine-1-carboxamide (PCN) through use of a PCN-negative mutant.

Figure 1 Root epidermis of a tomato seedling 3 days after inoculation. This region, located approximately 1 cm under the stem, is covered by *Pseudomonas fluorescens* strain WCS365. Numbers declined drastically down the root. Most bacteria are clearly covered by a mucigel (semi-transparent) layer. The bar represents 10 μm. (From Ref. 65.)

One of the first studies evaluating use of microbial antagonists for control of FORL under field conditions was performed in Florida by Marois et al. [66]. A mixture of three isolates of *Trichoderma harzianum*, one isolate of *Aspergillus ochraceus*, and one isolate of *Penicillium funiculosum* were applied to field plots. Disease incidence at harvest was 7% in antagonist-treated plots and 37% in untreated plots. The population of the pathogen in soil was also reduced from 600 propagules per gram to 200 propagules per gram in the antagonist-treated plots. Sivan and Chet [67] found that *Trichoderma harzianum* provided effective control of Fusarium crown rot in Israel. In the United States. *T. harzianum* and *Glomus intraradices* were found to control Fusarium crown and root rot in commercial production fields in Florida [68].

Additional work with fungal antagonists of FORL include evaluation of the mycoparasite *Pythium oligandrum*, which increased resistance to FORL in tomato [69]. When tomato plants were previously inoculated with *P. oligandrum*, the fungus

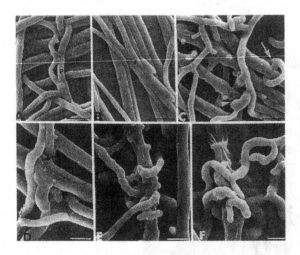

Figure 2 Scanning electron micrographs of *Pythium oligandrum* hyphae interacting with cells of *Fusarium oxysporum* f. sp. *radicis-lycopersici* in dual cultures. (A and B) Controls grown in pure culture. (A) *P. oligandrum* (P) and (B) *F. oxysporum* f. sp. *radicis-lycopersici* (F). Both fungi form a dense, branched mycelium (×1500; bar = 10 μm). (C–F) Hyphal interactions in dual cultures 2, 3, 4, or 5 days after inoculation, respectively. (C) Hyphae of both fungi appear closely intertwined with hyphae of *F. oxysporum* f. sp. *radicis-lycopersici*. Contact between the fungi is apparently established through a thick mucilage (arrowhead). Features of coiling are observed (arrow) (×1500; bar = 10 μm). (D) Slight wall deformations are seen along the areas of contact (arrow) (×2400; bar = 5 μm). (E) Early signs of collapse shown by wrinkled cell surface (×2400; bar = 5 μm). (F) Marked collapse and loss of turgor of some cells is observed (×2000; bar = 5 μm). (From Ref. 69.)

showed strong antagonism in the rhizosphere and in planta towards FORL (Fig. 2). In addition, inoculation with *P. oligandrum* induced structural and biochemical barriers in host tissue that adversely affected pathogen growth and development.

Fungal wall fragments such as glucan, chitin, or chitosan oligomers have been shown to be active inducers of plant defense responses also called elicitors [70,71]. Benhamou and Thériault [72] demonstrated that tomato plants treated with chitosan were protected against Fusarium crown and root rot. In further studies, Benhamou et al. [73] showed that seed coating with chitosan in combination with substrate amendment increased resistance of tomato seedlings to FORL attack. Resistance was correlated with restricted fungal growth in root tissue, decreased pathogen viability, and accumulation of deposits in host cells (Fig. 3). This work indicates that external application of chitosan stimulates the overall plant defense system and is capable of reducing disease incidence of important and aggressive soilborne pathogens such as *Fusarium* spp.

Further work to improve the control achieved with both biological agents and elicitors of plant defense responses such as chitosan has been performed. Combinations of chitosan and the endophytic bacteria *Bacillus pumilus*, a PGPR known to induce defense reactions in plants, were evaluated for activity and cytological response in tomato challenged with FORL [74]. Bacterial treatment induced enhanced physiological and biochemical changes at sites where the fungal pathogen attempted penetration of the host cells. Combination treatments of chitosan and *B. pumilus* were associated with restricted pathogen growth in root tissue, decreased pathogen viability, and accumulation of callose-enriched cell wall apositions on the inner cell wall surface in epidermal and outer cortical cells.

2. Fusarium Wilt (*Fusarium oxysporum* f. sp. *lycopersici*)

Symptoms of Fusarium wilt (FOL) include a stunting of infected seedling plants, drooping and downward curving of older leaves, and plant wilt and death. Symptoms often become apparent on older plants between blossoming to fruit maturation [36]. Early symptoms include chlorosis of older leaves, which often develops on only one side of the plant. As the chlorosis progresses the plant will wilt during the heat of the day [36]. The wilt progresses over a period of days until the plant dries up and completely collapses. Extensive vascular browning which extends up the stem and is noticable in the petiole scar is characteristic of this pathogen. Soil pH near 7.0 reduces disease severity. Fusarium wilt can be managed by increasing the ratio of nitrate to ammonium nitrogen fertilizers. Nitrates raise soil pH, which reduces pathogen growth and increases host resistance to phytotoxins like fusaric acid [55,56].

Natural suppression of Fusarium wilt has mainly been attributed to fluorescent pseudomonads and nonpathogenic *Fusarium oxysporum* isolates. Mechanisms involved in disease suppression include microbial antagonism during the

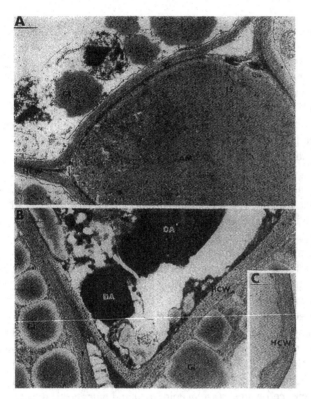

Figure 3 Transmission electron micrographs of tomato root tissues from chitosan-treated seeds (1 mg/mL) grown in chitosan-amended substrate (1 mg/mL) and inoculated with *Fusarium oxysporum* f. sp. *radicis-lycopersici*. (A) Labeling of β-1,3-glucans with a purified tobacco β-1,3 glucanase complexed to gold. The globules (GI) as well as the material (AM) filling the intercellular spaces are specifically and intensely labeled. Scattered gold particles are seen all over the host cell wall (HCW) (×24,000; bar = 0.5 μm). (B and C) Labeling of phenolic and ligninlike compounds with a purified laccase complexed to gold. (B) The globules (GI) are unlabeled while the dense aggregates (DA) are heavily labeled. The host cell wall (HCW) is decorated by a significant number of gold particles (×40,000; bar = 0.25 μm). (From Ref. 73.)

saprophytic phase of pathogen growth and induced resistance in the host during the pathogenic phase [75–77]. A nonpathogenic *F. oxysporum* isolate which both expressed microbial antagonism and induced resistance was more efficient suppressing Fusarium wilt in commercial tomato production settings than isolates that expressed only one mechanism [78].

Mao et al. [79] found that, in studies involving multiple soilborne pathogens, plant stand was increased to levels comparable to the noninfested control plants with the addition of combinations of the biocontrol agents *Gliocladium virens* and *Burkholderia cepacia*. In addition to *Fusarium oxysporum* f. sp. *lycopersici*, soilborne pathogens in that study included *Rhizoctonia solani*, *Pythium ultimum*, and *Sclerotium rolfsii*. When transplants were set out into field plots infested with multiple pathogens, the combination of biocontrol agents resulted in greater fruit yield in tomato than those obtained with either biocontrol agent alone [79]. In tests evaluating nonpathogenic *Fusarium* spp., *Trichoderma* spp., *G. virens*, *P. fluorescens*, and *B. cepacia*, for biological control potential, Larkin and Fravel [80] found that isolates of *F. oxysporum* and *F. solani* collected from a Fusarium wilt–suppressive soil provided the most consistent disease control. These isolates were also found to be effective in controlling Fusarium wilt diseases of other crops including watermelon and muskmelon. Other organisms tested also reduced disease but not as consistently as the nonpathogenic Fusarium isolates. *Penicillium oxalicum* applied to roots as a biological agent for FOL reduced disease in stem inoculated plants [81]. Due to inoculation techniques that maintained physical separation of pathogen and antagonist, disease control was attributed to induced resistance. A rapid method for evaluating biocontrol potential was developed using *Penicillium oxalicum* as a biological control agent for FOL [82]. This method consists of growing plants in flasks with nutrient solution. Biocontrol agents and pathogens are added to the solution. Typical disease symptoms were observed and biocontrol effects were clear. Nutrient solution consumption was correlated with disease parameters and proved to be an easy method for quantifying disease severity. Further work showed that the timing and application method of *Penicillium oxalicum* affected the level of suppression of FOL [83]. Application of the biocontrol agent to tomato seedlings in seedbeds, rather than to tomato seed, provided more effective disease suppression, which was maintained for 60–100 days after transplanting.

3.   Verticillium Wilt (*Verticillium dahliae*)

Verticillium wilt occurs in all tomato-growing regions but is favored by cool conditions and neutral to alkaline soils [84]. This fungus has a very wide host range, which includes many vegetables. *Verticillium dahliae* overwinters in plant debris or soil as microsclerotia and can remain viable for up to 30 years. This fungus is a poor competitor in soil and often invades plant tissue through wounds and nematode feeding sites [84]. Soil solarization has proven effective in reducing inoculum levels of *V. dahliae* while increasing populations of beneficial fungi including *Talaromyces flavus* [85]. Exposure of microsclerotia to sublethal temperatures during solarization, combined with increases in populations of thermophilic antagonists in soil, resulted in increased mortality of the pathogen and suppression of Verticillium wilt [86]. Cultural control measures for Verticillium wilt include resistant cultivars, crop rotation, and reducing nematode populations, which reduces infection sites.

4. *Rhizoctonia solani* Diseases

Rhizoctonia solani Kuhn is commonly found in soils and causes some type of disease on most cultivated plants. The fungus survives in soil and on dead plant material, both of which serve to disseminate the pathogen. *Rhizoctonia solani* causes a variety of diseases on tomato worldwide including damping-off, root rot, stem canker, stem rot, and fruit rot. Damping-off can occur either pre-emergence or postemergence in the greenhouse or field. Rhizoctonia root rot is often more severe when plants have sustained damage by root-knot nematodes or when under low-temperature stress. Providing plants with optimum growing conditions and preventing injury and incidence of root-knot nematodes will lessen susceptibility to *Rhizoctonia*. This pathogen is most aggressive under optimum soil moisture. Both dry and waterlogged soil conditions inhibit fungal growth [87].

Symptoms of damping-off include tip necrosis and/or reddish-brown lesions on the seedling. Postemergence symptoms include a dark constricted lesion at the soil line that causes the plant to fall over. Root rot lesions are distinct and dark in color. Basal stem canker or foot rot produces sunken reddish-brown lesions just below the soil line. Ripe tomatoes are more susceptible to fruit rot, which occurs under warm, wet conditions and produces a brown rot that may have alternating light and dark bands [87]. Fruit loss can be reduced by eliminating fruit contact with soil.

Asaka and Shoda [88] determined that iturin A and surfactin, two antibiotics produced by the biological control agent *Bacillus subtilis* RB14, play an important role in the suppression of damping-off caused by *R. solani*. They found that when plants were treated with culture broth of *Bacillus subtilis* RB14 without the pathogen *R. solani* present, the growth of tomato plants was the same as that of the control plants, indicating that the bacterium imparts no growth-enhancing activity to the plant. This study indicates that treatment of soil with the culture broth, cell suspension, or centrifuged culture broth is as effective as using the biological control agent.

5. *Pythium* Diseases

Pythium spp., including *P. aphanidermatum*, *P. myriotylum*, *P. arrhenomanes* Drechs., *P. ultimum* Trow, and *P. debaryanum* R. Hesse, cause several diseases of young tomatoes including seed rot, preemergence and postemergence damping-off, and stem rot. Symptoms include a soft, mushy rot of seed, which occurs before radicle emergence, dark-colored, water-soaked lesions affecting the entire seedling prior to emergence, and water-soaked lesions extending up the stem after seedling emergence. Use of high-quality seed and growing plants under optimal temperature, moisture, and nutritional conditions will reduce incidence of *Pythium* diseases. Excessive moisture and poor drainage should be avoided. Fruit rot can be lessened by avoiding fruit contact with soil [89].

A variety of antagonsistic miocroorganisms have been used as seed treatments to control *Pythium* damping-off [90–92]. Elad and Chet [93] found that competition for nutrients between germinating oospores of *P. aphanidermatum* and several rhizosphere bacteria was significantly correlated with disease suppression in greenhouse trials. Their results indicate that the presence of bacteria in the root zone of susceptible plants reduced the potential sites for establishment of *Pythium* along the roots.

Van Dijk and Nelson [94] showed that the bacterium *Enterobacter cloacae* can utilize seed exudate from a number of plant species, including tomato, as a sole carbon and energy source reducing the stimulatory effect that those compounds have on *Pythium ultimum* sporangia. This is an example of a biological control agent indirectly affecting a pathogen by having an impact on compounds produced by the host that are necessary for stimulation of the infective propagule.

Another potential biocontrol agent against pathogenic *Pythium* spp. is the nonpathogenic fungus *Pythium oligandrum*. The occurrence of large populations of this fungus has been correlated with soil suppressiveness to damping-off caused by other *Pythium* spp. [95]. Rey et al. [96] provided a detailed investigation of the interaction of *P. oligandrum* colonization and interaction with tomato roots. That study showed the tomato root tissue was extensively colonized by the fungus but showed no signs of necrotic symptoms (Fig. 4). Also, that study noted the accumulation of osmophilic or electron dense chemical compounds in invaded and reacting cells considered to be phenolic in nature (Fig. 4).

Tomato is a good candidate for use in exploring the genetic basis for host interactions with biocontrol agents. This is due to tomato being a self-pollinated, diploid plant that exists primarily as homogeneous, homozygous lines that have been extensively mapped for inherited traits. Smith et al. [97] found differences among tomato lines for both resistance to the pathogen and response to biological control. The fact that the two traits were independent indicates that it may be possible to combine them through breeding to improve disease suppression. Smith et al. [98] went on to identify three quantitative trait loci (QTL) in tomato associated with disease suppression by the biological control agent *Bacillus cereus*. Two of the QTL for disease suppression by *B. cereus* map to the same locations as QTL for other traits, which suggests that the host effect on biocontrol is mediated by different mechanisms.

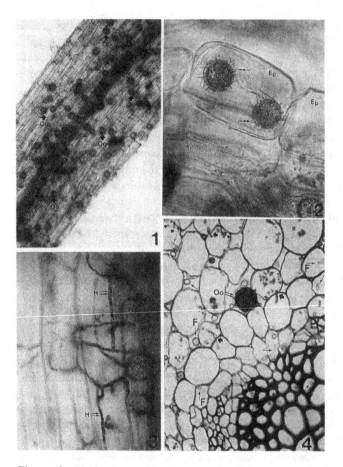

Figure 4 Light micrographs of tomato (cv. Prisca) root colonized by *P. oligandrum* 48–72 hours after inoculation. Ep: Epidermal cell; F: fungus; H: hypha; Oo: oogonium. (1) Many oogonia have developed over the root surface (arrows). Note that root invasion with *P. oligandrum* is not associated with a necrotic reaction (×26). (2) Oogonia present in an epidermal cell. Their external walls are ornamented with spines (double-headed arrows) (×260). (3) Hyphae (double arrows) on the root surface stained blue by the immuno-enzymatic treatment (oogonia are unstained). The hyphae form a loose mycelial network over the surface (×260). (4) All root tissues, including the vascular stele, are invaded by *P. oligandrum*. Host wall penetration is achieved by means of constricted hyphae (thick arrows). Numerous hyphae appear like empty shells (thin arrows). Unlike on the root surface, only few oogonia have developed within the root tissues (double arrows) (×65). (From Ref. 96.)

6. Southern Stem Blight or Stem Rot (*Sclerotium rolfsii*)

Southern blight is a widespread disease among vegetables in the southern United States. The first symptoms of this disease are wilting and chlorosis of the leaves. Warm temperatures and high moisture following dry periods favor this disease. To help control southern blight farmers should avoid planting in heavily infested fields, deep plow before planting, and reduce cultivation after planting. Studies in North Carolina showed that soil solarization combined with the biological control agent *Gliocladium virens* reduced disease incidence by 49% the first season after solarization and 60% the second season [99].

7. Late Blight (*Phytophthora infestans*)

Early symptoms of late blight are blackish-purple water-soaked lesions on leaf margins of lower leaves. Under humid conditions, a white spore–producing hyphal mass appears at the margin of the lesion on the lower leaf surface. Late blight can be introduced on transplants or can be introduced into a tomato field from infested potato or tomato fields as airborne spores. There are resistant cultivars available. Other control practices include isolating tomato and potato fields, using disease-free transplants, and destroying potato cull piles [1].

8. Buckeye Rot (*Phytophthora parasitica, P. capsici*, and *P. drechsleri*) and Phytophthora Root Rot (*Phytophthora parasitica* and *P. capsici*)

Buckeye rot occurs in high-humidity, high–soil moisture conditions worldwide. In the United States, the disease occurs most frequently in the southeast. Fruit symptoms include a brownish rot at the point of contact with soil. Buckeye rot lesions remain firm and smooth with a pattern of concentric rings. Fruit eventually decay, while the foliage remains unaffected. Warm, wet conditions favor this pathogen, which is spread by surface water and splashing rain. Heavy, poorly drained soils should be avoided [100]. *Phytophthora* spp. have a broad host range, including many weeds.

Phytophthora root rot is a major problem on tomatoes in California. Symptoms on roots include dry, water-soaked, dark brown lesions that may girdle the roots and cause extensive decay. Root rot is more severe in compacted, poorly drained soils. Dassi et al. [101] investigated the possible involvement of different pathogenesis-related (PR) protein families in the biocontrol of *P. parasitica* by the arbuscular mycorrhizal (AM) fungus *Glomus mosseae*. They found that although the bioprotective effect of the AM fungus towards the pathogen was evident, PR proteins belonging to five common families of compounds were not directly involved. Cordier et al. [102] used antibodies to label hyphae of *Phytophthora nicotianae* var. *parasitica* to distinguish it from hyphae of the AM fungus *Glomus mosseae* and found that precolonization by the AM fungus resulted in decreased root damage by the pathogen.

9. Timber Rot (*Sclerotinia sclerotiorum*)

Timber rot is also known as white mold or Sclerotinia stem rot. Symptoms first appear on the stem at flowering and may result in large portions eventually appearing bleached and dry [103]. Wide plant spacing and low plant density reduce disease development by *Sclerotinia sclerotiorum*. Sclerotia may survive in soil for 3–4 years. *Coniothyrium minitans* is a mycoparasite of *S. sclerotiorum* that is found only in sclerotia and does not appear to be a plant pathogen. Application of *C. minitans* onto crop debris infected with *S. sclerotiorum* has shown potential to reduce disease carry-over [104].

10. Bacterial Canker (*Clavibacter michiganensis* subsp. *michiganensis*)

Bacterial canker of tomato is a potentially devastating disease that occurs worldwide. Symptoms of canker include downward turning of lower leaves, marginal necrosis, wilting of leaflets, and systemic wilt of the plant. Elongated stem lesions that form cankers may or may not appear. Fruit symptoms may not appear but are distinctive, consisting of lesions with raised brown centers surrounded by a white halo and often referred to as birds-eye spots [105].

Clipping or pruning of direct-seeded or transplanted tomatoes can result in severe losses. The use of clean seed and transplants is the most effective way of controlling this pathogen. This pathogen overwinters in soil, plant debris, weeds, and volunteers and on tomato stakes. Rotation to a nonhost is advised if a field becomes infested with canker [105].

11. Bacterial Wilt (*Ralstonia solanacearum*)

Bacterial wilt is a serious disease of tomato in warm, temperate, subtropical, and tropical regions worldwide. This bacterium attacks more than 200 species of cultivated plants and weeds in 33 families. The occurrence of bacterial wilt in a field can range from sporadic to widespread and causes rapid plant death. Characteristic symptoms of this disease include a rapid wilt and death of the plant with no yellowing of the foliage [36]. Before wilting, plants may appear stunted. The pith near the crown is often dark colored and water soaked. Bacterial streaming may be evident upon cutting the stem at its base. Rapid wilting and death, lack of chlorosis, and pith decay distinguish bacterial wilt from Fusarium and Verticillium wilts.

The biology and epidemiology of bacterial wilt have been reviewed in detail by Hayward [106]. This bacterium can survive in the soil for extended periods of time without a host and enters the plant through any type of wound. Cultural controls include avoiding planting seedbeds on infested land. Rotation to nonsola-naceous crops may help but is only of limited effectiveness due to the wide host range of the pathogen. Movement of water, soil, and equipment from infested fields to noninfested fields should be avoided. Cultivars Venus and Saturn have exhibited resistance in the United States, but this resistance has not always held up in other areas. Resistant breeding lines developed outside the United States include Rodade, Scorpio, Redlands Summertaste, Redlander, Kewada, Rosita, Caribo, and Durable Shinburo [107].

The hypersensitive reaction (HR) is believed to be an important component of disease resistance. In phytopathogenic bacteria, *hrp* (hypersensitive reaction and pathogenicity) genes control the ability to cause disease and to elicit hyper-sensitive reactions on resistant plants. Research has led to the isolation of *hrp* mutants of pathovars of *Pseudomonas syringae* [108,109]. *Hrp*-mutants of *Ralstonia solanacearum* have been described as potential biological control agents of bacterial wilt of tomato. The ability of *Hrp*-mutants to protect against invasion by pathogens is correlated with its aggressiveness in invading and colo-nizing host tissue [110–112]. Mechanisms responsible for biological control by these organisms include bacteriocin-mediated antibiosis within the rhizosphere and plant tissues [112].

Studies performed to characterize the microhabitats of microbial communi-ties in soil indicate that microhabitats conducive for growth differed among strains of soil bacteria and that little competition existed between *R.solanacearum* and other soil bacteria when their microhabitats differed [113]. These results indicate that selection of potential biological control agents should include an understanding of their ecological suitability for survival in the habitats that the pathogen occupies.

C. Nematode Management

Nematodes often form disease complexes with other pathogens of tomato. Nema-tode infestation can cause damage to roots that can result in water stress and stunting and damage is often dependent on the population density of the pathogen and the host's ability to tolerate stress. Nematode-infested areas should be identi-fied and isolated to avoid spreading inoculum with machinery and irrigation wa-ter. Many vegetables including tomato, cole crops, beans, eggplant, cucumber, muskmelon, watermelon, honeydew, okra, and pepper are susceptible to root-knot nematodes (*Meloidogyne* spp.). This makes rotation difficult in vegetable production areas. Although rotation to nonhost crops for more than a year may reduce populations of root-knot nematodes it will not eliminate them. Rotation with asparagus, corn, onions, garlic, small grains, cahaba white vetch and 'nova' vetch, crotalaria, velvetbean, soybean, and ryegrass reduce numbers of root-knot nematode [114]. Crops grown in the southeastern United States that consistently reduce populations of *Meloidogyne* spp. include sorghum-sudangrass and bahiagrass [115]. French marigold (*Tagetes patula*), in a solid planting for a full season, has been reported to decrease some species of root-knot nematodes but is a host to the northern root-knot nematode.

There are some nematode-resistant tomato cultivars developed through traditional plant breeding. High soil temperature (>28–30°C) is a major limiting factor for the use of resistance in tomato to root-knot nematodes (*Meloidogyne* spp.) due to the breakdown in expression of the Mi gene resistance at high temperature. Also, there are resistance-breaking biotypes of *Meloidogyne* spp. associated with tomato. Evidence suggests that resistance genes other than the Mi gene exist in exotic germplasm [116]. Ammati et al. [117] found stability of root-knot nematode resistance under heat stress in several *Lycopersicon* genotypes. There are also efforts to develop genetically engineered plants that would function as trap crops by allowing the nematodes to enter the roots but not to reproduce.

Various materials have been used as soil amendments for nematode control [19]. Mechanisms responsible for nematode suppression may differ for each amendment used and can include release of toxic compounds such as hydrogen sulfide, organic acids, and ammonia. Addition of organic amendments to soil can also reduce nematode infestation by increasing populations of beneficial microorganisms. Application of chitin or collagen, which are components of the gelatinous matrix produced by *Meloidogyne* spp., nematode egg shells, and cuticular proteins, enhances the components of the microflora capable of utilizing those compounds and often results in reduction of pathogenic nematode populations [118–120]. Ground seed of castor, crotalaria, hairy indigo, and wheat were evaluated for effects on populations of root-knot nematode on tomato [121]. Crotalaria and hairy indigo added to soil at 2% almost completely suppressed egg mass production of both *M. javanica* and *M. incognita*. In these studies, levels of amendment rather than type of amendment had more effect on egg mass production.

Rhizobacteria have been extensively investigated as nematode antagonists for many years. Results using commercial formulations for nematode control have been variable [17]. Several rhizobacteria, including *Bacillus* spp. [122], fluorescent pseudomonads [123], and *Telluria chitinolytica* [124], have been shown to inhibit penetration of roots by *Meloidogyne* spp. and reduce galling. It has been hypothesized that these bacteria may interfere with chemotaxis by blocking receptors on the roots or by modifying host plant root exudates [122,123,125]. Oka et al. [126] found that the combination of a proteinaceous amendment, such as peptone, and an ammonia-producing bacterium reduced galling by root-knot nematode. Significant reduction in *M. incognita* gall incidence on tomato occurred after seed treatment with the three rhizobacteria *Bacillus cereus*, *Bacillus subtilis*, and *Pseudomonas* spp. [127]. An increase in seedling biomass, a reduction in galling caused by *M. incognita*, and a yield increase were observed with one or more of the bacterial treatments.

Bacillus thuringiensis (BT) is a gram-positive bacterium that has been widely used as a biological control agent for insects. BT has also been reported to be lethal to plant parasitic nematodes in vitro [128,129]. In field trials, the BT isolate CR-371 effectively controlled root-knot nematode on tomato and pepper when applied as a drench [130]. Populations of reniform (*Rotylenchulus reniformis*) and lesion (*Pratylenchus penetrans*) nematodes also decreased in response to application of BT indicating broad nematicidal activity. Due to the small size of the stylet orifice and other natural body openings in plant parasitic nematodes, it is likely that BT exotoxins are responsible for the nematicidal activity reported in these studies. A thermostable beta exotoxin from a BT strain has proven to be nematicidal to *Meloidogyne incognita* and other soil nematodes [128,129]. Delta endotoxins released into the soil upon lysis of bacterial cells have also been implicated in nematode control.

Pasteuria penetrans is an obligate mycelial endosprore-forming bacterial parasite of some parasitic nematodes including *Meloidogyne* spp. This bacterium was first described as a protozoan and named *Duboscquia penetrans* [131]. After closer observation using the electron microscope, it was determined to be a bacterium and was designated as *Bacillus penetrans* [132]. Further study of the organism revealed its branched, filamentous vegetative thallus, which resembled an actinomycete. This observation resulted in the renaming of the organism *Pasteuria penetrans* [133]. Spores from isolates that attack *Meloidogyne* spp. attach to the cuticle of second-stage juveniles in the soil and germinate after the juvenile enters the root and initiates feeding. The pathogen enters the body of the nematode by producing a germ tube, which penetrates the cuticle. The bacteria then produces branched microcolonies, which give rise to daughter colonies that proliferate throughout the body cavity of the nematode, interfering with normal growth and reproduction [134]. *Pasteuria penetrans* is an extremely specific obligate parasite that depends completely on its host for development. Experiments by Bird [135] evaluating the effect of parasitism by *P. penetrans* on development of *Meloidogyne javanica* indicated that a balance was struck between the parasite and host nematode, which caused a minimal disturbance in the feeding and functioning of the nematode while selectively destroying the nematode's ability to reproduce. Early trials by Stirling [136] confirmed that the incorporation of *P. penetrans* into soil prior to planting tomatoes reduced galling and soil populations of root-knot nematode to levels comparable to nematicide. In other studies, *Pasteuria penetrans* was found to have no effect on nematode populations, galling, and yield in the first crop but did reduce populations and galling in successive crops due to active multiplication in the soil [137]. *Pasteuria penetrans* is currently not commercially available due to difficulties in mass production of this obligate pathogen.

The fungus *Paecilomyces lilacinus*, a common soil hyphomycete, is known to parasitize nematode eggs [138,139]. Nematodes of the group Heteroderidae, which deposit their eggs in a gelatinous matrix, are more vulnerable to attack by egg destroying fungi such as *P. lilacinus* than the eggs of migratory parasites. *P. lilacinus* suppressed root galling, number of egg masses, and egg hatch in greenhouse experiments and increased yield of soybean in microplots in two consecutive years without reapplication of the fungus the second year [140]. In experiments on greenhouse tomatoes, plants inoculated with the fungus 4–6 days prior to nematode inoculation had significantly lower gall indices than those treated and inoculated at other times [141]. The integration of neem and other oil-cakes as organic amendments with the application of *P. lilacinus* increased parasitism of root-knot nematode females, egg masses, and eggs [142].

The endoparasitic fungus *Meria coniospora* has been reported to have efficacy against root-knot nematodes [143]. This fungus grows well in culture and produces abundant conidia on both infected nematodes and artificial substrates. *Meria coniospora* is very aggressive, colonizing 15 of 17 nematode species tested including *Meloidogyne javanica* and *M. incognita*. A *Scytalidium*-like fungus, isolated from black egg masses of *M. javanica* on tomato roots, lowered the hatch rate of juveniles in vitro [144]. Application of the fungus did not inhibit penetration of juveniles into tomato roots, but the nematode population in treated soil was lower than in nontreated soil after one generation of nematodes.

The fungus *Verticillium lecanii* is another potential biological control agent that has been extensively investigated for control of *Meloidogyne incognita* on tomato [145,146] and other nematode/host combinations [147–149]. Various formulations, including an alginate granule, of isolates of *V. lecanii* have been tested against root-knot nematode on tomato with inconsistent results. Root drench applications of this fungus did not increase its efficacy against *M. incognita* on tomato [146].

There has been substantial research on nematode-trapping fungi in recent years. These fungi trap the nematodes in hyphal rings, or with a network of sticky hyphal structures, and invade the body cavity, consuming its contents. Activities of these fungi may be influenced by soil pH, moisture, and temperature, which can limit their establishment and reproduction in agricultural soils. In general, these fungi have limited competitive ability and slow rates of multiplication resulting in inconsistent performance as biological agents in the field. Stirling et al. [150] investigated the effects of formulations on *Arthrobotrys dactyloides* as a control agent for *Meloidogyne javanica*. Formulations that had been subjected to solid phase incubation prior to drying consistently reduced numbers of *M.* juveniles in greenhouse experiments using field soil. In later field trials formulations of *A. dactyloides* applied at 220–440 kg/ha reduced the number of nematodes present in roots 4–8 weeks after planting [151]. It was concluded that formulations with greater biological activity are needed to achieve levels of control comparable to chemical nematicides.

Studies on the effects of soil microfauna on populations of the nematophagous fungi *Hirsutella rhossiliensis* and *Monacrosporium gephyropagum* indicate that enchytraeids did not reduce populations of formulated nematode trapping fungi [152]. However, these studies did indicate that organisms smaller that 20 μm do negatively affect establishment of these fungi in soils. Bacteria associated with *Arthrobotrys oligospora*, called nematophagous fungus helper bacteria (NHB), were found to enhance in vitro fungal activity against *Meloidogyne mayaguensis* resulting in better nematode control and improved plant growth [153].

Entomopathogenic nematodes are commonly used for the control of various insect pests. *Steinernema glaseri* and *S. carpocapsae* were studied by Bird and Bird [154] for effects on populations of *Meloidogyne javanica*. Both *Steinernema* and *Meloidogyne* are chemotactically attracted to CO_2, which is given off by root tips. It was found that the larger and more active *Steinernema* spp. outcompeted *M. javanica* for space and consequently reduced the number of egg-laying *Meloidogyne* females in tomato roots. It was also observed that the addition of *S. glaseri* seemed to stimulate plant growth and that several less concentrated applications were superior to a single concentrated dose for control of root-knot nematodes and enhanced plant growth. Lewis and Perez [155] found that tomato seedlings inoculated with *Steinernema feltiae* and *Heterorhabditis bacteriophora* had fewer *M. incognita* eggs than roots inoculated with *M. incognita* alone.

D. Management of Virus Diseases

There are approximately 30 plant viruses that infect tomato. Virus diseases are extremely difficult to control and can result in significant crop loss. Ten of the most important viruses of tomato are vectored by aphids: five by whiteflies, two by thrips, one by nematodes, and four by beetles, leafhoppers, or treehoppers [156]. Disease incidence and severity vary due to the complex relationship that exists among the virus, host, vector, and environment.

The most effective way to control viruses is to limit their spread. In general, tomato viruses are spread by infected seed or transplants and insect feeding. It is important to use certified virus-free seed or transplants and virus-resistant cultivars and to control weeds and insects around tomato fields. Weed control is especially important for control of tobacco etch virus, the primary source of which is infected solanaceous weeds [1]. Soaps and oils can be used to control aphids and thrips [1]. Reflective plastic mulches have been shown to delay the onset of some virus diseases compared to black plastic [157].

Tomato spotted wilt virus (TSWV) is one of the most economically destructive viruses of tomato, often causing severe infection rates of 50–90% in some vegetable crops [158]. This virus is found in regions with tropical climates and is vectored by thrips [159]. Leaves typically become bronze and develop dark spots. Plants may show evidence of stunting or irregular growth, and ripe fruit may have chlorotic ring spots, although symptoms can vary [160]. Approximately 800 species of plants in 80 families are susceptible to TSWV. Predominant vegetable and field crop hosts of TSWV include tomato, pepper, potato, tobacco, lettuce, and peanut [161]. TSWV is difficult to control due to the high reproductive rate and wide host range of the vector combined with the ability of thrips to develop resistance to insecticides. The Sw-5 resistance gene in tomato has provided some control of the virus [162] and can be combined with the use of thrips-proof mesh that provides a barrier against the vector [163]. Diez et al. [163] found that clean TSWV-resistant transplants performed best under the protection of thrips-proof mesh but that mesh can enhance infection and reduce yield if used in combination with susceptible cultivars under high disease pressure. Adkins [164] provides a more complete review of recent research on TSWV.

Tomato yellow leaf curl virus (TYLCV) is a whitefly (*Bemisia tabaci*)-transmitted virus that is extremely devastating on tomato. TYLCV is a geminivirus with a narrow host range affecting only a few species in six plant families, which include Compositae, Leguminosae, and Solanaceae [165]. While many hosts remain symptomless, tobacco and tomato are severely affected by TYLCV [166]. Symptoms include weakening and stunting of the entire plant and flower and fruit abscission. This virus can cause complete crop loss when infection occurs before flowering. TYLCV occurs in many tropical and subtropical regions and is spreading into new areas [166]. Use of resistant varieties is the most effective way to manage this pathogen; however, only partially resistant hybrids are commercially available [166]. In order to reduce sources of the virus, clean planting material must be used and plants that serve as virus reservoirs such as tobacco, beans, and volunteer tomatoes should be eliminated [167]. In order to control the transmission of the virus, the vector must be controlled. Biological control of *B. tabaci* in Mediterranean regions can be accomplished using the parasites *Encarsia formosa*, *Encarsia lutea*, and *Eretmocerus mundus* [168].

The ability of one virus strain to protect against infection by a second strain of the same virus is known as cross protection. Practical use of attenuated strains began 30 years ago and has become a widely used means of biological control of tobacco and tomato mosaic virus in greenhouse tomatoes [169]. More recently, alternative strategies have been developed using transgenic plants containing viral coat protein genes and replicase-associated genes [170]. Transgenic plants have been obtained that express the coat protein gene of TYLCV and show a high positive correlation between the presence of the coat protein and disease reduction [171].

Plant growth–promoting rhizobacteria have been reported to act as inducers of systemic resistance towards cucumber mosaic virus in tomato [172]. PGPR strains that induced protection in cucumber against the fungal pathogens *Colletotrichum orbiculare* and *Fusarium oxysporum* and the bacterial pathogen *Pseudomonas syringae* pv. *lachrymans*, reduced disease severity in cucumber mosaic cucumovirus (CMV)–susceptible tomato compared to the untreated control. This research also demonstrated that specific PGPR strains have the potential of protecting various crops against multiple and diverse pathogens.

Experiments conducted over 5 years in Alabama and Florida evaluating strains of PGPR for induction of resistance against CMV and whitefly-transmitted tomato mottle virus (ToMoV) demonstrated a reduction in the incidence of viral infection and an increase in tomato yield in PGPR-treated plants [173]. Tomato mottle virus is a problem for transplant and field production of tomato in west-central and southwest Florida [174]. These studies offer hope for control of insect-transmitted diseases, such as viruses, using PGPR-mediated induced resistance.

IV. SUMMARY

A greater understanding of the interaction between biocontrol agents, pathogens, environmental conditions, and host plants is imperative if successful strategies are to be developed to limit disease incidence without the use of chemical pesticides. This information is necessary to improve performance of biocontrol agents by lessening their chances of failure due to application during conditions unfavorable to their growth or under extremely high pathogen pressure that could be reduced by cultural practices. Successful use of biological control will require growers to adjust current crop management practices. The resulting integration of biological control practices into a multifaceted management program will require knowledge-based decisions by growers regarding ecological principals involving the environment, host, pathogen, and biocontrol agent. A current list of commercially available biological control agents is available through the USDA, ARS Biological Control of Plant Diseases website [175].

REFERENCES

1. M Peet. Sustainable practices for vegetable production in the south, 2000. http://www.cals.ncsu.edu/sustainable/peet/IPM/disease/d mgmt.html.
2. EC Tigchelaar. Botany and culture. In: JB Jones, JP Jones, RE Stall, TA Zitter, eds. Compendium of Tomato Diseases. St. Paul, MN: APS Press, 1991, pp 2–4.

3. JC Watterson. Seed production. In: JB Jones, JP Jones, RE Stall, and TA Zitter, eds. Compendium of Tomato Diseases. St. Paul, MN: APS Press, 1991, pp 4–5.
4. CD Stanley, CM Geraldson. Containerized transplant production. In: JB Jones, JP Jones, RE Stall, TA Zitter, eds. Compendium of Tomato Diseases. St. Paul, MN: APS Press, 1991, pp 5–8.
5. KJ Bradford. Manipulation of seed water relations via osmotic priming to improve germination under stress conditions. HortScience 21:1105–1112, 1986.
6. GE Harman, AG Taylor. Improved seedling performance by integration of biological control agents at favorable pH levels with solid matrix priming. Phytopathology 78:520–525, 1988.
7. A Sivan, I Chet. Biological control of *Fusarium* spp. in cotton, wheat and muskmelon by *Trichoderma harzianum*. Phytopathol Z 116:39–47, 1986.
8. Y Chang, Y Chang, R Baker, O Kleifeld, I Chet. Increased growth of plants in the presence of the biological control agent *Trichoderma harzianum*. Plant Dis 70:145–148, 1986.
9. WL Chao, EB Nelson, GE Harman, HC Hoch. Colonization of the rhizosphere by biological control agents applied to seeds. Phytopathology 76:60–65, 1986.
10. Y Hadar, GE Harman, AG Taylor. Evaluation of *Trichoderma koningii* and *T. harzianum* from New York soils for biological control of seed rot caused by *Pythium* spp. Phytopathology 74:106–110, 1984.
11. N Kokalis-Burelle, EN Rosskopf, RA Shelby, DO Chellemi, CS Vavrina. Field evaluation of amended transplant mixes and soil solarization for tomato and pepper production. Phytopathology 89:41, 1999.
12. A Sivan, O Ucko, I Chet. Biological control of Fusarium crown rot of tomato by *Trichoderma harzianum* under field conditions. Plant Dis 71:587–592, 1987.
13. S Nemec, LE Datnoff, J Strandberg. Efficacy of biocontrol agents in planting mixes to colonize plant roots and control root diseases of vegetables and citrus. Crop Prot 15:735–742, 1996.
14. S Nemec. Longevity of microbial biocontrol agents in a planting mix amended with *Glomus intraradices*. Biocontrol Sci Technol 7:183–192, 1997.
15. R Vargas-Ayela. Nematode population dynamics and microbial ecology in a rotation program with *Mucuna deeringiana*, and other crops: a biological control approach. PhD dissertation, Auburn University, Auburn, AL, 1995.
16. DP Breakwell, RF Turco. Nutrients and phytotoxic contributions of residue to soil in no-till continuous-corn ecosystems. Biol Ferti Soils 8:328–334, 1990.
17. GR Stirling. Biological Control of Plant Parasitic Nematodes: Progress, Problems and Prospects. Wallingford, Oxon, UK: CAB International, 1991.
18. HAJ Hoitink. Basis for the control of soilborne plant pathogens with composts. Annu Rev Phyopathol 24:93–114, 1998.
19. R Rodriguez-Kabana. Organic and inorganic nitrogen amendments to soil as nematode suppressants. J Nematol 18:129–135, 1986.
20. D Chun, JL Lockwood. Reduction of *Pythium ultimum*, *Thielaviopsis basicola*, and *Macrophomina phaseolina* populations in soil associated with ammonia generated from urea. Plant Dis 69:154–158, 1985.
21. FS Chew. Biological effect of glucosinolates. In: HG Cutler, ed. Biologically Active Natural Products. ACS Symp Ser 380. 1988, pp 155–181.

22. HS Mayton, C Olivier, SF Vaughn, R Loria. Correlation of fungicidal activity of *Brassica* species with allyl isothiocyanate production in macerated leaf tissue. Phytopathology 86:267–271, 1996.
23. Y Chen, A Gamliel, JJ Stapleton, T Aviad. Chemical, physical, and microbial changes related to plant growth in disinfested soil. In: J Katan, JE DeVay, eds. Soil Solarization. Boca Raton, Fl: CRC Press, 1991, pp 103–129.
24. RJ Cook, KF Baker. The Nature and Practice of Biological Control of Plant Pathogens. St. Paul, MN: APS Press, 1983, pp 1–539.
25. DA Pavlica, TS Hora, JJ Bradshaw, RK Skogerboe, R Baker. Volatile compounds from soil influencing activities of soil fungi. Phytopathology 68:758–765, 1978.
26. N Kokalis-Burelle, R Rodriquez-Kabana. Effects of pine bark extracts and pine bark powder on fungal pathogens, soil enzyme activity, and microbial populations. Biol Control 4:269–276, 1994.
27. N Kokalis-Burelle, R Rodriquez-Kabana. Changes in populations of soil microorganisms, nematodes, and enzyme activity associated with application of powdered pine bark. Plant Soil 162:169–175, 1994.
28. J Katan. Soil solarization. In: I Chet, ed. Innovative Approaches to Plant Disease Control. New York: John Wiley & Sons, 1987, pp 77–105.
29. A Gamliel, J Katan. Suppression of major and minor pathogens by fluorescent pseudomonads in solarized soil. Phytopathology 83:320–327, 1993.
30. A Gamliel, JJ Stapleton. Characterization of antifungal volatile compounds evolved from solarized soil amended with cabbage residues. Phytopathology 83:899–905, 1993.
31. JJ Stapleton, A Gamliel. Feasibility of soil fumigation by sealing soil amended with fertilizers and crop residues containing biotoxic volatiles. Proc. 24th Natl. Agricultural Plastics Congr Am Soc Plasticulture, Raleigh, NC, 1993, pp 200–205.
32. A Gamliel, JJ Stapleton. Effect of soil amendment with chicken compost or ammonium phosphate and solarization on pathogen control, rhizosphere microorganism and lettuce growth. Plant Dis 77:886–891, 1993.
33. DO Chellemi, SM Olsen, DJ Mitchell. Effects of soil solarization and fumigation on survival of soilborne pathogens of tomato in northern Florida. Plant Dis 78: 1167–1172, 1994.
34. CA Chase, TR Sinclair, DG Shilling, JP Gilreath, SJ Locascio. Light effects on rhizome morphogenesis in nutsedges (*Cyperus* spp.): implications for control by soil solarization Weed Science 46:575–580, 1998.
35. WR Stevenson, KL Pohronezny. Anthracnose. In: JB Jones, JP Jones, RE Stall, TA Zitter, eds. Compendium of Tomato Diseases. St. Paul, MN: APS Press, 1991, pp 9–10.
36. GW Simone. Disease control in tomato (*Lycopersicon esculentum*). In: GW Simone, RS Mullin, eds. 1998 Florida Plant Disease Management Guide Vol. 3: Fruit and Vegetables. 1998, pp 324–355. University of Florida, Gainesville, FL.
37. JP Jones. Early blight. In: JB Jones, JP Jones, RE Stall, TA Zitter, eds. Compendium of Tomato Diseases. St. Paul, MN: APS Press, 1991, pp 13–14.
38. AF Nash, RG Gardner. Tomato early blight resistance in a breeding line derived from *Lycopersicon hirsutum* PI126445. Plant Dis 72:206–209, 1988.
39. WR Jarvis. Managing diseases in greenhouse crops. Plant Dis 73:190–194, 1989.

40. Y Elad, J Köhl, NJ Fokkema. Control of infection and sporulation of *Botrytis cinerea* on bean and tomato by saprophytic bacteria and fungi. Europe. J Plant Pathol 100:315–336, 1994.
41. TM O'Neill, A Niv, Y Elad, D Shtienberg. Biological control of *Botrytis cinerea* on tomato stem wounds with *Trichoderma harzianum*. Eur J Plant Pathol 102:635–643, 1996.
42. AJ Dik, Y Elad. Comparison of antagonists of *Botrytis cinerea* in greenhouse-grown cucumber and tomato under different climatic conditions. Eur J Plant Pathol 105:123–127, 1999.
43. D Shtienberg, Y Elad. Incorporation of weather forecasting in integrated, biological-chemical management of *Botrytis cinerea*. Phytopathology 87:332–340, 1996.
44. RE Stall. Gray mold. In: JB Jones, JP Jones, RE Stall, TA Zitter, eds. Compendium of Tomato Diseases. St. Paul, MN: APS Press, 1991, pp 16–17.
45. E Fallik, J Klein, S Griberg, E Lomaniec, S Lurie, A Lalazar. Effect of postharvest heat treatment of tomatoes on fruit ripening and decay caused by *Botrytis cinerea*. Plant Dis 77:985–988, 1993.
46. M Mari, M Guizzardi, M Brunelli, A Folchi. Postharvest biological control of grey mould (*Botrytis cinerea* Pers.: Fr.) on fresh-market tomatoes with *Bacillus amyloliquefaciens*. Crop Prot 15:699–705, 1996.
47. JP Jones. Bacterial spot. In: JB Jones, JP Jones, RE Stall, TA Zitter, eds. Compendium of Tomato Diseases. St. Paul, MN: APS Press, 1991, pp 26–27.
48. JP Jones. Bacterial speck. In: JB Jones, JP Jones, RE Stall, TA Zitter, eds. Compendium of Tomato Diseases. St. Paul, MN: APS Press, 1991, p 27.
49. WG Bonn, RD Gitaitis, BH MacNeill. Epiphytic survival of *Pseudomonas syringae* pv. *tomato* on tomato transplants shipped from Georgia. Plant Dis 69:58–60, 1985.
50. RW Schneider, RG Grogan. Bacterial speck of tomato: sources of inoculum and establishment of a resident population. Phytopathology 67:388–394, 1977.
51. DR Smitley, SM McCarter. Spread of *Pseudomonas syringae* pv. *tomato* and role of epiphytic populations and environmental conditions in disease development. Plant Dis 66:713–717, 1982.
52. DA Cooksey, Reduction in infection by *Pseudomonas syringae* pv. *tomato* using a nonpathogenic, copper-resistant strain combined with a copper bactericide. Phytopathology 78:601–603, 1988.
53. JE Colin, Z Chafik. Comparison of biological and chemical treatments for control of bacterial speck of tomato under field conditions in Morocco. Plant Dis 70:1048–1050, 1986.
54. RE Stall. Bacterial stem rot. In: JB Jones, JP Jones, RE Stall, TA Zitter, eds. Compendium of Tomato Diseases. St. Paul, MN: APS Press, 1991, pp 27–28.
55. JP Jones, AW Engelhard, SS Woltz. Management of Fusarium wilt of vegetables and ornamentals by macro- and microelement nutrition. In: AW Engelhard, ed. Soilborne Plant Pathogens: Management of Diseases with Macro- and Microelements. St. Paul, MN:APS Press 1989, pp 18–32.
56. B Barna, ART Sarhan, Z Kiraly. The influence of nitrogen nutrition on the sensitivity of tomato plants to culture filtrates of *Fusarium* and to fusaric acid. Physiol Plant Pathol 23:257–263, 1983.
57. WH Elmer, Association between Mn-reducing root bacteria and NaCl applications

in suppression of Fusarium crown and root rot of asparagus. Phytopathology 85: 1461–1467, 1995.

58. AM Fenton, PM Stephens, J Crowley, M O'Callaghan, F O'Gara. Exploitation of gene(s) involved in 2,4-diacetylphloroglucinol biosynthesis to confer a new biocontrol capability to a *Pseudomonas* strain. Appl Environ Microbiol 58:3873–3878, 1992.

59. C Keel, DM Weller, A Natsch, G Défago, RJ Cook, LS Thomashow. Conservation of the 2,4-diacetylphloroglucinol biosynthesis locus among fluorescent *Pseudomonas* strains from diverse geographic locations. Appl Environ Microbiol 62:552–563, 1996.

60. C Keel, U Schnider, M Maurhofer, C Voisard, J Laville, U Burger, P Wirther, D Haas, G Défago. Suppression of root diseases by *Pseudomonas fluorescens* CHA0: importance of the bacterial secondary metabolite 2,4-diacetylphloroglucinol. Mol Plant-Microbe Interact 5:4–13, 1992.

61. D Cronin, Y Moenne-Loccoz, A Fenton, C Dunne, DN Dowling, F O'Gara. Role of 2,4-diacetylphloroglucinol in the interactions of the biocontrol pseudomonad F113 with the potato cyst nematode *Globodera rostochiensis*. Appl Environ Microbiol 63:1357–1361, 1997.

62. A Sharifi-Tehrani, M Zala, A Natsch, Y Moenne-Loccoz, G Biocontrol of soilborne fungal plant diseases by 2,4-diacetylphloroglucinol-producing fluorescent pseudomonads with different restriction profiles of amplified 16s rDNA. Eur J Plant Pathol 104:631–643, 1998.

63. BK Duffy, G Défago. Zinc improves biocontrol of Fusarium crown and root rot of tomato by *Pseudomonas fluorescens* and represses the production of pathogen metabolites inhibitory to bacterial antibiotic biosynthesis. Phytopathology 87: 1250–1257, 1997.

64. P M'Piga, RR Bélanger, TC Paulitz, N Benhamou. Increased resistance to *Fusarium oxysporum* f. sp. *radicis-lycopersici* in tomato plants treated with the endophytic bacterium *Pseudomonas fluorescens* strain 63-28. Physiol Mol Plant Pathol 50:301–320, 1997.

65. TFC Chin-A-Woeng, GV Bloemberg, AJ van der Bij, KMGM ban der Drigt, J Schripsema, B Kroon, RJ Scheffer, C Keel, RAHM Bakker, H-V. Tichy, FJ deBruijn, JE Thomas-Oates, BJJ Lugtenberg. Biocontrol by phenazine-1-carboxamide-producing *Pseudomonas chlororaphis* PCL1391 of tomato root rot caused by *Fusarium oxysporum* f. sp. *radicis-lycopersici*. Mol Plant-Microbe Interact 11:1069–1077, 1998.

66. JJ Marois, DJ Mitchell, RM Sonoda. Biological control of Fusarium crown rot of tomato under field conditions. Phytopathology 71:1257–1260, 1981.

67. A Sivan, I Chet. Integrated control of fusarium crown and root rot of tomato with *Trichoderma harzianum* in combination with methyl bromide or soil sterilization. Crop Prot 12:380–386, 1993.

68. LE Datnoff, S Nemec, K Pernezny. Biological control of Fusarium crown and root rot of tomato in Florida using *Trichoderma harzianum* and *Glomus intradices*. Biol Control 5:427–431, 1995.

69. N Benhamou, P Rey, M Cherif, J Hockenhull, Y Tirilly. Treatment with the mycoparasite *Pythium oligandrum* triggers induction of defense-related reactions in to-

mato roots when challenged with *Fusarium oxysporum* f. sp. *radicis-lycopersici.*
Phytopathology 87:108–122, 1997.

70. LA Hadwiger, C Chiang, S Victory, D Horovitz. The molecular biology of chitosan
in plant-pathogen interactions and its application in agriculture. G Skjak, BT An-
thonsen, P Sanford, ed. Chitin and Chitosan: Sources Chemistry, Biochemistry,
Physical Properties, and Applications. Amsterdam: Elsevier Applied Sciences,
1988, pp 119–138.

71. MG Hahn, P Bucheli, F Cervone, SH Doares, RA O'Neill, A Darvill, P Albersheim,
E Nester, T Kosuge. The roles of cell wall constituents in plant-pathogen interactions. In: Plant-Microbe Interactions. New York: McGraw-Hill, 1989, pp 131–138.

72. N Benhamou, G Thériault. Treatment with chitosan enhances resistance of tomato
plants to the crown and root rot pathogen *Fusarium oxysporum* f. sp. *radicis-
lycopersici.* Physiol Mol Plant Pathol 41:33–52, 1992.

73. N Benhamou, PJ Lafontaine, M Nicole. Induction of systemic resistance to
Fusarium crown and root rot in tomato plant by seed treatment with chitosan. Phytopathology 84:1432–1444, 1994.

74. N Benhamou, JW Kloepper, S Tuzun. Induction of resistance against Fusarium
wilt of tomato by combination of chitosan with an endophytic bacterial strain: ultrastructure and cytochemistry of the host response. Planta 204:153–168, 1998.

75. C Alabouvette, P Lemanceau. Natural suppressiveness of soils in management of
fusarium wilts. In: R Utkhede, VK Gupta, eds. Management of Soil-Borne Diseases. Ludhiana: Kalyani Publisher, 1996, pp 301–322.

76. J-G Fuchs, Y Moënne-Loccoz, G Défago. Nonpathogenic *Fusarium oxysporum*
strain Fo47 induces resistance to Fusarium wilt in tomato. Plant Dis 81:492–496,
1997.

77. J-G Fuchs, Y Moënne-Loccoz, G Défago. Ability of nonpathogenic *Fusarium oxysporum* Fo47 to protect tomato against Fusarium wilt. Biol Control 14:105–110,
1999.

78. BJ Duijff, D Pouhair, C Olivain, C Alabouvette, P Lemanceau. Implication of systemic induced resistance in the suppression of Fusarium wilt of tomato by *Pseudomonas fluorescens* WCS417r and by nonpathogenic *Fusarium oxysporum* Fo47.
Eur J Plant Pathol 104:903–910, 1988.

79. W Mao, JA Lewis, RD Lumsden, KP Hebber. Biocontrol of selected soilborne
disease of tomato and pepper plants. Crop Prot 17:535–542, 1998.

80. RP Larkin, DR Fravel. Efficacy of various fungal and bacterial biocontrol organisms for control of Fusarium wilt of tomato. Plant Dis 82:1022–1028, 1998.

81. A De Cal, S Pascual, P Melgarejo. Involvement of resistance induction by *Penicillium oxalicum* in the biocontrol of tomato wilt. Plant Pathol 46:72–79, 1997.

82. A De Cal, S Pascual, P Melgarejo. A rapid laboratory method for assessing the
biological control potential of *Penicillium oxalicum* against Fusarium wilt of tomato. Plant Pathol 46:699–707, 1997.

83. A De Cal, R García-Lepe, S Pascual, P Melgarejo. Effects of timing and method
of application of *Penicillium oxalicum* on efficacy and duration of control of Fusarium wilt of tomato. Plant Pathol 48:260–266, 1999.

84. KL Pohronezny. Verticillium wilt. In: JB Jones, Re Stall, TA Zitter, eds. Compendium of Tomato Diseases. St. Paul, MN: APS Press, 1991, pp 24–25.

85. EC Tjamos, DA Biris, EJ Paplomatas. Recovery of olive trees with Verticillium wilt after individual application of soil solarization in established olive orchards. Plant Dis 75:557–562, 1991.

86. EC Tjamos, DR Fravel. Detrimental effects of sublethal heating and *Talaromyces flavus* on microsclerotia of *Verticillium dahliae*. Phytopathology 85:388–392, 1995.

87. SM McCarter. *Rhizoctonia* diseases. In: JB Jones, JP Jones, RE Stall, TA Zitter, eds. Compendium of Tomato Diseases. St. Paul, MN: APS Press, 1991, pp 21–22.

88. O Asaka, M Shoda. Biocontrol of *Rhizoctonia solani* damping-off of tomato with *Bacillus subtilis* RB14. Appl Environ Biol 62:4081–4085, 1996.

89. SM McCarter. *Pythium* diseases. In: JB Jones, JP Jones, RE Stall, TA Zitter, eds. Compendium of Tomato Diseases. St. Paul, MN: APS Press, 1991, pp 20–21.

90. Y Hadar, GE Harman, AG Taylor, JM Norton. Effects of pregerminating of pea and cucumber seeds and of seed treatment with *Enterobacter cloacae* on rots caused by *Pythium* spp. Phytopathology 73:1322–1325, 1983.

91. CR Howell, RD Stipanovic. Suppression of *Pythium ultimum* induced damping-off of cotton seedlings by *Pseudomonas fluorescens* and its antibiotic pyoluteorin. Phytopathology 70:712–715, 1980.

92. A Sivan, Y Elad, I Chet. Biological control of *Pythium aphanidermatum* by a new isolate of *Trichoderma harzianum*. Phytopathology 74:498–501, 1984.

93. Y Elad, I Chet. Possible role of competition for nutrients in biocontrol of *Pythium* damping-off by bacteria. Phytopathology 77:190–195, 1987.

94. K Van Dijk, EB Nelson. Inactivation of seed exudate stimulants of *Pythium ultimum* sprorangium germination by biocontrol strains of *Enterobacter cloacae* and other seed-associated bacteria. Soil Biol Biochem 30:183–192, 1998.

95. FN Martin, JG Hancock. Association of chemical and biological factors in soils suppressive to *Pythium ultimum*. Phytopathology 76:1221–1231, 1986.

96. P Rey, N Benhamou, E Wulff, Y Tirilly. Interactions between tomato (*Lycopersicon esculentum*) root tissues and the mycoparasite *Pythium oligandrum*. Physiol Mol Plant Pathol 53:105–122, 1998.

97. KP Smith, J Handelsman, RM Goodman. Modeling dose-response relationships in biological control: partitioning host responses to the pathogen and biocontrol agent. Phytopathology 87:720–729, 1997.

98. KP Smith, J Handelsman, RM Goodman. Genetic basis in plants for interactions with disease suppressive bacteria. Proc Natl Acad Sci 96:4786–4790, 1999.

99. JB Ristaino, KB Perry, RD Lumsden. Effect of solarization and *Gliocladium virens* on sclerotia of *Sclerotium rolfsii*, soil microbiota, and the incidence of southern blight of tomato. Phytopathology 91:1117–1124, 1991.

100. WR Stevenson. Buckeye rot and Phytophthora root rot. In: JB Jones, JP Jones, RE Stall, TA Zitter, eds. Compendium of Tomato Diseases. St. Paul, MN: APS Press, 1991, p 11.

101. B Dassi, E Dumas-Gaudot, S Gianinazzi. Do pathogenesis-related (PR) proteins play a role in bioprotection of mycorrhizal tomato roots towards *Phytophthora parasitica*? Physiol Mol Plant Pathol 52:167–183, 1998.

102. C Cordier, S Gianinazzi, V Gianinazzi-Pearson. Colonization patterns of root tis-

sues by *Phytophthora nicotianae* var. *parasitica* related to reduced disease in mycorrhizal tomato. Plant Soil 185:223–232, 1996.

103. KL Pohronezny. White Mold. In: JB Jones, Re Stall, TA Zitter, eds. Compendium of Tomato Diseases. St. Paul, MN: APS Press, 1991, pp 24–25.

104. M Gerlagh, JM Whipps, SP Budge, HM Goossen-van de Geijn. Efficiency of isolates of *Coniothyrium minitans* as mycoparasites of *Sclerotinia sclerotiorum*, *Sclerotium cepivorum*, and *Botrytis cinerea* on tomato stem pieces. Eur J Plant Pathol 102:787–793, 1996.

105. RD Gitaitis. Bacterial canker. In: JB Jones, JP Jones, RE Stall, TA Zitter, eds. Compendium of Tomato Diseases. St. Paul, MN: APS Press, 1991, pp 25–26.

106. AC Hayward. Biology and epidemiology of bacterial wilt caused by *Pseudomonas solanacearum*. Annu Rev Phytopathol 29:65–87, 1991.

107. SM McCarter. Bacterial wilt. In: JB Jones, JP Jones, RE Stall, TA Zitter, eds. Compendium of Tomato Diseases. St. Paul, MN: APS Press, 1991, pp 28–29.

108. PB Lindgren, RC Peet, NJ Paopoulos. Gene cluster of *Pseudomonas syringae* pv. *phaseolicola* controls pathogenicity on bean plants and hypersensitivity on nonhost plants. J Bacteriol 168:512–522, 1986.

109. F Niepold, D Anderson, D Mills. Cloning determinants of pathogenesis from *Pseudomonas syringae* pathovar *syringae*. Proc Natl Acad Sci 82:406–410, 1985.

110. A Trigalet, D Demery. Invasiveness in tomato plants of Tn5-induced avirulent mutants of *Pseudomonas solanacearum*. Physiol Mol Plant Pathol 28:423–430, 1986.

111. A Trigalet, D Trigalet-Demery. Use of avirulent mutants of *Pseudomonas solanacearum* for the biological control of bacterial wilt of tomato plants. Physiol Mol Plant Pathol 36:27–38, 1990.

112. P Frey, P Prior, C Marie, A Koutoujansky, D Trigalet-Demery, A Trigalet. Hrp mutants of *Pseudomonas solanacearum* as potential biocontrol agents of tomato bacterial wilt. Appl Environ Microbiol 60:3175–3181, 1994.

113. K Toyota, M Kimura. Growth of the bacterial wilt pathogen *Pseudomonas solanacearum* introduced into soil colonized by individual soil bacteria. Soil Biol Biochem 28:1489–1494, 1996.

114. R Rodriguez-Kabana, GH Canullo. Cropping systems for the management of phytonematodes. Phytoparasitica 20:211–224, 1992.

115. R McSorley, RN Gallaher. Nematode changes and forage yields of six corn and sorghum cultivars. J Nematol (Suppl) 23:673–677, 1991.

116. M Ammati, IJ Thomason, PA Roberts. Screening *Lycopersicon* species for new genes imparting resistance to root-knot nematodes (*Meloidogyne* spp.). Plant Dis 69:112–115, 1985.

117. M Ammati, IJ Thomason, HE McKinney. Retention of resistance to *Meloidogyne incognita* in *Lycopersicon* genotypes at high soil temperature. J Nematol 18:491–495, 1986.

118. R Rodriguez-Kabana, D Boube, RW Young. Chitinous materials from blue crab for control of root-knot nematode. I. Effect of urea and enzymatic studies. Nematropica 19:53–74, 1989.

119. Y Spiegel, I Chet, E Cohn. Use of chitin for controlling plant-parasitic nematodes II. Mode of action. Plant Soil 98:337–345, 1987.

120. Y Spiegel, E Cohn. Chitin is present in gelatinous matrix of *Meloidogyne*. Rev Nematol 8:184–185, 1985.
121. JR Rich, GS Rahi. Suppression of *Meloidogyne javanica* and *M. incognita* on tomato with ground seed of castor, crotalaria, hairy indigo, and wheat. Nematropica 25:159–164, 1995.
122. JO Becker, E Zavaleta-Mejia, SF Colbert, MN Schroth, AR Reinhold, JG Hancock, SD Van Gundy. Effects of rhizobacteria on root-knot nematodes and gall formation. Phytopathology 78:1466–1469, 1988.
123. M Oostendorp, RA Sikora. In vitro interrelationships between rhizosphere bacteria and *Heterodera schactii*. Rev Nematol 13:269–274, 1990.
124. JP Bowman, LI Sly, AC Hayward, Y Spiegel, E Stackebrandt. *Telluria mixta* (*Pseudomonas mixta* Bowman, Sly and Hayward, 1988) gen. nov., comb. nov., and *Telluria chitinolytica* sp. *nov.*, soil-dwelling organisms which actively degrade polysaccharides. Int J Systemat Bacteriol 43:120–124, 1993.
125. Y Spiegel, E Cohn, S Galper, E Sharon, I Chet. Evaluation of a newly isolated bacterium *Pseudomonas chitinolytica* sp. nov., for controlling the root-knot nematode *Meloidogyne javanica*. Biocon Sci Technol 1:115–125, 1991.
126. Y Oka, I Chet, Y Spiegel. Control of the root knot nematode *Meloidogyne javanica* by *Bacillus cereus*. Biocontrol Sci Technol 3:115–126, 1993.
127. S Hoffmann-Hergarten, MK Gulati, RA Sikora. Yield response and biological control of *Meloidogyne incognita* on lettuce and tomato with rhizobacteria. J Plant Dis Prot 105:349–358, 1998.
128. SSSV Prasad, KVBR Tilak, KG Gollakote. Role of *Bacillus thuringiensis* var. *thuringiensis* on the larval survivability and egg hatching of *Meloidogyne* spp. the causative agent of root knot disease. J Invert Path 20:377–378, 1972.
129. CM Ignoffo, VH Dropkin. Deleterious effects of the thermostable toxin of *Bacillus thuringiensis* on species of soil-inhabiting, myceliophagous, and plant-parasitic nematodes. J Kansas Entomol Soc 50:394–398, 1977.
130. BM Zuckerman, MB Dicklow, N Acosta. A strain of *Bacillus thuringiensis* for control of plant-parasitic nematodes. Biocon Sci Technol 3:41–46, 1993.
131. G Thorne. *Duboscqia penetrans*, n. sp. (*Sporozoa, Microsporidia, Nosematidae*), a parasite of the nematode *Pratylenchus pratensis* (de Man) Fililjev. Proc Helminthol Soc Wash 7:51–53, 1940.
132. R Mankau. *Bacillus penetrans* n. comb. causing a virulent disease of plant-parasitic nematodes. J Invertebr Pathol 26:333–339, 1975.
133. RM Sayre, MP Starr. *Pasteuria penetrans* (ex Thorne, 1940) nom.rev., comb.n.sp.n., a mycelial and endospore-forming bacterium parasitic in plant parasitic nematodes. Proc Helminthol Soc Wash 52:149–165, 1985.
134. RM Sayre, WP Wergin. Bacterial parasite of a plant nematode: morphology and ultrastructure. J Bacteriol 129:1091–1101, 1977.
135. AF Bird. The influence of the actinomycete, *Pasteuria penetrans*, on the host-parasite relationship of the plant-parasitic nematode, *Meloidogyne javanica*. Parasitology 93:571–580, 1986.
136. GR Stirling. Biological control of *Meloidogyne javanica* with *Bacillus penetrans*. Phytopathology 74:55–60, 1984.
137. M Eddaoudi, M Bourijate. Comparative assessment of *Pasteuria penetrans* and

three nematicides for the control of *Meloidogyne javanica* and their effect on yields of successive crops of tomato and melon. Fundam Appl Nematol 21:113–118, 1998.

138. G Godoy, R Rodriguez-Kabana, G Morgan-Jones. Parasitism of eggs of *Heterodera glycines* and *Meloidogyne arenaria* by fungi isolated from cysts of *H. glycines*. Nematropica 12:111–119, 1982.

139. MT Dunn, RM Sayer, A Correll, WR Wergin. Colonization of nematode eggs by *Paecilomyces lilacinus* (Thom) Samson as observed with scanning electron microscope. Scan Elect Microsc 3:1351–1357, 1982.

140. B Dube, GC Smart, Jr. Biological control of *Meloidogyne incognita* by *Paecilomyces lilacinus* and *Pasteuria penetrans*. J Nematol 19:222–227, 1987.

141. J Román, A Rodríguez-Marcano. Effect of the fungus *Paecilomyces lilacinus* on the larval population and root knot formation of *Meloidogyne incognita* in tomato. J Agric Univ Puerto Rico 69:159–167, 1984.

142. TA Khan, SK Saxena. Integrated management of root knot nematode *Meloidogyne javanica* infecting tomato using organic materials and *Paecilomyces lilacinus*. Bioresource Technol 61:247–250, 1997.

143. HB Jansson, A Jeyaprakash, BM Zuckerman. Control of root-knot nematodes on tomato by the endoparasitic fungus *Meria coniospora*. J Nematol 17:327–329, 1985.

144. Y Oka, I Chet, Y Spiegel. A fungal parasite of *Meloidogyne javanica* eggs: evaluation of its use to control the root-knot nematode. Biocontrol Sci Technol 7:489–497, 1997.

145. SLF Meyer. Effects of a wild type strain and a mutant strain of the fungus *Verticillium lecanii* on *Meloidogyne incognita* populations in greenhouse studies. Fundam Appl Nematol 17:563–567, 1994.

146. SLF Meyer. Evaluation of *Verticillium lecanii* strains applied in root drenches for suppression of *Meloidogyne incognita* on tomato. J Helminthol Soc Wash 65:82–86, 1998.

147. SLF Meyer, RJ Meyer. Effects of a mutant strain and a wild type strain of *Verticillium lecanii* on *Heterodera glycines* populations in the greenhouse. J Nematol 27:409–417, 1995.

148. SLF Meyer, RJ Meyer. Greenhouse studies comparing strains of the fungus *Verticillium lecanii* for activity against the nematode *Heterodera glycines*. Fund Appl Nematol 19:305–308, 1996.

149. SLF Meyer, RN Huettel. Application of a sex pheromone, pheromone analogs, and *Verticillium lecanii* for management of *Heterodera glycines*. J Nematol 28:36–42, 1996.

150. GR Stirling, LJ Smith, KA Licastro, LM Eden. Control of root-knot nematode with formulations of the nematode-trapping fungus *Arthrobotrys dactyloides*. Biol Control 11:224–230, 1998.

151. GR Stirling, LJ Smith. Field tests of formulated products containing either *Verticillium chlamydosporium* or *Arthrobotrys dactyloides* for biological control of root-knot nematodes. Biol Control 11:231–239, 1998.

152. BA Jaffee. Enchytraeids and nematophagous fungi in tomato fields and vineyards. Phytopathology 89:398–406, 1999.

153. RM Duponnois, B Amadou, T Mateille. Effects of some rhizosphere bacteria for the biocontrol of nematodes of the genus *Meloidogyne* with *Arthrobotrys oligospora*. Fundam Appl Nematol 21:157–163, 1998.

154. AF Bird, J Bird. Observations on the use of insect parasitic nematodes as a means of biological control of root-knot nematodes. Int J Parasitol 16:511–516, 1986.

155. EE Lewis, EE Perez. Effect of entomopathogenic nematodes on movement and egg production of *Meloidogyne incognita*. J Nematol 32:442, 2000.

156. TA Zitter. Diseases caused by viruses. In: JB Jones, JP Jones, RE Stall, TA Zitter, eds. Compendium of Tomato Diseases. St. Paul, MN:APS Press, 1991, pp 31–42.

157. JE Brown, JM Dangler, FM Woods, MD Henshaw, WA Griffey, MS West. Delay in mosaic virus onset and aphid vector reduction in summer squash grown on reflective mulches. HortScience 28:895–896, 1993.

158. JW Moyer, T German, JL Sherwood, D Ullman. An update on tomato spotted wilt virus and related tospoviruses. APSnet Feature at http://www/scisoc.org/feature/tospovirus/Top.html, 1999.

159. TL German, DE Ullman, JW Moyer. Tospoviruses: diagnosis, molecular biology, phylogeny, and vector relationships. Annu Rev Phytopathol 30:315–348, 1992.

160. TA Zitter. Tomato spotted wilt. In: Compendium of Tomato Diseases. JB Jones, JP Jones, RE Stall, TA Zitter, eds. St. Paul, MN: APS Press, 1991, p 40.

161. R Goldbach, D Peters. Possible causes of the emergence of tospovirus diseases. Sem Virol 5:113–120, 1994.

162. JJ Cho, DM Custer, SH Brommonschenkel, SD Tanksley. Conventional breeding: host-plant resistance and the use of molecular markers to develop resistance to tomato spotted wilt virus in vegetables. Acta Hort 431:367–378, 1996.

163. MJ Diez, S Rosello, F Nuez, J Costa, A Lacasa, MS Catala. Tomato production under mesh reduces crop loss to tomato spotted wilt virus in some cultivars. Hortscience 34:634–637, 1999.

164. S Adkins. Tomato spotted wilt virus—positive steps towards negative success. Mol Plant Pathol 1:151–158, 2000.

165. S Cohen, Y Antignus. Tomato yellow leaf curl virus (TYLCV), a whitefly-borne geminivirus of tomatoes. Adv Dis Vector Res 10:259–288, 1994.

166. B Picó, MJ Díez, F Nuez. Viral diseases causing the greatest economic losses to the tomato crop. II. The tomato yellow leaf curl virus—a review. Sci Hort 67:151–196, 1996.

167. S Cohen, J Kern, I Harpez, R Ben-Joshep. Epidemiological studies of the tomato yellow leaf curl virus (TYLCV) in the Jordan Valley, Israel. Phytoparasitica 16:259–270, 1988.

168. K Natarajan. Natural enemies of *Bemisia tabaci* (Gennadius) and effect of insecticides on their activity. J Biol Control 4:86–88, 1990.

169. L Broadbent. Epidemiology and control of tomato mosaic virus. Annu Rev Phytopath 14:75–96, 1976.

170. JH Fitchen, RN Beachy. Genetically engineered protection against viruses in transgenic plants. Annu Rev Microbiol 47:739–763, 1993.

171. T Kunik, R Salomon, D Zamir, N Navot, M Zeidan, I Michelson, Y Gafni, H Czosnek. Transgenic tomato plants expressing the tomato yellow leaf curl virus capsid protein are resistant to the virus. Bio/Technology 12:121–129, 1994.

172. GS Raupach, L Liu, JF Murphy, S Tuzun, JW Kloepper. Induced systemic resistance in cucumber and tomato against cucumber mosaic cucumovirus using plant growth–promoting rhizobacteria (PGPR). Plant Dis 80:891–894, 1996.
173. GW Zehnder, C Yao, JF Murphy, ER Sikora, JW Kloepper, DJ Schuster, JE Polston. Microbe induced resistance against pathogens and herbivores: Evidence of effectiveness in agriculture. In: AA Agrawal, S Tuzun, E Bent, eds. Induced Plant Defenses Against Pathogens and Herbivores St. Paul, MN: APS Press, 1999, pp 335–355.
174. JE Polston, E Hiebert, RJ McGovern, PA Stansly, DJ Schuster. Host range of tomato mottle virus: a new geminivirus infecting tomato in Florida. Plant Dis 77: 1181–1184, 1993.
175. http://www.barc.usda.gov/psi/bpdl/bpdl.html.

12
Biological Disease Control in the Production of Apple

Mark Mazzola
Agricultural Research Service, U.S. Department of Agriculture, Wenatchee, Washington

I. INTRODUCTION

Apple is by no means a fundamental crop for human subsistence, and in many instances it could be considered a superfluous addition to the diet. As such, production historically has been dominated by countries with economic systems that could provide investment income to establish orchards and a populace with sufficient disposable income to purchase such a nonessential food item. As production costs have become primary and transportation costs have waned as elements contributing to the economic viability of production systems, there has been a shift in the relative importance in the regions of apple production.

The production of apples is common to most regions of the world that possess a temperate climate. Through the mid-1980s the Soviet Union was the largest producer of apples, followed by the United States and China, both of which produced in the order of 3.5 million metric tons per year [1]. Although production has remained nearly flat in most of Europe, apple production in other regions has climbed significantly and has been associated with changes in varietal composition of that production. Chinese apple production has increased dramatically during the past 20 years and now approaches or exceeds 20 million metric tons annually [2]. During this expansion in apple production, a consistent reduction in the mainstream varieties, including Red and Golden Delicious along with Cox's Orange Pippin, has been observed, with a corresponding increase in market share for newer varieties such as Fuji, Gala, and Braeburn.

Due to the availability of resources and difficulty in breeding pest resistance into horticulturally acceptable cultivars, chemical measures have been the dominant form of pest control employed in apple production systems. However, impending and potential regulatory issues are driving interest in the development of alternative control measures in production aspects ranging from the use of preplant soil fumigants for the control of apple replant disease to the application of fungicides for the control of postharvest rots. Another element leading the drive toward formulating alternative pest control alternatives has been the absence of effective chemical control options. Control of fire blight has traditionally relied on use of the antibiotic streptomycin, and control of gray mold was in large part dependent on use of benzimidazole fungicides. In both instances, the development of resistant pathogen populations and the absence of chemical controls that are as effective as these predecessors has stimulated interest in the formulation of integrated disease control systems.

Biological disease control would appear to have significant potential, in terms of both environmental and economic issues, for incorporation into organic and conventional apple production systems. Concern over the impact of chemical pesticides used in production agriculture on environmental quality and human health has, in some cases, led to a reevaluation of specific products by regulatory agencies or, in the case of methyl bromide, an eventual ban in use. In addition, recent difficult economic conditions have led to the adoption of organic farming systems by some apple growers due to the potential for greater financial returns. In the 1999 growing season approximately 2200 acres of apple orchard were certified organic in Washington State. During the same period over 3500 acres were in transition from conventional production practices to be certified organic within 3 years [3]. Obviously, such systems rely more heavily on alternative pest control practices and will benefit greatly from the development and implementation of effective biological disease control measures.

Biological control of insect pests has proven quite successful in apple. However, with a few notable exceptions (e.g., fire blight and postharvest diseases of apple), in comparison to annual crop production systems, there has been a relative dearth of study on biological control as a disease management practice in apple. In part, this may stem from the perennial nature of the crop and the long-term nature of studies to investigate the efficacy of such control measures in these systems.

Perennial cropping systems provide unique opportunities and impediments to the use of biological measures in the control of plant diseases. As perennial cropping systems such as apple are established through the use of clonal stock rather than seed, these systems may possess multiple opportunities for the introduction of biological agents during the process of plant establishment. Several opportunities will exist in the nursery environment for the application of microbial inoculants such as biocontrol rhizobacteria or mycorrhizal fungi. This is particularly applicable in those nursery operations where rootstocks and other propagative materials are initially established in fumigated soils. Such an environment will likely possess diminished microbial competition, and thus success in establishing the introduced microbial inoculant will be enhanced. An additional opportunity for introduction of biological control agents exists at the time of orchard establishment and can take the form of direct soil incorporation or application to the root system of planting stock.

Without question, a major obstacle to the effective utilization of biological measures in apple production systems is the extended duration of the protection period often required to achieve successful disease control. Implementation of biological measures for the control of root diseases of established perennial production systems provide unique barriers not typically found in annual production systems. One such barrier is the postplanting introduction and establishment of the biocontrol agent at the potential point of pathogen invasion. While such agents could plausibly be introduced each year at planting in annual production systems, successful introduction of a biological agent into the rhizosphere of a plant with a long established microflora is not probable.

Control of aerial pathogens in perennial production systems in large part is analogous to those methods and measures employed in annual production systems. The protection period varies with the host-pathogen system, but it does not differ significantly with regards to foliar pathogens from annual cropping systems. A primary exception to this statement would be a variety of canker diseases and pathogens, which induce systemic infections. In such an instance, the pathogen is dormant during a portion of the year but becomes active during some period of each successive growing season. Again, this may be a hidden opportunity such as those instances where applications during a dormant period may provide effective control and the need to suppress an actively growing and reproducing pathogen is made unnecessary. On the other hand, it is more often the case that pathogens causing perennial cankers are internal to the host and effective biological control will be difficult if not impractical.

In this chapter, the application of biological measures for the control of some specific diseases of apple will be discussed. A wide disparity in development of such methods exists among disease systems. Thus, the depth to which individual diseases are discussed will vary in a similar manner.

II. FIRE BLIGHT

Fire blight is caused by the bacterium *Erwinia amylovora* and is a serious constraint to pome fruit production, including apple, in many regions of the world [4]. The bacterium is capable of damaging virtually all plant tissues, causing symptoms ranging from flower necrosis to stem girdling near the graft union of trees. As a result, losses can range from reductions in the current year crop to the death of mature trees in a single growing season. An artificially induced, but certainly economic, concern is the market access barrier E. amylovora presents for export of fruit from countries where the disease has been reported to countries where the disease is absent. E. amylovora is reputedly the most intensively investigated bacterial plant pathogen [5]. During the past three decades some advancement, particularly in the identification of active biological agents, has been attained in the development of suitable measures for control of fire blight. Control continues to be dependent upon an integrated management approach employing strategies ranging from strict quarantine in regions where the bacterium is not established to the prodigious use of antibiotics.

A first line of defense in the control of fire blight is the management of inoculum potential in the orchard through the elimination of overwintering cankers during the dormant season and removal of active lesions during the growing season [6]. Chemical control of fire blight has been dependent upon the use of the antibiotics streptomycin and oxytetracycline or the application of copper compounds. However, resistance to streptomycin among populations of E. amylovora is common in many production areas where this compound has been employed [7–9]. As a result, oxytetracycline and in some instances copper sprays have been substituted. Control achieved with these measures is less effective, and potential for the development of resistance to these compounds is evident [7], if not likely, based on history of other bacterial pathogens [10,11]. Though other antibiotics may prove effective for the control of fire blight [12], the current concern over the impact of agricultural uses of antibiotics and resistance development in human pathogens may preclude commercial application.

Resistance to E. amylovora is known to exist within apple germplasm [13,14]. However, the use of such germplasm has not been extensive due to difficulty in incorporating desired horticultural traits into such material. The recent shift away from traditional cultivars such as Red Delicious toward more susceptible varieties such as Fuji and Gala has intensified the problem. Rootstocks resistant to E. amylovora have also been identified [15,16], but some of the more common dwarfing rootstocks employed, such as M26, are among the most susceptible to fire blight. Recent advances in the transformation of apple hold significant promise for over coming past deficiencies in merging resistance with desirable horticultural traits in apple [17,18].

A. Biological Control with Bacterial Agents

Biological control of fire blight through the application of epiphytic bacteria has been extensively studied, and numerous recent reviews have addressed the topic [19–21]. Although progress has been made in developing control practices for the management of fire blight, the inadequacies of such measures cited above point to significant opportunities for the employment of alternative strategies. One such element is the incorporation of biological control into an integrated approach to the management of fire blight. Control of the blossom blight phase of the disease is generally thought to be integral to effective management of fire blight [21] as this serves as the primary means of entry to the host, and bacterial ooze emanating from blossom infections is a source of inoculum contributing to the secondary phase of disease development. As a result, attempts to develop agents for the biological control of fire blight have focused on this phase of the disease cycle.

Although significant obstacles continue to exist, biological control of the blossom blight phase of fire blight has been obtained in the field [21–24], resulting in the development of commercial biocontrol products (BlightBan A540; a.i. *Pseudomonas fluorescens* strain A506, Plant Health Technologies, Boise, ID), and has been utilized as an alternative or a complementary measure to the application of antibiotics for disease control [25,26]. In the realm of biological control research, such a history can only be looked upon as a success, even when considering the associated limitations of this control practice, which require continued study to overcome.

Biological control of fire blight has focused on the use of bacterial agents for suppression of epiphytic populations of *E. amylovora* that develop on blossoms. This is a critical step in obtaining effective biological control of fire blight as populations of *E. amylovora* that develop on flower stigmas are subsequently transported to the hypanthia where infection generally occurs [27]. Thus, bacteria possessing the ability to effectively colonize flower stigmas are likely to be superior agents for biocontrol of fire blight, regardless of the mechanism employed by the agent.

The majority of studies on the biological control of fire blight have focused on the use of the bacterial ephiphytes *Pantoea agglomerans* (*Erwinia herbicola*) and fluorescent *Pseudomonas* spp. The initial motivation for focus on these groups is not clear, though the frequent isolation of *P. agglomerans* in concert with *E. amylovora* from infected blossoms appears to have been a stimulus [28]. The focus on these bacterial groups has been warranted based on subsequent studies on the composition and antagonistic potential of apple and pear blossom and leaf microflora. In large scale in vitro screening tests, isolates from these groups of bacteria were among the most effective in suppressing disease development [29]. Likewise, representative isolates from *Pantoea* and *Pseudomonas* were more effective in limiting populations of *E. amylovora* on the stigma of crab apple blossoms [24] relative to isolates from other bacterial genera including *Bacillus* and *Arthrobacter*. However, in field settings, *P. agglomerans* does not appear to be a major component of the microflora of healthy apple or pear blossoms [30–32]. In contrast, fluorescent *Pseudomonas* spp. are a large component of the microflora resident to the flowers of apple and pear. Regardless, numerous studies have demonstrated the efficacy of strains from both of these bacterial groups for control of fire blight when applied to blossoms prior to colonization by *E. amylovora* [23–25,29].

B. Mechanisms of Biological Control

Bacterial strains applied for the biological control of fire blight employ several different mechanisms ranging from preemptive colonization of susceptible tissues to direct suppression of *E. amylovora* through the production of various metabolites including antibiotics. Preemptive exclusion is an important, and in some instances the dominant, mechanism by which bacterial antagonists suppress populations of *E. amylovora*. Although *P. fluorescens* strain A506 inhibited colonization of the stigma by *E. amylovora* when applied prior to inoculation of the pathogen, strain A506 failed to provide control when co-inoculated with the pathogen [33]. It was proposed that the observed inhibition was the result of preemptive utilization of a growth-limiting resource, though other factors including the induced cessation of nectar secretion were not excluded.

A similar mechanism has also been reported to contribute to the inhibition of *E. amylovora* by various isolates of *P. agglomerans*. Scanning electron microscopy studies have demonstrated that these bacteria colonize the same sites on the stigmatic surfaces of apple [34] and hawthorn [35]. When *P. agglomerans* (= *E. herbicola*) strain HL9N13 was applied to hawthorn flowers 24 hours prior to inoculation with *E. amylovora*, the growth rate and final population of the pathogen were suppressed, resulting in a significant reduction in disease incidence [35]. In contrast to *P. fluorescens* strain A506, strain HL9N13 was able to reduce populations of *E. amylovora* applied prior to or when co-inoculated with the pathogen. This indicates that this strain of *P. agglomerans* prevented *E. amylovora* colonization of the stigma by preemptive and competitive occupation of colonization sites. Once occupation of the site was attained, the ability of HL9N13 to limit multiplication of the pathogen was attributed to the depletion of a growth-limiting resource.

In vitro inhibition of *E. amylovora* is a characteristic common to numerous isolates of *P. agglomerans* [36,37]. Several studies have investigated antimicrobial compounds produced by strains of *E. herbicola* and have demonstrated that these compounds are not uniform with regards to class (summarized in Ref. 19). The same strain may have the potential to produce multiple antibiotics [38]. As in other systems, gene disruption studies have demonstrated a definitive role of antibiotic production in the ability of certain *P. agglomerans* strains to provide biological control of *E. amylovora*. Vanneste et al. [39] demonstrated that an antibiotic-deficient derivative of Eh252 generated through transposon mutagenesis was less effective than the parental strain in controlling fire blight. Although antibiotic production contributes to biological control of *E. amylovora*, the fact that some level of disease control is maintained indicates that other mechanisms, including preemptive and competitive exclusion of the pathogen, function in the efficacy of these *P. agglomerans* strains.

Recent studies have begun to address the development of alternative strategies for the effective deployment of agents for the biocontrol of fire blight. As occupation of the niche preferred by the pathogen is intrinsic to effective control of fire blight, particular attention has been placed on the use of novel derivatives of the pathogen as potential biological control agents.

One suggested strategy is the introduction of genes from *E. herbicola* Eh252 involved in peptide antibiotic production into a nonpathogenic derivative of *E. amylovora* [40]. Alternatively, the application of avirulent mutants of the pathogen as biological control agents has been proposed [41]. Faize et al. [42] examined the effect of a Hrp regulatory mutant and a secretory mutant of *E. amylovora* for the ability to provide control of fire blight. It was found that the regulatory mutant, but not the secretory mutant, suppressed multiplication of the pathogen. It was suggested that control was achieved by induction of a host defense response as the regulatory mutant stimulated two known plant defense response enzymes but the secretory mutant had no such impact.

C. Integrated Strategies Utilizing Biological Control Agents

The application of streptomycin for the control of fire blight has been an effective means of disease control, but its continued utility is in question due to widespread resistance development in the pathogen population [7–9]. Likewise, bacterial antagonists typically fail to provide complete disease control under field conditions, thus integrated strategies that employ the use of biocontrol agents in concert with antibiotics would appear to be a desirable disease management strategy. Such a scheme would require that the antagonist exhibit resistance to either of the two antibiotics commonly employed for the control of fire blight—streptomycin and oxytetracycline. Stockwell et al. [26] observed no difference in recovery of the antagonists *P. fluorescens* A506 and a streptomycin-resistant mutant of *P. agglomerans* strain C9-1 from apple blossoms when streptomycin was applied 2 and 7 days after the antagonists. Although application of oxytetracycline had a negative impact on both bacterial strains, the impact was negligible if antibiotic application was delayed until 7 days after application of the antagonists. Studies conducted with *P. fluorescens* A506 in concert with antibiotic applications suggest that the control achieved is likely to be additive in nature [25]. Although integration of these control practices appears promising, effective use will continue to require attention to the timing of antagonist application, particularly in regions where oxytetracycline is the only chemical alternative due to the presence of a streptomycin-resistant pathogen population.

III. APPLE REPLANT DISEASE

Apple replant disease is widespread and has been documented in all of the major fruit-growing regions of the world [43]. Replant disease of apple has been attributed to a variety of biotic and abiotic factors, but the fact that other fruit trees planted in the same soil grow normally [44] and that soil pasteurization [45,46] or fumigation [47,48] dramatically improves plant growth demonstrate that this disease is primarily a biological phenomenon rather than the result of abiotic factors. Replant disease is the primary biological impediment to the establishment of an economically viable orchard on sites previously planted to apple. This disease is becoming an increasingly important problem as economics dictate shorter orchard rotations and the availability of land suitable for orchard establishment, but not previously planted to apple, becomes limited.

A variety of agents, some known pathogens and others generally considered to possess a saprophytic habit, have been implicated as causal agents of apple replant disease. The lesion nematode (*Pratylenchus* spp.) has been reported to have a major role in apple replant disease in the eastern United States [47,49], British Columbia [50], and Australia [51] but appears to have a less significant role in the Pacific Northwest of the United States [52,53]. Several studies have suggested a role for soilborne fungi in the etiology of apple replant disease [46,48,54,55]. In a recent systematic study, a fungal complex was reported as the dominant cause of apple replant disease [53]. Other studies have suggested the possible involvement of bacteria [56] or actinomycetes [57,58] as potential causal agents of replant disease, but the data implicating these organisms in the disease etiology are tenuous.

A. Biological Control of the Causal Pathogen Complex

Contradictory information concerning the primary biological factors that contribute to apple replant disease has been a major impediment to the development of biologically based measures for disease control. As multiple agents generally contribute to disease development, failure to identify or target all components of the causal pathogen complex will invariably lead to failure of a biological, or any other, control strategy. In practice, development of biological controls for apple replant disease has generally focused on the use of agents with activity against individual elements of the pathogen complex with the goal of integrating such agents with other control practices. Such an approach is likely to be the only feasible manner to effectively use biological agents for the control of apple replant disease.

A number of studies have demonstrated benefits resulting from rootstock applications of bacteria to the subsequent growth of apple. These studies have investigated both plant growth–promoting [59,60] and disease-suppressive rhizobacteria [61–63]. Unfortunately, with few exceptions, these studies have been observational in nature and mechanistic studies have been lacking. In general these initial notable observations have not received subsequent investigation to determine modes of action, efficacy under field conditions or feasibility as commercial measures for disease control.

A diversity of bacterial species have been identified that suppress individual causal elements of apple replant disease and enhance growth of apple in replant orchard soils. Biological control of *Phytophthora cactorum*, which contributes to replant disease [53] and can cause crown and collar rot of apple [1], has been reported in response to application of *Enterobacter aerogenes* [64]. In the field, such treatments resulted in increased tree growth and fruit yield [65]. When applied to infected tissues *Pseudomonas* spp. strain 3A17 was found to suppress the expansion of lesions induced by *P. cactorum* [61]. In vitro inhibition of *P. cactorum* by strain 3A17 was attributed to siderophore production by the bacterium.

Bacillus subtilis strain EBW-4 was reported to have potential for control of apple replant disease [62]. However, in apple seedling bioassays conducted in the greenhouse, this bacterium enhanced growth of apple in pasteurized but not nonpasteurized orchard replant soil when applied as a soil drench [62]. In a field trial, the bacterium enhanced shoot growth of M.26 rootstock in replant orchard soil, but whether this was due to a growth-promotion effect or disease control was not ascertained. EBW-4 has been show to enhance tree growth in subsequent field trials [66], but in the same trials soil fumigation failed to improve tree growth. In the absence of analyses to determine pathogen suppression, these findings reinforce the concept that growth promotion rather than disease control was achieved in this instance.

Pseudomonas putida strain 2C8 was originally isolated from the roots of apple grown in soil that had been in continuous wheat monoculture prior to orchard establishment [63]. This bacterial isolate was found to inhibit in vitro growth of each element of the fungal complex reported to incite replant disease in Washington State. In greenhouse trials, incorporation of strain 2C8 into the soil profile significantly enhanced growth of M.26 rootstock in multiple apple replant soils [63]. In field trials, root-dip application of strain 2C8 to Gala on M.26 at planting resulted in an initial significant increase in shoot growth relative to nontreated trees, but this difference was not maintained through the end of the initial growing season (Table 1). Fumigation with methyl bromide, but not root application of strain 2C8, significantly increased trunk diameter of Gala on M.26 apple trees grown at two replant orchard sites (Table 2). Although strain 2C8 effectively controlled root infection by *Rhizoctonia* sp., root colonization by *Cylindrocarpon destructans*, *Phytophthora cactorum* sp., and *Pythium* spp. was not suppressed by application of this biocontrol strain. This may explain the lack of a sustained positive growth response at these field sites.

Table 1 Average Shoot Length of Gala on M.26 Rootstock Established in Replant Soils During Initial Year of Tree Growth[a]

| | Shoot length (cm)[b] | | |
Treatment	27 June, 1998	13 July, 1998	30 August, 1998
Control	22.6a	24.2a	44.0a
MeBr fumigation	26.9b	30.4b	58.4b
Pseudomonas putida 2C8	27.2b	30.7b	43.6a

[a] Studies were conducted at the CV orchard, Orondo, WA.
[b] Means in the same column followed by the same letter are not significantly ($p = 0.05$) different based on Fisher's protected LSD test.

Caesar and Burr [59] identified two fluorescent pseudomonads and an enteric bacterium possessing the ability to promote growth of apple in replant soils. Each of these bacteria demonstrated broad-spectrum antibiosis in vitro against a number of fungi found to infect apple roots. However, the fact that these isolates also promoted growth of apple in nonreplant soils suggested that the enhanced growth did not solely result from disease control, but that other growth-promotion effects were induced by the bacterial strains. Enhanced growth in replant soils in response to application of these rhizobacteria was associated with a reduction in root infection by *Cylindrocarpon destructans*, a fungal pathogen known to contribute to apple replant disease [49,53].

Adequate development of endomycorrhizal relationships is essential to the normal growth and development of apple [67–69]. Trees exhibiting symptoms of apple replant disease typically are deficient in development of mycorrhizal associations [70,71]. A number of studies have reported enhanced growth of apple

Table 2 Increase in Trunk Diameter of Gala on M.26 Apple Trees After 2 Years of Growth in Replant Soils

| | Trunk diameter increase (mm)[a] | |
Treatment	CV orchard[b]	WVC orchard
Control	11.6a	8.1a
MeBr fumigation	14.1b	12.9b
Pseudomonas putida 2C8	10.3a	11.1ab

[a] Means in the same column followed by the same letter are not significantly ($p = 0.05$) different based on Fisher's protected LSD test.
[b] Studies were conducted at the CV and WVC orchards located in Orondo and Wenatchee, WA, respectively.

in replant soils in response to root inoculation with a specific VA mycorrhizal isolate [72]. However, as is common in studies of this disease syndrome, the absence of sufficient microbial analyses prevents a determination as to whether this was the result of disease control or enhanced plant nutrition.

B. Phytomanagement of Resident Microbial Antagonists

The use of cover crops in orchard management systems is not an uncommon practice, although its application varies widely. Cover crops have been used as a means to enhance subsequent nutrition of orchards established on replant sites [73], as well as a cultural practice to suppress populations of the lesion nematode in replant soils prior to replanting [74,75]. Crops related to nematode suppression have been chosen on the basis of nonhost status or known nematicidal activity. Use of cover crops in orchard systems as a means to exploit natural biological control of the ecosystem through the enhanced activity of resident soil microbial antagonists has not received extensive study.

Habitat manipulation through the maintenance of certain cover crops in orchard ecosystems has been examined extensively as a means to manage various insect pests [76,77]. Various cover crops have been used to harbor distinctive complexes of natural enemies of orchard insect pests [78,79]. Recent studies examined the use of wheat as a cover crop to enhance populations of resident soil microbial antagonists that suppress activity of plant pathogenic fungi in apple replant soils. Wheat was selected due to the observation that a wheat field soil was suppressive to *Rhizoctonia* root rot of apple caused by an introduced isolate of *R. solani* AG 5, but that soil from the same site planted to apple for 3 or more years (replant soil) was conducive to disease development [80]. The fluorescent pseudomonad population from wheat-field soil was dominated by isolates of *Pseudomonas putida* that suppressed in vitro growth of various apple root pathogens. In contrast, *Pseudomonas fluorescens* bv. III and *Pseudomonas syringae* dominated the fluorescent pseudomonad population from replant soil and the rhizosphere of apple seedlings grown in these soils. In general the fluorescent pseudomonad isolates from apple replant soils lacked in vitro antifungal activity and failed to provide biocontrol of *R. solani* AG 5.

In greenhouse trials, cultivation of replant soils with any of three wheat cultivars was found to enhance growth of apple, and the increase in plant growth resulted from a suppression in root infection by *Pythium* and *Rhizoctonia* spp. [81]. The wheat cultivar Penawawa was superior to Eltan and Rely in ability to enhance subsequent growth of apple seedlings in orchard replant soils. The rhizosphere of apple seedlings grown in Penawawa cultivated replant soils possessed a fluorescent pseudomonad population that had a higher proportion of *P. putida* isolates than did seedlings grown in the same soil cultivated to either Eltan or Rely. These and additional studies [82] suggest that alterations in composition of the resident fluorescent pseudomonad community, at least in part, contribute to the reduction in disease severity achieved through cultivation of replant soils with wheat prior to planting apple.

IV. POWDERY MILDEW

Powdery mildew of apple is caused by the biotrophic fungus *Podosphaera leucotricha*. The disease is common wherever apple is grown and can be a significant commercial problem in every stage of apple development, from reducing the growth of nursery stock to causing fruit russetting [1]. In apple, the pathogen overwinters in dormant buds, and infected buds may fail to develop new shoots during the subsequent growing season. Conidia developing from overwintering mycelium serve as the primary source of inoculum, and secondary spread of the pathogen is incited by inoculum that develops from infection of young leaves, blossoms or fruit. Ascospores of *P. leucotricha* are not believed to have a major role in the disease cycle [1]; however, based on observations in the transport of *Podosphaera clandestina* cleistothecia [83], it is possible that wind dispersal of ascocarps may be involved in the long-distance transport of the pathogen.

Biological control of powdery mildews of various crop species has received extensive examination. To date, all agents that have been reported to provide biological control of powdery mildews have been fungal in nature. As powdery mildews are biotrophs and typically do not require exogenous nutrients for germination and initial penetration, control through competition for nutrients is not a viable strategy. Likewise, as exposure of the pathogen on the leaf surface after spore germination is limited, control through antibiosis is not likely to be a suitable mechanism for disease control. As such, the greatest attention has been focused on the use of mycoparasites for the suppression of sporulation and dissemination of powdery mildews. These include *Ampelomyces quisqualis* [84–86], *Sporothix flocculosa* [87], *Tilletiopsis* spp. [88,89], and *Verticillium lecanii* [90]. The mechanism of biocontrol for all of these fungi, with the exception of *Tilletiopsis* spp., has been established as hyperparisitism as these fungi possess the ability to colonize the mycelium of powdery mildews and produce reproductive structures. Antibiosis has been suggested as the mechanism by which *Tilletioipsis* spp. inhibit sporulation of the cucumber powdery mildew [91].

The mycoparasite *A. quisqualis* isolate A-10 has been released as a commercial product (AQ10TM) for the biological control of grape powdery mildew. Several recent studies have examined the efficacy of this product for the biological control of powdery mildew of apple. Application of AQ10 using various spray schedule intervals failed to provide control of foliar infection and fruit russetting by *P. leucotricha* in field trials conducted in Washington [92]. Although apparently ineffective when utilized alone, the biofungicide has provided marginal improvement of powdery mildew control when used in a well-integrated schedule that employs effective fungicides or sulfur [93].

V. POSTHARVEST DISEASES

Apples account for a large proportion of the total fruit volume that is stored for extended periods under controlled atmospheric conditions. Storage periods extending for upwards of 10 months is not uncommon, resulting in fruit being subjected to extended attack by numerous postharvest pathogens. Conservative estimates suggest that losses attributable to postharvest decay of apples exceed $4.4 million per year in the United States [94]. The major postharvest diseases of apple include blue mold, bull's-eye rot, grey mold, and Mucor rot, incited, respectively, by *Penicillium* spp., *Pezicula malicorticis*, *Botrytis cinerea*, and *Mucor piriformis*. Although upwards of 11 different species of *Penicillium* have been implicated as causal agents of blue mold, *P. expansum* is by far the most common and economically important species [1].

In most instances, postharvest pathogens gain access to susceptible apple tissues and cause infection via entry through fruit surface wounds that are generated through the process of harvesting and postharvest handling. Alternatively, some pathogens access susceptible tissues through natural openings such as lenticels [1], or decay may be initiated in the sinus between the calyx and core cavity [95]. Even a relatively small volume loss resulting from the activity of postharvest pathogens can result in significant economic losses due to the accumulated monetary inputs that have occurred during the growing, harvesting, and storing of this crop.

Attention to fruit-handling practices in the field and during storage to reduce mechanical and physical injuries and management of controlled atmospheric conditions can be effective techniques in reducing postharvest disease. However, these methods do not ensure effective protection of stored fruit, and therefore, postharvest disease control has traditionally been reliant on the use of synthetic fungicides [96]. The effective use of such chemicals in control of postharvest diseases of apple has become increasingly less reliable due to the development of resistant populations among the target pathogens [97,98]. Perhaps even more intensely than their use in the control of preharvest diseases, there exists considerable concern over the use of synthetic fungicides on fruits and vegetables in a postharvest setting in general due to perceived or actual hazards to human health, as well as the environment.

As the consumption of fruits and vegetables can be a source of direct ingestion of fungicides, the use of biological methodologies for the control of postharvest diseases of apple has enjoyed significant attention, and even a modicum of success, over the past two decades. Biological control of postharvest diseases in general has been viewed more positively in the commercial sector due to the perception of a greater potential for success than other systems. This is primarily a function of the fact that such a setting possesses a controlled environment that can be effectively managed to benefit the activity and survival of introduced biocontrol agents.

A. Biocontrol Agents

The study and use of biological agents for the control of postharvest diseases of apple was delayed in respect to other plant disease systems but has received significant attention during the past two decades. The primary source of biological agents for the control of postharvest pathogens has been the microbial community resident to the phyllosphere and fruit surface of apple [99–101]. Bacteria and yeasts or yeast-like organisms have been the most commonly employed agents for the biological control of postharvest diseases of apple.

Several studies have identified bacterial agents for the control of postharvest diseases of apple. A saprophytic strain of *Pseudomonas syringae* initially found to control postharvest rots of pear, including gray mold, blue mold, and Mucor rot [102], was subsequently shown to provide biological control of all three pathogens on wounded Red and Golden Delicious apples in co-inoculations with the individual pathogens [103]. The strain (*P. syringae* ESC-11) is currently registered for postharvest application to apples and is marketed as the product Bio-save 110 (formerly Bio-save 11) by EcoScience Corp. (Langhorne, PA). An isolate of *Burkholderia cepacia* provided biological control of blue mold and gray mold on Golden Delicious apples [104]. *Bacillus subtilis* applied to wounded apples reduced fruit rot caused by *B. cinerea*, *P. exapansum*, and *P. malicorticis* [105].

Successful biological control of numerous postharvest diseases of fruits and vegetables has been obtained through the application of yeasts. In apple, a diversity of yeasts have been studied for the biological control of postharvest decays. Control of *Penicillium expansum*, *Botrytis cinerea*, and *Mucor* spp. on apple has been reported in several studies using the yeasts *Candida guilliermondii* [106], *Candida oleophila* [107], *Cryptococcus laurentii* [100], *Kloeckera apiculata* [106], and *Sporobolomyces roseus* [108]. *Rhodotorula glutinis* provided biological control of bull's-eye rot on Golden Delicious apples [105]. The yeast *Candida oleophila* strain 182 has been commercialized as the postharvest biofungicide Aspire (Ecogen, Inc., Langhorne, PA) for the control of *Botrytis* spp. and *Penicillium* spp.

While biological control of postharvest diseases has been obtained through the use of individual isolates, the application of strain mixtures has been promoted. Fruit is submerged in dump tanks during the sizing and packing process to reduce damage to fruit, and the solutions contained in these tanks invariably harbor spores of various potential postharvest pathogens. In such an environment, it is likely that the control of multiple pathogens will be necessitated, yet, relative to other control methods including synthetic chemicals, biological control agents typically possess a relatively narrow spectrum of activity. It has been shown in certain soilborne disease systems that the effective control of one pathogen in a complex often leads to increased plant damage due to the increased activity of other pathogens that compete in the same niche [53,109]. Thus, it is plausible that in the postharvest setting control of a single pathogen may simply favor another. The application of multiple agents, which in concert expand the range of biocontrol activity, is a means to avert such a circumstance.

Janisiewicz [110] found that a combination of an isolate of *Pseudomonas* sp. in combination with *Acremonium breve* gave complete control of *B. cinerea* and *P. expansum* on apple. A co-application involving the bacterial antagonist *P. syringae* and the yeast *S. roseus* applied in equal biomass provided control of blue mold that was superior to that obtained by treatment with the individual agents applied separately using a biomass equivalent to that of the mixture [111]. A mixture of two *Aureobasidium pullulans* strains and an isolate of *Rhodotorula glutinis* was superior to any of the strains applied individually in controlling decay caused by *B. cinerea*, *P. expansum*, or *P. malicorticis* [105].

B. Mechanisms of Biological Control

Several of the same mechanisms implicated in the biological control of fire blight have been shown to function in the ability of agents to provide biological control of postharvest pathogens of apple. Mechanisms reported to contribute to biological control include antibiosis, competitive exclusion, induced resistance in the host, and production of hydrolytic enzymes.

Suppression of *B. cinerea* and *P. expansum* by *Burkholderia cepacia* is reported to involve, in part, the production of the antibiotic pyrrolnitrin [112]. Isolates of *Bacillus subtilis* have been employed for the control of numerous postharvest diseases of tree fruits [113,114]. Production of iturin peptides, which possess a wide spectrum of antifungal activity, was determined to be the primary mode of action of this bacterium [115].

Without question, competition is the most often cited mechanism in the biological control of postharvest diseases, particularly in the interaction between biocontrol yeasts and target fungal pathogens. The competitive interaction may take different forms. Roberts [100] cited the preemptive utilization of carbon or nitrogen sources as a plausible mechanism in the control of the gray mold pathogen *B. cinerea* by *Cryptococcus laurentii*. Nutrient competition has also been cited as the operative mechanism in the control of blue and gray mold through application of *Sporobolomyces roseus* [108]. An additional competitive mechanism of biological control is the physical exclusion of the pathogen by yeasts that rapidly colonize wound sites [100], and this preemptive exclusion may operate by formation of a yeast EPS layer over the infection court [116].

Direct parasitism of fungal plant pathogens is a well-documented mechanism in the biological control of various soilborne diseases by numerous species of *Trichoderma* [117]. In contrast, there have been few reports of direct parasitism functioning as a mechanism in the biological control of postharvest pathogens of apple. Wisniewski et al. [118] observed direct attachment of the yeast biocontrol agent *Pichia guilliermondii* to mycelium of *B. cinerea*. This yeast was found to produce high levels of β-1,3-glucanase activity in the presence of cell walls of several fungal plant pathogens, and upon removal of the yeast cells from *B. cinerea* hyphae degradation of cell wall was observed. Jijakli and Lepoivre [101] found that the activity of exo-β-1,3-glucanase production by *Pichia anomala* was enhanced in vitro when cell wall preparations of *B. cinerea* were supplied as the sole carbon source. The β-1,3-glucanase was found to inhibit germ tube growth of *B. cinerea* and induced leakage of cytoplasm and cell swelling.

β-1,3-Glucanase was detected on fruit treated with the biocontrol yeast and application of a *B. cinerea* cell wall preparation to fruit treated with *P. anomala* enhanced disease suppression. Although the relative role of this mechanism to disease control has yet to be determined, these studies suggest that cell wall–degrading enzymes may make a significant contribution to the suppression of postharvest fungal pathogens.

Induced resistance is a well-documented phenomenon that operates through the activation of the host plant's own defense apparatus. This response is common to a great number of crop-pathogen systems and can be elicited in a number of different manners including the application of biological and chemical agents [119,120]. Exploitation of this phenomenon in the control of postharvest diseases has primarily focused on the use of various chemical and environmental applications as a means to elicit the resistance response in harvested fruits and vegetables [121]. Droby and Chalutz [122] suggested that such a mechanism may operate in the control of *Penicillium digitatum* by *P. guilliermondii*. They observed that application of the yeast to wounds on different fruits resulted in enhanced ethylene production, though production of ethylene by the yeast was not detected when cultured in vitro. As ethylene is reported to have a role in induction of the resistance processes, it was proposed that induction of host resistance may operate in the disease suppression resulting from yeast applications.

C. Postharvest Biological Control in a Preharvest Setting

The integration of other strategies with biological measures for the control of postharvest diseases has received considerable attention. Such strategies may include the management of nutritional conditions to favor the antagonist [123], the use of synthetic fungicides in concert with biocontrol agents [124], and altering fruit storage conditions to ensure survival and activity of introduced biocontrol agents [125]. Genetic modification of biocontrol agents to enhance their activity is also a commonly raised option, though the general public concern over the use of genetically modified organisms will likely preclude the implementation of this alternative in the short term. Until rather recently, little attention has been given to earlier phases in the crop production cycle that could significantly impact control of postharvest diseases of apple.

There has been increasing interest in determining the practicality of applying such agents in the field prior to harvest as a means to control postharvest pathogens [126]. This interest emanates from the fact that infection of fruit by postharvest pathogens may occur in the field prior to harvest [127] or as a result of wounds created during harvest and handling but preceding the movement of fruit into dump tanks where biocontrol agents have traditionally been applied. The presence of biocontrol agents on the fruit surface during this period could enhance control of initial infections that often occur in the field prior to the application of agents in the packing house. This could result in a reduction in inoculum production and further limit potential spread of the pathogen through the storage facility. Such an approach would be dependent upon the prolonged survival of the biocontrol agent at sufficient populations after application.

Biological control of postharvest diseases of apple in a preharvest setting does appear to be a plausible option based on the limited studies conducted to date. Application of either of two *Trichoderma harzianum* strains to the apple cultivar Aroma in the field significantly reduced symptoms of bull's-eye rot when apples were stored at 4°C for up to 100 days [128]. Leibinger et al. [105] applied combinations of the antagonists *Aurobasidium pullans*, *Rhodotorula glutinis*, and *Bacillus subtilis* to Golden Delicious apple in the field in an attempt to control the postharvest pathogens *B. cinerea*, *P. expansum*, and *P. malicorticis*. A combination containing two isolates of *A. pullans* and an isolate of *R. glutinis* was as effective as the chemical fungicide used in the trial in suppressing postharvest infections and fruit rot. These findings suggest that application of certain biocontrol agents that are a characteristic element of the resident microflora of apple may be useful in reducing infection in the field and subsequent infections during fruit storage.

VI. CONCLUSIONS

Biological control of certain target pests is currently a viable option in most apple production systems in both the preplant and the postplant settings. Although the use of biological control by commercial producers has focused primarily on management of insect pests, several factors including consumer preference, government regulation, and pathogen resistance to chemical pesticides will lead to an increased role for biological control in management of apple diseases. At present, the most significant opportunities for successful use of biological control remain in the management of fire blight and postharvest diseases as commercial biocontrol products are currently available. This will likely remain the case in the future due to a lack of suitable acceptable alternatives for management of streptomycin-resistant *E. amylovora* and the general unwillingness of consumers and regulators to allow for continued use of chemical treatments on edible plant products.

In all probability, the use of biological control for preharvest applications will necessitate integration with other management practices. Research that is centered on the ecology of orchard ecosystems will be integral to the successful implementation of biological control as a component of a systems approach to disease management. This has been clearly demonstrated for the use of biological control in the management of fire blight in apple. Efforts to develop an integrated system employing antibiotics with biological agents for control of fire blight were refined through extensive study of the phenology and ecology of the interaction between host, pathogen, and biocontrol agent [25,26].

The management of orchard ground covers shows significant promise as a means to induce or support natural biological control of insect pests, weeds, and pathogens. The effective use of such a practice awaits extensive investigation of the biological complexity intrinsic to these ecosystems. As an example, it has been demonstrated that induction of a microbial community suppressive to re-plant disease of apple in response to growing wheat occurs in a wheat cultivar-specific manner [81; M. Mazzola, unpublished data]. Is the use of such an approach compatible with insect pest management, and if so, are the impacts uniform across wheat cultivars? Obviously, the questions to be raised are numerous and include aspects beyond pest management, including possible adverse impacts arising from competition and altered soil nutrient status. Thus, it is apparent that extensive ecological research will be required in order to achieve optimum utilization of the biological resources resident to orchard ecosystems for managing diseases of apple.

REFERENCES

1. AL Jones, HS Aldwinckle. Compendium of Apple and Pear Diseases. St. Paul, MN: APS Press, 1990.
2. AD O'Rourke. Trends in world apple production and marketing. Good Fruit Grower 47:46–50, 1996.
3. D Granatstein, P Dauer. Trends in organic tree fruit production in Washington state. Center for Sustaining Agriculture and Natural Resources Report No. 1, Washington State University, Wenatchee, WA, 2000.

4. T Van der Zwet, HL Keil. Fire blight: A bacterial disease of rosaceous plants. Agriculture Handbook 510. Washington, DC: USDA Science and Education Administration, 1979.

5. MN Schroth, SV Thomson, DC Hildebrand, WJ Moller. Epidemiology and control of fire blight. Annu Rev Phytopathol 12:389–412, 1974.

6. HS Aldwinckle, SV Beer. Fire blight and its control. Hortic Rev 1:423–474, 1979.

7. JE Loper, MD Henkels, RG Roberts, GG Grove, MJ Willet, TJ Smith. Evaluation of streptomycin, oxytetracycline, and copper resistance of *Erwinia amylovora* isolates from pear orchards in Washington State. Plant Dis 75:287–290, 1991.

8. SV Thomson, SC Gouk, JL Vanneste, CN Hale, RG Clark. The presence of streptomycin resistant strains of *Erwinia amylovora* in New Zealand. Acta Hortic 338: 223–230, 1993.

9. PS McManus, AL Jones. Epidemiology and genetic analysis of streptomycin-resistant *Erwinia amylovora* from Michigan and evaluation of oxytetracyline for control. Phytopathology 84:627–633, 1994.

10. DA Cooksey. Genetics of bactericide resistance in plant pathogenic fungi. Annu Rev Phytopathol 28:201–219, 1990.

11. EL Palmer, AL Jones. Tetracycline-resistance determinants present in Michigan apple orchards. Phytopathology 87:S74, 1997.

12. RA Spitko, M Alvarado. Use of agry-gent (gentamicin sulfate) as a control material for fire blight of apple in the U.S. Acta Hortic 489:629, 1999.

13. TM Thomas, AL Jones. Severity of fire blight on apple cultivars and strains in Michigan, Plant Dis 76:1049–1052, 1992.

14. HS Aldwinckle, HL Gustafson, PL Forsline. Evaulation of the core subset of the USDA apple germplasm collection for resistance to fire blight. Acta Hortic 489: 269–272, 1999.

15. JW Travis, JL Rytter, KD Hickey. The susceptibility of apple rootstocks and cultivars to *Erwinia amylovora*. Acta Hortic 489:235–241, 1999.

16. TL Robinson, JN Cummins, SA Hoying, WC Johnson, HS Aldwinckle, JL Norelli. Orchard performance of fire blight-resistant Geneva apple rootstocks. Acta Hortic 489:287–294, 1999.

17. K Ko, SK Brown, JL Norelli, K Düring, HS Aldwinckle. Construction of plasmid binary vector for enhanced fire blight resistance in apple. Phytopathology 87:S53, 1997.

18. JL Norelli, E Borejsza-Wysocka, MT Momol, JZ Mills, A Grethel, HS Aldwinckle, K Ko, SK Brown, DW Bauer, SV Beer, AM Abdul-Kader, V Hanke. Genetic transformation for fire blight resistance in apple. Acta Hortic 489:295–296, 1999.

19. JL Vanneste. Honey bees and epiphytic bacteria to control fire blight, a bacterial disease of apple and pear. Biocontrol News Inform 17:67N–78N, 1996.

20. M Wilson. Biocontrol of aerial plant diseases in agriculture and horticulture: current approaches and future prospects. J Indust Microbiol Biotechnol 19:188–191, 1997.

21. KB Johnson, VO Stockwell. Management of fire blight: a case study in microbial ecology. Annu Rev Phytopathol 36:227–248, 1998.

22. SV Thomson, MN Schroth, WJ Moller, WO Reil. Efficacy of bactericides and saprophytic bacteria in reducing colonization and infection of pear flowers by *Erwinia amylovora*. Phytopathology 66:1457–1459, 1976.

23. JL Vanneste, J Yu. Biological control of fire blight using *Erwinia herbicola* Eh252 and *Pseudomonas fluorescens* A506 separately or in combination. Acta Hortic 411: 351–353, 1996.

24. PL Pusey. Laboratory and field trials with selected microorganisms as biocontrol agents for fire blight. Acta Hortic 489:655–661, 1999.

25. SE Lindow, G McGourty, R Elkins. Interactions of antibiotics with *Pseudomonas fluorescens* A506 in the control of fire blight and frost injury of pear. Phytopathology 86:841–848, 1996.

26. VO Stockwell, KB Johnson, JE Loper. Compatibility of bacterial antagonist of *Erwinia amylovora* with antibiotics used for fire blight control. Phytopathology 86: 834–840, 1996.

27. SV Thomson. The role of the stigma in fire blight infections. Phytopathology 76: 476–482, 1986.

28. RN Goodman. In vitro and in vivo interactions between components of mixed bacterial cultures isolated from apple buds. Phytopathology 55:217–221, 1965.

29. RJ McLaughlin, RG Roberts. Laboratory and field assays for biological control of fire blight in d'anjou pear. Acta Hortic 338:317–319, 1993.

30. C Manceau, J-C Lalande, G Lachaud, R Chartier, J-P Paulin. Bacterial colonization of flowers and leaf surface of pear trees. Acta Hortic 272:73–76, 1990.

31. LP Kearns, CN Hale. Incidence of bacteria inhibitory to *Erwinia amylovora* from blossoms in New Zealand apple orchards. Plant Pathol 44:918–924, 1995.

32. VO Stockwell, RJ McLaughlin, MD Henkels, JE Loper, D Sugar, RG Roberts. Epiphytic colonization of pear stigmas and hypanthia by bacteria during primary bloom. Phytopathology 89:1162–1168, 1999.

33. M Wilson, SE Lindow. Interactions between the biological control agent *Pseudomonas fluorescens* A506 and *Erwinia amylovora* in pear blossoms. Phytopathology 83:117–123, 1993.

34. MJ Hattingh, SV Beer, EW Lawson. Scanning electron microscopy of apple blossoms colonized by *Erwinia amylovora* and *E. herbicola*. Phytopathology 76:900–904, 1986.

35. M Wilson, HAS Epton, DC Sigee. Interactions between *Erwinia herbicola* and *E. amylovora* on the stigma of hawthorn blossoms. Phytopathology 82:914–918, 1992.

36. Y Ophir, SV Beer. Antibiotic producing Israeli strains of *Erwinia herbicola* and control of fire blight. Phytopathology 83:1230, 1993.

37. RS Wodzinski, JP Paulin. Frequency and diversity of antibiotic production by putative *Erwinia herbicola* strains. J Appl Bacteriol 76:603–607, 1994.

38. CA Ishimaru, EJ Klos, RR Brubaker. Multiple antibiotic production by *Erwinia herbicola*. Phytopathology 78:746–750, 1988.

39. JL Vanneste, J Yu, SV Beer. Role of antibiotic production by *Erwinia herbicola* Eh252 in biological control of *Erwinia amylovora*. J Bacteriol 174:2875–2796, 1992.

40. JL Vanneste, J Yu, DA Cornish, MD Voyle, M Melbourne. Designing a biological control of fire blight: expression of a new peptide antibiotic gene in *Erwinia amylovora* and of the harpin gene in *Erwinia herbicola*. Acta Hortic 489:669–670, 1999.

41. M Tharaud, J Laurent, M Faize, JP Paulin. Fire blight protection with avirulent mutants of *Erwinia amylovora*. Microbiology 143:625–632, 1997.
42. M Faize, M Tharaud, MN Brisset, JP Paulin. Protective effect of HRP mutants of *Erwinia amylovora* against a virulent strain of the pathogen: expression in several biological systems and preliminary mechanistic studies. Acta Hortic 489:635–637, 1999.
43. JA Traquair. Etiology and control of orchard replant problems: a review. Can J Plant Pathol 6:54–62, 1984.
44. BM Savory. Specific replant diseases causing root necrosis and growth depression in perennial fruit and plantation crops. Research Review No. 1. Commonw. Bur. Hortic. and Plantation Crops, E. Malling, England.
45. H Hoestra. Replant diseases of apple in the Netherlands. Ph.D. thesis, Meded. Landbouwhogesch. Wageningen, the Netherlands.
46. BA Jaffe, GS Abawi, WF Mai. Fungi associated with roots of apple seedings grown in soil from an apple replant site. Plant Dis 66:942–944, 1982.
47. WF Mai, GS Abawi. Controlling replant diseases of pome and stone fruits in northeastern United States by preplant fumigation. Plant Dis 65:859–864, 1981.
48. JT Slykhuis, TSC Li. Response of apple seedlings to biocides and phosphate fertilizers in orchard soils in British Columbia. Can J Plant Pathol 7:294–301, 1985.
49. BA Jaffe, GS Abawi, WF Mai. Role of soil microflora and *Pratylenchus penetrans* in apple replant disease. Phytopathology 72:247–251, 1982.
50. RS Utkhede, TC Vrain, JM Yorston. Effects of nematodes, fungi, and bacteria on the growth of young apple trees grown in apple replant disease soils. Plant Soil 139:1–6, 1992.
51. SR Dullahide, GR Stirling, A Nikulin, AM Stirling. The role of nematodes, fungi, bacteria, and abiotic factors in the etiology of apple replant problems in the Granite Belt of Queensland. Aust J Exp Agric 34:1177–1182, 1994.
52. RP Covey, Jr., NR Benson, WA Haglund. Effect of soil fumigation on the apple replant disease in Washington. Phytopathology 69:684–686, 1979.
53. M Mazzola. Elucidation of the microbial complex having a causal role in the development of apple replant disease in Washington. Phytopathology 88:930–938, 1998.
54. PG Braun. The combination of *Cylindrocarpon lucidum* and *Pythium irregulare* as a possible cause of apple replant disease in Nova Scotia. Can J Plant Pathol 13: 291–297, 1991.
55. GWF Sewell. Effects of *Pythium* species on the growth of apple and their possible causal role in apple replant disease. Ann Appl Biol 97:149–169, 1981.
56. JA Bunt, D Mulder. The possible role of bacteria in relation to the apple replant disease. Meded Fac Landbouwwet Rijksuniv Gent 38:1381–1385, 1973.
57. SW Westcott, III, SV Beer, HW Israel. Interactions between actinomycete-like organisms and young apple roots grown in soil conducive to apple replant disease. Phytopathology 77:1071–1077, 1987.
58. G Otto, H Winkler. Colonization of rootlets of apple seedlings from replant soils by actinomycetes and endotrophic mycorrhiza. Acta Hortic 324:53–59, 1993.
59. AJ Caesar, TJ Burr. Growth promotion of apple seedling rootstocks by specific strains of rhizobacteria. Phytopathology 77:1583–1588, 1987.
60. AJ Caesar. Effects, characterization, and formulation of plant growth-promoting

rhizobacteria of apple from New York soils. PhD thesis, Cornell University, Ithaca, NY.

61. WJ Janisiewicz, RP Covey. Biological control of collar rot caused by *Phytophthora cactorum*. Phytopathology 73:822, 1983.

62. RS Utkhede, TSC Li. Evaluation of *Bacillus subtilis* for potential control of apple replant disease. J Phytopathol 126:305–312, 1989.

63. M Mazzola. Towards the development of sustainable alternatives for the control of apple replant disease in Washington. In: Proc. Int. Res. Conf. Methyl Bromide Alternatives and Emissions Reductions, MBAO, Fresno, CA, 1998, pp 8.1–8.3.

64. RS Utkhede, EM Smith. Biological and chemical treatments for control of *Phytophthora* crown and root rot caused by *Phytophthora cactorum* in a high density aple orchard. Can J Plant Pathol 13:267–270, 1991.

65. RS Utkhede, EM Smith. Effectiveness of dry formulations of *Enterobacter agglomerans* for control of crown and root rot of apple trees. Can J Plant Pathol 19:397–401, 1997.

66. RS Utkhede, EM Smith. Development of biological control of apple replant disease. Acta Hortic 363:129–134, 1994.

67. B Mosse. Growth and chemical composition of mycorrhizal and non-mycorrhizal apples. Nature 179:922–924, 1957.

68. NR Benson, RP Covey. Response of apple seedlings to zinc fertilization and mycorrhizal inoculation. HortScience 11:252–253, 1976.

69. RP Covey, BL Koch, HJ Larsen. Influence of vesicular-arbuscular mycorrhizae on the growth of apple and corn in low-phosphorous soil. Phytopathology 71:712–715, 1981.

70. FL Caruso, BF Neubauer, MD Begin. A histological study of apple roots affected by replant disease. Can J Bot 67:742–749, 1989.

71. Čatská, H Taube-Baab. Biological control of replant problems. Acta Hortic 363: 115–120, 1994.

72. LN Bhardwaj. Vesicular-arbuscular mycorrhizal colonization and population dynamics on young apple seedlings grown in replant disease soil. In: Proc. 4th Int'l Symp. on Replant Problems, Budapest, Hungary, 1996, p 15.

73. AR Biggs, TA Baugher, AR Collins, HW Hogmire, JB Kotcon, M Glenn, AJ Sexstone, RE Byers. Growth of apple trees, nitrate mobility and pest populations following a corn versus fescue crop rotation. Am J Alternative Agric 12:162–172.

74. DH MacDonald, WF Mai. Suitability of various cover crops as hosts for the lesion nematode (*Pratylenchus penetrans*). Phytopathology 53:730–731, 1963.

75. EA Merwin, WC Stiles. Root lesion nematodes, potassium deficiency, and prior cover crops as factors in apple replant disease. J Am Hortic Sci 114:724–728, 1989.

76. WL Tedder. Insect management in deciduous orchard ecosystems: habitat manipulation. Environ Manage 7:29–34, 1983.

77. RL Bugg, C Waddington. Using cover crops to manage arthropod pests of orchards: a review. Agric Ecosystems Environ 50:11–28, 1994.

78. S Haley, EJ Hogue. Ground cover influence on apple aphid, Aphis pomi DeGeer (Homoptera: Aphididae), and its predators in a young apple orchard. Crop Prot 9: 225–230, 1990.

79. W Liang, M Huang. Influence of citrus orchard ground cover plants on arthropod communities in China: a review. Agric Ecosystems Environ 50:29–37, 1994.
80. M Mazzola. Transformation of soil microbial community structure and *Rhizoctonia*-suppressive potential in response to apple roots. Phytopathology 89:920–927, 1999.
81. M Mazzola, Y-H Gu. Impact of wheat cultivation on microbial communities from replant soils and apple growth in greenhouse trials. Phytopathology 90:114–119, 2000.
82. M Mazzola, DM Granatstein, DC Elfving, K Mullinix, Y-H Gu. Cultural management of microbial community structure to enhance growth of apple in replant soils. Phytopathology 91: (in press), 2001.
83. G Grove, R Boal, W Duplag. Wind dispersal of cleistothecia of *Podosphaera clandestina* in eastern Washington. Phytopathology 89:S30, 1999.
84. WR Jarvis, K Slingsby. The control of powdery mildew of greenhouse cucumber by water sprays and *Ampelomyces quisqualis*. Plant Dis Rep 61:728–730, 1977.
85. A Stzjenberg, S Galper, S Mazar, N Lisker. *Ampelomyces quisqualis* for biological and integrated control of powdery mildews in Israel. J Phytopathol 124:285–295, 1989.
86. SP Falk, DM Gadoury, RC Pearson. Partial control of grape powdery mildew by the mycoparasite *Ampelomyces quisqualis*. Plant Dis 79:483–490.
87. WR Jarvis, LA Shaw, JA Traquair. Factors affecting antagonism of cucumber powdery mildew by *Stephanoascus flocculosus* and *S. rugulosus*. Mycol Res 92:162–165, 1989.
88. HC Hoch, R Provvident. Mycoparasitic relationships: cytology of the *Sphaerotheca fuliginea-Tilletiopsis* sp. interaction. Phytopathology 69:359–362, 1979.
89. EJ Urquhart, JG Menzies, ZK Punja. Growth and biological control activity of *Tilletiopsis* species against powdery mildew (*Sphaerotheca fuliginea*) on greenhouse cucumber. Phytopathology 84:341–351, 1994.
90. MA Verhaar, T Hijwegen, JC Zadoks. Glasshouse experiments on biocontrol of cucumber powdery mildew (*Sphaerotheca fuliginea*) by the mycoparasites *Verticillium lecanii* and *Sporothrix rugulosa*. Biol Control 6:353–360, 1996.
91. T Hijwegen. Effect of culture filtrate of seventeen fungicolous fungi on sporulation of cucumber powdery mildew. Neth J Plant Pathol 95:95–98, 1989.
92. GG Grove, RJ Boal. Apple powdery mildew control trials using the mycoparasite *Ampelomyces quisqualis* (AQ10). F & N Tests 52:7, 1997.
93. KS Yoder, AE Cochran, II, WS Royston, Jr., SW Kilmer, JE Scott. Integrated fungicide schedules for suppression of powdery mildew and other diseases on idared apple. F. & N. Tests 53:43–44, 1998.
94. DA Rosenberger. Recent research and changing options for controlling postharvest decays of apples. In: Proc. Harvesting, Handling, and Storage Workshop. Northeast Reg. Agric. Eng. Serv. Publ. NRAES-112. 14 Aug. 1997. Cornell University, Ithaca, NY, 1997.
95. RA Spotts, RJ Holmes, WS Washington. Factors affecting wet core rot of apples. Australas Plant Pathol 17:53–57, 1988.
96. JW Eckert, JM Ogawa. The chemical control of postharvest diseases: deciduous fruit, berries, vegetables and root/tuber crops. Annu Rev Phytopathol 26:433–469, 1988.

97. DA Rosenberger, FW Meyer. Postharvest fungicides for apples: development of resistance to benomyl, vinclizolin, and iprodion. Plant Dis 65:1010–1013, 1981.

98. RA Spotts, LA Certvantes. Populations, pathogenicity, and benomyl resistance of *Botrytis* spp., *Penicillium* spp., and *Mucor piriformis* in packinghouses. Plant Dis 70:106–108, 1986.

99. WJ Janisiewicz. Postharvest biological control of blue mold on apples. Phytopathology 77:481–485, 1987.

100. RG Roberts. Postharvest biological control of gray mold of apple by *Cryptococcus laurentii*. Phytopathology 80:526–530, 1990.

101. MH Jijakli, P Lepoivre. Characterization of an exo-β-1, 3-glucanase produced by *Pichia anomala* strain K, antagonist of *Botrytis cinerea* on apples. Phytopathology 88:335–343, 1998.

102. WJ Janisiewicz, A Marchi. Control of storage rots on various pear cultivars with a saprophytic strain of *Pseudomonas syringae*. Plant Dis 76:555–560, 1992.

103. SN Jeffers, TS Wright. Comparison of four promising biological control agents for managing postharvest diseases of apples and pears. Phytopathology 84:1082, 1994.

104. WJ Janisiewicz, J Roitman. Biological control of blue mold and gray mold on apple and pear with *Pseudomonas cepacia*. Phytopathology 78:1697–1700, 1988.

105. W Leibinger, B Breuker, M Hahn, K Mendgen. Control of postharvest pathogens and colonization of the apple surface by antagonistic microorganisms in the field. Phytopathology 87:1103–1110, 1997.

106. RJ McLaughlin, CL Wilson, S Droby, R Ben-Arie, E Chalutz. Biological control of postharvest diseases of grape, peach and apple with the yeasts *Kloeckera apiculata* and *Candida guilliermondii*. Plant Dis 76:470–473, 1992.

107. J Mercier, CL Wilson. Colonization of apple wounds by naturally occurring and introduced *Candida oleophila* and their effect on infection by *Botrytis cinerea* during storage. Biol Control 4:138–144, 1994.

108. WJ Janisiewicz, DL Peterson, R Bors. Control of storage decay of apples with *Sporobolomyces roseus*. Plant Dis 78:466–470, 1994.

109. K Xi, JHG Stephens, SF Hwang. Dynamics of pea seed infection by *Pythium ultimum* and *Rhizoctonia solani*: effects of inoculum density and temperature on seed rot and pre-emergence damping-off. Can J Plant Pathol 17:19–24, 1995.

110. WJ Janisiewicz. Biocontrol of postharvest diseases of apples with antagonist mixtures. Phytopathology 78:194–198, 1988.

111. WJ Janisiewicz, B Bors. Development of a microbial community of bacterial and yeast antagonists to control wound-invading postharvest pathogens of fruits. Appl Environ Microbiol 61:3261–3267, 1995.

112. WJ Janisiewicz, L Yourman, J Roitman, N Mahoney. Postharvest control of blue mold and gray mold of apples and pears by dip treatment with pyrrolnitrin, a metabolite of *Pseudomonas cepacia*. Plant Dis 75:490–494, 1991.

113. PL Pusey, CL Wilson. Postharvest biological control of stone fruit brown rot by *Bacillus subtilis*. Plant Dis 68:753–756, 1984.

114. RS Utkhede, PL Sholberg. In vitro inhibition of plant pathogens by *Bacillus subtilis* and *Enterobacter aerogenes* and in vivo control of two postharvest cherry diseases. Can J Microbiol 32:963–967, 1986.

115. CD McKeen, CC Reilly, PL Pusey. Production and partial characterization of anti-

fungal substances antagonistic to *Monilinia fructicola* from *Bacillus subtilis*. Phytopathology 76:136–139, 1986.

116. S Droby, E Chalutz, I Chet. Possible role of glucanase and extracellular polymers in the mode of action of yeast antagonists of postharvest diseases. Phyparasitica 2:167, 1993.

117. I Chet. Mycoparasitism-recognition, physiology and ecology. In: RR Baker, PE Dunn, eds. New Directions in Biological Control: Alternatives for Suppressing Agricultural Pests and Diseases. New York: Alan R. Liss Inc., 1990, pp 725–733.

118. ME Wisniewski, C Biles, S Droby, R McLaughlin, C Wilson, E Chalutz. Mode of action of the postharvest biocontrol yeast, *Pichia guillermondii*. I. Characterization of attachment to *Botrytis cinerea*. Physiol Mol Plant Pathol 39:245–258, 1991.

119. J Kuc. Plant immunization and its applicability for disease control. In: I Chet, ed. Innovative Approaches to Plant Disease Control. New York: John Wiley & Sons, 1987, pp 255–274.

120. H Kessman, T Staub, C Hofmann, T Maetzke, J Herzog, E Ward, S Uknes, J Ryals. Induction of systemic acquired disease resistance in plants by chemicals. Annu Rev Phytopathol 32:439–459, 1994.

121. M Forbes-Smith. Induced resistance for the biological control of postharvest diseases of fruit and vegetables. Food Aust 51:382–385, 1999.

122. S Droby, E Chalutz. Mode of action of biocontrol agents of postharvest diseases. In: CL Wilson, ME Wisniewski, eds. Biological Control of Postharvest Diseases. Theory and Practice. Boca Raton, FL: CRC Press, 1994, pp 365–389.

123. WJ Janisiewicz, J Usall, B Bors. Nutritional enhancement of biocontrol of blue mold on apples. Phytopathology 82:1364–1370, 1992.

124. A Tronsmo. Biological and integrated controls of *Botrytis cinerea* on apple with *Trichoderma hazianum*. Biol Control 1:59–62, 1991.

125. PL Pusey. Enhancement of biocontrol agents for postharvest diseases and their integration with other control strategies. In: CL Wilson, ME Wisniewski, eds. Biological Control of Postharvest Diseases. Theory and Practice. Boca Raton, FL: CRC Press, 1994, pp 77–88.

126. JL Smilanick. Stategies for the isolation and testing of biocontrol agents. In: CL Wilson, ME Wisniewski, eds. Biological Control of Postharvest Diseases. Theory and Practice. Boca Raton, FL: CRC Press, Inc., 194, pp 25–41.

127. AR Biggs. Detection of latent infections in apple fruit with paraquat. Plant Dis 79:1062–1067, 1995.

128. A Tronsmo. *Trichoderma hazianum* in biological control of fungal diseases. In: R Hall, ed. Principles and Practices of Managing Soilborne Plant Pathogens. St. Paul, MN: APS Press, 1996, pp 213–236.

13
Biological Control of Postharvest Diseases of Citrus Fruits

Ahmed El Ghaouth
University of Nouakchott, Nouakchott, Mauritania, Africa

Charles L. Wilson and Michael Wisniewski
Agricultural Research Service, U.S. Department of Agriculture, Kearneysville, West Virginia

Samir Droby
Agricultural Research Organization, The Volcani Center, Bet Dagan, Israel

Joseph L. Smilanick
Agricultural Research Service, U.S. Department of Agriculture, Parlier, California

Lise Korsten
University of Pretoria, Pretoria South Africa

I. INTRODUCTION

Worldwide, postharvest losses of fruits and vegetables including citrus have been estimated at 25%, much of which is due to fungal and bacterial infections. In developing countries, postharvest losses are often severe due to the lack of adequate handling and refrigerated storage facilities. Postharvest decay of citrus fruits can be traced to infections that occur either between flowering and fruit maturity or during harvesting and subsequent handling and storage [43]. In the former case, preharvest infections are mainly caused by fungal pathogens such as *Phytophthora* species, *Colletotrichum gloeosporioides*, *Botrytis cinerea*, *Diplodia natalensis*, *Phomopsis citri*, and *Alternaria citri*. Stem-end infections caused by *Diplodia*, *Phomopsis*, and *Alternaria* spp. remain quiescent until the fruit becomes senescent during prolonged storage [43,67]. Infections initiated by *Phytophthora* species occur during wet periods before harvest, and *B. cinerea* infections can occur in the orchard and during storage. On the other hand, postharvest infections that occur through surface wounds inflicted during harvest and subsequent handling are mainly caused by wound pathogens such as *Penicillium digitatum*, *Penicillium italicum*, *Geotrichum citri-aurantii* (= *G. candidum*), and *Trichoderma viride*. Among the wound pathogens, green mold (*P. digitatum*) and blue mold (*P. italicum*) account for most of the decay of citrus fruits worldwide. Sour rot caused by *Geotrichum citri-aurantii* is the most rapidly spreading postharvest disease and can be disastrous on fruit stored at temperatures above 10°C. Sour rot is most prevalent on lemon fruit, and the causal yeast-like fungus grows very slowly below 10°C.

Although preharvest quiescent infections are difficult to control, postharvest decay of citrus fruits are reduced by avoiding injury to the fruit during harvest and subsequent handling, stringent sanitation practices, and maintaining the natural resistance of the fruit through the use of hormones, cold storage, and elimination of ethylene from the storage environment [43,102]. These beneficial practices, however, are usually not sufficient to completely protect harvested commodities from infection. Currently, synthetic fungicides such as imazalil and thiabendazole are the primary means of controlling postharvest diseases of citrus fruits [42].

The global trend, however, is shifting towards reduced pesticide use in agriculture in general and in postharvest in particular. Pesticide residues on fruit and vegetables are a major concern to consumers and to the fruit and vegetable industry. With growing health and environmental concerns over pesticide disposal and residue levels on fresh commodities, the development of fungicide-resistant strains of postharvest pathogens, and the deregistration of some of the more effective fungicides has generated a growing interest in the development of safer alternatives that are effective and pose no risk to human health and the environment. Currently, several promising biological approaches that include antagonistic microorganisms, compounds of natural origin or that are generally recognized as safe, and induced resistance have been proposed as potential alternatives to synthetic fungicides for postharvest disease control. Among the proposed alternatives, development of antagonistic microorganisms has been the most studied, and substantial progress has been made in this area [117].

The fundamental basis, the potential, and limitations of these strategies will be presented with special references on citrus fruits. For more extensive coverage of biological approaches for the control of postharvest decay, the reader is referred to previously published reviews [47,48,62,69,117,118,123].

II. MICROBIAL BIOCONTROL AGENTS

In recent years, research on the use of microbial biocontrol agents for the control of postharvest diseases of fruits has gained considerable attention and has moved from the laboratory to commercial application [117]. From these efforts, a large source of information regarding the use of microbial biocontrol agents to control postharvest diseases is now available [62,117]. The selection of putative microbial biocontrol agents has been based mainly on the ability of antagonists to rapidly colonize fruit surfaces and wounds, outcompete the pathogen for nutrients, and survive and develop under a wide range of temperature conditions. A simple and reliable screening technique for antagonists has been developed utilizing the wound site as a selective medium [116]. Utilizing these procedures and other comparable protocols, several antagonistic bacteria, yeasts, and filamentous fungi have been isolated and shown to protect a variety of harvested commodities, including citrus and pome fruit, against postharvest decay [26,28,36,50,55,59,61, 75,92].

In laboratory studies, microbial antagonists *Debaryomyces hansenii* and *Pichia guilliermondii* were shown to be effective in reducing decay of oranges, lemons, and grapefruits caused by *Penicillium digitatum* and *Penicillium italicum* [26,36,38,78]. Both antagonists also reduced the incidence of sour rot of lemon caused by *Geotrichum candidum*, a disease that is not controlled by any of the fungicides currently registered for postharvest use [26]. A reduction of decay of citrus fruit caused by *Penicillium digitatum* was also reported with other antagonistic yeasts such as *Saccharomyces cerevisiae* [29], *Candida sake* [9], *Candida famata* [8], *Candida oleophila* [26,40,78], and *Candida saitoana* [51,52]. *Candida saitoana*, when used alone on early-season oranges and lemons, resulted in 40% less decay than in the water-treated control [51,52]. On late season oranges and lemons, however, the decay incidence following *Candida saitoana* treatment alone was approximately 20% less than in the water-treated control [51,52].

Control of green mold caused by *Penicillium digitatum* was also obtained on citrus fruit with several *Pseudomonas* spp. [23,57,58,103] and *Bacillus subtilis* [7,100]. In 1995, two *Pseudomonas syringae* strains were approved by the U.S. Environmental Protection Agency (EPA) for postharvest use on citrus fruit. Reduction of green mold was also reported on citrus fruit treated with fungal antagonists including *Aureobasidium pullulans* [115], *Myrothecium rorodum* Streuden [5], and *Trichoderma viride* [16].

The success of some of these microbial antagonists in laboratory and large-scale studies has generated the interest of several agro-chemical companies in the development and promotion of postharvest biological products for control rots of fruits and vegetables. A number of microbial antagonists have been patented and evaluated for commercial use as a postharvest treatment. Currently, four antagonistic microorganisms—two yeasts, *Candida oleophila* and *Crypto-coccus albidus*, and two strains of the bacterium *Pseudomona syringae*—are commercially available under the trade names ASPIRE, YieldPlus, and BIO-SAVE-110, respectively.

A. Mode of Action

Although the biocontrol activity of antagonistic bacteria and yeasts has been demonstrated on a variety of commodities, including pome and citrus fruits, the mode of action of the microbial biocontrol agents has not been fully elucidated. In the case of bacterial antagonists, it has been suggested that their biocontrol activity may be in part associated with the production of antibiotics. The bacterial antagonist *Pseudomonas syringae*, which controls *Penicillium* molds of citrus fruit and gray molds of pome fruit, produces an antibiotic, syringomycin, that inhibits the germination of *Penicillium digitatum* [24,25]. When applied as a wound treatment, syringomycin reduced decay of lemons and oranges caused by *Penicillium digitatum*. Bull et al. [25] characterized syringomycin production of two of these strains, ESC-10 and ESC-11, by the generation of syringomycin deficient mutants by the disruption of syringomycin synthesis gene syrB by lacZ. Beta-galactosidase activity, a measure of syringomycin production by the mutant strains, did not increase when lemon or orange albedo extracts were added to the ESC-11 mutant, but did increase when added to the ESC-10 mutant. This provides some evidence that syringomycin may have an active role in control of green mold, because ESC10 more effectively controlled green mold than ESC11, and that this antibiotic may be present at the infection court. Conclusive experiments evaluating the role of this antibiotic have not been done.

Earlier work characterized the role of the antibiotic pryrrolnitrin, produced by *Burkolderia (Pseudomonas) cepacia*, on the suppression of postharvest pathogens. Both the bacterium and its antibiotic were shown to control decay of apples, pears, and citrus fruit [60,103]. This antibiotic may have a role in suppression of decay by this antagonist, but conclusive experiments quantifying its role have not been done. Smilanick and Dennis-Arrue [103] found that *Penicillium digitatum* isolates with very high levels of pyrrolnitrin resistance were still partially controlled on citrus fruit by *Pseudomonas cepacia* application. In other host-pathogen combinations, some pyrrolnitrin-production–deficient mutants retained all of their biocontrol capacity [68] or lost some or all of their effectiveness [86]. Reduction of postharvest decay by an antibiotic produced by a microbial antagonist has also been reported on peaches [53]. The antibiotic, iturin, is produced by the antagonist, *Bacillus subtilis*.

In addition to producing antifungal compounds, the bacterial antagonists *Bacillus subtilis*, *Pseudomonas cepacia*, and *Pseudomonas syringae* are able to outcompete pathogens for nutrients and space in fruit wounds, thus suggesting a multicomponent mode of action. In citrus and pome fruits, *Pseudomonas cepacia,* *Pseudomonas syringae,* and *Enterobacter cloacae* were shown to rapidly and extensively colonize fruit wounds [55,60,62,103,104,121], thereby possibly preventing the establishment of a nutritional base for the pathogen. The implication of nutrient and space competition in the mode action of bacterial antagonists is indirectly supported by the fact that their biocontrol activity is dependent on the concentration of the antagonist propagules and is partially or completely reversed by the addition of exogenous nutrients [62,121].

On the other hand, antagonistic yeasts have been selected mainly on the basis that they do not produce an antibiotic in vitro. Competition for nutrients and space is believed to be the major component of the complex mode of action of antagonistic yeasts [36,39,50,122]. In citrus fruits, the biocontrol potential of *Pichia guilliermondii* at the wound site could be easily reversed by the addition of exogenous nutrient [36]. Application of a low amount of glucose to citrus wounds that had been treated with *P. guilliermondii* and challenge-inoculated with *Penicillium italicum* increased the incidence of decay [36]. Involvement of nutrient competition in the mode of action of microbial antagonists is also indirectly supported by the close relationship between the performance of an antagonist and fruit physiology.

In most commodities, the protective effect of antagonists often declines with an increase in fruit ripeness. With lemon fruits, the ability of *Pseudomonas cepacia* to prevent lesion development caused by *Penicillium digitatum* was shown to diminish with an increase in fruit maturity as indicated by changes in tissue coloring. *Pseudomonas cepacia* was shown to be more effective in controlling decay on green lemons than on yellow lemons [103]. A similar phenomenon also was reported on oranges and lemons where the microbial antagonist *Candida saitoana* was shown to be more effective in controlling decay on early-season fruits than on late-season fruits [51,52]. Implication of the host physiology in biocontrol activity of microbial antagonists was also observed in pome fruits treated with antagonistic yeasts and bacteria [62,92]. The decline in biocontrol activity of microbial antagonist with the onset of ripening is believed to be due to the increase in nutrient availability as a result of biochemical changes associated with ripening.

In most reports on biological control of postharvest diseases of fruits, a quantitative relationship has been demonstrated between the concentration of the antagonist applied in the wound and the efficacy of the biocontrol agent [36,60,62,79,121], thus indicating also a role for space competition. On orange and pome fruit, the biocontrol activity of microbial antagonists has been shown to increase with increasing levels of microbial antagonists and decreasing levels of the pathogen inoculum [26,62,79,92]. In general, microbial antagonists are most effective in controlling postharvest decay when applied at 10^8 cfu/mL [36,50,61,79,92]. Often no control of decay was observed when antagonistic yeasts were applied at 10^5 cfu/mL.

The efficacy of microbial agents is also dependent on their ability to multiply in fruit wounds and reach a critical level to avoid the establishment of the pathogen in the fruit wounds. In fruit wounds, the antagonistic yeasts *Pichia guilliermondii*, *Debaryomyces hansenii*, *Cryptococcus laurentii*, *Aureobasidium pullulans*, *Candida* spp., and *Sporobolomyces roseus* have all been shown to multiply abundantly and form a matrix of yeast cells that cover the cell layers ruptured during wounding [36,38,50,51,55,61,92,101,112]. When applied in citrus fruit wounds, the antagonist population often shows an initial increase after 6 hours and within 24 hours; the population size usually increases by nearly 10-fold and stabilizes thereafter [23,36,38,51,52]. A comparable growth pattern has also been observed with antagonistic bacteria and other yeasts on pome fruit [50,61,62,101]. This may result in a reduction of available nutrients and thus a reduction in pathogen spore germination and hyphal development. With a mutant strain of *Candida oleophila* that exhibited no biocontrol activity, its population size in fruit wounds remained unchanged with time [122,124].

Beside nutrient and space competition, antagonistic yeasts have been shown to directly parasitize major postharvest pathogens [39,50,122]. *Pichia guilliermondii* and *Debaryomyces hansenii*, when co-cultured with *Botrytis cinerea*, strongly attached to hyphae of *B. cinerea*, causing swelling and, in extreme cases, complete disruption of the hyphal-wall structure. Hyphal surfaces appeared concave and partial degradation of the fungal-wall was observed at attachment sites [122]. Attachment was blocked when the yeast cells or the pathogen hyphae were exposed to compounds that affect protein integrity or respiratory metabolism, thus indicating a lectin-type recognition [122]. The partial degradation of *B. cinerea* cell walls by *Pichia guilliermondii* was attributed to its tenacious attachment to hyphal walls in conjunction with its production of β-1,3-glucanase [122]. *Pichia guilliermondii* produced high levels of β-1,3-glucanase activity when co-cultured on various carbon sources or on cell walls of several fungal pathogens [122]. A similar attachment pattern has been observed with antagonistic yeast *Candida saitoana* when co-cultured with *B. cinerea* [50]. When *B. cinerea* was allowed to grow for 16 hours before adding *C. saitoana*, the yeast cells attached tightly to *Botrytis* hyphae despite extensive rinsing with distilled water. The same attachment pattern also was observed in fruit wounds. Yeast cells attached to the hyphal cell walls of *Botrytis cinerea* caused swelling and the disruption of hyphal wall structure [50].

III. INDUCED RESISTANCE

Strengthening the endogenous defense capabilities of plants has been advanced as a promising strategy for crop protection, and several active microbial and chemical elicitors have been identified and shown to protect a variety of plants

[70–72,95,97,113]. Upon infection or treatment with elicitors, plant tissue often reacts by activating a highly coordinated biochemical and structural defense system that helps ward off the spread of pathogens [17,31,34]. Induced defense reactions can be restricted to tissue close to the site of the stimulus or can be expressed systematically throughout the tissue. Site-restricted biochemical and structural defense responses include the rapid death of cells referred to as a hypersensitive reaction, the reinforcement of the cell wall by deposition of lignin, callose, and hydroxyproline-rich glycoproteins, and the accumulation of phytoalexins [10,12,20]. Host defense responses expressed systematically involve the synthesis and accumulation of proteinase inhibitors and antifungal glucanohydrolases such as chitinase, chitosanase, and β-1,3-glucanase [15,17,31,34].

Among the diverse biochemical defense responses, glucanohydrolases such as chitinase, chitosanase, and β-1,3-glucanase have received considerable attention because they are considered to play a major role in induced resistance of plants against invading pathogens [15,95,97,99]. Chitinase and β-1,3-glucanase are known to hydrolyze fungal cell walls, and in combination they have been shown to inhibit the in vitro growth of several pathogenic fungi [96,98]. The insertion of a chitinase gene in tobacco and canola plants was shown to enhance disease resistance [19]. In several plant-pathogen interactions, the induction and accumulation of PR proteins including chitinase and β-1,3-glucanase is often correlated with the onset of induced resistance [70,71,95,111,113].

Phytoalexins, antimicrobial secondary metabolites, are synthesized de novo in tissue area close to the site where infection occurs or the elicitor is applied [10,34]. Phytoalexins show a wide diversity in structure and biogenetic origin and are predominately phenylpropanoids, isoprenoids, and acetylenes. Their implication in disease resistance has dominated most studies of active defense responses and a large body of evidence supports a disease resistance function for phytoalexins [10,34]. Activating biochemical defense responses in harvested tissue through prestorage treatment with innocuous elicitor(s) soon enough and in sufficient magnitude could be a promising strategy of enhancing disease resistance and consequently prolonging fruit storage life. The systemic nature and persistence of antifungal hydrolases in plant tissue upon elicitation could be of significant importance in retarding the resumption of quiescent infections, which typically become active when tissue resistance declines.

In recent years, considerable attention has been paid to induced resistance in harvested commodities as an important manageable form of fruit protection. This interest has been generated by basic and applied research, which demonstrate that fruits can be rendered resistant by artificially turning on their natural defense mechanisms [118]. The reported control of postharvest diseases of pome and citrus through prestorage treatment with fungal cell wall fragments [1], chitosan [44–46], and UV-C light [81,110] bears witness to the validity of induced resistance as a novel and promising approach for postharvest disease control.

A. UV-C radiation

Among the elicitors, non-ionizing UV-C (190–280 nm) radiation has been the most extensively studied, and substantial progress has been made in this area [37,81,93,110]. Prestorage treatment of a variety of commodities, including citrus fruit, with UV-C radiation has been shown to reduce decay, and this has been attributed to UV-mediated induction of disease-resistance rather than to its germicidal effect [81,93,110]. In oranges and lemons, UV-C treatment triggered a gradual development of tissue resistance to infection by *Penicillium digitatum* that coincide with the induction of phenylalanine ammonia lyase (PAL) activity, a key enzyme in the phenylpropanoid pathway, and the accumulation of the phytoalexin scoparone [27,37,93]. In grapefruit, PAL activity showed a transient increase in the first hours after UV treatment, peaking at 24 hours after treatment [27,37]. Immunoblottng analysis using citrus chitinase and β-1,3-glucanase antibodies showed that UV irradiation of citrus fruit induced the accumulation of a 25 kDa chitinase protein [87].

Lers et al. [73], using differential display, isolated and cloned a cDNA representing an mRNA that accumulated in grapefruit peel upon UV irradiation. Sequence analysis revealed that this cDNA represents a gene encoding for an isoflavone reductase–like protein. In UV-C–treated grapefruit, the onset of the resistance was often found to coincide the maximal accumulation of the isoflavone reductase–like protein transcript. Induction of disease resistance by UV-C treatment also has been reported in soybean [18], peach fruit [110], and carrot [81] and attributed to the induction and accumulation of phytoalexins. Mercier et al. [81] showed a relationship between the level 6-methoxymellen and the resistance of carrot slices to infection by *Botrytis cinerea* and *Sclerotinia sclerotiorum*. Concentration of 6-methoxymellen in UV-treated carrot exceeded the ED_{50} necessary for the inhibition of *Botrytis cinerea*.

UV-C–mediated resistance is highly dependent on the physiological maturity of the fruit. In grapefruit, oranges, and lemons, UV-C treatment appeared more effective in reducing decay on early-season fruits than on late-season fruits [27,37]. Similar observations were reported in tomato fruit where UV treatment was more effective in reducing Rhizopus soft rot in mature green and breaker stage tomato than in the ripe red tomatoes [74]. In carrot, UV-induced resistance against *B. cinerea* was shown to decrease with the increased physiological age of the tissue [81]. This was probably associated with the decreased ability of tissue to synthesize inhibitory compounds in response to UV treatment.

In most commodities tested, UV treatment triggered a gradual development of nonpersistent tissue resistance. Challenge inoculation of citrus with *P. digitatum* at different times following UV-C treatment showed that the resistance occurred within 24–48 hours after UV treatment and decreased afterward [27]. In citrus fruit stored at 20°C, the reduction of lesion development in response to UV-

C treatment lasted for up to 7 days, after which resumption of lesion development occurred [37]. A similar trend was observed in lemon fruit [13,14], indicating the need for a lag period for the establishment of a resistance state.

B. Bioactive Elicitors

Several bioactive compounds derived from plants and microbes have been shown to induce disease resistance [35,70,76,91]. Induction of disease-resistant responses has been observed with chitosan application in a variety of harvested commodities. Chitosan, a β-1,4-glucosamine polymer that is found as a natural constituent in the cell wall of many fungi, is produced from chitin of arthropod exoskeletons that has been deacetylated to provide sufficient free amino groups to render the polymer readily soluble in diluted organic acids. Chitosan is known to interfere with the growth of a wide range of fungi [3,44–46,56] and activate defense mechanisms in plant tissues [35,44,46,63]. This dual activity gives chitosan great potential as an antifungal preservative for fresh horticultural commodities.

In laboratory tests, chitosan treatment controlled postharvest decay caused by *Botrytis cinerea*, *Penicillium expansum*, *Penicillium digitatum*, and *Penicillium italicum* on a variety of fruit [49]. The control of decay by chitosan appears to be due to the interplay between the antifungal and eliciting properties of chitosan. In vitro, chitosan was shown to be inhibitory to major postharvest pathogens, presumably by interfering with fungal membranes [44,45]. The inhibitory effect of chitosan has also been observed in planta. On bell pepper, chitosan treatment adversely affected the potential of *Botrytis cinerea* to initiate infection and caused severe cellular damage to invading hyphae [46]. In addition to interfering directly with fungal growth, chitosan induced the accumulation of β-1,3-glucanase and chitinase and elicited the formation of various structural defense barriers in several postharvest commodities [44,46]. Expression of defensive reactions by chitosan treatment seems to be implicated in the restriction of fungal infection. This was indirectly supported by the fact that ingress of the pathogen was limited to epidermal cells ruptured during wounding [46].

Plant growth regulators (jasmonic acid and gibberelic acid), signal transduction compound (salicylic acid), and polypeptides such as harpin have been shown to induce disease resistance in several harvested commodities [2,41,76,91]. Prestorage treatment of lemons fruits and celery with gibberellic acid reduced the incidence of decay and delayed the breakdown of constitutive inhibitors such as marmesin in celery [2]. Induction of the accumulation of preformed antifungal compounds was also observed in avocado fruits stored under high CO_2 [89]. The increased levels of diene in CO_2-treated avocado fruit is believed to be due to an increase in concentration of epicatechin, an endogenous inhibitor of lipoxygenase, a key enzyme in the metabolism of the antifungal diene.

Jasmonates are believed to play an important role as signal molecules in plant defense responses. They have been shown to accumulate in plant tissue in response to elicitor treatment [35,83] and to activate putative defense genes [54,76,125]. Jasmonates have been shown to protect potato and tomato plant from infection [30] and reduce decay of harvested commodities such strawberry, pepper, grapefruit, and roses [41,82,80]. At a concentration of 10 μmol/L methyl jasmonate and jasmonic acid effectively reduced decay of cold-stored grapefruit caused by *Penicillium digitatum* [41]. This concentration also reduced the percentage of fruits exhibiting chilling injury symptoms after 6 weeks of storage at 2°C and 4 additional days at 20°C. Since no direct effect of jasmonates on the fungus was found, it is suggested that jasmonates probably reduced postharvest decay indirectly by enhancing the natural resistance of the fruit at high and low temperatures.

C. Microbial Elicitors

Induction of disease resistance following treatment with microbial inducers has been demonstrated in a number of crops and shown to provide a protection against a wide range of pathogens [66,70,71,113]. In harvested commodities, the induction of disease resistance by microbial antagonists has been inferred but not clearly demonstrated since no attempt was made to separate the antagonistic activity of the yeasts from the fruit-mediated disease suppression [39,50]. Antagonistic yeasts have been shown to induce several biochemical defense responses in harvested commodities. Treatment of lemon wounds with *Pichia guilliermondii* was shown to enhance the production of the phytoalexin scoparone [93]. Arras [8] showed that scoparone accumulation could be 19 times higher when the antagonist *Candida famata* was inoculated 24 hours prior to *Penicillium digitatum* and only four times higher if inoculated 24 hours after the pathogen. Induction of defense responses by antagonistic yeast was also reported in apple fruit. The antagonist yeast *Candida saitoana* was shown to induce chitinase and cause deposition of papillae along host cell in apple tissue [50].

In apple wounds *Aureobasidium pullulans* caused a transient increase in β-1,3-glucanase, chitinase, and peroxidase activities starting 24 hours after treatment and reaching maximum levels 48 and 96 hours after treatment. Wounding also triggered an increase in β-1,3-glucanase, chitinase, and peroxidase activity; however, the level of increase was markedly lower than that detected in yeast-treated fruit [59]. Induction of disease-resistance was also reported in avocado fruits treated with nonpathogenic endophytic mutant of *Colletotrichum magna* [90]. The mutant strain was shown to penetrate the peel of avocado fruits and prevent anthracnose symptoms caused by *Colletotrichum gloeosporioides* by activation of various host-defense responses.

Although the implication of induced disease resistance in the mode of action of microbial antagonists remains to be determined, the observed accumulation of chitinase, β-1,3-glucanase, and scoparone in yeast-treated tissue suggests a putative involvement of these biochemical defense responses in the biocontrol activity of the antagonists.

It is quite possible that yeast-mediated defense reactions played a supporting role in restricting fungal spread. The ability of the antagonistic yeast to outcompete the pathogen for nutrient and space can be anticipated to render the pathogen more susceptible to host antifungal hydrolases and secondary metabolites.

D. Prestorage Heating

Prestorage heating holds potential as a nonchemical method for control of postharvest diseases by directly inhibiting pathogen growth, activating the natural resistance of the host, and slowing down the ripening process. Control of postharvest diseases by heat treatment has been reported for a number of fruits and vegetables [13,14,32,65]. On citrus fruit, control of postharvest pathogens by a brief hot water treatment has been evaluated repeatedly and was in common commercial use before fungicides were introduced [43]. It is particularly effective for the control of brown rot, caused by *Phytopthora* spp. [67], and partially successful for the control of green and blue molds because the temperature of the hot water required to completely control these diseases occasionally injures fruit [88]. Hot water is less effective for stem-end decay or sour rot, because these infections are more resistant to heat than the fruit. Suppression of green mold occurs only at high humidity during thermal treatments [21]. The risk of sour infections by *Geotrichum citri-aurantii* may increase because this fungus grows rapidly at this temperature [43].

In South Africa and Israel heat treatment of citrus fruit has been implemented successfully on a commercial scale [33,88]. De Villiers et al. [33] showed that packhouse dipping of fruit at temperatures of 36 and 40°C for 1–5 minutes effectively reduced postharvest decay of citrus while maintaining fruit quality. Porat et al. [88] showed that exposure of grapefruit to hot water brushing treatment for a short period of time (20 s) at temperatures of 56, 59, and 62°C reduced decay development caused by *Penicillium digitatum*. Hot water brushing treatment at 56°C for 20 seconds reduced development of natural decay on different citrus cultivars such as Minneola tangerines, Shamouti oranges, and Star Ruby grapefruit by 45–55% after 6 weeks of storage at the appropriate temperature. Heat treatment can have other positive impacts as well. Rodov and coworkers [94] showed that brief hot water treatments improve the resistance of citrus fruit to subsequent chilling injury; presumably, storage temperatures could be reduced, and this would affect the benefits of delayed senescence and further reductions in decay losses.

Beside directly interfering with the pathogens, heat treatment has been shown to induce disease resistance in harvested tissue. In lemons it induced the accumulation of the phytoalexin scoparone and increased tissue resistance to infection [13,14]. In spite of interesting possibilities emerging with prestorage heating, the sensitivity of many harvested crops to heat treatment and the energy required for the treatments may prove to be a liability.

IV. NATURAL COMPOUNDS

A wide variety of plant- and animal-derived compounds are known to be fungicidal, and some have been shown to be effective in reducing postharvest decay of fruit and vegetables [47]. Essential oils, volatile substances, and extracts from various plants have been shown to inhibit radial growth of major postharvest pathogens such as *Botrytis cinerea*, *Penicillium expansum*, and *Monolinia fructicola* [120]. There have been several attempts to use essential oils and volatiles of plant origin for the control of postharvest decay [77,85,114,120]. Reduction of decay by extracts and volatiles from plants have been reported in citrus fruit [105], strawberry [85,114], sweet cherry [77], and apple fruit [109]. Limited exposure of fruit to benzaldehyde, acetaldehyde, or 2-nonanone reduced the incidence of decay without causing tissue damage [77,114]. In addition, fumigation with acetaldehyde appeared to enhance the quality attributes of several postharvest commodities [84,85].

A common natural product, ethanol, was evaluated by Smilanick and coworkers [105]. A brief immersion in a solution containing ethanol at 10% wt/vol or more at 44°C or higher reduced green mold incidence by 80% or more on lemons without significant injury to the fruit. However, the effectiveness of ethanol treatment was less than that of a similar brief immersion in heated sodium carbonate (3% wt/vol) solution, particularly when the fruit were held in storage long periods, and it has not been implemented commercially.

Reduction of postharvest decay of fruits was also observed with organic acid preservatives (sorbate, acetate, and benzoate) and medium-chain fatty acids [4,6]. For instance, treatment of citrus fruits with sorbic acid was shown to be effective in controlling green mold caused by *Penicillium digitatum* [43], and, when heated, it will partially control sour rot caused by *Geotrichum citri-aurantii* [64]. The inhibitory effect of organic acid preservatives is highly dependent upon their dissociation constant. Since the pKa value of most acids is between 3 and 5, they are generally most effective at low pH values. The inhibitory effect of organic acids has been attributed to the depression of intracellular pH and the alteration of membrane permeability.

Some active oxygen species are potent biocides, and two, hydrogen peroxide and ozone, have been evaluated for postharvest use on citrus fruit. Immersion in hydrogen peroxide solutions (5–15% wt/vol) achieved significant but modest reduction in green mold incidence on lemons, but the risk of injury to the fruit was high [106]. Ozone, applied either in air or water, does not stop infections from developing from inoculated wounds on citrus fruit. Ozone at low concentrations (0.3–1 ppm or less) in the storage room atmosphere does, however, retard the sporulation of many pathogens on the surface of lesions that do develop, and control of sporulation is of some value to reduce spore contamination in storage rooms and packing houses [108]. Ozone also oxidizes ethylene, and ethylene reduces storage life by accelerating senescence that exacerbates decay problems. Some commercial facilities have implemented the use of ozone, and more practical applications of ozone use will develop as ozone generation equipment improves and regulatory issues about its use are resolved [106].

V. ENHANCEMENT OF BIOCONTROL EFFICACY

Although the currently proposed biological control approaches, including the use of antagonistic microorganisms, natural fungicides, and induced resistance, have been shown to reduce postharvest diseases, each comes with limitations that can affect their commercial applicability. When used as a stand-alone treatment, none of the advanced biological control approaches has been shown clearly to consistently offer an economically sufficient level of disease control that would warrant its acceptance as a viable alternative to synthetic fungicides. For instance, induced resistance via prestorage treatment with elicitors provides only temporary protection, and the level of control is intricately affected by the physiological status of tissue. The ability of harvested commodities to respond to elicitor treatment, thereby becoming resistant to infection, declines with the onset of fruit ripening. This may represent a major setback since the resistant state of the fruit to decay could not be maintained or activated during the ripening stage. A stage when the fruit undergoes the most desirable biochemical changes from the consumer's standpoint. Furthermore, there is the possibility that the deliberate activation defense reaction may result in an accumulation of undesirable compounds especially secondary metabolites. As to plant- and animal-derived bioactive compounds, their performance level is generally lower than synthetic fungicides, and instances of phytotoxicity have been reported especially with essential oils and volatile compounds.

Currently available microbial antagonists confer only a protective effect that diminishes with ripening and provide no control of previously established infections [51,52,92,103,104,119]. Under commercial conditions, antagonistic microorganisms often confer a level of control lower than synthetic fungicides [22,40,119]. In large-scale tests on Navel and Valencia oranges, biocontrol products such as ASPIRE and BIOSAVE-100 often provide a level of control equivalent to synthetic fungicides only when combined with a low doses of synthetic fungicides [22,40]. For instance, *Candida oleophila* in combination with 200 mg/ mL of thiabendazole controlled citrus decay at the level equivalent to the commercial fungicide treatment and reduced the variability often observed when using the antagonistic yeast alone [22,40].

Similar results were reported on apple fruits treated with a combination of *Cryptococcus infirmo-miniatus* with 264 mg/mL thiabendazole [28]. The inability of antagonists alone to offer the level of control comparable to synthetic fungicides has been attributed in part to their inability to eradicate incipient infections that occur during and after harvest [22,38,40,119]. Since infection of fruit can occur either prior to or during harvest and subsequent handling, biological products are expected to protect wounds and also eradicate previously established infections in a manner similar to synthetic fungicides.

Recent attempts to overcome the variable performance and augment the efficacy of existing biological approaches led to the development of a combination of complementary biological approaches for additive and/or synergistic effects. For instance, $CaCl_2$ has been shown to enhance the performance of several antagonistic yeasts in controlling postharvest decay of a variety of fruits. On pome fruit, the addition of $CaCl_2$ was shown to increase the protective effect of some antagonistic yeasts and also greatly reduce the populations of yeasts required to give effective control [79,124]. The potential of $CaCl_2$ as an additive, however, has not been studied extensively in citrus because the interference of $CaCl_2$ at the recommended concentration (2–4%) with packhouse waxes. The biocontrol activity of microbial antagonists was also shown to be enhanced by carbonate salts [51,52,107]. Smilanick and coworkers [107] showed that control of green mold on oranges was maximized when dip treatments in sodium carbonate or bicarbonate were followed by the application of *Pseudomonas syringae* strain ESC10, the active ingredient in the postharvest biological control BioSave™ products. Similar results were also observed with lemon fruits pretreated with sodium carbonate prior to treatment with *Candida saitoana* [51,52].

Biocontrol activity of microbial antagonistic on citrus and pome fruit was also shown to be augmented by addition of nitrogenous (L-asparagine and L-proline) compounds and 2-deoxy-D-glucose, a sugar analog [51,52,61]. When applied in fruit wounds before inoculation, the combination of *C. saitoana* with a low dose of 0.2% (w/v) 2-deoxy-D-glucose was more effective in controlling decay of apple, orange, and lemon caused by *B. cinerea*, *P. expansum*, and *P. digitatum* than either *C. saitoana* or 2-deoxy-D-glucose alone [51,52]. The level of control obtained from the combination of *C. saitoana* with 2-deoxy-D-glucose on lemon and orange fruit was similar to that of imazalil, a common fungicide with worldwide usage. The effectiveness of the combination of *C. saitoana* with 0.2% 2-deoxy-D-glucose appears to stem from the interplay of the biological activity of *C. saitoana* and the antifungal property of 2-deoxy-D-glucose.

A similar protective effect was also reported in apple and pear fruit treated with a combination of *Pseudomonas syringae* or *Sporobolus roseus* and 2-deoxy-D-glucose when applied before inoculation [61]. In addition to protecting citrus and apple fruit from infection, the combination of *C. saitoana* with 2-deoxy-D-glucose was also effective against infections established up to 24 hours before treatment. When applied within 24 hours after inoculation, the combination of *C. saitoana* with 0.2% 2-deoxy-D-glucose resulted in a level of control of green mold of oranges and lemons equivalent to imazalil treatment. No apparent control of blue mold of apple and mold of oranges and lemons was observed when *C. saitoana* or 2-deoxy-D-glucose were applied within 24 hours after inoculation of the fruit with *Penicillium digitatum*. The observed curative activity of the combination of *C. saitoana* with 2-deoxy-D-glucose represents a substantial improvement over existing microbial biocontrol products.

Recently, we have developed a biocontrol product termed "a bioactive coating" that consists of a unique combination of an antagonistic yeast with chemically modified chitosan [51,52]. This combination makes it possible to exploit the antifungal property of chemically modified chitosan and the biological activity of the antagonist yeast. In laboratory studies, the biocontrol activity of *C. saitoana* against decay of apple, lemon, and orange caused by *Botrytis cinerea*, *Penicillium expansum*, and *Penicillium digitatum* was markedly enhanced by the addition of glycolchitosan [51,52]. Under semi-commercial conditions, the bioactive coating was superior to *C. saitoana* and glycolchitosan in controlling decay of oranges (Washington navel, Valencia, Pineapple, and Hamlin) and Eureka lemons, and the control level was equivalent to that with imazalil [51,52]. On apple fruit, depending on the apple variety used, the bioactive coating was comparable or superior to thiabendazole in reducing decay. Unlike *C. saitoana*, which showed poor performance on late-season fruit, the bioactive treatments offered consistent control of decay on Washington navel oranges and Eureka lemons in early and late seasons.

Enhancement of microbial biocontrol agents has been also reported with physical additives such as curing and heat treatments [11,58,110], ultraviolet light [110], and modified or controlled atmosphere (MA/CA) and cold storage [75,112]. The efficacy of *Pichia guilliermondii* against *Penicillium digitatum* increased when orange fruits were stored at optimal low storage temperature under controlled atmosphere [75]. Integrating UV-C radiation with antagonistic yeast was shown to enhance the performance of the yeast and provide a level of control equivalent to synthetic fungicides [110]. Huang et al. [58] demonstrated that biocontrol of green mold using *Pseudomonas glathei* could be enhanced when heat was applied to retard conidia germination of *Penicillium digitatum* while simultaneously stimulating bacterial multiplication.

VI. CONCLUSIONS

Among the proposed alternatives, development of microbial antagonists has received most of the attention, and presently a number of microbial antagonists are commercially available for the control of postharvest decay. If we are to maximize, however, the biocontrol potential of microbial antagonists, a more fundamental understanding of their mode of action, the microecology, their compatibility of postharvest commercial practices, and the effect of host physiology on their biological activity is imperative. Similarly, the potential of naturally occurring biocides as specific biofungicides for the control of postharvest decay will depend on our ability to develop safe and effective formulations that have no undesirable effects on the quality attributes of harvested products.

Success with the induction of resistance in several harvested commodities by treatment with biotic and abiotic elicitors suggests potential for this technology for the reduction of postharvest diseases. More effective methods of harnessing the endogenous defense capabilities of fruits and vegetables through prestorage treatment with elicitors should ultimately emerge as we learn more about [1] the biological activity of elicitors, [2] the regulation of defense genes associated with the induced resistance in harvested tissue and their role in resistance, [3] the signal transduction pathways that link the host perception with the expression of defense genes required to ward off infection, and [4] the temporal regulation of defense response genes in relation with the physiological changes that occur in harvested tissue. As more is learned about the function and regulation of disease resistance genes in harvested tissue, it will be possible to engineer the expression of complex resistance genes in postharvest commodities. Since most harvested commodities are more susceptible to infection as they approach the ripening stage, the use of developmentally regulated promoters to drive the expression of defense genes could be a useful approach. This could confer disease resistance at specific stages when the fruits are most susceptible to infection. Such an approach will be possible once organ-specific and developmentally regulated promoters are available.

As we learn more about the fundamental basis underlying the protective effect of microbial antagonists, bioactive compounds, and induced resistance, more effective methods of formulating, applying, and combining complementary biological approaches for additive and/or synergistic effects will emerge. So far the results obtained with the different combination of biological products demonstrate the potential of this multifaceted approach as a viable alternative to synthetic fungicides. Such biological strategies should also be expected to have greater stability and effectiveness than approach utilizing single biological agents. The complexity of the mode of action displayed by combined alternatives should make the development of pathogen resistance more difficult and present a more highly complex disease-deterrent barrier than an approach relying on a single biological agent.

REFERENCES

1. NK Adikaram, AE Brown, TR Swinburne. Phytoalexin induction as a factor in the protection of *Capsicum annum* L. fruits against infection by *Botrytis cinerea* Pers. J Phytopathol 122:267, 1988.
2. U Afek, N Aharoni, S Carmeli. Increased celery resistance to pathogens during storage and reducing high-risk psoralen concentration by treatment with GA_3. J Am Soc Hort Sci 120:562–565, 1995.
3. CR Allan, LA Hadwiger. The fungicidal effect of chitosan on fungi of varying cell wall composition. Exp Mycol 3:285–287, 1979.
4. AB Al Zaemey, N Magan, AK Thompson. Studies on the effect of fruit-coating polymers and organic acids on growth of *Colletotrichum musae* in vitro and on postharvest control of anthracnose of bananas. Mycol Res 12:1463–1468, 1993.
5. DJ Appel, R Gees, MD Coffey. Biological control of the postharvest pathogen *Penicillium digitatum* on Eureka lemons. Phytopathology 12:1595, 1988.
6. DK Arora, KG Mukerji, EH Marth. Handbook of Applied Mycology. Vol. 3. New York: Marcel Dekker, 1991.
7. G Arras, G D'Hallewin. In vitro and in vivo control of *Penicillium digitatum* and *Botrytis cinerea* in citrus fruit by *Bacillus subtilis* strains. Agric Mediter 124:56–61, 1994.
8. G Arras. Mode of action of an isolate of *Candida famata* in biological control of *Penicillium digitatum* in orange fruits. Postharvest Biol Technol 8:191–198, 1996.
9. G Arras, P Sanna, V Astone. Biological control of *Penicillium italicum* of citrus fruits by *Candida sake* and calcium salt. Proc 49th Int Symposium Crop Prot 62: 1071–1078, 1997.
10. JA Bailey, JW Mansfield. Phytoalexins. New York: John Wiley & Sons, 1982.
11. R Barkai-Golan, JP Douglas. Postharvest heat treatment of fresh fruits and vegetables for decay control. Plant Dis 75:1085–1090, 1991.
12. RC Beier. Natural pesticides and bioactive compounds in foods. Rev Environ Contam Toxicol 113:47–137, 1990.
13. S Ben-Yohoshua, B Shapiro, JJ Kim, J Sharoni, S Carmeli, Y Kashman. Resistance of citrus fruit to pathogens and its enhancement by curing. In: R Goren, K Mendel, eds. Proceedings of the Sixth International Citrus Congress, Tel Aviv, Israel. Philadelphia: Balaban Publishers, 1988, pp 1371–1379.
14. S Ben-Yehoshua, V Rodov, JJ Kim, S Carmeli. Preformed and induced antifungal materials of citrus fruits in relation to the enhancement of decay resistance by heat and ultraviolet treatments. J Agric Food Chem 40:1217–1221, 1992.
15. T Boller. Hydrolytic enzymes in plant disease resistance. In: T Kosuge, EW Nester, eds. Plant-Microbe Interactions. Vol. 2. New York: Macmillan, 1987, pp 385–402.

16. AD Borras, RV Aguilar. Biological control of *Penicillium digitatum* by *Trichoderma viride* on postharvest citrus fruits. Intl J Food Microbiol 11:179–184, 1990.

17. DJ Bowles. Defense-related proteins in higher plants. Annu Rev Biochem 59:873–907, 1990.

18. MA Bridge, WL Klarman. Soybean phytoalexin, hydroxyphaseollin, induced by ultraviolet irradiation. Phytopathology 63:606–609, 1972.

19. K Broglie, I Chet, M Holliday, R Cressman, P Biddle, S Knowlton, CJ Mauvais, R Broglie. Transgenic plants with enhanced resistance to the fungal pathogen *Rhizoctonia solani*. Science 254:1194–1197, 1991.

20. GE Brown. Host defenses at the wound site on harvested crops. Phytopathology 79:1381–1384, 1989.

21. GE Brown. Development of green mold in degreened oranges. Phytopathology 63:1104–1107, 1973.

22. GE Brown, M Chambers. Evaluation of biological products for the control of postharvest diseases of Florida citrus. Proc Fla State Hort Soc 109:278–282, 1996.

23. CT Bull, JP Stack, JL Smilanick. *Pseudomonas syringae* strains ESC-10 and ESC-11 survive in wounds on citrus and control green and blue molds of citrus. Biol Control 8:81–88, 1997.

24. CT Bull, ML Wadsworth, KN Sorenson, J Takemoto, R Austin, JL Smilanick. Syringomycin E produced by biological agents controls green mold on lemons. Biol Control 12:89–95, 1998.

25. CT Bull, ML Wadsworth, TD Pogge, TT Le, SK Wallace, JL Smilanick. Molecular investigations into the mechanisms in the biological control of postharvest diseases of citrus. IOBC Bull 21:1–6, 1998.

26. E Chalutz, CL Wilson. Postharvest biocontrol of green and blue mold and sour rot of citrus by *Debaryomyces hansenii*. Plant Dis 74:134–137, 1990.

27. E Chalutz, S Droby, CL Wilson, M Wisniewski. UV-induced resistance to postharvest diseases of citrus fruit. J Phytochem Photobiol 15:367–374, 1992.

28. T Chand-Goyal, RA Spotts. Biological control of postharvest diseases of apple and pear under semi-commercial and commercial conditions using three saprophytic yeasts. Biol Control 10:199–206, 1997.

29. LH Cheah, TB Tran. Postharvest biocontrol of *Penicillium* rot of lemons with industrial yeasts. Proceedings of the 48th New Zealand Plant Protection Conference, 1995, pp 155–157.

30. Y Cohen, U Gisi, T Niderman. Local and systemic protection against *Phytophthora infestans* induced in potato and tomato plants by jasmonic acid and jasmonic methyl ester. Phytopathology 83:1054–1062, 1993.

31. DB Collinge, AJ Slusarenko. Plant gene expression in response to pathogens. Plant Mol Biol 9:389–410, 1987.

32. HM Couey. Heat treatment for control of postharvest diseases and insect pest of fruit. Hortscience 24:198–201, 1989.

33. EE De Villiers, K Van Dyk, SH Swart, JH Smith, L Korsten. Potential alternative decay control strategies for South African citrus packhouses. Proceedings of the 8th Congress of the International Society of Citriculture, 1996, pp 410–414.

34. RA Dixon, M Harrison. Activation, structure, and organization of genes involved in microbial defense in plants. Adv Genet 28:165–180, 1990.

35. SH Doares, T Syrovets, EW Weiler, CA Ryan. Oligogalacturonides and chitosan activate plant defense genes through the octadecanoid pathway. Proc Natl Acad Sci USA 92:4095–4098, 1995.

36. S Droby, E Chalutz, CL Wilson, ME Wisniewski. Characterization of the biocontrol activity of *Debaryomyces hansenii* in the control of *Penicillium digitatum* on grapefruit. Can J Microbiol 35:794–800, 1989.

37. S Droby, E Chalutz, B Horev, L Cohen, V Gaba, CL Wilson, ME Wisniewski. Factors affecting UV-induced resistance in grapefruit against the green mold decay caused by *Penicillium digitatum*. Plant Pathol 42:418–424, 1993.

38. S Droby, R Hofstein, CL Wilson, ME Wisniewski, B Fridlender, L Cohen, B Weiss, A Daus, D Timar, E Chalutz. Pilot testing of *Pichia quilliermondii*: a biocontrol agent of postharvest diseases of citrus fruit. Biol Control 3:47–52, 1993.

39. S Droby, E Chalutz. Mode of action of biocontrol agents for postharvest diseases. In: CL Wilson, ME Wisniewski, eds. Biological Control of Postharvest Diseases of Fruits and Vegetables—Theory and Practice. Boca Raton, FL: CRC Press, 1994, pp 63–75.

40. S Droby, A Cohen, B Weiss, B Horev, E Chalutz, H Katz, M Keren-Tzur, A Shachnai. Commercial testing of Aspire: a yeast preparation for the biological control of postharvest decay of citrus. Biol Control 12:97–100, 1998.

41. S Droby, R Porat, L Cohen, B Weiss, B Shapirom, S Philosoph-Hadas, S Meir. Suppressing green mold decay in grapefruit with postharvest jasmonate application. J Am Hort Sci 124:184–188, 1999.

42. JW Eckert, JM Ogawa. The chemical control of postharvest diseases: deciduous fruits, berries, vegetables and roots/tuber crops. Annu Rev Phytopath 26:433–469, 1988.

43. JW Eckert, IL Eaks. Postharvest disorders and diseases of citrus fruits. In: W Reuther, EC Calavan, GE Carman, eds. The Citrus Industry. Berkeley: University of California Press, 1989, pp 179–260.

44. A El Ghaouth, J Arul, J Grenier, A Asselin. Antifungal activity of chitosan on two postharvest pathogens of strawberry fruits. Phytopathology 82:398–402, 1992.

45. A El Ghaouth, J Arul, A Asselin, N Benhamou. Antifungal activity of chitosan on post-harvest pathogens: induction of morphological and cytological alterations in *Rhizopus stolonifer*. Mycol Res 96:769–779, 1992.

46. A El Ghaouth, J Arul, C Wilson, N Benhamou. Ultrastructural and cytochemical aspects of the effect of chitosan on decay of bell pepper fruit. Physiol Mol Plant Pathol 44:417–432, 1994.

47. A El Ghaouth, C Wilson. Biologically based technologies for the control of postharvest diseases. Postharvest News Inf 6:5–11, 1995.

48. A El Ghaouth. Biologically based alternatives to synthetic fungicides for the control of postharvest diseases. J Indust Microbiol Biotechnol 19:160–162, 1997.

49. A El Ghaouth. Manipulation of defense systems with elicitors to control postharvest diseases. ACIAR Proc 80:131–135, 1998.

50. A El Ghaouth, C Wilson, M Wisniewski. Ultrastructural and cytochemical aspect of the biocontrol activity of *Candida saitoana* in apple fruit. Phytopathology 88:282–291, 1998.

51. A El-Ghaouth, J Smilanick, E Brown, A Ippolito, CL Wilson. Application of *Can-*

dida saitoana and glycolchitosan for the control of postharvest diseases of apple and citrus fruit under semi-commercial conditions. Plant Dis 84:243–248, 2000.

52.   A El-Ghaouth, J Smilanick, M Wisniewski, CL Wilson. Improved control of apple and citrus fruit decay with a combination of *Candida saitoana* with 2-deoxy-D-glucose. Plant Dis 84:249–253, 2000.

53.   RC Gueldner, CC Reilly, PL Pusey, CE Costello, RF Arrendale, RH Cox, DS Himmelsbach, FG Crumley, HG Cutler. Isolation and identification of iturins as antifungal peptides in biological control of peach brown rot with *Bacillus subtilis*. J Agric Food Chem 36:366–370, 1988.

54.   H Gundlach, MJ Muller, TM Kutchan, MH Zenk. Jasmonic acid is a signal transducer in elicitor-induced plant cell cultures. Proc Natl Acad Sci USA 89:2389–2393, 1992.

55.   ML Gullino, C Aloi, M Palitto, D Benzi, A Garibaldi. Attempts at biocontrol of postharvest diseases of apple. Med Fac Landbouw Rijksuiv Gent 56:195, 1991.

56.   S Hirano, C Itakura, H Seino, Y Akiyama, I Nonaka, N Kanbara, T Kawakami. Chitosan as an ingredient for domestic animal feeds. J Agric Food Chem 38:1214–1217, 1990.

57.   Y Huang, BL Wild, SC Morris. Postharvest biological control of *Penicillium digitatum* decay on citrus fruit by *Bacillus pumilus*. Ann Appl Biol 120:367–372, 1992.

58.   Y Huang, BJ Deverall, SC Morris. Postharvest control of green mold on oranges by a strain of *Pseudomonas glathei* and enhancement of its biocontrol by heat treatment. Postharvest Biol Technol 3:129–137, 1995.

59.   A Ippolito, A El-Ghaouth, CL Wilson. Control of postharvest decay of apple fruit by *Aurobasidium pullulans* and induction of defense responses. Postharvest Biol 19:265–272, 2000.

60.   W Janisiewicz, L Yourman, J Roitman, N Mahoney. Postharvest control of blue mold and gray mold of apples and pears by dip treatment with pyrrolnitrin, a metabolite of *Pseudomonas cepacia*. Plant Dis 75:490–494, 1991.

61.   W Janisiewicz. Enhancement of biocontrol of blue mold with nutrient analog 2-deoxy-D-glucose on apples and pears. Appl Environ Microbiol 60:2671–2676, 1994.

62.   W Janisiewicz. Biocontrol of postharvest diseases of temperate fruits: challenges and opportunities. In: J Boland, LD Kuykendall, eds. Plant-Microbe Interactions and Biological Control. New York: Marcel Dekker, 1998, pp 171–198.

63.   FD Kendra, D Christian, LA Hadwiger. Chitosan oligomers from *Fusarium solani/* pea interactions, chitinase/β-glucanase digestion of sporelings and from fungal wall chitin actively inhibit fungal growth and enhance disease resistance. Physiol Mol Plant Pathol 35:215–223, 1989.

64.   H Kitagawa, K Kawada. Effect of sorbic acid and potassium sorbate on the control of sour rot of citrus fruits. Proc Fla State Hort Soc 97:133–135, 1984.

65.   DJ Klein, S Lurie. Postharvest heat treatment and fruit quality. Postharvest News Inf 2:15–19, 1991.

66.   JW Kloepper, S Tuzun, JA Kuo. Proposed definitions related to induced disease resistance. Biocontrol Sci Technol 2:349–351, 1992.

67.   LJ Klotz. Color Handbook of Citrus Diseases. University of California, Berkeley, 1973.

68. J Kraus, JE Loper. Lack of evidence for a role of antifungal metabolite production by *Pseudomonas fluorescens* Pf-5 in biological control of *Pythium* damping-off of cucumber. Phytopathology 82:264–271, 1992.

69. L Korsten, EE De Villiers, FC Wehner, JM Kotze. A review of biological control of postharvest diseases of subtropical fruits. ACIAR Proc 50:172–185, 1994.

70. J Kuc. Immunization for the control of plant disease, In: D Homby, ed. Biological Control of Soil-borne Plant Pathogens. Oxfordshire, UK: C.A.B International, 1990, pp 355–373.

71. J Kuc, N Strobel. Induced resistance using pathogens and nonpathogens. In: E Tjamos, G Papavisas, eds. Biological Control of Plant Diseases. New York: Plenum Press, 1992, pp 295–303.

72. K Lawton, L Friedrich, M Hunt, K Weymann, H Kessmann, T Staub, J Ryals. Benzothiadiazole induces disease resistance in *Arabidopsis* by activation of the systemic acquired resistance signal transduction pathway. Plant J 10:71–82, 1996.

73. A Lers, S Burd, E Lomnaniec, S Droby, E Chalutz. The expression of grapefruit gene encoding an isoflavone reductase-like protein is induced in response to UV irradiation. Plant Mol Biol 36:847–856, 1998.

74. J Liu, C Stevens, VA Khan, JY Lu, CL Wilson, O Adeyeye, MK Kabwe, PL Pusey, E Chalutz, T Sultana, S Droby. Application of ultraviolet-C light on storage rots and ripening of tomatoes. J Food Prot 56:868–872, 1993.

75. S Lurie, S Droby, L Chalupowicz, E Chalutz. Efficacy of *Candida oleophila* strain 182 in preventing *Penicillium expansum* infection of nectarine fruits. Phytoparasitica 23:231–234, 1995.

76. GD Lyon, T Reglinski, AC Newton. Novel disease control compounds: the potential to immunize plants against infections. Plant Pathol 44:407–427, 1995.

77. J Mattheis, R Roberts. Fumigation of sweet cherry (*Prunus avium* 'Bing') fruit with low molecular weight aldehydes for postharvest decay control. Plant Dis 77: 810–814, 1993.

78. RG McGuire. Application of *Candida guilliermondii* in commercial citrus coating for biocontrol of *Penicillium digitatum* on grapefruits. Biol Control 3:1–7, 1994.

79. RJ McLaughlin, ME Wisniewski, CL Wilson, E Chalutz. Effect of inoculum concentration and salt solutions on biological control of postharvest diseases of apple with *Candida* sp. Phytopathology 80:456–461, 1990.

80. SS Meir, S Philosoph-Hadas, S Lurie, S Droby, G Akerman. Reduction of chilling injury in stored avocado, grapefruit, and bell pepper by methyl jasmonate. Can J Bot 74:870–874, 1998.

81. J Mercier, J Arul, R Ponnampalam, M Boulet. Induction of 6-methoxymellein and resistance to storage pathogens in carrot slices by UV-C. J Phytopathol 137:44–55, 1993.

82. HE Moline, JG Buta, RA Saftner, JL Maas. Comparison of three volatile natural products for the reduction of postharvest decay in strawberries. Adv Strawberry Res 16:43–48, 1997.

83. H Nojiri, M Sugimori, H Yamane, Y Nishimura, A Yamada, N Shibuya, O Kodama, N Murofushi, T Omori. Involvement of jasmonic acid in elicitor-induced phytoalexin in suspension-cultured rice cells. Plant Physiol 110:387–392, 1996.

84. O Paz, HW Janes, BA Prevost, C Frenkel. Enhancement of fruit sensory quality

by postharvest applications of acetaldehyde and ethanol. J Food Sci 47:270–276, 1991.

85. E Pesis, I Avissar. Effect of postharvest application of acetaldehyde vapour on strawberry decay, taste and certain volatiles. J Sci Food Agr 52:377–385, 1990.

86. WF Pfender, J Kraus, JE Loper. A genomic region from *Pseudomonas fluorescens* Pf-5 required for pyrrolnitrin production and inhibition of *Pyrenphora triticirepentis* in wheat straw. Phytopathology 83:1223–1228, 1993.

87. R Porat, A Lers, S Dori, L Cohen, B Weiss, A Daus, CL Wilson, S Droby. Induction of chitinase and β-1,3-endoglucanase proteins by irradiation and wounding in grapefruit peel tissue. Phytoparasitica 27:233–238, 1999.

88. R Porat, A Daus, B Weiss, L Cohen, E Fallik, S Droby. Reduction of postharvest decay in organic citrus fruit by a short hot water brushing treatment. Postharvest Biol Technol 18:151–157, 2000.

89. D Prusky, RA Plumbey, I Kobiler. Modulation of natural resistance of avocado fruits to *Colletotrichium gloesporiodes* by CO_2 treatment. Physiol Mol Plant Pathol 39:325–334, 1991.

90. D Prusky, S Freeman, RJ Rodrigues, NT Keen. A nonpathogenic mutant strain of *Colletotrichium magna* induces resistance to *C. gloesporiodes* in avocado fruits. Mol Plant-Microbe Interact 7:326–333, 1994.

91. D Qui, ZM Wei. Effect of messenger on gray mold and other fruit rot diseases. Phytopathology 90:62, 2000.

92. RG Roberts. Biological control of gray mold of apple by *Cryptococcus laurentii*. Phytopathology 80:526–530, 1990.

93. V Rodov, S Ben-Yehoshua, R Albaglis, D Fang. Accumulation of phytoalexins scoparone and scopoletin in citrus fruits subjected to various postharvest treatments. Acta Hort 381:517–523, 1994.

94. V Rodov, S Ben-Yeoshua, R Albagi, DQ Fang. Reducing chilling injury and decay of stored citrus fruit by hot water dips. Postharvest Biol Technol 5:119–127, 1995.

95. J Ryals, U Neuenschwander, M Willits, A Molina, HY Steiner, M Hunt. Systemic acquired resistance. Plant Cell 8:1809–1819, 1996.

96. A Schlumbaum, F Mauch, U Vogeli, T Boller. Plant chitinases are potent inhibitors of fungal growth. Nature 324:365–367, 1986.

97. M Schroder, K Hahlbrock, E Kombrink. Temporal and spatial patterns of β-1,3-glucanase and chitinase induction in potato leaves infected by *Phytophthora infestans*. Plant J 2:161–172, 1992.

98. MB Sela-Buurlage, AS Ponstein, B Bres-Vloemans, LO Melchers, P Van den ELzen, BJC Cornelissen. Only specific tobacco chitinases and β-1,3-glucanases exhibit antifungal activity. Plant Physiol 101:857–863, 1993.

99. L Sequeira. Induced resistance: physiology and biochemistry. In: R Alan, ed. New Directions in Biological Control Alternatives for Suppressing Agricultural Pests and Diseases. New York: Liss Inc, 1990, pp 663–678.

100. V Singh, BJ Deverall. *Bacillus subtilis* as a control agent against fungal pathogens of citrus fruit. Trans Br Mycol Soc 83:487–490, 1984.

101. PA Shefelbine, RG Roberts. Population dynamics of *Cryptococcus laurentii* in wounds in apple and pear fruit stored under ambient or controlled atmospheric conditions. Phytopathology 80:1020, 1990.

102. RL Shewleft. Postharvest treatment for extending the shelf life of fruits and vegetables. Food Technol 40:70–80, 1986.

103. JL Smilanick, R Dennis-Arrue. Control of green mold of lemons with *Pseudomonas* species. Plant Dis 76:481–485, 1992.

104. JL Smilanick, R Denis-Arrue, JR Bosch, AR Gonzales, DJ Henson, WJ Janisiewicz. Biocontrol of postharvest brown rot of nectarines and peaches by *Pseudomonas* species. Crop Prot 12:513–520, 1993.

105. JL Smilanick, DA Margosan, DJ Henson. Evaluation of heated solutions of sulfur dioxide, ethanol, and hydrogen peroxide to control postharvest green mold of lemons. Plant Dis 79:742–747, 1995.

106. JL Smilanick, C Crisosto, F Mlikota. Postharvest use of ozone for decay control. Perishables Handling Qu 99:10–14, 1999.

107. JL Smilanick, DA Margosan, F Mlikota, J Usall, IF Michael. Control of citrus green mold by carbonate and bicarbonate salts and the influence of commercial postharvest practices on their efficacy. Plant Dis 83:139–145, 1999.

108. DH Spalding. Effects of ozone atmospheres on spoilage of fruits and vegetables after harvest. Agricultural Research Service, Washington, DC: USDA Marketing Research Report No. 801, 1968.

109. GJ Stadelbacher, Y Prasad. Postharvest decay control of apple by acetaldehyde vapor. J Am Soc Hort Sci 99:364–369, 1974.

110. C Stevens, VA Kahn, JY Lu, CL Wilson, A El Ghaouth, E Chalutz, S Droby. Low dose UV-C light as a new approach to control decay of harvested commodities. Recent Res Develop Plant Pathol 1:155–169, 1996.

111. L Sticher, B Mauch-Mani, JP Metraux. Systemic acquired resistance. Annu Rev Phytopathol 35:235–270, 1997.

112. D Sugar, RG Roberts, RJ Hilton, TL Reghetti, EE Sanchez. Integration of cultural methods with yeast treatment for control of postharvest decay in pear. Plant Dis 78:791–795, 1994.

113. LC Van Loon, PA Bakker, MJ Pieterse. Systemic resistance induced by rhizosphere bacteria. Annu Rev Phytopathol 36:453–483, 1998.

114. SF Vaugh, GF Spencer, S Shasha. Volatile compounds from raspberry and strawberry fruit inhibit postharvest decay fungi. J Food Sci 58:793–796, 1993.

115. CL Wilson, E Chalutz. Postharvest biological control of *Penicillium* rots of citrus with antagonistic yeasts and bacteria. Sci Hort 40:105–112, 1989.

116. CL Wilson, ME Wisniewski, S Droby, E Chalutz. A selection strategy for microbial antagonists to control postharvest diseases of fruits and vegetables. Sci Hort 40:105–112, 1993.

117. CL Wilson, ME Wisniewski. Biological Control of Postharvest Diseases of Fruits and Vegetables—Theory and Practice. Boca Raton, FL: CRC Press, 1994.

118. CL Wilson, A El Ghaouth, S Droby, E Chalutz, C Stevens, JY Lu, VA Kahn, J Arul. Potential of induced resistance to control postharvest diseases of fruits and vegetables. Plant Dis 78:837–844, 1994.

119. CL Wilson, ME Wisniewski, A El Ghaouth, S Droby, E Chalutz. Commercialization of antagonistic yeasts for the biological control of postharvest diseases of fruits and vegetables. J Indust Microbiol Biotechnol 46:237–242, 1996.

120. CL Wilson, A El Ghaouth, J Solar, J, M Wisniewski. Rapid evaluation of plant

extracts and essential oils for fungicidal activity against *Botrytis cinerea*. Plant Dis 81:204–210, 1997.

121. ME Wisniewski, CL Wilson, W Hershberger. Characterization of inhibition of *Rhizopus stolonifer* germination and growth by *Enterobacter cloacae*. Can J Bot 67: 2317–2323, 1989.

122. ME Wisniewski, C Biles, S Droby, R McLaughlin, CL Wilson, E Chalutz. Mode of action of the postharvest biocontrol yeast, *Pichia guilliermondii*. I. Characterization of the attachment to *Botrytis cinerea*. Physiol Mol Plant Pathol 39:245–258, 1991.

123. ME Wisniewski, CL Wilson. Biological control of postharvest diseases of fruits and vegetables. Recent Adv Hort 27:94–98, 1992.

124. ME Wisniewski, S Droby, E Chalutz, Y Eilam. Effect of Ca^{+2} and Mg^{+2} on *Botrytis cinerea* and *Penicillium expansum* in vitro and on the biocontrol activity of *Candida oleophila*. Plant Pathol 44:1016–1024, 1995.

125. Y Xu, PL Chang, D liu, ML Narashima, KG Raghothama, PM Hasegawa, RA Bressan. Plant defense genes are synergistically induced by ethylene and methyl jasmonate. Plant Cell 6:1077–1085, 1996.

14

Biological Control of Turfgrass Diseases

W. Uddin and G. Viji
Pennsylvania State University, University Park, Pennsylvania

Management of turfgrass diseases is one of the most difficult challenges that turf managers routinely experience in the production and maintenance of high-quality turf in golf courses, landscapes, sod production, and athletic fields. An effective disease management program involves significant efforts and costs especially in the golf course industry. Traditionally, the turfgrass industry has been heavily dependent on fungicides as the primary means of disease control. Although most fungicides labeled for selected diseases usually provide satisfactory controls for a number of diseases, this often does not offer all the solutions for disease management. Application of certain fungicides can often result in new sets of problems in the development of disease control strategies. Some of the most undesired effects of fungicide applications are development of resistance in the pathogen population, nontarget effects of the particular compound used, and the phytotoxicity. While use of fungicides remains a significant component of a turfgrass health management system, there is a growing trend in the turfgrass industry towards promoting environmental stewardship. Significant advances have been made in the golf course industry in the development of educational programs for turf managers that entail the adaptation of various principles and practices in integrated turfgrass disease management.

There has been tremendous interest in the development of various biological control strategies that can be incorporated into a broader integrated disease management spectrum. Because of the positive attributes that fungicides currently hold, it is unlikely that biological control methods could be used as an alternative to replace fungicides in the turfgrass industry in the foreseeable future. However, biological control is increasingly considered as part of an integrated disease management approach by the turfgrass industry for a number of reasons that include:

Augmenting fungicide use

Reduction of nontarget effects and increased microbial diversity in soil

Improved formulation, shelf life, and methods of delivery of biocontrol
   agents

Enhancement of root and foliar growth of turf

Good public relations for the golf course industry and recreational facilities

There are four general approaches to the biological control of foliar and
root diseases of turfgrasses:

1. Application of specific microbial inoculants on turfgrass foliage or incorporation of the antagonists into the soil
2. Application of complex microbiota in natural organic substrates (organic amendment) into the soil to increase microbial diversity
3. Use of organic fertilizers and biostimulants to enhance root growth
4. Inoculation of potential endophytes in turf species for establishing disease resistance

I. MICROBIAL INOCULANTS IN DISEASE SUPPRESSION

Several genera and species of bacteria have been reported to be effective biological control agents for root and foliar diseases of turfgrasses. The mechanisms
by which pathogens are suppressed include antibiosis, competition, parasitism,
predation, and induced resistance [6]. Antagonism is further defined as "active
opposition" that results from the production of substances by one organism that
are toxic to other organisms, such as production of antibiotics that causes lysis
or death of the latter [99] and, additionally, from competition for food, oxygen,
and space [6]. Siderophores produced by bacteria deprive pathogens of iron, thus
suppressing their growth [121]. The ability of microorganisms to lyse pathogens
by degrading cell wall components can also serve as a powerful tool for biological
control [67].

   Microbial inoculants for turfgrass disease control consist of beneficial microorganisms that are part of the microbial community found in the turfgrass soil,
rhizosphere, thatch, or plant residue [110]. The goals of using microbial inoculants in the management of turfgrass diseases are [1] to isolate the organisms
from their natural substrate in the soil or aboveground plant parts, [2] to identify
their antagonistic or competitive abilities against pathogens, and [3] to increase
their population in the laboratory using artificial media and introduce them into
the soil or on the foliage of turf to inhibit the growth and reproduction of the
pathogen. Once they are introduced to the soil or turf canopy, they must not only
survive and adapt to the new environment, but also continuously multiply and
remain viable throughout the growing season, especially during the period when

pathogens remain active. The high number of biocontrol agents in the soil or phyllosphere of turf is of paramount importance in achieving the desired level of disease control [51]. The shelf life of the microbial inoculant is a major factor that influences the efficacy of an inoculant product. The product should possess a reasonably long shelf life—turf managers are accustomed to the extended shelf lives of synthetic chemical preparations, which can be used for the entire growing season or longer.

In addition to competing with other soil microbes, antagonists require optimum environmental conditions such as temperature, moisture, pH, and other soil factors for their growth and proliferation. Survival of the antagonists and their ability to tolerate direct exposure to environmental stresses including ultraviolet (UV) radiation and low water availability largely influence their success or failure in field conditions. Pigments produced by bacteria are considered to be responsible for increasing tolerance to UV radiation [5]; various *Pseudomonas* species produce siderophores, which are considered to be important for tolerance to UV exposure. Strains of *Pseudomonas aeruginosa* are known to produce siderophores and salicylic acid, and their beneficial effects on plant growth are pronounced even when plants are subjected to suboptimal conditions and unfavorable climatic conditions [27].

Several genera and species of bacteria, fungi, and actinomycetes serve as expedient inoculants for various turfgrass diseases. Evaluation of the effectiveness of microbial inoculants for disease suppressiveness begins with in vitro laboratory assays, followed by growth chamber, greenhouse, and field studies, provided the organisms remain effective throughout the series of tests. Only a few laboratory and greenhouse studies end up being tested in the field; therefore, information on efficacy for turfgrass disease control in the field is much more limited.

Certain microbial agents have also been used in the management of turfgrasses for improved availability of fertilizers for plants [88], reduction of thatch layer [66], and control of insects [119]. Application of *Xanthomonas campestris* pv. *poaannua* as a biological agent for control of weeds such as annual bluegrass has been tested. Some biotypes of *X. campestris* have afforded up to 82% control of annual bluegrass [53,131]. Ecosoil Systems (San Diego, CA) is developing this bacterium as a biocontrol agent for annual bluegrass control for golf courses. However, there has been an increased interest in the use of microbial inoculants for turf disease management as they are produced and marketed in various formulations with improved delivery systems. Success in the identification of new organisms that exhibit significant control of various root and foliar diseases in the past two decades has also contributed to the rising interest in biological control in the turfgrass industry. In 1997, Bio-Trek®22G, marketed by Wilbur-Ellis Co., Fresno, CA, was the first product approved by the U.S. Environmental Protection Agency (EPA) and registered for use in turfgrass disease management.

A. Bacterial Antagonists

Bacterial antagonists are the most widely used microbes in a turfgrass management system. Among the various species tested against turfgrass pathogens, *Pseudomonas* species and members of the Enterobacteriaceae are significantly more efficient than heterotrophic bacteria.

1. *Pseudomonas* spp.

Several *Pseudomonas* species are known to inhibit the growth of a number of fungal pathogens and suppress turfgrass diseases. One of the most widely recognized bacteria currently used in the turfgrass industry for its biocontrol potential is a *Pseudomonas aureofaciens* strain commonly known as TX-1 [116]. This bacterium was originally isolated from the thatch layer of a creeping bentgrass from Texas. The antifungal activity of this bacterium is reportedly due to its metabolite, phenazine-1 carboxylic acid, which is effective against *Sclerotinia homoeocarpa* (dollar spot), *Bipolaris sorokiniana* (spring leaf spot), and *Magnaporthe poae* (summer patch) [91]. This bacterium, marketed under the trade name Spot-less, has gained tremendous popularity among golf course superintendents in several regions of the United States because of its suppressive quality for turfgrass diseases when delivered through the Bioject® system (Ecosoils, Inc., San Diego, CA). Formulations of *Burkholderia casidae*, *Bacillus subtilis*, and *Pseudomonas fluorescens* have also been shown to suppress brown patch, although the level of suppression afforded did not equal the commercially acceptable quality standards for golf course turfs and putting greens in particular [28]. The efficacy of certain fungicides has been reported to be increased by application of *P. aureofaciens* to turf, where disease is managed by specific fungicides. Application of *P. aureofaciens* to turf has been shown to significantly increase dollar spot control achieved using the fungicides cyproconazole and propiconazole [114]. A number of subscribers in the turfgrass industry claim that timely application of *P. aureofaciens* through the Bioject system initiated early in spring prevents dollar spot from recurring during the season.

   Pseudomonas fluorescens is another species that has good biological control potential in turfgrass disease management. Strains of *P. fluorescens* are known to be antagonistic to *Sclerotinia homoeocarpa* and to have disease-suppressive activity for dollar spot of Kentucky bluegrass and creeping bentgrass (*Agrostis palustris* Huds.) [4,45,98]. In a study, preventive application of several *P. fluorescens* strains on Kentucky bluegrass (*Poa pratensis* L.) provided significant suppression of dollar spot [45]. A strain of antibiotic-producing *P. fluorescens* has also been shown to inhibit mycelial growth of *S. homoeocarpa* in vitro and also significantly suppressed the disease on Kentucky bluegrass and creeping bentgrass infested with the fungus [98]. Antagonism and disease suppression by *P. fluorescens* with respect to other foliar diseases of turfgrasses are inconclusive. Several strains of *P. fluorescens* were reported to be highly antagonistic to *Drechslera poae* and *D. dictyoides*, providing good control of spring leaf spot or melting out of Kentucky bluegrass [98]. However, some strains of *P. fluorescens* did not provide satisfactory control for spring leaf spot of Kentucky bluegrass caused by *Bipolaris sorokiniana*, which is closely related to *D. poae* [45]. Although strains of *P. fluorescens* were reported to be antagonistic to *M. poae*, significant suppression of summer patch disease of Kentucky bluegrass by this bacteria was observed in greenhouse tests, but not in field studies [113]. However, strains of *P. fluorescens* have been reported to be effective in the suppression of take-all patch (Ophiobolus patch) caused by *Gaeumannomyces graminis* var. *avenae* in the greenhouse [125] and field studies [7], conforming with earlier reports of antagonism and disease suppression by fluorescent pseudomonads on take-all patch of wheat [22,99,106,107,120,124]. Species of fluorescent pseudomonads appear to play a major role in the remission of the central zone of take-all patch in take-all decline phenomenon in turf. A study on creeping bentgrass greens indicated that the percentage of fluorescent *Pseudomonas* species were progressively higher in the rhizosphere bacterial population beginning from the asymptomatic area at the edge of the turf to the center of the patch where regrowth of the turf was in progress [65,101]. Among the population of fluorescent *Pseudomonas* species quantified, the number of antagonists for *G. graminis* var. *avenae* was reportedly higher in the rhizosphere of the recolonized zone of the take-all patch than in the asymptomatic area outside the patch; however, no details on the identity of these fluorescent pseudomonads were provided. In a different study, significant reduction of take-all patch on Penncross and Penneagle creeping bentgrasses was achieved 2 weeks after inoculation with three unidentified strains of *Pseudomonas* [64].

Pseudomonas aeruginosa, which is a plant growth-promoting bacterium [47] and a siderophore [46] and salicylic acid producer [13], also has great potential for the control of turfgrass diseases. Strains of *P. aeruginosa* isolated from spent mushroom substrate have been found to be highly antagonistic to *Pyricularia grisea*, *Sclerotinia homoeocarpa*, *Rhizoctonia solani*, *R. cerealis*, and *Fusarium culmorum* in vitro [117]. Additionally, these strains of *P. aeruginosa* provided satisfactory disease control when evaluated for efficacy and timing of application to control gray leaf spot (blast disease) of perennial ryegrass turf [118]. In this controlled environmental chamber study, bacterial isolates provided a level of control that was comparable to that of propiconazole, but significantly lower than that of azoxystrobin. All three application intervals tested at 1-, 3-, and 7-day intervals prior to pathogen inoculation with the bacterial strains provided significant disease suppression (Table 1). A comparison of disease severity for plants maintained in controlled-environment chambers and an outdoor setting showed no significant differences in the efficacy of *P. aeruginosa* under those

Table 1 Effects of Timing of Application of *Pseudomonas aeruginosa* (Days Prior to Inoculation with *Pyricularia grisea*) on Incidence and Severity of Gray Leaf Spot on Perennial Ryegrass Turf

	Incidence (%)			Severity (0–10 index)		
Treatment	Day 7	Day 3	Day 1	Day 7	Day 3	Day 1
B12	34.5 b	31.8 b	26.4 b	3.1 b	3.0 b	2.5 bc
B15	35.7 b	32.5 b	30.8 b	3.3 b	2.9 b	2.7 b
B38	33.6 b	27.1 b	25.9 b	3.1 b	2.8 b	2.7 b
Propiconazole	22.8 c	26.3 b	8.4 c	2.6 bc	2.5 bc	2.3 bc
Azoxystrobin	12.3 d	10.6 c	8.5 c	2.1 c	1.9 c	1.7 c
Control	72.0 a	72.9 a	75.2 a	7.3 a	7.6 a	7.9 a

Numbers followed by different letters within a column are significantly different according to Student-Newman-Keul's test ($p \leq 0.05$).

conditions. Further, the biocontrol activity of the bacterial isolates was not compromised by exposure to UV radiation.

Other pseudomonads such as *P. lambergii* and *P. putida-fluorescens* have also been reported to be antagonistic to certain turfgrass pathogens. Strains of *P. lambergii* were found to be highly antagonistic to *B. sorokiniana*, reducing disease severity of dollar spot and melting-out of Kentucky bluegrass when sprayed with the bacterial suspension [45]. Antagonism and suppression of the disease by *P. putida-fluorescens* has been documented only for take-all patch in turfgrasses [123–125]. In these studies, a significant inhibition of mycelial growth in vitro and a substantial reduction of take-all patch of creeping bentgrass were observed after inoculation of bacteria and the pathogen into the sterilized soil. Complete recovery of the infected turf in 6 weeks clearly indicates the highly antagonistic nature of *P. putida-fluorescens* to *G. graminis* var. *avenae* and persistence of the bacterium in the soil for several weeks [124]. Several species of *Pythium* are important pathogens causing foliar blight or root dysfunction in turfgrasses. Not much has been explored about biocontrol of *Pythium*-induced diseases by pseudomonads as in other diseases discussed earlier; however, one report indicated that a few strains of unidentified *Pseudomonas* species isolated from thatch and soil were suppressive to *Pythium aphanidermatum*, which causes foliar blight of turfgrasses [79].

2. Enterobacter cloacae

Enterobacter cloacae is one of the most important microbial inoculants tested in depth for its biological control properties against a number of soilborne pathogens [29,38,41,69,76,81,96,97,108]. It is an effective microbial antagonist with high

Table 2 Suppression of Dollar Spot on a Creeping Bentgrass/Annual Bluegrass Putting Green with Strains of *Enterobacter clocae*[a]

Strain	1988		1989		1989	
	Spots per plot[b]	Control[c] (%)	Rating 1[d]	Control (%)	Rating 2[d]	Control (%)
Untreated	41	—	5.0	—	5.0	—
Uninoculated CMS[e]	35	14.6	3.6	28.0	5.2	0.0
EcCT-501	15	63.4	2.8	44.0	5.4	0.0
EcH-1	28	31.7	4.8	4.0	4.0	20.0
E1	31	24.4	4.4	12.0	3.0	40.0
E6	39	4.9	4.0	20.0	4.4	12.0
E1-R6	23	43.9	NT	—	NT	—
E6-R8	38	7.3	NT	—	NT	—
EcH-1-R8	NT[f]	—	3.8	24.0	4.6	8.0
EcCT-501-R3	NT	—	3.8	24.0	4.6	8.0
LSD ($p = 0.05$)	18		1.6		1.9	

[a] Inoculated into a mixture consisting of 25% cornmeal and 75% fine sand. Applied as a topdressing at the rate of 465 cm^3/m^2.
[b] Number of infection centers per plot area. Ratings assessed 64 days after application.
[c] Based on a percentage of the disease severity in untreated plots.
[d] Rating scale: 1 = 10% of plot area diseased; 10 = 100% of plot area necrotic. Rating 1 taken 32 days after application; rating 2 taken 55 days after application.
[e] CMS = Cornmeal/sand mixture.
[f] NT = Not tested.
Source: Ref. 76.

disease suppressiveness against several turfgrass diseases. Reduction of dollar spot in bentgrass putting greens by use of *E. cloacae* as a microbial inoculant in soil was clearly demonstrated in a study in New York [76]. In this study, application of cornmeal and sand mixtures fortified with strains of *E. cloacae* on creeping bentgrass putting greens naturally infested with *Sclerotinia homoeocarpa* provided satisfactory control of dollar spot (Table 2). Monthly application of a strain of the bacterium provided up to 63% disease control, which was comparable to the fungicides iprodione and propiconazole. Disease suppression was reportedly more effective when *E. cloacae* was applied preventively rather than curatively. Persistence of this organism in thatch was outstanding, as indicated by continuous suppression of the disease for up to 2 months and the high recovery rates of the organism for over 3 months following the application. As a single species, *E. cloacae* may have a much broader biological control spectrum than several other bacterial inoculants for control of turfgrass diseases. Strains of *E. cloacae* have also been shown to be antagonistic to two other turfgrass fungal pathogens, *Pythium aphanidermatum* and *Magnaporthe poae*, which belong to two distinct phyla, Oomycota and Ascomycota, respectively. Tissue culture plate assays and controlled environment chamber studies indicated that strains of *E. cloacae* significantly suppressed Pythium foliar blight of perennial ryegrass [79]. In the tissue plate assays, disease suppression provided by strains of *E. cloacae* was significant on 3-day-old ryegrass seedlings, whereas disease severity on plants generated from *E. cloacae*-treated seeds and sand medium did not differ from that of the seedlings from noninoculated controls. In controlled environment chambers, one of the *E. cloacae* strains used in tissue plate assays provided control of Pythium foliar blight on 7- to 10-week-old ryegrass similar to that of the fungicide metalaxyl and noninoculated control plants. There was significant variability among the strains of *E. cloacae* in this study, and it was hypothesized that the variability might have been due to suppression specificity among enterobacterial antagonists of *Pythium* species. The suppression specificity of *E. cloacae* may be explained by a direct relationship between growth inhibition of *Pythium* with binding of *E. cloacae* cells to hyphae of *Pythium ultimum* [84]. The same strains of *E. cloacae* that provided satisfactory control for Pythium foliar blight [79] have been shown to be effective against *Magnaporthe poae*, which causes summer patch disease of Kentucky bluegrass [113]. In this study, little or no inhibition of *M. poae* by strains of *E. cloacae* was observed; however, in a growth chamber study up to 31% control of summer patch was provided by the bacterium. In the field experiments, one strain provided up to 56% control of the disease. As in Pythium blight, there was significant variability among the strains of bacterium in suppression of summer patch.

3. Other Bacteria

Other bacteria that provide various levels of disease suppression in a turfgrass system include *Xanthomonas maltophilia*, *X. campestris*, *Serratia marcescens*, *Serratia* species, *Streptomyces* species, and *Bacillus* sp. Significant suppression of summer patch disease of Kentucky bluegrass by strains of *X. maltophilia*, species of *S. marcescens*, and *Bacillus* sp. [54,55] and spring leaf spot of tall fescue (*Festuca arundinacea* Schreb.) by strains of *Stenotrophomonas maltophilia* (= *Xanthomonas maltophilia*) [130] have been well documented. In one study, suppression of summer patch by strains of *X. maltophilia* was outstanding, providing over 70% control of the disease [55]. Timing of application of the bacteria to turfgrass rhizosphere was also significant (Table 3). No suppression of the disease was evident when the bacterial suspension was applied during the time of planting after inoculation with *M. poae*; however, significant reduction of the disease was observed when bacterial suspension was applied 1–4 weeks after planting the grass. Suppression of the disease was influenced by bacterial concentration. Delivery of the suspension at 10^9 or 10^{10} cfu/mL provided significantly less disease control than did 10^8 cfu/mL. Persistence of *X. maltophilia* in soil may be of great practical value in summer patch disease management, as indicated by

Table 3 Effect of Timing of Bacterial Application on Suppression of Summer Patch on Kentucky Bluegrass cv. Baron

		Disease rating[b]	
	Inoculation time[a]	Xanthomonas maltophilia	Serratia marcescens
Expt. 1	0	3.1	2.9
	1	2.1*	2.5*
	2	2.2*	2.8
	3	2.4*	2.3*
	4	2.0*	1.9*
	Untreated check	3.2	3.2
Expt. 2	0	2.6	2.9
	1	1.7*	2.1
	2	1.8*	2.1
	3	1.9*	1.9
	4	1.6*	1.7*
	Untreated control	2.7*	2.9

[a] 0 = time of planting; 1 = 1 wk after planting; 2 = 2 wk after planting; 3 = 3 wk after planting; 4 = 4 wk after planting; untreated control = fungal-inoculated plants not treated with bacteria.
[b] Disease ratings based on a 9-point logarithmic scale.
* Significant difference compared to untreated control plants.
Source: Reprinted from Soil Biol Biochem 27:1479–1487 (1995), Kobayashi et al. Isolation of the chitinolytic bacteria *Xanthomonas maltophilia* and *Serratia marcescens* as biological control agents for summer patch disease of turfgrass. With permission from Elsevier Science.

recovery of the bacterium from nonrhizosphere and rhizosphere soils 35 days after infestation of the soil. Although monitoring of *X. maltophilia* population dynamics in the soil beyond a 35-day study period was apparently not conducted in this study, such a monitoring system for an extended period would reveal valuable information regarding the fitness of the bacterium in a turfgrass system. In the same study, the efficacy of *S. marcescens* in suppressing summer patch of Kentucky bluegrass was relatively less than that of *X. maltophilia*. Additionally, recurrence of the disease was greater in turf treated with *S. marcescens* than that with *X. maltophilia*. Suspension of *S. marcescens* at a concentration of 10^9 cfu/mL provided the best control of summer patch. Persistence of *S. marcescens* in soil appeared to be higher than *X. maltophilia*, as indicated by recovery assays of nonrhizosphere and rhizosphere soils. Despite the lower efficacy of *S. marcescens* than *X. maltophilia*, application of *Serratia* spp. in summer patch disease control may still be of practical value. This is clearly demonstrated by achievement of over 80% disease control following drenching of the soil with strains of *Serratia* sp. in summer patch–affected Kentucky bluegrass turf [54]. This study also confirms an earlier report that this antagonist showed adaptation and persistence in the soil. *Bacillus* sp. also provided over 60% disease control, but, its persistence in soil was reportedly poor.

Phylloplane bacteria have also been implicated as antagonists of plant pathogens [24,30,60]. Strains of *X. campestris* isolated from phylloplane of perennial ryegrass (*Lolium perenne* L.) have been reported to exhibit biocontrol properties against *Drechslera dictyoides* that causes leaf blight (net blotch) of perennial ryegrass [4]. The inhibitory effects of *X. campestris* on *D. dictyoides* were indicated by delayed germination of conidia, reduced rate of germ-tube elongation, and plasmolysis of hyphae of *D. dictyoides* cultures. A substantial reduction in the number of necrotic blades of perennial ryegrass was also observed. Filamentous bacteria such as species of *Streptomyces* have also been shown to suppress various turfgrass diseases such as spring leaf spot, dollar spot, brown patch, and Pythium root rot [45,94,102].

B. Fungal Antagonists

Fungal antagonists are also important biological control agents of turfgrass diseases. In recent years, there has been increasing interest in the utilization of fungi as potential biocontrol agents in the turfgrass industry. Disease-suppressive antagonists include fungi from a wide range of phyla. While some are phylogenetically distant from their target turfgrass pathogens, others are closely related to the pathogens, which often turns out to be avirulent strains of the same species of a specific pathogen. Among all the disease-suppressive fungi previously identified, two extensively studied inoculants for the biological control of turfgrass diseases are *Typhula phacorrhiza* [10,12,58,59,73,83] and species of *Trichoderma* [35,36,39,62,63,73,86,92,93,103]. Additionally, *Acremonium* sp., *Fusarium heterosporum*, *Gaeumannomyces* spp., *Gliocladium virens*, and *Laetiseria* spp. have been shown to suppress various turfgrass diseases [11,26,32,33,39, 42,43,126–129].

1. *Typhula phacorrhiza*

Introduction of microbial inoculants with extended periods of survival in the turfgrass system has been a major focus of research in the management of gray snow mold caused by *Typhula incarnata* and *T. ishikariensis*. In such cool season diseases, use of fungicides is generally limited to preventive application prior to snowfall as the pathogens are active during the period of extended snow cover. However, use of microbial inoculants in management of cool season turfgrass diseases has been a challenge because of the importance of survivability of the newly introduced organisms when subjected to low-temperature environment in soil and the turf canopy. In the early 1980s, a low-temperature–tolerant fungus, *T. phacorrhiza*, closely related to gray snow mold pathogens, isolated from the thatch layer of Kentucky bluegrass was effectively used to suppress the disease on creeping bentgrass [12]. In this study on creeping bentgrass greens, application of wheat grains colonized by *T. phacorrhiza* showed up to 70% reduction in gray snow mold disease severity. Percent reduction of the disease seemed to vary with the density of *T. phacorrhiza*–infested wheat grains. A grain density of 100 g/m^2 provided 44% disease reduction, whereas 200 g/m^2 reduced the disease up to 70% on golf greens. In another study, application of *T. phacorrhiza*–infested cereal grains to gray snow mold–affected turf provided significant reduction in gray snow mold severity, sped up recovery rate of the affected grass, increased the number of sclerotia of the antagonist, and decreased the number of sclerotia of the pathogens in thatch [58]. Additionally, studies on interactions between hyphae of *T. phacorrhiza* and isolates of *T. incarnata* and *T. ishikariensis* suggested that the mechanism of disease suppression by *T. ishikariensis* was not hyperparasitism or cellular lysis induced by an antibiotic or by hyphal contact [12]. A nutrient-mediated reduction in the suppression of growth of *T. incarnata* and *T. ishikariensis* on culture medium exposed to *T. phacorrhiza* further suggested that the mechanism of suppression was possibly due to competition for nutrient and colonization of the substrate.

2. *Trichoderma* Species

Species of *Trichoderma* are antagonistic to a number of foliar and root pathogens of turfgrass and are known to be to effective biocontrol agents against various turfgrass diseases [39,40,62,63,86]. One of the most extensively studied *Trichoderma* species is *T. harzianum*. Currently, *T. harzianum* is a commercially available biocontrol agent as Bio-Trek 22G. This fungus is antagonistic to several turfgrass pathogens and has strong rhizosphere competence [40]. Studies have shown that commercial formulations of *T. harzianum* provide significant control of brown patch, dollar spot, and Pythium root diseases of bentgrass [62,63]. Application of *T. harzianum* (10^7 cfu/g soil) to soil of creeping bentgrass greens provided disease control up to 68, 50, and 87% in turf affected by dollar spot, brown patch, and Pythium root rot, respectively. Further field studies on the suppression of dollar spot indicated that monthly application of granular or peat-based formulations of *T. harzianum* reduced initial disease severity by as much as 71% and delayed disease development up to 30 days on bentgrass greens. One of the most desirable attributes of this organism is its ability to persist in the rhizosphere of creeping bentgrass. Monthly application of *T. harzianum* has been effective in maintaining populations at levels of nearly 10^6 cfu/g of thatch and soil. The fungus was also shown to overwinter at population levels between 10^5 and 10^6 cfu/g of soil and thatch which was adequate to achieve a significant level of biological control. A significant increase in the population of *Trichoderma* spp. in soil and thatch was observed after the application of *T. harzianum* to turf. A similar trend in increase of species of *Trichoderma* population after the application of *T. harzianum* was observed in bentgrass–annual bluegrass mix golf greens [92]. In another study, a significant suppression of brown patch on creeping bentgrass (80% control) was achieved following the application of co-nidial suspension of *Trichoderma* sp. to turf. Isolates of *T. harzianum* are also antagonistic to *Typhula incarnata*, the gray snow mold pathogen. Severe inhibi-tion of the germination of *T. incarnata* in soil by *T. harzianum* has been well documented [39]. Another species of *Trichoderma* that is a potentially good bio-control agent for turfgrass diseases is *T. hamatum* [36,93]. The fungus is highly antagonistic to isolates of *Pythium apanidermatum* and *P. ultimum*, the pathogens of foliar blight disease. The mechanism of suppression of the foliar blight patho-gens by *T. hamatum* appears to involve hyperparasitism [93]. Species of *Trichod-erma* such as *T. konigii* and *T. viridae* are also potentially effective biocontrol agents against certain turfgrass pathogens. The antagonism of these fungi to the gray snow mold pathogen, *T. incarnata*, is extremely high with complete inhibi-tion of sclerotial germination on potato-dextrose agar, sterilized soil, and natural soil [39].

3. Other Fungi

A number of avirulent strains of turfgrass pathogens such as *Rhizoctonia solani*, binucleate *Rhizoctonia* species, *Rhizoctonia*-like fungi, and *Gaeumannomyces graminis* and its anamorph *Phialophora* species are suppressive to turfgrass diseases caused by their virulent counterparts and other unrelated fungi [11,26,32,33,112,126,128,129]. A significant reduction of severity of Rhizoctonia blight (brown patch) of tall fescue, normally maintained as tall-cut turf, has been achieved by application of the mycelia of the binucleate *Rhizoctonia* to tall fescue plants [129]. Suppression of the disease was reportedly associated with reduced growth of *Rhizoctonia solani* in the presence of the binucleate isolate. A similar suppression of Rhizoctonia blight was also observed on close-cut bentgrass turf maintained as putting greens [11]. Similarly, several avirulent strains of take-all pathogen, *G. graminis* var. *graminis*, *G. graminis* var. *tritici*, *G. graminis* var. *avenae*, *Phialophora radicola* var. *radicola*, and *P. radicola* var. *graminicola* have been shown to be suppressive to the virulent strains. A study in Australia indicated that a complete control of take-all patch (Ophiobolus patch) on bent-grass was achieved following the incorporation of these avirulent fungi into soil at the time of planting the grass [26]. Suppression of take-all patch appears to vary with volumes of take-all suppressive soil and the strain of *G. graminis* from which the suppressive soil was developed. Control of the disease by suppressive soil at 20 mm depth developed from the mycelia of *G. graminis* var. *avenae* was greater than that at 5 or 10 mm depth. There was an obvious trend in the effective-ness of disease suppression with increase in depth up to 20 mm, the level that provided almost complete control of the disease. Suppressive soil developed from

G. graminis var. *graminis* also provided similar control; however, it did not appear to be as effective as the soil developed from *G. graminis* var. *avenae*. Further, a high frequency of recovery of avirulent strains of *P. radicicola* from numerous golf and bowling greens have also been reported in a survey conducted in four states in Australia. In a study in the United Kingdom, a high population of *P. radicicola* was reportedly present in grasslands, and it was hypothesized that the population build-up in high pH soils restricted the growth of the take-all pathogen, thus leading to less prevalence of the disease [26]. Other fungi that are potentially effective biocontrol agents for turfgrass diseases are *Acremonium* sp., *Fusarium heterosporum*, *Gliocladium virens*, *Laetiseria arvalis*, and *Actinomycetes* [33,42,43,111,112].

II. APPLICATION OF COMPOSTS AS DISEASE-SUPPRESSIVE SOIL AMENDMENTS

Composts, sewage sludge, and organic fertilizers are three primary organic wastes that have disease-suppressive characteristics, protecting plants from pathogens. Although composts have been used to control fruit and vegetable diseases, their application in the management of turfgrass diseases has been extensively investigated only in recent years. Because recreational turfgrasses are subjected to extensive wear and tear, they are made vulnerable to various diseases and soil compaction. These problems can be addressed without chemical application by using composts as top dressings and soil amendments. Composts, when formulated, are rich in beneficial microbes and nutrients, thereby stimulating turf growth and development, in addition to increasing resistance to the most common diseases, such as dollar spot and brown patch. The physical, chemical, and biological properties of composts, as well as their maturity level and composting method, affect suppression of diseases. The activity of antagonists in compost is also dependant on the nutrients present [49]. Composted yard waste trimmings are high in K, P, Ca, Mg, total carbon-to-nitrogen ratio, and nitrate nitrogen concentration [34]. Immature composts are not suppressive to pathogens, but as composts mature, they are fully recolonized by mesophilic bacteria, heterotrophic fungi, as well as actinomycetes, which consistently induce suppression to diseases [15–17,21,34,50,56]. Various oligotrophic *Pseudomonas* species in composts are known to be effective root colonizers of *Pythium ultimum* [57], while *Bacillus subtilis* are effective is supressing *Pythium* species and various other soilborne pathogens [89,90]. Recent studies have shown that *Pseudomonas aeruginosa* isolated from spent mushroom substrate are effective in controlling gray leaf spot of perennial ryegrass when tested in controlled environmental chamber experiments [117,118]. These isolates afforded significant reduction of the disease when applied as foliar sprays. Another study has indicated that strains of actinomycetes isolated from *Pythium*-suppressive composts were effective in reducing Pythium root rot in greenhouse conditions [111]. In addition, fungal antagonists such as species of *Trichoderma* are also suppressive to *Rhizoctonia solani* [82].

Suppression of specific diseases by the antagonists depend on the effective colonization of specific biocontrol agents against a particular pathogen [56,57,74,75,77,78,82,90]. Various investigators have shown that elevated levels of microbial activity result in increased competition between the compost-inhabiting microbial populations and the pathogen *P. graminicola*, since components of root exudates are essential for the germination of *Pythium* propagules and mycelial growth [17,37,70,115]. Application of composts to turf promotes the population of antagonistic microorganisms that interfere with the activities of the pathogenic fungi [71], and turf managers are increasingly interested in using composts as top dressing applications based on its proven success in suppressing turf diseases.

Alleviating soil compaction is a persistent problem in landscape management as compacted soil impedes turf establishment, thus making it vulnerable to diseases. Traditional methods such as reseeding or resodding are expensive, and in recent years turf managers have started using composts amended with wood chips or aged crumb rubber in order to alleviate soil compaction, improve root penetration and turf establishment, and enhance resistance to diseases. A research study conducted at a golf course in Colorado Springs, Colorado, has shown that turf grown in soil amended with compost requires 30% less water, fertilizer, and chemicals [1]. It has also been observed that applications of composts on putting greens in late fall suppresses Typhula blight in addition to protecting the surface from winter ice and freezing damage [71].

The use of various kinds of composts as starting material has yielded promising results and suppression of turf diseases including Typhula blight, Rhizoctonia blight, Pythium blight, Pythium root rot, red thread, summer patch, necrotic ring spot, and other diseases (Table 4). The preventive and curative applications of compost as top dressings provide disease control on putting greens of creeping bentgrass and annual bluegrass mix to a level comparable to fungicide iprodione in reducing dollar spot disease [80]. Dollar spot was also found to be significantly reduced with application of composted materials such as sewage sludge [23]. A study has shown that composted municipal sludge applied as top dressings was effective in suppression of brown patch on tall fescue [87]. Amending sand-based greens with municipal biosolid compost, brewery sludge compost, or peat is known to suppress Pythium root rot [72]. These amendments afforded complete control of the disease 6 months after incorporation, and the suppression was retained for up to 4 years. Composted municipal biosolid topdressing fortified with *Trichoderma hamatum* and *Flavobacterium balustinum* have also been shown to significantly suppress dollar spot on creeping bentgrass [36]. Several diseases caused by soilborne pathogens are suppressed in containers amended with com-

Table 4 Amendments for Turfgrass Disease Control

Amendment	Diseases controlled	Maximum level of control observed (%)[a]
Municipal and industrial sludges		
Activated sewage sludge	Dollar spot	99
Composted municipal biosolids	Brown patch	42
	Dollar spot	40
	Pythium root rot	63
	Red thread	51
	Typhula blight	70
Composted brewery sludge	Brown patch	25
	Dollar spot	15
	Pythium root rot	68
	Red thread	36
	Typhula blight	70
Uncomposted natural organic fertilizers		
Animals and plant meals	Brown patch	75
	Dollar spot	74
	Necrotic ring spot	96
	Pythium root rot	56
	Red thread	57
	Typhula blight	0
Animal manures		
Composted cow or horse manure	Brown patch	25
	Dollar spot	73
	Pythium root rot	31
	Red thread	9
	Typhula blight	55
Composted poultry litter	Brown patch	75
	Dollar spot	55
	Necrotic ring spot	86
	Pythium root rot	94
	Red thread	79
	Typhula blight	15
Other		
Composted yard waste	Brown patch	39
	Dollar spot	5
	Red thread	0
Spent mushroom compost	Brown patch	25
	Dollar spot	0
	Red thread	0
Reed-sedge peat	Pythium root rot	68

[a] Percentages represent the maximum values published. Considerable variation in suppressiveness exists among different compost feedstocks, different batches of the same feedstock, and at different sites.
Source: Ref. 73.

posted hardwood bark [14,25,68,109]. Species of *Trichoderma* and *Gliocladium* are known to be most effective for control of soilborne pathogens in bark compost, while in addition a variety of bacterial antagonists such as *Pseudomonas putida*, *Flavobacterium balustinum*, and *Xanthomonas maltophilia* also play a role [49].

With the increase in biological control of turfgrass diseases in recent years, composted organic amendments using feedstocks and food waste have also been investigated for disease suppressiveness. Composts from a variety of feedstocks have been tested to suppress seedling and root diseases of creeping bentgrass caused by *Pythium graminicola* [21]. In field experiments, these composts afforded a significant level of suppression even when disease pressure was high. Populations of actinomycetes, fluorescent pseudomonads, and fungi were found to be increased in animal manures [3]. Besides these types of composts, natural organic fertilizers such as Ringer Compost Plus, Ringer Greens Restore, and Sustane significantly suppress the severity of dollar spot [80]. A similar study conducted on creeping bentgrass (*Agrostis palustris*) green and Kentucky bluegrass (*Poa pratensis*) over a 3-year period showed that Ringer Greens Super and Ringer Turf Restore significantly suppressed dollar spot [61]. In contrast, in another study it was observed that amendments with organic fertilizers did not influence the growth rate, nitrogen uptake, or suppression of brown patch on Rebel tall fescue and Tifway Bermudagrass. Turf Restore amended with *Trichoderma viride*, *Bacillus* spp., and actinomycetes was not beneficial when tested in the greenhouse for a period of 72 days [88]. The reason for the contradictory results between these two studies may be the insufficient period of time allowed in the latter.

Efforts to commercialize the controlled process of inoculating composts with specific antagonists and market them as a pest control product is underway [48]. The success of utilization of composts depends on the consistency of the level of disease suppression achieved. A major challenge in the use of composts is the variation in the disease control capacity with respect to different batches, various environmental conditions, and sites tested. Although numerous assays have been developed to reduce variability in the physical and chemical properties of composts [52], methods to assess microbial aspects and disease suppressiveness have not yet been determined.

III. ENDOPHYTES IN TURFGRASS SYSTEMS

Endophytes are long considered to be mutualistic with their host plants; hence, the term "defensive mutualism" has been used [19]. Most endophytes belong to phylum Ascomycota in the family Clavicipitaceae, and they are common in many grasses [20,122]. Tall fescue and perennial ryegrass (*Lolium perenne* L.) are the most widely known cool season turfgrass species that host endophytes. Only certain genera of these fungi, such as *Epichloë* and *Neotyphodium* (syn.

Acremonium), associated with cool season grasses, are transmitted by seeds. However, species of *Balansia* found in warm season grasses are not seed-transmitted [95]. Therefore, development of endophyte-enhanced turfgrass varieties is more practical in tall fescue and perennial ryegrass than warm season grasses, and most research related to the effects of endophytes on host fitness involves endophyte-associated tall fescue and perennial ryegrass [44,104]. Grasses with endophytes are reportedly more aggressive [2,9,110], and they resist attack by insect and fungal plant pathogens more effectively than noninfected grasses [18,100,105]. A study on Chewings fescue, creeping fescue, and hard fescue turf indicated that the grasses inoculated with endophytes developed significantly less dollar spot disease than noninoculated grasses [18]. It has been also suggested that specific endophyte isolates from fine fescue have a high degree of inhibition against *Sclerotinia homoeocarpa* (dollar spot) and *Pyricularia grisea* (gray leaf spot) [95]. Further, in laboratory studies, chemical extracts from endophyte-associated Chewings fescue seeds strongly inhibited the growth of *S. homoeocarpa*, suggesting a chemical basis for resistance in endophyte-associated turf. Antifungal activity by several species of *Neotyphodium* (syn. *Acremonium*) on a number of turfgrass pathogens, including *Bipolaris sorokiniana* (Helminthosporium leaf spot), *Colletotrichum graminicola* (foliar anthracnose/basal rot), *Limonomyces roseipellis* (pink patch), *Rhizoctonia zeae* (Rhizoctonia sheath spot), and *R. cerealis* (yellow patch), has also been documented [44,104]. Work on the use of endophytes in turfgrass disease management is still in the early stages of development. Turfgrasses hold an advantage over forage grasses in the use of endophytes in plant health management as there is no animal toxicity concern especially in golf courses and athletic fields.

IV. ROLES OF PLANT GROWTH PROMOTERS

Extensive research on the use of plant growth–promoting bacteria (PGPR) in enhancing plant growth and yields of many commercial and agricultural crops has been conducted [8,31,85]; however, research on use of PGPRs on turf has not been conducted. These organisms have been shown to serve as plant growth promoters by enhancing indole acetic acid synthesis in plants and promoting root development. The bacterium *Azosporillum brasilense* is currently being evaluated by university researchers and certain commercial organizations for its effect on turfgrass growth and development and protection against certain pathogens.

V. THE FUTURE

Apart from these alternatives to control turf diseases, recent advances in biotechnology have led to the development of transgenic turf species and genetically engineered biocontrol agents that exhibit fitness to survive and establish in adverse environmental conditions. Efficient management of beneficial microbes will provide solutions to management of foliar and root diseases and help maintain healthy turf. Creeping bentgrass cultivars that tolerate the nonselective herbicide glufosinate (Finale) and cultivars with improved resistance to dollar spot and brown patch are being developed by researchers at Rutgers and Michigan State University [28]. Genetically engineered and transformed turf cultivars that will require smaller amounts of pesticides and fertilizers are likely to be commercially available in the foreseeable future.

REFERENCES

1. Anonymous. Innovative uses of compost erosion control, turf remediation, and landscaping. Circular EPA530-F-97-043, 1997.
2. M Arechaveleta, CW Bacon, CS Hoveland, DE Radcliffe. Effect of the tall fescue endophyte on plant response to environmental stress. Agron J 81:83–90, 1989.
3. IP Aryanatha, R Cross, DI Guest. Suppression of *Phytophthora cinnamomi* in potting mixes amended with uncomposted and composted animal manures. Phytopathology 90:775–782, 2000.
4. B Austin, CH Dickinson, M Goodfellow. Antagonistic interactions of phylloplane bacteria with *Drechslera dictyoides* (Drechsler) Shoemaker. Can J Microbiol 23: 710, 1977.
5. JB Bahme, MN Scroth. Spatial-temporal colonization patterns of a rhizobacterium on underground organs of potato. Phytopathology 77:1093–1100, 1987.
6. KF Baker, RJ Cook. Biological Control of Plant Pathogens. San Francisco, CA: W. H. Freeman and Company, 1974, p. 431.
7. NA Baldwin, AL Capper, DJ Yarham, eds. Evaluation of biological agents for the control of take-all patch (*Gaeumannomyces graminis*) of fine turf. In: ABR Beemster, ed. Developments in Agricultural and Managed-Forest Ecology. Amsterdam: Elsevier Science Publishers, 1991, pp. 231–235.
8. Y Bashan, H Levanony. Current status of *Azospirillum* inoculation technology: *Azospirillum* as a challenge for agriculture. Can J Microbiol 36:591–605, 1990.
9. DP Belesky, WC Stringer, NS Hill. Influence of endophyte and water regime upon tall fescue accessions. I. Growth characteristics. Ann Bot 63:495–503, 1989.
10. LL Burpee. Interactions among low-temperature-tolerant fungi: prelude to biological control. Can J Plant Path 16:247–250, 1994.
11. LL Burpee, LG Goulty. Suppression of brown patch disease of creeping bentgrass by isolates of non-pathogenic *Rhizoctonia* spp. Phytopathology 74:692–694, 1984.
12. LL Burpee, LM Kaye, LG Goulty, MB Lawton. Suppression of gray snow mold on creeping bentgrass by an isolate of *Typhula phacorriza*. Plant Dis 71:97–100, 1987.
13. S Buysens, K Heungens, J Poppe, M Hofte. Involvement of pyochelin and pyroverdin in suppression of Pythium-induced damping-off of tomato by *Pseudomonas aeruginosa* 7NSK2. Appl Environ Microbiol 62:865–871, 1996.

14. W Chen, HAJ Hoitink. Interactions between thermophilic fungi and *Trichoderma hamatum* in suppression of Rhizoctonia damping-off in a bark compost-amended with container medium. Phytopathology 78:836–840, 1990.

15. W Chen, HAJ Hoitink, AF Schmitthenner. Factors affecting suppression of pythium damping-off in container media amended with composts. Phytopathology 77:755–760, 1987.

16. W Chen, HAJ Hoitink, LV Madden. Microbial activity and biomass in container media predicting suppressiveness to damping-off caused by *Pythium ultimum*. Phytopathology 78:1447–1450, 1988.

17. W Chen, HA, Hoitink, AF Schmitthenner, O Tuovinen. The role of microbial activity in suppression of damping-off caused by *Pythium ultimum*. Phytopathology 78: 314–322, 1988.

18. BB Clarke, JF White Jr., CR Funk Jr., S Sun, DR Huff, RH Hurley. Enhanced resistance to dollar spot in endophyte-infected fine fescues. Plant Dis. In press.

19. K Clay. Fungal endophytes of grasses: a defensive mutualism between plants and fungi. Ecology 69:10–16, 1988.

20. K Clay, A Leuchtmann. Infection of woodland grasses by fungal endophytes Mycologia 81:805–811, 1989.

21. CM Craft, EB Nelson. Microbial properties of composts that suppress damping-off and root rot of creeping bentgrass caused by *Pythium graminicola*. Appl Environ Microbiol 62:1550–1557, 1996.

22. RJ Cook, AD Rovira. The role of bacteria in the biological control of *Gaeumannomyces graminis* by suppressive soils. Soil Biol Biochem 8:269–273, 1976.

23. RN Cook, RE Engel, S Bachelder. A study on the effect of nitrogen carriers on turfgrass disease. Plant Dis Rep 48:254–255, 1964.

24. JE Crosse. Interactions between saprophytic and pathogenic bacteria in plant disease. In: Ecology of Leaf Surface Microorganisms. London: Academic Press, 1971, pp. 283–290.

25. GE Daft, HA Poole, HAJ Hoitink. Composted hardwood bark: a substitute for stem sterilization and fungicide drenches for control of poinsettia crown and root rot. Hort Sci 14:185–187, 1979.

26. JW Deacon. Factors affecting occurrence of the Ophiobolus patch disease of turf and its control by *Phialophora radicicola*. Plant Pathol 22:149–155, 1973.

27. G De Meyer, and M Hofte. Salicylic acid produced by the rhizobacterium *Pseudomonas aeruginosa* 7NSK2 induces resistance to leaf infection by *Botrytis cinerea* on bean. Phytopathology 87:588–593, 1997.

28. PH Dernoeden. Creeping Bentgrass Management: Summer Stress, Weeds, and Selected Maladies. Chelsea, MI: Ann Arbor Press, 2000, p. 133.

29. Y Elad, R Baker. The role of competition for iron and carbon is suppression of chlamydospore germination of *Fusarium* spp. by *Pseudomonas* spp. Phytopathology 75:1053–1059, 1985.

30. AK Fraser. Growth restrictions of pathogenic fungi on the leaf surface. In: Ecology of Leaf Surface Microorganisms. London: Academic Press, 1971, pp. 529–535.

31. T Gamo. *Azospirillum* spp. from crop roots: a promoter of plant growth. Jpn Agric Res Qty 24:253–259, 1991.

32. LJ Giesler, GY Yuen, ML Craig. Evaluation of fungal antagonists against *Rhizoctonia solani* in tall fescue. Biol Cult Tests Cont Plant Dis 8:123, 1993.

33. DM Goodman, LL Burpee. Biological control of dollar spot disease on creeping bentgrass. Phytopathology 81:1438–1446, 1991.
34. ME Grebus, ME Watson, HAJ Hoitink. Biological, chemical and physical properties of composted yard trimmings as indicators of maturity and plant disease suppression. Compost Sci Utiliz 2:57–71, 1994.
35. ME Grebus, ME Watson, HAJ Hoitink, PR Harder, J Troll. Antagonism of *Trichoderma* spp. to sclerotia of *Typhula incarnata*. Plant Dis Rep 57:924–926, 1995.
36. ME Grebus, JW Rimelspach, HAJ Hoitink. Control of dollar spot of turf with biocontrol agent-fortified compost topdressings. Phytopathology 85:1166, 1995.
37. Y Hadar, R Mandelbaum. Suppression of *Pythium aphanidernatum* damping-off in container media containing composted liquorice roots. Crop Prot 5:88–92, 1986.
38. Y Hadar, GE Harman, AG Taylor, JM Norton. Effects of pregermination of pea and cucumber seeds and of seed treatment with *Enterobacter cloacae* on rots caused by *Pythium* spp. Phytopathology 73:1322–1325, 1983.
39. PR Harder, J Troll. Antagonism of *Trichoderma* spp. to sclerotia of *Typhula incarnata*. Plant Dis Rep 57:924–926, 1973.
40. GE Harman, C-T Lo. The first registered biological control product for turf disease: Bio Trek 22G. Turfgrass Trends 5(5):8–14, 1996.
41. GE Harmon, Y Hadar. Biological control of *Pythium* species. Seed Sci Technol 11:893–906, 1983.
42. RA Haygood, AR Mazur. Evaluation of *Gliocladium virens* as biocontrol agent for dollar spot on bermudagrass. Phytopathology 80:435, 1990.
43. RA Haygood, JF Walter. Biological control of brown patch of centipedegrass with *Gliocladium virens* in a growth chamber. Biol Cult Tests Cont Plant Dis 5:86, 1991.
44. NS Hill, DP Belesky, WC Stringer. Competitiveness of tall fescue (*Festuca arundinacea* Schreb.) as influenced by endophyte (*Acremonium coenophialum* Morgan-Jones and Gams). Crop Sci 31:185–190, 1991.
45. CF Hodges, DA Campbell, N Christians. Potential biocontrol of *Sclerotinia homoeocarpa* and *Bipolaris sorokiniana* on the phylloplane of *Poa pratensis* with strains of *Pseudomonas* spp. Plant Pathol 43:500–506, 1994.
46. M Hofte, S Buysens, N Koedam, P Cornelis. Zinc affects siderophore-mediated high affinity iron uptake systems in the rhizosphere *Pseudomonas aeruginosa* 7NSK2. Biometals 6:85–91, 1993.
47. M Hofte, KY Seong, E Jurkrvitch, W Verstraete. Pyoverdin production by plant growth beneficial *Pseudomonas* strain 7NSK2: Ecological significance in soil. Plant Soil 130:249–258, 1991.
48. HAJ Hoitink. Marketing compost as a pest control product. BioCycle May:65–67, 1995.
49. HAJ Hoitink, PC Fahy. Basis for the control of soil borne plant pathogens with composts. Ann Rev Phytopathol 24:93–114, 1986.
50. HAJ Hoitink, HM Keener, CR Krause. Key steps to successful composting. BioCycle 34(8):30–33, 1993.
51. B Horvath, J Vargas JR., Biological control: it's a number game. Golf Course Manage 68:55–58, 2000.
52. Y Inbar, Y Chen, Y Hadar, HAJ Hoitink. New approaches to compost maturity. Biocycle 31:64, 1990.

53. BJ Johnson. Biological control of annual bluegrass with *Xanthomonas campestris* pv. Poannua in Bermudagrass. Hort Sci 29:659–662, 1994.
54. DY Kobayashi, NEH El-Barrad. Selection of bacterial antagonists using enrichment cultures for the control of summer patch disease in Kentucky bluegrass. Microbiology 32:106–110, 1996.
55. DY Kobayashi, M Guglielmoni, BB Clarke. Isolation of the chitinolytic bacteria *Xanthomonas maltophilia* and *Serratia marcescens* as biological control agents for summer patch disease of turfgrass. Soil Biol Biochem 27:1479–1487, 1995.
56. GA Kuter, HAJ Hoitink, W Chen. Effects of municipal sludge compost curing time on suppression of *Pythium* and *Rhizoctonia* diseases of ornamental plants. Plant Dis 72:751–756, 1988.
57. OCH Kwok, PC Fahy, HAJ Hoitink, GA Kuter. Interactions between bacteria and *Trichoderma hamatum* in suppression of Rhizoctonia damping-off in bark compost media. Phytopathology 77:1206–1212, 1987.
58. MB Lawton, LL Burpee. Effects of rate and frequency of application of *Typhula phacorrhiza* on biological control of Typhula blight of creeping bentgrass. Phytopathology 80:70–73, 1990.
59. MB Lawton, LL Burpee, LG Goulty. Influence of several factors on the biocontrol of gray snow mold of turfgrass by *Typhula phacorrhiza*. Phytopathology 77:119, 1987.
60. C Leben, GC Daft. Influence of an epiphytic bacterium on cucumber anthracnose, early blight of tomato and northern leaf blight of corn. Phytopathology 55:760–762, 1965.
61. L Liu, T Hsiang, K Carey, JL Eggens. Microbial populations and suppression of dollar spot disease in creeping bent grass with inorganic and organic amendments. Plant Dis 29:144–147, 1995.
62. C-T Lo, EB Nelson, GE Harman. Biological control of turfgrass diseases with a rhizosphere competent strain of *Trichoderma harzianum*. Plant Dis 80:736–741, 1996.
63. C-T Lo, EB Nelson, GE Harman. Improved the biocontrol efficacy of *Trichoderma harzianum* for controlling foliar phases of turf diseases by spray applications. Plant Dis 81:1132–1138, 1997.
64. P Lucas, A Sarniguet. Screening in the greenhouse of some treatments to control take-all patch on turfgrass. Phytopathology 81:1198, 1991.
65. P Lucas, A Sarniguet, S Lelarge. Bacterial populations related to the progress of the disease in take-all patches on turf grass. In: ABM Beemster, GJ Bollen, M Gerlagh, MA Ruissen, B Schippers, A Tempel, eds. Biotic Interactions and Soilborne Diseases. Amsterdam: Elsevier Publishing Company, 1991, pp. 264–270.
66. CF Mancino, M Barakat, A Maricic. Soil and thatch microbial populations in 80% sand: 20% peat creeping bentgrass putting green. Hort Sci 28:189–191, 1993.
67. R Mitchell, M Alexander. The mycolytic phenomenon and biological control of *Fusarium* in soil. Nature 190:109–110, 1961.
68. AM Moustafa, HAJ Hoitink. Suppression of Pythium damping-off of tomato in hardwood bark compost. Proc Ann Phytopathol Soc 4:173, 1977.
69. EB Nelson. Biological control of Pythium seed rot and pre-emergence damping-off of cotton with *Enterobacter cloacae* and *Erwinia herbicola* applied as seed treatments. Plant Dis 75:140–142, 1988.

70. EB Nelson. Exudate molecules initiating fungal responses to seeds and roots. Plant Soil 129:61–73, 1990.
71. EB Nelson. Biological control of diseases on golf course turf. USGA Green Section Record Mar/Apr 11–14, 1992.
72. EB Nelson. Enhancing turfgrass disease control with organic amendments. Turfgrass Trends 5(6):1–15, 1996.
73. EB Nelson. Biological control of turfgrass diseases. Golf Course Manage 65(7): 60–69, 1997.
74. EB Nelson, HAJ Hoitink. Factors affecting the suppression of *Rhizoctonia solani* in container media. Phytopathology 72:275–279, 1982.
75. EB Nelson, HAJ Hoitink. The role of microorganisms in the suppression of *Rhizoctonia solani* in container media amended with composted hardwood bark. Phytopathology 73:274–278, 1983.
76. EB Nelson, CM Craft. Introduction and establishment of strains of *Enterobacter cloacae* in golf course turf for the biological control of dollar spot. Plant Dis 75: 510–514, 1991.
77. EB Nelson, CM Craft. Suppression of Pythium root rot with top-dressings amended with composts and organic fertilizers. Biol Cult Tests Plant Dis 7:104, 1991.
78. EB Nelson, CM Craft. Suppression of dollar spot with top-dressings amended with composts and organic fertilizers. Biol Cult Tests Cont Plant Dis 6:93, 1991.
79. EB Nelson, CM Craft. A miniaturized and rapid bioassay for the selection of soil bacteria suppressive to Pythium blight of turfgrasses. Phytopathology 82:206–210, 1992.
80. EB Nelson, CM Craft. Suppression of dollar spot on creeping bentgrass and annual bluegrass turf with compost-amended topdressings. Plant Dis 76:954–958, 1992.
81. EB Nelson, AP Maloney. Molecular approaches for understanding biological control mechanisms in bacteria: Studies of the interaction of *Enterobacter cloacae* with *Pythium ultimum*. Can J Plant Pathol 14:106–114, 1992.
82. EB Nelson, GA Kuter, HAJ Hoitink. Effects if fungal antagonists and compost age on suppression of Rhizoctonia damping-off in container media amended with composted hardwood bark. Phytopathology 72:275–279, 1983.
83. EB Nelson, LL Burpee, MB Lawton. Biological Control of Turfgrass Diseases. In: A Leslie, ed. Integrated Pest Management of Turfgrass (and Ornamentals). Chelsea, MI: Lewis Publishers Inc., 1994, pp. 409–427.
84. EB Nelson, WL Chao, JM Norton, GT Nash, GE Harman. Attachment of *Enterobacter cloacae* to hyphae of *Pythium ultimum*: possible role in the biological control of Pythium pre-emergence damping off. Phytopathology 76:327–335, 1986.
85. Y Okon, CA Labandera-Gonzalez. Agronomic applications of *Azospirillum*. An evaluation of 20 years worldwide field inoculation. Soil Boil Biochem 26:1551–1601, 1994.
86. AL O'Leary, DJ O'Leary, SH Woodhead. Screening potential bioantagonists against pathogens of turf. Phytopathology 78:1593, 1988.
87. NR O'Neill. Plant pathogenic fungi in soil/compost mixtures. Research for small farms. U.S. Dep. Agric Res Serv Misc Publ 1422, 1982, pp. 285–287.
88. CH Peacock, PF Daniel. A comparison of turfgrass response to biologically amended fertilizers. Hort Sci 27:883–884, 1992.

89. CG Phae, M Shoda. Expression of the suppressive effect of *Bacillus subtilis* on phytopathogens in inoculated composts. J Ferment Bioeng 70:409–414, 1990.

90. CG Phae, M Sasaki, M Shoda, H Kubota. Characteristics of *Bacillus subtilis* isolated from composts suppressing phytopathogenic microorganisms. Soil Sci Plant Nutr 36(4):575–586, 1990.

91. JF Powell, JM Vargas, MG Nair. Management of dollar spot on creeping bentgrass with metabolites of *Pseudomonas aureofaciens* (TX-1). Plant Dis 84:19–24, 2000.

92. ZK Punja, RG Grogan, T Unruh. Comparative control of *Sclerotium rolfsii* on golf greens in northern California with fungicides, inorganic salts, and *Trichoderma* spp. Plant Dis 66:1125–1128, 1982.

93. C Rasmussen-Dykes, WM Brown Jr. Integrated control of Pythium blight on turf using metalaxyl and *Trichoderma hamatum*. Phytopathology 72:976, 1982.

94. HM Reuter, GL Schumann, ML Matheny, RT Hatch. Suppression of dollar spot (*Sclerotinia homoeocarpa*) and brown patch (*Rhizoctonia solani*) on creeping bentgrass by an isolate of *Streptomyces*. Phytopathology 81:124, 1991.

95. MD Richardson, JF White Jr., FC Belanger. The use of endophytes to improve turfgrass performance. In: M Sticklen, M Kenna, eds. Turfgrass Biotechnology. Ann Arbor, MI: Ann Arbor Press, 1998, pp. 97–107.

96. DP Roberts, CJ Sheets, JS Hartung. Evidence for proliferation of *Enterobacter cloacae* on carbohydrates in cucumber and pea spermosphere. Can J Microbiol 38: 1128–1134, 1992.

97. DP Roberts, NM Short Jr, AP Maloney, EB Nelson, DA Schaff. Role of colonization in biocontrol: studies with *Enterobacter cloacae*. Plant Sci 101:83–89, 1994.

98. F Rodriguez, WF Pfender. Antibiosis and antagonism of *Sclerotinia homeocarpa* and *Drechslera poae* by *Pseudomonas fluorescens* Pf-5 in vitro and in planta. Phytopathology 87:614–621, 1997.

99. AD Rovira, GB Wildermuth. The nature and mechanism of suppression. In: MJC Asher, PJ Shipton, eds. Biology and Control of Take-all. New York: Academic Press, 1981, pp. 385–414.

100. DD Rowan, DL Gaynor. Isolation of feeding deterrents against Argentine stem weevil from ryegrass infected with the fungal endophyte *Acremonium loliae*. J Chem Ecol 12:647–658, 1986.

101. A Sarniguet, P Lucas. Evaluation of populations of flourescent pseudomonads related to decline of take-all on turfgrass. Plant Soil 145:11–15, 1992.

102. GL Schumann, HM Reuter. Suppression of dollar spot with wheat bran topdressings. Biol Cult Tests Cont Plant Dis 7:113, 1993.

103. L Segall. Marketing compost as a pest control product. BioCycle, May:65–67, 1995.

104. MR Siegel, GCM Latch, MC Johnson. Fungal endophytes of grasses. Ann Rev Phytopathol 25:293–315, 1987.

105. MR Siegel, GCM Latch, LP Bush, FF Fannia, DD Rowan, BA Tapper, CW Bacon, MC Johnson. Fungal endophyte-infected grasses: alkaloid accumulation and aphid response. J Chem Ecol 16:3301–3315, 1990.

106. RW Smiley. Antagonists of *Gaeumannomyces graminis* from rhizoplane of wheat in soils fertilized with ammonium or nitrate nitrogen. Soil Biol Biochem 10:169–174, 1978.

107. RW Smiley. Wheat-rhizoplane pseudomonads as antagonists of *Gaeumannomyces graminis*. Soil Biol Biochem 11:371–376, 1979.

108. B Sneh, M Dupler, Y Elad, R Baker. Chlamydospore germination of *Fusarium oxysporum* f. sp. *cucumerinum* as affected by fluorescent and lytic bacteria from Fusarium suppressive soil. Phytopathology 74:1115–1124, 1984.

109. CT Stephens, LJ Herr, HAJ Hoitink, AF Schimitthenner. Control of Rhizoctonia damping-off by the use of composted hardwood bark. Plant Dis 65:796–797, 1981.

110. M Sticklen, M Kenna. Turfgrass Biotechnology. Ann Arbor, MI: Ann Arbor Press, 1998, p. 256.

111. CT Stockwell, EB Nelson, CM Craft. Biological control of *Pythium graminicola* and other soil borne pathogens of turfgrass with actinomycetes from composts. Phytopathology 84:1113, 1994.

112. EM Sutker, LT Lucas. Biocontrol of *Rhizoctonia solani* in tall fescue turfgrass. Phytopathology 77:1721, 1987.

113. DC Thompson, BB Clarke, DY Kobayashi. Evaluation of bacterial antagonists for reduction of summer patch symptoms in Kentucky bluegrass. Plant Dis 80:856–862, 1996.

114. SS Vaiciunas, PR Majumdar, LP Tredway, BB Clarke. Control of dollar spot with chemical and biological fungicides. Fungic Nematicide Tests 54:502, 1999.

115. K van Dijk. Seed exudate stimulant inactivation by *Enterobacter cloacae* and its involvement in the biological control of *Pythium ultimum*. M.S. thesis, Cornell University, Ithaca, NY, 1994.

116. JM Vargas. Biological control: a work in progress. Golf Course Manage 68:55–58, 1999.

117. G Viji, W Uddin, CP Romaine. Evaluation of bacterial antagonists from spent mushroom substrate for suppression of turfgrass diseases. Phytopathology 90:S81, 2000.

118. G Viji, W Uddin, CP Romaine. Efficacy and timing of application of *Pseudomonas aeruginosa* for control of ryegrass blast. Phytopathology 90:S80, 2000.

119. MG Villani, What's new in turfgrass insect pest management products: focus on biological controls. Turfgrass Trends 4:1–6, 1995.

120. DM Weller, RJ Cook. Suppression of take-all of wheat by seed treatments with fluorescent pseudomonads. Phytopathology 73:463–469, 1983.

121. DM Weller, WJ Howie, RJ Cook. Relationships between in vitro inhibition of *Gaeumannomyces graminis* var *triticii* and suppression of take-all of wheat by fluorescent pseudomonads. Phytopathology 78:1094–1100, 1988.

122. JF White Jr. Endophyte-host association in forage grasses. XI. A proposal concerning origin and evolution. Mycologia 80:442–446, 1988.

123. PTW Wong, R Baker. Control of wheat take-all and Ophiobolus patch of Agrostis turfgrass by fluorescent pseudomonads from a Fusarium-suppressive soil. Phytopathology 71:1008, 1981.

124. PTW Wong, R Baker. Suppression of wheat take-all and Ophiobolus patch by fluorescent pseudomonads from a Fusarium-suppressive soil. Soil Biol Biochem 16:397–403, 1984.

125. PTW Wong, R Baker. Control of wheat take-all and Ophiobolus patch of turfgrass

by fluorescent pseudomonads. In: CA Parker, AD Rovira, KJ Moore, PTW Wong, and JF Kollmorgen, eds. Ecology and Management of Soilborne Plant Pathogens, 1985, pp. 151–153.

126. PTW Wong, TR Siviour. Control of Ophiobolus patch in Agrostis turf using a virulent fungi and take-all suppressive soils in pot experiments. Ann Appl Biol 92: 191–197, 1979.

127. PTW Wong, DJ Worrand. Preventive control of take-all patch of bentgrass turf using triazole fungicides and *Gaeumannomyces graminis* var. *graminis* following soil fumigation. Plant Prot Qty 4:70–72, 1989.

128. GY Yuen, ML Craig. Reduction in brown patch severity by a binucleate *Rhizoctonia* antagonist. Biol Cult Tests Cont Plant Dis 7:114, 1992.

129. GY Yuen, ML Craig, LJ Giesler. Biological control of *Rhizoctonia solani* on tall fescue using fungal antagonists. Plant Dis 78:118–123, 1994.

130. Z Zhang, GY Yuen. Biological control of *Bipolaris sorokiniana* on tall fescue by *Stenotrophomonas maltophilia* strain C3. Phytopathology 89:817–822, 1999.

131. T Zhou, JC Neal. Annual bluegrass (Poa annua) control with *Xanthomonas campestris* pv poa annua in New York State. Weed Technol 9:173–177, 1995.

15

Implementation of Biological Control of Plant Diseases in Integrated Pest-Management Systems

X. B. Yang
Iowa State University, Ames, Iowa

Luis del Rio
North Dakota State University, Fargo, North Dakota

The reduction of fungicides available for use in agriculture in the United States [1] is a point of concern for producers. The primary concern about the use of fungicides is the production of carcinogenic residues in food. As a result, the use of biological control and other ecologically based pest-management practices has become one of the driving forces in the recent development of pest sciences. Despite the amount of resources and time invested in this area, biological control is still far from our early expectation that it would play a major role in integrated pest management (IPM). In this chapter we discuss the role of biological control in IPM and the theoretical and practical considerations of implementation of biological control in IPM.

I. IPM AND BIOLOGICAL CONTROL

IPM is the use of a combined set of strategies and practices aimed to keep pests at levels under the economic damage threshold. The term IPM, coined by entomologists in the mid-1960s, was developed in order to reduce the risks associated with excessive use of pesticides and to preserve the ecological balance [2]. As pointed out by Jacobson [3], this approach presumed that pests, like insects, were easy to quantify and detect—two factors that did not apply to most of the known plant pathogens. From the view of systems science, IPM is a system dealing with hosts, pathogens, environments, and socioeconomic components. Biological control is one element of the system.

Many definitions have been proposed to describe "biological control" in plant pathology (see Chapter 1). All of them imply an interaction between microorganisms and the host.

II. FACTORS LIMITING BIOLOGICAL CONTROL AS SOLO MEASURES IN PRODUCTION

The factors that limit development and use of biological control measures as major or solo measures for the control of plant diseases at a commercial level can be divided into two groups: biological and ecological aspects of control and aspects associated with economical and social interests.

A. Biological Considerations

Many potentially good biocontrol agents cannot be moved from the experimental phase to a commercialization phase due to impractical dosage recommendations, limited or inconsistent control efficacy, inadequate delivery systems, and incompatibility with current production methods. A better understanding of the ecological and epidemiological relationship between microorganisms and new delivery systems that will carry fungal strains with enhanced fungicide resistance will help close the gap between experimental results and commercial use of biopesticides.

An example of how combining knowledge of the biology and the ecology of a parasite and its host can help solve the problem of unrealistic dosages was presented by Adams and Fravel [6]. *Sporidesmium sclerotivorum* is a parasite that attacks the sclerotia of many species of the genus *Sclerotinia*. In a previous study, Adams and Ayers [7] had demonstrated the great parasitic ability of *S. sclerotivorum* and its potential as a biocontrol agent of *Sclerotinia minor*, the causal agent of lettuce drop. They achieved their goal by spraying spores of *S. sclerotivorum* on the soil surface and then incorporating them at a rate of 2300 kg/ha. Adams and Fravel [6] considered the natural distribution of sclerotia in the soil and the fact that most of the newly formed sclerotia were clustered in plant residues and decided to apply the spore suspension directly on infected crop residues before their incorporation in the soil. By doing so, the probability of *S. sclerotivorum* contacting its sclerotial host increased dramatically. Economical control levels were achieved with as little as 2 kg of inoculum per hectare. The same principle was successfully applied by del Rio [8] to control Sclerotinia stem rot of soybean. It was demonstrated that "timing and placement of the biocontrol agent are often more important to the success of control than the population size of the biocontrol agent'' [9].

Specificity in the relationship between antagonists contributes to the stability of the ecosystem. However, too much specificity can inhibit overall protection of plants in other environments. Papaya ringspot virus (PRSV), considered a major pathogen in papaya-producing areas, can be successfully controlled by cross protection using the mild strain PRSV HA 5-1 in Hawaii [10]. However, plants protected with the Hawaiian strain failed to protect against 11 other serologically related strains of the virus from diverse geographical regions [11].

New delivery methods and fungicide-resistant biocontrol agents are being developed. Longer shelf life is required to make biopesticides attractive to farmers or to industry. Germination of conidia of *Coniothyrium minitans*, formulated as Contans (Prophyta), remains over 70% for up to one year when stored at 5°C. However, if no refrigeration is provided, germination will sharply decline after 4 months. Under the same circumstances fungicides have a shelf life of a few years. The combination of peat with a previous formulation of *Trichoderma harzianum*–based wheat bran [12] has extended the shelf life of this mycoparasite to one year at 23°C [13]. In the same way, delivery of *Trichoderma koningii* as a seed coat on pea resulted in improved activity against *Pythium ultimum* compared to soil incorporation of the parasite [14].

Although the use of fungicide-resistant strains as biocontrol agents is now standard [15], development of such strains is not always an easy task and may require the use of mutational agents [16,17] or genetic engineering [18,19]. It is not surprising, then, that more than two thirds of the biopesticides available worldwide contain bacteria or yeasts as active ingredients [20].

B. Theoretical Consideration for Implementation Success

The chance of the success of implementing a biological control measure in IPM is theoretically related to the establishment and growth of the population of a biological control agent. Of the two mechanisms utilized by biological control agents, competing with resources for growth [21] and parasitism [22], the probabilities of success with implementation are determined by different factors. For the first case, the ability to compete with pathogens, such as *Rhizoctonia* and *Fusarium*, growth is important, and the chance of success is determined by the available resources of a system. For a pathosystem with limited resources, the more competitive the biological control agents are, the more likely they will succeed, as indicated by a low infection level. When resources for growth are unlimited, the use of competitive-type agents in such a system would likely be unsuccessful. The relationship between available resources and competition is illustrated in Figure 1. The amount of infection in a competitive environment is a function of the amount of available resources. If there is no competition for

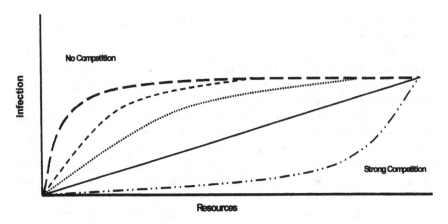

Figure 1 Theoretical relationship between level of infections and amount of resource for growth of a plant pathogen and a biological control agent in a competitive environment.

resources from biological control agents, the pathogen population builds up quickly and, consequently, the infection level increases exponentially. When the competition for resources is strong and the pathogen is not a good competitor with the control agent, the pathogen population will have a reduced growth rate and, consequently, a relative lower inoculum density. Correspondingly, the level of infection increases slowly with the availability of resource, indicating a higher chance of success.

The resource-infection relationship can be further used to determine the relationship between probability of successful implementation and competitiveness of biological control agents. In a pathosystem with limited resources available for population growth, the probability of successfully implementing a biological control agent to reduce disease infection is an exponential relationship. The greater the competitiveness, the higher the disease reduction is likely to be. When the competitiveness reaches a level where the pathogen population can no longer compete for resources to grow, maximum control will be achieved.

C. Social, Practical, and Economic Aspects

A combination of economics, lack of appropriate protocols, and complex government regulations makes the registration of a biopesticide a difficult process. After the discovery of a new potential candidate for biological control, exhaustive ecological and biological studies are conducted. Studies are first conducted under controlled environments, and later in small experimental fields. After several years and hundreds of thousands of dollars, some excellent candidates may be ready for larger-scale field trials; however, most of these will not be scaled-up. On the researcher's side, the lack of funding and the generalized idea that scale-up trials are not their responsibility partly explain this problem [23]. On the industry side, the small market potential typical of biocontrol agents, due to the specificity of its antagonism, present an inconvenience that may be impossible to overcome [24]. It has been argued that scientists usually do not know about the registration process and of the agencies involved [25]. The Environmental Protection Agency (EPA) regulates the registration process for all pesticides intended for use in agriculture in the United States. Under the Federal Insecticide, Fungicide, and Rodenticide Act (FIFRA) regulated by the EPA, biologically based pesticides are considered in the same category as chemical pesticides and, therefore, must submit results of extensive toxicological tests as a requisite for their registration. Tests are conducted in specialized laboratories and may cost up to $200,000. Each agent and each intended use of that agent is subject to a separate registration. If toxicological tests designed to evaluate chemical pesticides are also used for biocontrol agents, the practice cannot be justified, considering the specificity of most biocontrol agents and the small niche in which they will be used as commercial products [30]. A lengthy registration period can discourage small companies [31], while late release of a biopesticide due to excessive regulatory procedures could result in loss of the marketing window for those that persevere [32].

Some of the problems identified in this section can be solved by a cooperative effort between the partners involved. Efforts towards a more friendly registration procedure are already being developed by the EPA. However, more decisive support from commodity groups and funding agencies should be provided for long-term projects working to scale up potential biopesticides.

1. Production Systems

Implementation of biological control practices in an IPM system varies from crop to crop. The practice depends to a great extent on the intensity of the production system in which the control is applied, the feasibility of its implementation, the availability of alternate practices, and the value of the crop to be protected. Cropping systems that rely heavily on pesticide use, like vegetables and ornamentals, will experience more problems with fungicide-resistant strains of plant pathogens and increasing disease control costs. The adoption of biological alternatives to chemical control in these systems will be greatly influenced by how effectively they compare to the chemical option, including degree of control and cost, and by how well they fit into production systems [4]. Less intense production systems, like those for corn, soybeans, and small grains, also have a smaller margin of profitability. Therefore, farmers in these domains may be more interested in longer-term solutions to their pest problems than in immediate but certainly more expensive and temporary ones. A sample of measures more easily adopted by producers in this domain includes crop rotation, monoculture, and use of green manure [5]. Perennial crops offer the greatest chance for development of IPM practices because of their more stable and undisrupted ecosystems.

Crops can be a significant limiting factor to the sustainability of the population of a biological control agent. To maintain the population of a control agent at equilibrium, the agroecosystem must have conditions favorable for agent survival and growth. The survival rate is less important than the growth rate for a perennial crop, than in a rotational cropping system. In the latter system, focusing on the winter period, the equilibrium is a function of overwintering capability, or winter survival. For a perennial crop, which shares a greater similarity with undisturbed ecosystems, the biological control agent would be able to maintain a population similar to that in a natural ecosystem. Implementation of a biological control in such a system has greater chance of success, as we illustrate with examples below. The yearly occurrence of a disease and the lagging effects of biological control on the disease are summarized in Figure 2. The more favorable the winter conditions to a control agent are, the higher the survival rate and greater the sustainability of its population may be.

An opposite case is exemplified by annual rotation cropping systems that have multiple crops. Such systems are highly perturbed/managed by farming practices. Uncertainty arises as to the sustainability of the population of biological control agents when rotation takes place. In a rotation system, the population of

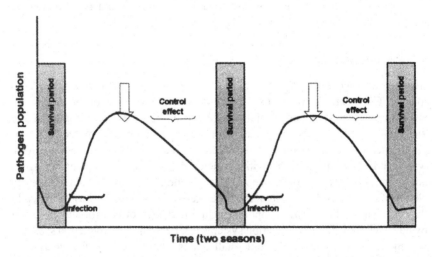

Figure 2 Multiseasonal dynamics of pathogen populations and timing and lagging effects of biological control agents on the pathogen population.

plant pathogens may build up in a favorable year, to be followed by an increase in the biological control agent. Subsequent changes in crops in the system will be followed by a rapid decline of the pathogen population and associated biological control agent. Due to the time lag effects, the biological control population will always be behind its host population and, therefore, will barely sustain its population above threshold levels over years. However, the population of a biological control agent that competes with the pathogen for resources may not be reduced by rotation as much as a parasitical agent.

2. The IPM Threshold and Implementing Biological Control

The IPM threshold is an important dimension in determining the success of implementing a biological control as a major measure in a production system. When a biological control is taken as a major measure, it may be effective in suppressing the development of the pathogen population from a view of population dynamics. However, the level of suppression may not be practically acceptable because it may not be below an economic threshold. This is especially true for high-value horticultural crops that have low tolerant threshold values for disease damage. In high-value crops, few successful examples of biological control measures have been demonstrated as major means in managing preharvest diseases. The lagging growth of the biological control agent compared to its pathogen population predetermines that the pathogen population will always reach a threshold level ahead of the control agent.

For example, *Sclerotinia sclerotiorum*, which has a wide host range, attacking both low- and high-value crops, can be reduced by *Sporidesmium sclerotivorum*. The fungus produces large amounts of conidial spores that infect the sclerotia of *S. sclerotiorum*. As a result, the infected sclerotia lose their ability to germinate and produce apothecia. This biological control agent was tested in California for controlling lettuce drop caused by *S. sclerotiorum* with limited success because minor infections can destroy the whole value of a lettuce head. While used for controlling Sclerotinia stem rot of soybean, a low-value crop, this control agent appears to have greater potential for large-scale application [8]. It is now being studied for the control of Sclerotinia rot of other low-value crops, such as of sunflower (C. Martinson, personal communication).

III. COMMERCIAL EXAMPLES OF BIOCONTROL PRACTICES AS SOLO MEASURES

A few diseases are currently controlled on a commercial level by a single biological control practice or agent. These include crown gall in grapes and other fruit trees, citrus tristeza on citrus, papaya ringspot on papaya, and chestnut blight. In the first three cases, the practice does not interfere with regular cropping practices. In the latter, a specific control measure was added to normal operational practices.

Biological control of crown gall is, without doubt, the best known and most successful example of a commercially implemented biological control measure. Control of *Agrobacterium tumefaciens* by its relative *Agrobacterium radiobacter* strain K84 [33] is achieved by immersing the roots of seedlings into a bacterial suspension before transplanting them in the fields. *A. radiobacter* actively colonizes the roots of the treated seedling, preventing infection by *A. tumefasciens*. *A. radiobacter* produces a bacteriocin called agrocin, which is thought to be responsible for the antagonistic affect [34]. K84 does not control all species of *Agrobacterium* that produce galls in plants, but new strains are being studied that may confer greater protection [35,36]. Commercial formulations of *A. radiobacter* include Galltroll-A, Nogall, Diegall, and Norbac 84C [20].

Citrus tristeza is caused by a closterovirus [37] that is transmitted by aphids. Since its arrival in Brazil, citrus tristeza virus has posed a threat to the citrus industry by killing millions of trees in less than 20 years. In 1961, a program to use a cross-protection virus was initiated [38]. After exhaustive screening, six mild strains of the virus were identified as good candidates for the program. Artificial inoculation of these strains produced mild symptoms on plants and reduced yield, but also protected the plants from a more severe strain. It is thought that the mild strain interferes with reproduction of the more severe one [39]. Currently, citrus plantations in South Africa, Australia, and Brazil are protected by this biocontrol method [40]. The same principle has been successfully applied to protect papaya plantations in Hawaii against papaya ringspot virus [10,41] for several years.

Chestnut blight is a devastating disease caused by the fungus *Cryphonectria parasitica*. Wind- and insectborne mycelium or ascospores are responsible for dissemination of this pathogen in the field [42]. Mixtures of hypovirulent strains are packed in tubes and inserted in holes around an active canker [42]. Canker healing occurs within the next few years. Although hypovirulence has worked very well in Europe, efforts to use it in the United States have failed. Among probable explanations for this are the more susceptible nature of the chestnut species predominant in America, a higher variability of the pathogen population, which makes the identification of compatible strains more difficult, and the differences between naturally occurring dsRNA from Europe and America [42].

Another example of a commercial biocontrol agent used as an alternative method for disease management is Kodiak. Kodiak (*Bacillus subtilis* strain GB03) is used for the prevention of Rhizoctonia root rot in peanuts and cotton and is applied as a seed treatment. Currently more than 60% of the cotton crop planted in the United States is treated with Kodiak [43,50]. Root colonization by *B. subtilis* GB03 reduces disease incidence and contributes to a healthier root development. *B. subtilis* is applied to the seeds along with fungicides [44,45].

IV. BIOLOGICAL CONTROL AS AN IMPLEMENT OF IPM

The IPM philosophy embraces sustainability and reduced environmental impact of agricultural production systems. Biological control in many ways is at the center of this philosophy. The pathogen-specific "silver bullet" approach [25] should give way to a more holistic approach, where plant health is preserved by a combination of cultural practices that enhance microbial diversity. These cultural practices can include the use of "silver bullets" if necessary. Some of the disadvantages of biological control agents mentioned earlier in this chapter, like complete control levels and specificity of control, can be overcome by the comnation of several other methods. The integration of different disease-control methods will prevent or reduce the possibility of disease outbreaks.

The dynamic nature of a pathosystem also predetermines the implementation of biological control measures in an IPM program. For some systems, a built-in control measure is effective and sustainable, and for other systems short-term inputs may be more effective. In many pathosystems, the frequency of destructive epidemics is low. Systems with a low frequency of pest outbreaks are economically efficient and sustainable if a built-in disease- or pest-management component is the basis for preventative pest control. For instance, in the United States and China, destructive epidemics of wheat rusts occur in approximately 4-year cycles as influenced by El Niño events [26]. From 1949 to 1990 only four major outbreaks occurred [27]. Management of diseases cannot depend on chemical control due to their unpredictable nature and cannot be justified by low crop price. Resistance has been effectively used as a built-in and preventative disease-management component in the pathosystem.

In pathosystems with equilibrium maintained by built-in pest-management components, we can classify those components into two groups: balancers and controllers. The controllers are single factors that have critical and major roles in system dynamics, e.g., an effective disease-resistance gene built in to the host crops. The effects of balancers on equilibrium are collective and secondary, with many balancers interacting through different pathways. Often a group of balancers function collectively to sustain the balance of the system over a longer period of time. Few examples of biological control used for disease management are below this level. However, many biological control agents are naturally effective balancers in agroecosystems. In a system that has a low disease epidemic frequency, implementation of a biological control measure aimed at developing a built-in preventative component is ecologically feasible and practical. However, because the population of a control agent often cannot be established early enough at a threshold level, as discussed previously, biological control measures would be less likely to play a solo major role, but would rather assume a complementary role in balancing the system.

A. Resistant Varieties

This is a built-in component of an IPM system; the best way to control a pest is through the use of resistant or tolerant cultivars. Resistant cultivars not only contribute less pollution to the ecosystem, they do it at a very low cost. Traditional breeding programs have produced a number of cultivars that express different levels of resistance to a number of plant pathogens, but with the aid of molecular techniques, new avenues for production of resistant materials have opened. Identification of resistance genes with the aid of linked molecular markers is currently underway in a number of crops. Transgenic cultivars with pathogen-derived resistance, like the papaya cultivars SunUp and UH Rainbow, which have the gene encoding for production of coat protein from PRSV [10], are being released. New varieties with enhanced ability to foster general or selective microbial colonization of its roots, or phylloplane, will also be available in the future [28]. Non-transgenic cultivars with resistance to several pathogens are also available in a large number of crops. The combined use of selected cultivars and biocontrol agents can provide better disease control than the use of any of them alone [29].

B. Solarization

Solarization is the destruction or decimation of populations of microorganisms in the soil by heat build-up [46]. A synergistic effect can be achieved [47], resulting in higher levels of protection. Chemical fumigation alone can produce the opposite result, allowing for a rapid colonization of treated areas by the pathogen intended to be eradicated [48].

C. Chemicals

Fungicide resistance is now considered a desired characteristic for a biological control agent [15], and therefore incompatibility with fungicides will no longer be a limitation for the adoption of biocontrol agents in the future. At the same time, improved activity has been observed when certain biopesticides are used in combination with paraffin oil [29,49]. The use of bacterial strains in seed treatments provides another avenue to circumvent the fungicide incompatibility issue [50].

D. Cultural Practices

Numerous cultural practices have been described as having a positive impact on the reduction of disease levels in different crops and production systems. The combined use of biopesticides and cultural practices that enhance the overall health of the plants and favor growth of natural enemies will increase the efficacy of the biopesticides [51]. The implementation of these methods in a production system [52] may result in healthier crops and a less polluted environment.

V. SYSTEMS APPROACH FOR IPM IMPLEMENTATION

A systems approach has been used in the quantitative study of all biological sciences, including plant pathology. The approach is a powerful tool for the implementation of biological control in integrated pest management because this approach was used to develop the theoretical basis of IPM framework. The systems approach originated from the study of the dynamics of large, complicated systems. The approach allows an understanding of such systems, which were traditionally incomprehensible quantitatively with conventional approaches [55]. The analytical power of systems science has been greatly enhanced by combining computer science with system simulation. Now this approach is being used to study almost any large system, such as ecosystems and engineering systems. Importantly, the application of different techniques of a systems approach enables us to predict quantitatively the effects of individual components on a system. In a complicated IPM system, this predictive power is very useful in understanding the function of individual biological control measures/agents.

Like any ecosystem, a pathosystem is an open system consisting of numerous interacting components. Epidemiologists have used a systems approach to develop simulation disease models and used them to evaluate disease-control strategies in different IPM systems. A recent example is the nematode epidemiology model by Been et al. [56] and Been and Schomaker [57] for potato nematode management in the Netherlands. In disease management, the function of a resistant host as one component in a pathosystem has been evaluated using this approach. Similarly, other modeling approaches can be applied to study the control efficacy of a biological control agent and how the control agent being one component interacts with many biotic and abiotic components in a complicated system [58,59]. With this approach, we can address the following questions quantitatively: (1) how a biological control agent interacts with the host plant, (2) how the agent interacts with plant pathogens, and (3) how parameters of the agent (competitiveness or dose-response) determine the probability of implementation success. We can also explore the effects of important environmental variables such as temperature, moisture, and soil pH on control efficacy. Such study enables us to predict the future development of and increase the success of implementing a control measure into IPM.

Sustainability of control agents, defined here as the ability to maintain the population level of the control agent, is a key factor in the efficacy of biological controls in IPM implementation. In a natural unperturbed ecosystem, the population equilibrium of a biological control agent is a function of survival and growth [54,55]. The sustainability of the control agent in a pathosystem is a function of the following multifactors: the growth and survival rates of the agent and the environmental conditions in which the specific biological control is implemented. Figures 3 illustrates these relationships with simplified schemes.

To have sustainable control of pathogen populations, maintaining a population of biological control agents around equilibrium over a long term with minimum inputs is desired. Little is known about the long-term dynamic of a pathosystem, especially with regard to biological control. Periodicity of climate affects disease epidemics, consequently influencing the population sustainability of biological control agents in IPM implementation. Disease fluctuations have been attributed to climate cycles, and epidemiological studies have clearly shown a link between climate cyclic patterns and disease outbreak patterns [26]. Long-term climate cycles influence the sustainability of parasitic biological control agents as measured against IPM threshold levels of certain diseases. Long-term records of agricultural systems show the fluctuating nature of plant diseases, an indication that plant pathogen populations are very dynamic as well. Wheat rust in China, for example, has a cycle of 3–4 years, associated with El Niño events [27,53]. Since the population of a control agent builds up in association with the periodic outbreaks of pathogen populations, the cyclical nature of climate may

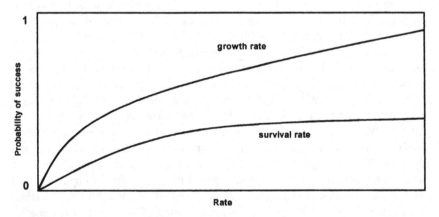

Figure 3 Theoretical relationship between probability of sustainability and biocontrol agent growth rate or survival rate. Growth rate has a more pronounced effect compared with survival rate.

produce a cyclical activity for biological control agents and, consequently, influence the sustainability of biological control when implemented in an IPM system.

ACKNOWLEDGMENT

The authors are grateful to Rick Hartman for assistance in the preparation of this manuscript.

REFERENCES

1. NN Ragsdale, HD Sisler. Social and political implications of managing plant diseases with decreased availability of fungicides in the United States. Annu Rev Phytopathol 32:545–557, 1994.
2. RF Smith, R van den Bosch. Integrated control. In: W Kilgore, R Doutt, eds. Pest Control: Biological, Physical, and Selected Chemical Methods. New York: Academic Press, 1967, pp 295–340.
3. BJ Jacobsen. Role of plant pathology in integrated pest management. Annu Rev Phytopathol 35:373–391, 1997.
4. EB Nelson. Current limits to biological control of fungal phytopathogens. In: DK Arora, B Rai, KG Mukerji, GR Knudsen, eds. Handbook of Applied Mycology. Vol. I: Soil and Plants. New York: Marcel Dekker, 1991, pp 327–355.
5. MN Schroth, JG Hancock. Soil antagonists in IPM systems. In: MA Hoy, DC Herzog, eds. Biological Control in Agricultural IPM Systems. New York: Academic Press, 1985, pp 415–431.
6. PB Adams, DR Fravel. Economical biological control of *Sclerotinia* lettuce drop by *Sporidesmium sclerotivorum*. Phytopathology 80:1120–1124, 1990.
7. PB Adams, WA Ayers. Biological control of Sclerotinia lettuce drop in the field by *Sporidesmium sclerotivorum*. Phytopathology 72:485–488, 1982.
8. LE del Rio. Biological control of Sclerotinia stem rot of soybean with *Sporidesmium sclerotivorum*. PhD dissertation, Iowa State University, Ames, IA, 1999.
9. DR Fravel, CA Engelkes. Biological management. In: CL Campbell, DL Benson, eds. Epidemiology and Management of Root Diseases. Berlin: Springer-Verlag, 1994, pp 292–308.
10. D Gonsalves. Control of papaya ringspot virus in papaya: a case study. Annu Rev Phytopathol 36:415–437, 1998.
11. PF Tennant, C Gonsalves, KS Ling, M Fitch, R Manshardt, JL Slightom, D Gonsalves. Differential protection against Papaya Ringspot Virus isolates in coat protein gene transgenic papaya and classically cross-protected papaya. Phytopathology 84: 1359–1366, 1994.
12. E Hadar, I Chet, Y Henis. Biological control of *Rhizoctonia solani* damping-off with wheat bran culture of *Trichoderma harzianum*. Phytopathology 69:64–68, 1979.

13. A Sivan, Y Elad, I Chet. Biological control of *Pythium aphanidermatum* by a new isolate of *Trichoderma harzianum*. Phytopathology 74:498–501, 1984.
14. EB Nelson, GE Harman, GT Nash. Enhancement of Trichoderma-induced biological control of Pythium seed rot and pre-emergence damping-off of peas. Soil Biol Biochem 20:145–150, 1988.
15. CL Wilson, ME Wisniewski. Biological control of postharvest diseases of fruits and vegetables: an emerging technology. Annu Rev Phytopathol 27:425–441, 1989.
16. GC Papavizas, JA Lewis, TH Abd-El Moity. Evaluation of new biotypes of *Trichoderma harzianum* for tolerance to benomyl and enhanced biocontrol capabilities. Phytopathology 72:126–132, 1982.
17. N Ossanna, S Mischke. Genetic transformation of the biocontrol fungus *Gliocladium virens* to benomyl resistance. Appl Environ Microbiol 56:3052–3056, 1990.
18. CK Hayes. Improvement of *Trichoderma* and *Gliocladium* by genetic manipulation. In: ES Tjamos, ed. Biological Control of Plant Diseases. New York: Plenum Press, 1992, pp 277–286.
19. T Katan, MT Dann, GC Papavizas. Genetics of fungicides resistance in *Talaromyces flavus*. Can J Microbiol 30:1079, 1984.
20. DR Fravel, RP Larkin. Availability and application of biocontrol products. Biol Cult Tests 11:1–7, 1996.
21. RM Atlas, R. Bartha. Microbial Ecology, Fundemantals and Applications. 4th ed. New York: AW Longman, 1998, pp 60–98.
22. GW Bruehl. Soilborne Plant Pathogens. London: Macmillan, 1987, pp 221–237.
23. RJ Cook. Making greater use of introduced microorganisms for biological control of plant pathogens. Annu Rev Phytopathol 31:53–80, 1993.
24. DE Mathre, RJ Cook, NW Callan. From discovery to use. Traversing the world of commercializing biocontrol agents for plant disease control. Plant Dis 83:972–983, 1999.
25. RS Utkhede. Potential and problems of developing bacterial biocontrol agents. Can J Plant Path 18:455–462, 1996.
26. Scherm H, Yang XB. Interannual variations in wheat rust development in China and the United States in relation to the El Nino/Southern Oscillation. Phytopathology 85:970–976, 1995.
27. Yang XB, and Zeng SM. Detecting patterns of wheat stripe rust pandemics in time and space. Phytopathology 82:571–576, 1992.
28. KP O'Connell, RM Goodman, J Handelsman. Engineering the rhizosphere: expressing a bias. Trends Biotechnol 14:83–88, 1996.
29. AJ Dik, MA Verhaar, RR Belanger. Comparison of three biological control agents against cucumber powdery mildew (*Sphaerotheca fuliginea*) in semi-commercial scale greenhouse trials. Eur J Plant Path 104:413–423, 1998.
30. RJ Cook. Assuring the safe use of microbial biocontrol agents: a need for policy based on real rather than perceived risks. Can J Plant Pathol 18:439–445, 1996.
31. JM Lynch. Environmental implications on the release of biocontrol agents. In: EC Tjamos, GC Papavizas, RJ Cook, eds. Biological Control of Plant Diseases. New York: Plenum Press, 1992, pp 389–398.
32. JV Cross, DR Polonenko. An industry perspective on registration and commercialization of biocontrol agents in Canada. Can J Plant Path 18:446–454, 1996.

33. LW Moore, G Warren. *Agrobacterium radiobacter* strain 84 and biological control of crown gall caused by *Agrobacterium tumefaciens*. Annu Rev Phytopathol 17: 163–179, 1979.

34. JE Slota, SK Farrand. Genetic isolation and physical characterization of pAgK84, the plasmid responsible for agrocin 84 production. Plasmid 8:175–186, 1982.

35. TJ Burr, C Bazzi, S Sule, L Otten. Biology of *Agrobacterium vitis* and the development of disease control strategies. Plant Dis 82:1288–1297, 1998.

36. XA Pu, RN Goodman. Tumor formation by *Agrobacterium tumefasciens* is suppressed by *Agrobacterium radiobacter* HLB-2 on grape plants. Am J Enol Vitic 44: 249–254, 1993.

37. RG Weber, A Granoff. Encyclopedia of Virology. Vol. I London: Academic Press, 1994, pp 242–248.

38. AS Costa, GW Muller. Tristeza control by cross protection: a US-Brazil cooperative success. Plant Dis 64:538–541, 1980.

39. REF Mathews. Plant Virology. 3rd ed. New York, Academic Press, 1991, pp 503–509.

40. SM Garnsey. Systemic disease. In: LW Timmer, LW Duncan, eds. Citrus Health Management. St. Paul, MN: APS Press, 1999, pp 95–106.

41. SD Yeh, D Gonsalves, HL Wang, R Namba, RJ Chiu. Control of papaya ringspot virus by cross protection. Plant Dis 72:375–380, 1988.

42. U Heiniger, D Rigling. Biological control of chestnut blight in Europe. Annu Rev Phytopathol 32:581–599, 1994.

43. PA Backman, M Wilson, JF Murphy. Bacteria for biological control of plant diseases. In: NA Rechcigl, JE Rechcigl, eds. Environmentally Safe Approaches to Crop Disease Control. Boca Raton, FL: Lewis Publishers, 1997, pp 95–109.

44. C Alabouvette, P Lemanceau, C Steinberg. Biological control of Fusarium wilts: opportunities for developing a commercial product. In: R Hall, ed. Principles and Practice of Managing Soilborne Plant Pathogens. St Paul, MN: APS Press, 1996, pp 192–212.

45. HAJ Hoitink, LV Madden, MJ Boehm. Relationships among organic matter decomposition level, microbial species diversity, and soilborne disease severity. In: R Hall, ed. Principles and Practices of Managing Soilborne Plant Pathogens. St. Paul, MN: APS Press, 1996, pp 237–249.

46. JJ Stapleton, JE de Vay. Soil solarization: a natural mechanism of integrated pest management. In: R Reuveni, ed. Novel Approaches to Integrated Pest Management. Boca Raton, FL: Lewis Publishers, 1995, pp 309–322.

47. A Sivan, Y Elad, I Chet. Integrated control of fusarium crown and root rot of tomato with *Trichoderma harzianum* in combination with methyl bromide or soil solarization. Crop Prot 12:380–386, 1993.

48. RC Rowe, JD Farley, DL Coplin. Air-borne spore dispersal and recolonization of steamed soil by *Fusarium oxysporum* in tomato greenhouses. Phytopathology 67: 1513–1517, 1977.

49. RR Belanger, C Labbe, WR Jarvis. Commercial-scale control of rose powdery mildew with a fungal antagonist. Plant Dis 78:420–424, 1994.

50. PM Brannen, DS Kenney. Kodiak—a successful biological-control product for sup-

pression of soil-borne plant pathogens of cotton. J Ind Microb Biotechnol 19:169–171, 1997.

51. DF Jensen, RD Lumsden. Biological control of soil borne pathogens. In: R Alajes, ML Gullino, JC van Lenteren, Y Elad. Integrated Pest and Disease Management in Greenhouse Crops. Boston: Kluwer Academic Publishers, 1999, pp 319–352.

52. JB Ristaino, SA Johnston. Ecologically based approaches to management of Phytophthora blight on bell pepper. Plant Dis 83:1080–1089, 1999.

53. XB Yang. Analysis of the long-term dynamics of stem and leaf rusts of wheat in North America using a time-series approach. J Phytopathol 143:651–657, 1995.

54. DO TeBeest, XB Yang, C Cisar. The status of biological control of weed with fungal pathogens. Annu Rev Phytopath 30:547–567, 1992.

55. XB Yang, DO TeBeest. Stability analysis of host-pathogen interactions in relation to weed biological control. Biol Control 2:266–271, 1992.

56. TH Been, CH Schomaker, JW Seinhoust. An advisory system for the management of potato cyst nematodes (*Globodera* spp.). In: FJ Gommers, PW Th Maas, eds. Nematology from Molecule to Ecosystem. Wildervank, The Netherlands: Husman 1995, pp 305–322.

57. TH Been, CH Schomaker. Development and evaluation of sampling methods for fields with infestation foci of potato cyst nematodes (*Globodera rostochiensis* and *G. pallida*). Phytopathology 90:647–656. 2000.

58. XB Yang. A general model for biological control. In: LJ Francl, DA Neher, eds. Exercises in Plant Disease Epidemiology. St. Paul, MN: APS Press, 1997, pp 132–137.

59. ME Hochberg. The potential role of pathogens in biological control. Nature 337:262–265. 1989.

16

Biocontrol Agents in Signaling Resistance

L. C. van Loon and Corné M. J. Pieterse
Utrecht University, Utrecht, The Netherlands

I. DISEASE SUPPRESSION THROUGH INDUCED SYSTEMIC RESISTANCE

The mechanisms by which biological control agents suppress disease comprise competition for nutrients, notably iron, production of antibiotics, and secretion of lytic enzymes, as well as inducing resistance in the plant [1,2]. The former three mechanisms act primarily on the pathogen by decreasing its activity, growth, and/or survival and require the biocontrol agent and the pathogen to be in close proximity. Because microorganisms with biocontrol properties and soilborne pathogens are both attracted to the rhizosphere, where root exudates and lysates provide a nutritious environment, antagonism between biocontrol agents and pathogens occurs locally. Such interactions may be favored by both the biocontrol agent and the pathogen growing preferentially over anticlinal walls of root epidermal cells [3]. However, pathogenic fungi such as *Fusarium oxysporum* f.sp. *lini* and f.sp. *raphani* grow towards root apices and penetrate through the tips that emerge virtually sterile from beneath the root cap (4; H. Steijl, T. van Welzenis, J. van den Heuvel, and L. C. van Loon, unpublished). Hence, those Fusarium wilts are among the diseases most difficult to control effectively and reliably by antagonistic microorganisms.

Induced resistance is a plant-mediated mechanism and can only be demonstrated unequivocally when the biocontrol agent and the pathogen never contact each other. Thus, induced resistance acts at a distance and can protect plants systemically. This elicitation of a systemically enhanced defense capacity is variously denoted as systemic acquired resistance (SAR) or induced systemic resistance (ISR) and can result from stimulation by either biotic or abiotic agents. SAR/ISR is relatively easily demonstrable with rhizobacteria as biocontrol agents that remain confined to the roots but protect plants against disease caused by a foliar pathogen. It is most difficult to prove when biocontrol fungi tend to colonize plant tissues systemically, and special measures are necessary to ensure that contact between the biocontrol agent and the pathogen is avoided. Systemic resistance induced by nonpathogenic, root-colonizing bacteria was first conclusively demonstrated in 1991 by van Peer et al. [5] in carnation and Wei et al. [6] in cucumber. Carnation cuttings were rooted on a substratum of rock wool and bacterized by pouring a suspension of a rifampicin-resistant derivative of *Pseudomonas fluorescens* strain WCS417 (WCS417r) over the roots. One week later, plants were stem-inoculated with *F. oxysporum* f.sp. *dianthi* between the first and second pair of leaves. As a result of root colonization by WCS417r, the number of wilted plants of the carnation cv. Pallas was reduced on an average from about 50% to 20% and in a single experiment with cv. Lena from 69% to 38%. Strain WCS417r could not be isolated from stem tissue, indicating that protection was plant-mediated [5]. Cucumber was protected against anthracnose, caused by *Colletotrichum orbiculare*, after seeds had been bacterized with 6 of 94 rhizobacterial strains tested. Treatment with these strains reduced both the number and the size of anthracnose lesions after challenge inoculation of the foliage with the fungus. None of the bacterial strains was recovered from surface-disinfected petioles on the day of challenge with *C. orbiculare*, clearly suggesting that systemic resistance had been induced [6].

In cucumber, systemic resistance against anthracnose could also be induced by several isolates of nonpathogenic fungi that were isolated from zoysiagrass rhizospheres. The fungal isolates were introduced into autoclaved potting medium in which surface-disinfected seeds were sown. After 21 days the seedling leaves were inoculated with *C. orbiculare*. The rhizosphere fungi did not colonize the aerial portions of the plant, indicating that the disease suppression observed was the result of induction of systemic resistance [7,8]. In other investigations ISR was demonstrated by using plants with split-root systems in which the biocontrol agent and the pathogen remain physically separated from each other. Colonization of one part of the root system of watermelon by selected isolates of nonpathogenic *F. oxysporum* protected the plants against Fusarium wilt when the other part was challenge inoculated with *F. oxysporum* f.sp. *niveum* [9]. Under similar conditions, nonpathogenic *F. oxysporum* isolate Fo47 and *Penicillium oxalicum* suppressed Fusarium wilt in tomato by inducing systemic resistance against pathogenic *F. oxysporum* f.sp. *lycopersici* [10,11]. Cordier et al. [12] likewise demonstrated that the arbuscular mycorrhizal fungus *Glomus mosseae* induces systemic resistance in tomato against *Phytophthora parasitica*. Also, the fungal biocontrol agents *Trichoderma harzianum* and *Trichoderma hamatum*, which are capable of antagonizing sensitive pathogenic fungi by producing antibi-

otics and lytic enzymes, have been reported to induce systemic resistance in tomato, lettuce, pepper, bean, and tobacco against grey mold, caused by *Botrytis
cinerea* [13], and in radish against leaf spot caused by *Xanthomonas campestris*
pv. *armoraciae* [14].

Most research on microbially induced systemic resistance has been conducted with rhizobacteria, however. Apart from carnation and cucumber, rhizobacteria-mediated ISR has been conclusively demonstrated in Arabidopsis, bean,
radish, tobacco and tomato and shown to be effective against fungi, bacteria,
viruses, and insects. Where investigated, ISR was shown to require a threshold
population of the biocontrol agent, above which little further increase in protection is evident. Moreover, a time interval is necessary between application of the
inducer and the onset of protection of the plant, indicating that the plant needs
time to reach the induced resistant state [2].

The ability of nonpathogenic rhizobacteria and fungi to induce systemic
resistance appears to be fairly common but by no means general. Resistance-
inducing biocontrol agents must express a specific determinant that is recognized
by plant roots and elicits a response that culminates in an enhanced defensive
capacity throughout the entire plant. Such systemic induction requires not only
local perception of the stimulus, but also its transport or the transport of a mobile
signal that is generated as a result of local stimulus perception. Moreover, the
mobile signal, in turn, must be transduced at the sites where resistance is expressed (Fig. 1). Phenomenologically, the situation resembles both the wound
response and pathogen-induced SAR [15,16], except that in those responses plant
tissues start reacting to various forms of cellular damage, whereas in biocontrol
agent–induced systemic resistance generally no deleterious effects of the inducing microorganism on the plant are visible.

II. HISTOLOGY OF THE INDUCED RESISTANT STATE

Histological investigations at the light and electron microscopic levels have attempted to relate the induced resistant state to tissue alterations that might provide
clues as to the mechanism of ISR. Adventitious roots from *Agrobacterium rhizo-
genes*–infected pea (Ri T-DNA transformed roots) were inoculated in vitro with
Bacillus pumilis strain SE34, a bacterium that protects cotton roots against *F.
oxysporum* f.sp. *vasinfectum* [17]. Whereas in nonbacterized roots the pea pathogen, *F. oxysporum* f.sp. *pisi*, multiplied abundantly through much of the tissue
including the vascular stele, in prebacterized roots pathogen growth was restricted
to the epidermis and the outer cortex. In these prebacterized roots typical host
reactions included strengthening of epidermal and cortical cell walls and deposition of wall appositions containing large amounts of callose as well as phenolic
compounds. No induction of visible host defense reactions was evident in bacte-

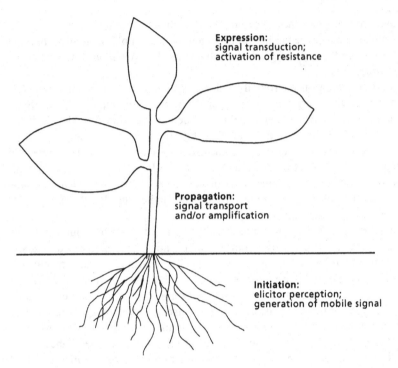

Figure 1   Representation of the stages in systemic resistance induced by rhizobacteria or nonpathogenic soilborne fungi.

rized pea roots before challenge with the pathogen, when a large number of bacteria had grown on the root surface and displayed the ability to colonize some intercellular spaces in the epidermis and the outer cortex. Because differences in defensive responses became evident only after challenge, it appears that the induced state constitutes a sensitization of the plant to respond more effectively to pathogen attack [18,19].

Essentially similar results were obtained when in vitro Ri T-DNA transformed pea roots were treated with the biocontrol bacterium *P. fluorescens* strain 63-28R and challenged with *Pythium ultimum* [20] or when in vivo tomato plants were bacterized with the same bacterial strain and subsequently infected with *F. oxysporum* f.sp. *radicis-lycopersici* [21]. Noninfected, 63-28R–treated tomato plants showed no symptoms, and their root systems appeared healthy. Bacteria grew actively at the root surface and colonized some host epidermal cells and intercellular spaces in the outermost root tissues. Apparent preservation of cell wall architecture, as shown by regular patterns of cellulose and pectin over the walls adjacent to the areas of bacterial colonization, provided evidence that wall-degrading lytic enzymes were not, or only very slightly, produced by the endophytic bacteria once inside the plant. A slight accumulation of electron-opaque substances along the epidermal cell walls and/or within the invaded intercellular spaces, as well as slight deposition of β-1,3-glucan in the host cell walls, pointed to some discrete host reactions that resemble defense responses. It thus appears that the perception of the bacteria by the roots leads to a minimal activation of wall-associated defenses, effectively priming the plant to respond faster and to a higher extent upon subsequent challenge inoculation.

Similar observations suggested that the mycoparasitic fungus *Pythium oligandrum* also has the potential to induce plant defense reactions in tomato roots challenged with *F. oxysporum* f.sp. *radicis-lycopersici* [22,23]. *Pythium oligandrum* was able to penetrate the root epidermis without extensive host wall degradation and, subsequently, ramified in all root tissues by inter- and intracellular growth. This implied that at least small amounts of cell wall hydrolytic enzymes, such as pectinases and cellulases, were produced to locally weaken or loosen the host cell walls, thereby facilitating spread into root tissues. However, the invading hyphae were structurally altered as evidenced by the frequent occurrence of empty fungal shells in the root tissues. Moreover, *Pythium* ingress in the root tissues was associated with host metabolic changes, leading to the elaboration of structural barriers at sites of potential fungal penetration. These observations indicate that the plant reacted defensively to invasion by *Pythium oligandrum*.

Similar formation of physical and chemical barriers at sites of potential fungal entry were detected in cucumber plants that reacted more rapidly and more efficiently to infection by pathogenic *Pythium ultimum* when pretreated with the endophytic biocontrol bacterium *Serratia plymuthica* strain R1GC4 [24]. A nonpathogenic strain of *F. oxysporum*, able to induce systemic resistance in tomato, likewise triggered typical defense reactions, such as wall appositions, intercellular plugging, and intracellular osmiophylic deposits [25]. Both *Pseudomonas corrugata* strain 13c and *Pseudomonas aureofaciens* strain 63-28 increased levels of phenylalanine ammonia-lyase (PAL), peroxidase, and polyphenoloxidase in cucumber roots, which peaked 2–4 days after root treatment. Similar experiments were done with split-root systems, which demonstrated that the induction of higher enzyme levels was systemic. Analysis of the peroxidase isoenzymes showed that the bacteria specifically induced an increase in one acidic isoform [26]. Similarly, upon root inoculation with the biocontrol agent *Trichoderma harzianum*, peroxidase and chitinase activities increased in both the roots and the leaves of treated cucumber seedlings, indicating that the fungus stimulated defense reactions in the plant [27]. In contrast, no histological differences were apparent in tomato stems in which resistance was induced by *Penicillium oxalicum*, as evidenced by reduced cambial loss upon challenge inoculation with *F. oxysporum* f.sp. *lycopersici* [28]. Similarly, no cell reactions were observed in mycorrhizal tomato root systems, but upon challenge with *Phytophthora parasitica*, accumulation of phenolics and plant cell defense responses were augmented [12].

III. ROLE OF SALICYLIC ACID IN INDUCED RESISTANCE

A. Pathogen-Induced SAR

Phenotypically, biocontrol agent–induced systemic resistance and pathogen-induced SAR are similar in that both develop as a result of the interaction of the plant with a microorganism and the resulting enhanced defensive capacity is expressed both locally and systemically against a broad spectrum of attacking organisms. Pathogen-induced SAR is a general phenomenon that occurs as a result of limited infection by a pathogen, notably in incompatible interactions that lead to a hypersensitive reaction. However, necrosis is not a prerequisite for pathogen-induced resistance, indicating that the signaling pathways leading to tissue necrosis, on the one hand, and induced resistance, on the other hand, are distinct. Nevertheless, hypersensitive necrosis contributes to the level of SAR attained [29]. Pathogen-induced SAR is triggered by elicitors that are involved either in specific gene-for-gene resistance or in the nonspecific elicitation of defense reactions. As a result, salicylic acid (SA) is produced as a signal and induces the resistant state. The role of SA has been established on the grounds that exogenous application of SA or its functional analogues 2,6-dichloroisonicotinic acid (INA) and benzothiadiazole (BTH) mimic pathogen-induced SAR by inducing resistance in many plant species towards the same broad spectrum of pathogens [16,30]. Endogenous accumulation of SA occurs both locally and, at lower levels, systemically concomitant with the development of SAR. Moreover, plants transformed with the *NahG* gene do not accumulate SA and do not develop SAR in response to biological or chemical inducers of SAR. The *NahG* gene encodes SA-hydroxylase, which converts SA into catechol, a product that does not induce resistance. The experiments with *NahG*-transformed plants indicate that SA is an essential signaling molecule in SAR induced by avirulent pathogens. Moreover, NahG-containing plants are more susceptible to a variety of fungal, bacterial, and viral pathogens [31].

SA can be synthesized from either cinnamic acid [32] or chorismic acid [33], and the regulation of its synthesis has not been clarified. Neither its mode(s) of action nor the molecular mechanism of resistance induction has been elucidated [34]. Although SA may be transported from locally infected leaves, it does not appear to be the primary long-distance signal for systemic induction [35,36]. Nevertheless, its presence is required for SAR to be expressed. Indeed, local application of SA to individual plant leaves does not lead to systemic induction of resistance unless SA can be transported out of the leaf to other plant parts

[37]. However, when applied to roots SA appears to be readily absorbed and transported throughout the plant, and acquired resistance is manifested systemically. Exceedingly low concentrations of SA applied to the roots of radish and bean have been reported to induce systemic resistance against Fusarium wilt and grey mold, respectively [38,39]. The higher levels of SA in plants expressing SAR suggest that an amplifying step is required for induced resistance to be manifested. Although exogenous application of SA can mimic the induction of resistance by pathogens, additional plant signals must play a role and the nature of the endogenous mobile signal is still unknown.

Associated with SAR triggered by either endogenous regulators or exogenous application of SA is the accumulation of pathogenesis-related proteins (PRs). PRs have been defined as plant proteins that are induced in pathological and related situations [40] and currently comprise 14 protein families [41]. Induction of at least some families of PRs occurs invariably in plants with necrotizing infections. Particularly appearance of the SA-inducible PR-1 type proteins is generally taken as a marker of the induced resistant state. Some of the PR families are β-1,3-glucanases and chitinases and are capable of hydrolyzing fungal cell walls. Other PRs have less well characterized antimicrobial activities or unknown functions. The association of PRs with SAR suggests an important contribution of these proteins to the enhanced defensive capacity of induced tissues. However, none seem to act against viruses, which, nevertheless, are also effectively protected against.

B. Rhizobacterium-Mediated ISR

To determine whether biocontrol agent–induced resistance is not only phenotypically but also mechanistically similar to pathogen-induced SAR, induced plants can be analyzed for the presence of specific PRs, and induction of resistance can be compared in untransformed and *NahG*-transformed plants. As described above, some biocontrol agents elicit increases in PR-type enzymic activities, suggestive of a pathway involving SA similar to that triggered by avirulent pathogens. At least 8 of the 10 major PRs induced in tobacco in response to pathogens causing hypersensitive necrosis were found in plants grown in soil containing *Pseudomonas fluorescens* CHA0, which suppresses necrotic lesion formation on leaves inoculated with tobacco necrosis virus (TNV) [42]. Similarly, when transgenic tobacco plants were assayed for induction of β-glucuronidase (GUS) activity under the direction of the tobacco PR-1a promoter, out of 10 rhizobacterial strains previously demonstrated to induce systemic resistance in cucumber, *Bacillus pumilis* strain T4, *Pseudomonas putida* strain 89B-61, *Serratia marcescens* strain 90-166, and *Burkholderia gladioli* strain IN-26 significantly enhanced GUS activity. The ability to enhance GUS activity was associated with reduction of symptoms of wildfire disease caused by *Pseudomonas syringae* pv. *tabaci* upon root bacterization by any of these four strains. Neither enhanced GUS activity nor reduction of symptoms of wildfire disease were noted with three control bacterial strains [43]. However, increased GUS activity was observed locally when the four inducing strains were infiltrated in leaves, and only strain T4 increased GUS activity in the leaves when it was applied as a soil drench. In contrast, *Enterobacter asburiae* strain JM-22, which did not induce resistance in cucumber when applied as a soil drench, did induce protection against wildfire disease in tobacco and also induced GUS activity in the leaves. Thus, the systemic induction of resistance against wildfire disease by application of any of the five strains to plant roots was associated with systemically enhanced GUS expression by two of these strains only.

Local induction of plant defense responses by rhizosphere bacteria or fungi has been observed upon leaf infiltration or stem inoculation, whereas upon root treatment such metabolic changes are often not apparent. Thus, a *Pseudomonas aureofaciens* strain induced symptoms of a hypersensitive response on bean cotyledons [44]. Similarly, out of 15 rhizobacteria tested, 9 strains of the genus *Pseudomonas* and two *Serratia* strains induced a hypersensitive response and the production of phytoalexins in wounded bean cotyledons and hypocotyls [45]. Inoculation of bean hypocotyls with nonpathogenic binucleate *Rhizoctonia* species induced systemic resistance against pathogenic *Rhizoctonia solani* or *Colletotrichum lindemuthianum* but also increased peroxidase, β-1,3-glucanase, and chitinase activities [46]. Such observations indicate that at least some rhizobacterial and fungal species have limited pathogenic activity when applied to aboveground plant parts.

No accumulation of PRs was found in either leaves or roots of radish with systemic resistance induced by *P. fluorescens* WCS417r [47,48], and neither were PR mRNAs or proteins apparent in Arabidopsis leaves upon root bacterization with the same rhizobacterial strain [49,50]. Similarly, *Pseudomonas aeruginosa* strain 7NSK2–induced resistance in tobacco was not associated with PR-1 expression [51]. Together, these results suggest that some rhizobacterial strains induce resistance by a SA-dependent mechanism, whereas others act independently of SA. This conclusion has been corroborated by experiments with SA-nonaccumulating NahG plants, even though the situation has turned out to be complex. *P. fluorescens* WCS417r, as well as *Pseudomonas putida* WCS358r, induced systemic resistance in NahG Arabidopsis to the same level as in untransformed plants, demonstrating that SA is not involved in ISR elicited by these strains [49,50]. In contrast, *P. aeruginosa* 7NSK2–induced systemic resistance was abolished in NahG tobacco [51] and tomato [52], indicating that ISR triggered by this rhizobacterial strain does depend on in planta SA accumulation, even though the marker gene for SA-dependent SAR, PR-1, is not activated. Conversely, *Serratia marcescens* 90-166, which activated the PR-1a promoter in the leaves of transgenic PR-1a-GUS tobacco [43], induced systemic resistance against wildfire disease equally well in NahG and in untransformed tobacco, ruling out an involvement of SA in the induction [53]. These results clearly indicate that the presence or absence of *PR* gene activation by these rhizobacterial strains is not a reliable parameter for determining whether the mechanism of resistance induction is SA-dependent or SA-independent.

Direct measurements of SA in leaves of tobacco plants grown in soil containing resistance-inducing *P. aeruginosa* 7NSK2 mutant KMPCH (which is deficient in production of the siderophores pyoverdin and pyochelin but can produce SA [see below]), increased free SA levels about 50% over the level in control plants, while amounts of bound SA were similar. A similar increase in free SA was detected in the rhizosphere of bean plants grown in soil with KMPCH. These increases seem sufficient to activate the SAR pathway [39], even though they are too low to activate PR gene expression [49,54]. In cucumber split-root systems, when *P. corrugata* 13 and *P. aureofaciens* 63–28 induced systemic resistance against *Pythium aphanidermatum*, SA accumulated up to sixfold in bacterized root parts and up to fourfold in distant roots on the opposite side. However, in this system exogenously applied SA failed to induce local or systemic resistance against a challenge infection by *Pythium* in planta [55]. This suggests that SA is not a primary causative factor in ISR in cucumber against *Pythium*.

IV. BACTERIAL DETERMINANTS OF ISR

A. Structural Components and Metabolic Compounds

On the one hand, the rhizobacterial strain should produce one or more ISR-eliciting compounds; on the other hand, the plant should possess a matching receptor and an inducible defense pathway downstream of it that activates the induced resistant state upon recognition. In the systemic resistance induced by *P. fluorescens* strains WCS374 and/or WCS417 in carnation, radish, tomato, and Arabidopsis, heat-killed bacteria, bacterial cell walls, or purified outer membrane lipopolysaccharide (LPS) were as effective in inducing systemic resistance as were live bacteria [50,56–58]. These observations provided original proof that ISR was not the result of a bacterial metabolite that might be transported through the plant and affect the activity of a pathogen upon contact. Rather, specific bacterial components must be perceived by the plant. In radish, the bacterial LPS appeared to act as an inducing determinant. In agreement with this finding, bacterial mutants that lacked the O-antigenic side chain of the LPS (OA$^-$) did not induce resistance in radish. Cell wall preparations of WCS417r likewise induced resistance in Arabidopsis. However, WCS417 OA$^-$ still induced resistance, indicating that another determinant of WCS417r also acts as an inducer of resistance in Arabidopsis. Similarly, *P. putida* WCS358, as well as its LPS, triggered ISR, but WCS358 OA$^-$ retained the capacity to induce systemic resistance. Therefore, like WCS417r, WCS358r must possess additional inducing determinants. Because LPS may be contaminated by flagella, purified flagella of WCS358r were also tested. Like LPS, the flagella induced resistance, and so did a mutant lacking flagella (P. A. H. M. Bakker, B. W. M. Verhagen, and I. van der Sluis, unpublished). Collectively, these results indicate that flagella, while capable of eliciting ISR, were unlikely to be the sole determinant in LPS preparations, and the O-antigen of LPS must be perceived by the plant independently from the perception of the flagella.

The O-antigenic side chain of LPS consists of repeated oligosaccharide moieties. Such oligosaccharides may resemble fungal elicitors that activate plant defenses [59] and may be perceived by similar types of receptors that are likely to be present in the plasma membrane of the root cells. So far, putative receptors for bacterial LPS have not been characterized in plants, and the mechanism of perception and coupling to induced resistance signaling remains unclear. In contrast, plant cells have been shown to possess a highly sensitive perception system for bacterial flagellins, the major structural protein of the flagella. Recognition occurs through perception by the plant of the most highly conserved domain within the N-terminus of the protein [60]. Sensitivity to flagellin in Arabidopsis is associated with the expression of a putative receptor kinase containing leucine-rich repeats and sharing structural and functional homologies with known plant resistance genes and with components involved in the innate immune system of mammals and insects [61]. Such results imply that the perception of bacterial flagella by plant roots might directly activate a signaling pathway involved in activating resistance responses. Bacterial components involved in antagonizing pathogens in the rhizosphere might also act in inducing resistance. Lytic enzymes, particularly glucanases, can liberate endogenous elicitors from plant cell walls, as has been shown for, e.g., β-1,3-glucanase in soybean [62]. The biocontrol strains P. fluorescens 89B-61 and Enterobacter asburiae JM22 were endophytic in cotton, and although they did not induce marked cellular alterations upon internal colonization, they did hydrolyze wall-bound cellulose [63]. However, no bacteria that trigger ISR and remain confined to the root surface have been shown to act through this mechanism.

Antibiotics can be toxic not only to pathogenic fungi and bacteria, but also to plant cells. Such toxicity may cause localized necrosis, leading, in turn, to systemically induced resistance. A transformant of P. fluorescens CHA0 that overproduced pyoluteorin and 2,4-diacetylphloroglucinol protected tobacco roots significantly better against black root rot, caused by Thielaviopsis basicola, than the wild-type strain, but drastically reduced the growth of the tobacco plants and was also toxic to sweet corn and cress [42,64]. There was no correlation between the sensitivity of the pathogens to the antibiotics and the degree of disease suppression by the overproducing strain. It seems, therefore, that resistance was induced, indicating that plants that are sensitive to antibiotics may be induced as a result of their toxic action. Rhizobacteria and fungi may also produce plant hormones than can affect plant growth and development as well as plant responses [65]. Notably, ethylene reduces root elongation while stimulating production of phytoalexins, synthesis of PRs, and strengthening of plant cell walls upon elicitation of plant defense responses [66,67].

B. Iron-Regulated Metabolites

Iron availability not only influences microbial antagonism by determining the extent of competition through siderophore production, but has also been found to be a major factor in the induction of systemic resistance by rhizobacteria. Pyoverdin siderophore production was implicated in the induction of systemic resistance against TNV in tobacco by *P. fluorescens* CHA0. The level of resistance induced was partially abolished when plants were bacterized with the siderophore-minus (sid⁻) mutant CHA400 instead of with CHA0 [42]. Purified pseudobactin siderophore of *P. putida* WCS358 induced resistance in Arabidopsis, but its sid⁻ mutant was still as effective as the wild type, confirming the presence of multiple resistance-inducing determinants in this strain. In radish, the pseudobactin siderophore of *P. fluorescens* WCS374r, but not of *P. putida* WCS358r or *P. fluorescens* WCS417r, induced resistance against *F. oxysporum* f. sp. *raphani*. The sid⁻ mutant of WCS374 still induced resistance, however, allegedly as a result of LPS acting as the inducing determinant. In the presence of ample available iron, when siderophore production is suppressed, the OA⁻ mutant did not induce resistance. In contrast, under conditions of low iron availability the OA⁻ mutant induced the same level of systemic resistance as the wild-type strain, and the level of resistance attained was greater than under iron-sufficient conditions (Fig. 2). This enhanced resistance induction was also evident in the sid⁻ mutant, suggesting that an additional iron-regulated factor might be involved. Similar results were obtained for strain WCS417, suggesting that the same situation might apply [38].

Under iron-limited conditions, certain rhizobacterial strains such as *P. fluorescens* CHA0, can produce SA as an additional siderophore [68,69]. Thus, it has to be considered that resistance-inducing bacteria may either activate SA-dependent induced resistance by triggering the SAR pathway in the plant, or themselves secrete the signaling compound SA in the rhizosphere. The induction of systemic resistance in tobacco to TNV by strain CHA0 might be fully explained by the production of SA by this strain, the more so because CHA0 also induced PRs in tobacco. Moreover, root colonization of the plants by either the wild-type strain or its sid⁻ mutant CHA400 caused up to fivefold increases in SA in the leaves [70]. Introduction of the SA-biosynthetic gene cluster *pchDCBA* from *P. aeruginosa* strain PAO1 [33] under the control of a strong constitutive promoter into strain CHA0 increased the production of SA in the rhizosphere of

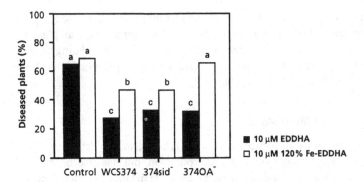

Figure 2   Disease incidence in radish grown in soil containing *Fusarium oxysporum*
f.sp. *raphani* without rhizobacteria (control) or in the presence of *Pseudomonas fluorescens* WCS374 or its pseudobactin siderophore-minus (sid⁻) or O-antigenic side chain
lacking (OA⁻) mutants, under iron limitation (10 μM EDDHA added) or with ample iron
supply (10 μM EDDHA complexed with excess Fe). Different letters indicate statistically
significant differences of $p \leq 0.05$.

tobacco but did not further increase protection against TNV. In contrast, introduction of the genes into the non–SA-producing strain P3 made this poor biocontrol
agent an effective inducer. Although NahG plants were not tested, these results
are consistent with bacterially produced SA acting as the inducing factor [70].

The induction of systemic resistance in bean against *Botrytis cinerea* and
in tobacco against tobacco mosaic virus (TMV) by *P. aeruginosa* 7NSK2 was
found to be iron-regulated. Under iron limitation 7NSK2 produces three siderophores: pyoverdin, pyochelin, and SA [71]. Both pyoverdin-negative mutants and
mutants lacking both pyoverdin and pyochelin induced resistance in bean and
tobacco, whereas mutants deficient in SA production did not. Bacterization of
NahG tobacco plants with either the wild-type strain or its mutants did not induce
systemic resistance to TMV, demonstrating that the resistance induced by 7NSK2
in tobacco, and probably also in bean, is dependent on bacterially produced SA
[51,72]. It appeared that nanogram amounts of SA produced by the rhizobacteria
are already sufficient to trigger the plant-mediated resistance response, probably
through the induction of SA biosynthesis in the plant [39]. However, 7NSK2-
induced resistance was not associated with PR-1 expression [51]. Because SA is
a precursor of pyochelin, it cannot be excluded that bacterially produced SA is
converted to pyochelin and that pyochelin acts as the inducing determinant.

A comparable situation may exist in the systemic resistance induced by
P. fluorescens WCS374r against *F. oxysporum* f.sp. *raphani* in radish. Both *P.
fluorescens* WCS374r and WCS417r are capable of producing SA in vitro under
iron-limiting conditions, whereas *P. putida* WCS358r is not. These capacities
correlate well with their ability to induce systemic resistance in radish [38,73].

However, no accumulation of PRs was apparent in either roots or leaves of bacterized plants. Cloning of the SA-biosynthetic locus of WCS374r revealed it to contain four open reading frames with homologies to bacterial isochorismate synthase, 2,3-dihydroxybenzoate AMP-lyase, histidine decarboxylase, and chorismate pyruvate-lyase (*pmsCBAE*), respectively. This SA-biosynthetic locus was found to be co-expressed with genes involved in the synthesis of the siderophore pseudomonine, which contains both a salicylate and a histamine moiety [74]. It is possible that in the rhizosphere histidine is provided in root exudates, whereupon SA is quickly incorporated into pseudomonine. Because under those conditions no free SA would be secreted by the bacterium, it can be assumed that SA is not involved in the induction of systemic resistance in radish by WCS374r. Similarly, SA produced by the biocontrol strain *Serratia marcescens* 90-166 is not the primary determinant of ISR in cucumber and tobacco. This strain, while capable of producing SA under low-iron conditions, induced resistance to wildfire disease in both untransformed and NahG tobacco. Bacterial mutants that did not produce detectable amounts of SA retained ISR-eliciting activity against *C. orbiculare* in cucumber, and an ISR-minus mutant still produced SA in vitro [53]. A siderophore may be involved, because fertilization of the plants with ferric iron significantly reduced the level of ISR in cucumber to *C. orbiculare*.

V. *ARABIDOPSIS THALIANA* AS A MODEL TO STUDY RHIZOBACTERIA-MEDIATED INDUCED SYSTEMIC RESISTANCE

A. Signal Transduction

1. SA-Independent Signaling

At least in those instances where no accumulation of PRs is associated with ISR and ISR was fully maintained in NahG plants, an SA-independent pathway must be operative. The existence of such a SA-independent pathway was first demonstrated in Arabidopsis, which was adopted as a model to study the signaling pathway(s) involved in rhizobacteria-mediated ISR [49]. In this model *P. fluorescens* WCS417r was used as the inducing agent, because this strain had been demonstrated to trigger ISR in several plant species, e.g., carnation [5], radish [73], and tomato [11]. Colonization of Arabidopsis roots by WCS417r bacteria was found to protect against different types of pathogens, including the bacterial leaf pathogens *Pseudomonas syringae* pv. *tomato* DC3000 and *Xanthomonas campestris* pv. *armoraciae*, the fungal root pathogen *Fusarium oxysporum* f.sp. *raphani*, and the oomyceteous leaf pathogen *Peronospora parasitica*. WCS417r induced ISR without systemic activation of *PR* genes (49,50; J. Ton and C. M. J. Pieterse, unpublished) and without a concomitant increase in SA levels [75].

368   van Loon and Pieterse

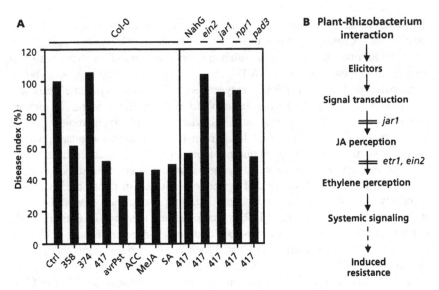

Figure 3 (A) Relative disease index of Arabidopsis ecotype Columbia (Col-0), transformant NahG and mutants treated with different bacterial strains and chemical inducers of systemic resistance and challenged with virulent *Pseudomonas syringae* pv. tomato. (B) Signal-transduction pathway leading to rhizobacteria-mediated induced systemic resistance in Arabidopsis.

Moreover, NahG plants developed normal levels of ISR after colonization of the roots by ISR-inducing WCS417r [49] (Fig. 3A) or WCS358r [50]. Similarly, the SA induction–deficient mutants *sid1-1* and *sid2-1* [76] expressed normal levels of induced resistance in response to ISR-inducing rhizobacteria (C. M. J. Pieterse, unpublished), providing compelling evidence that, in contrast to pathogen-induced SAR, rhizobacteria-mediated ISR in Arabidopsis functions independently of SA. As a consequence, SA-dependent resistance responses were not activated upon bacterization of the roots with WCS417r [54]. Therefore, ISR-expressing plants are unlikely to be protected against pathogens that are resisted exclusively by an SA-dependent defense mechanism. Indeed, WCS417r-mediated ISR was found to be ineffective against turnip crinckle virus (J. Ton, unpublished), which is a pathogen that is resisted by Arabidopsis ecotype Dijon through a defense response that is dependent on SA but not on jasmonic acid (JA) and ethylene [77].

2. Jasmonic Acid– and Ethylene-Dependent Signaling

Studies on mutants and transgenics of Arabidopsis and other plant species, such as tobacco and soybean, revealed that besides SA, JA and ethylene play a key role in the regulation of plant defenses, because blocking the response to either of these hormonal compounds can render plants more susceptible to certain pathogens and even insects. For instance, mutants that are affected in JA biosynthesis or signaling are more susceptible to pathogens such as *Pythium mastophorum* [78], *Pythium irregulare* [79], and *P. syringae* pv. *tomato* [80], as well as to insect herbivory [81,82]. Similarly, ethylene-insensitive tobacco plants transformed with a mutant *etr1-1* gene from Arabidopsis lost their ability to resist the soilborne pathogen *Pythium sylvaticum* [83]. Similar results were obtained with ethylene-insensitive soybean plants, which developed more severe symptoms in response to *Septoria glycines* and *Rhizoctonia solani* [84]. Furthermore, ethylene-insensitive Arabidopsis mutants exhibit enhanced susceptibility to the necrotrophic fungal pathogens *Alternaria brassicicola* and *Botrytis cinerea* [85,86] and to the bacterial leaf pathogens *Erwinia carotovora* pv. *carotovora* [87] and *P. syringae* pv. *tomato* [80].

To investigate the possible role of JA and ethylene in the signaling pathway controlling rhizobacteria-mediated ISR in Arabidopsis, Pieterse et al. [80] tested the JA response mutant *jar1-1* and the ethylene response mutants *etr1-1* and *ein2* for their ability to express ISR. Both types of mutants were unable to develop ISR against *P. syringae* pv. *tomato* in response to bacterization of the roots with WCS417r (Fig. 3A), indicating that both JA and ethylene are involved in the ISR signaling pathway. Like treatment with WCS417r, application of methyl jasmonate (MeJA) or the ethylene precursor 1-aminocyclopropane-1-carboxylate (ACC) was effective in inducing resistance against *P. syringae* pv. *tomato* in wild-type (Fig. 3A), as well as NahG plants. MeJA-induced protection was blocked in *jar1-1* and *etr1-1* plants, whereas ACC-induced protection was affected in *etr1-1* plants, but not in *jar1-1* plants, indicating that components from the JA response act upstream of the ethylene response in the ISR pathway [80] (Fig. 3B). Whereas in rhizobacterially mediated ISR JA and ethylene appear to act sequentially, a concerted action of JA and ethylene has been described for other defense-related responses. For instance, JA and ethylene are acting together in activating proteinase inhibitor gene expression in tomato in response to wounding [88]. Similarly, induction of the plant defensin gene *Pdf1.2* in Arabidopsis requires concomitant activation of both the JA- and ethylene-response pathway [89].

To further investigate the roles of JA and ethylene in the ISR signaling pathway, the levels of these signaling molecules were determined in plants upon root bacterization. In plants grown in soil containing WCS417r, neither the JA content nor the level of ethylene evolution was altered in systemically resistant leaves [75,90]. Also, at the site of application of ISR-inducing rhizobacteria the levels of JA and ethylene did not change [75,90], indicating that rhizobacteria-mediated ISR is not based on the induction of changes in the biosynthesis of either of these signal molecules.

By using the *Lox2* co-suppressed transgenic line S-12, additional evidence was obtained that an increase in JA production is not required for the induction or expression of ISR. Transgenic S-12 plants, which are affected in the production of JA in response to wounding [91] and pathogen infection [75], developed normal levels of ISR in response to treatment with WCS417r [75], demonstrating that plants are capable of expressing ISR in the absence of increased JA levels. These results seem to suggest that the JA and ethylene dependency of ISR is based on an enhanced sensitivity to these hormones, rather than on an increase in their production.

3. The Role of NPR1 in ISR Signaling

Although pathogen-induced SAR and rhizobacteria-mediated ISR follow distinct signaling pathways in Arabidopsis, they are both blocked in the regulatory mutant *npr1-1* (for *n*onexpresser of *PR* genes) [80,92] (Fig. 3A). NPR1 (also called NIM1 or SAI1) was originally discovered as a key regulatory protein that functions downstream of SA in the SAR pathway [92–94]. Recently, several research groups provided evidence that, upon induction of SAR, NPR1 activates *PR-1* gene expression by physically interacting with a subclass of basic leucine zipper protein transcription factors that bind to promoter sequences required for SA-inducible *PR* gene expression [95–97]. Elucidation of the sequence of ISR-signaling events revealed that NPR1 also functions downstream of the JA and ethylene response in the ISR pathway [80]. This suggests that NPR1 is required not only for the SA-dependent expression of *PR* genes that are activated during SAR, but also for the JA- and ethylene-dependent activation of so far unidentified enhanced defensive responses resulting from rhizobacteria-mediated ISR. The mechanism underlying the divergence of the SAR and the ISR pathway downstream of NPR1 is not known. Possibly, interactions of pathway-specific proteins with NPR1 are involved.

4. Application of Rhizobacteria to Roots or Leaves Triggers the Same ISR Pathway

Similar to root application, infiltration of leaves with ISR-inducing WCS417r bacteria induces protection against *P. syringae* pv. *tomato* in noninfiltrated leaves [49]. To test whether infiltration of leaves with ISR-inducing rhizobacteria triggers the same signaling pathway as root application, Arabidopsis genotypes Columbia (Col-0), NahG, *jar1-1*, *etr1-1*, and *npr1-1* were tested for their ability to express ISR against *P. syringae* pv. *tomato* after pressure infiltrating three lower leaves with ISR-inducing WCS417r bacteria. Leaf infiltration and root application of WCS417r were similarly effective in eliciting ISR in wild-type Col-0 plants. SA-nonaccumulating NahG plants also developed a statistically significant level of ISR after leaf induction. In contrast, mutants *jar1-1*, *etr1-1*, and *npr1-1* did not express ISR after infiltration of the leaves with ISR-inducing WCS417r bacteria [75]. Moreover, infiltration of three lower leaves per plant with WCS417r or WCS358r resulted in a significant level of protection against *P. syringae* pv. *tomato* in the nontreated leaves, whereas WCS374r did not induce resistance. These results are in full agreement with those obtained after application of WCS417r, WCS358r, or WCS374r bacteria to the roots [50,80] (Fig. 3A) and demonstrate that ISR-inducing rhizobacteria trigger the same systemic signaling pathway when applied to either roots or leaves.

5. ISR Requires Ethylene-Dependent Signaling at the Site of Induction

Knoester et al. [90] tested a set of well-characterized Arabidopsis mutants that are affected at different steps in the ethylene-signaling pathway for their ability to express ISR. None of the mutants developed ISR against *P. syringae* pv. *tomato* after treatment of the roots with WCS417r, confirming that an intact ethylene-signaling pathway is required for the expression of ISR. Mutant *eir1-1*, which is insensitive to ethylene in the roots but not in the shoots, was able to mount ISR when WCS417r was infiltrated into the leaves, but not when the bacteria were applied to the roots. If ethylene signaling were required only for the systemic expression of ISR at the site of challenge inoculation, *eir1-1* plants should develop normal levels of ISR in the leaves after application of WCS417r to the roots. However, this was not the case. Therefore, one can postulate that ethylene signaling is required at the site of application of the inducer and may be involved in the generation or translocation of the systemically transported ISR signal [90] (Fig. 3B).

B. Molecular-Genetic Analysis

1. Search for ISR-Related Genes in *Arabidopsis*

The state of pathogen-induced SAR is characterized by the concomitant activation of a set of *PR* genes. In SAR-expressing plants, *PR* gene products accumulate systemically to levels from 0.3 to 1% of the total mRNA and protein content [98]. However, although some PRs possess antimicrobial activity, a causal relationship between accumulation of PRs and the broad-spectrum resistance characteristic of SAR has never been convincingly demonstrated [99]. Of many defense-related genes tested in Arabidopsis (e.g., the well-characterized SA-inducible genes *PR-1*, *PR-2*, and *PR-5*, and JA- and/or ethylene-responsive genes *Lox1*, *Lox2*, *Atvsp*, *Pdf1.2*, *Hel*, *ChiB*, and *Pal1*, none were found to be upregulated in plants expressing ISR, either locally in the roots, or systemically in the leaves [54] (Fig. 4). Moreover, neither standard differential screening of a cDNA library of WCS417r-induced plants nor 2D-gel analysis of proteins from induced and noninduced plants yielded significant differences [100]. Thus, in contrast to SAR, the onset

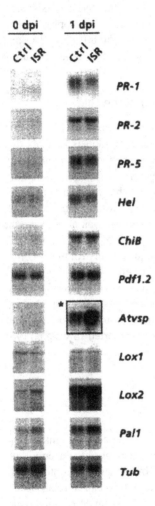

Figure 4 Expression of defense-related genes in ISR-expressing Arabidopsis leaf tissue after challenge inoculation with *P. syringae* pv. *tomato*. ISR was induced by growing the plants in soil containing WCS417r. Leaves of control-treated plants and leaves expressing ISR were harvested 0 and 1 day after inoculation (dpi). The box with the asterisk shows potentiation of the *Atvsp* gene in ISR-expressing tissue.

of ISR is not associated with major changes in gene expression. Nevertheless, ISR-expressing plants are clearly more resistant to different types of pathogens. Therefore, plants must possess as yet undiscovered defense-related gene products that contribute to their broad-spectrum disease resistance.

In a search for ISR-related genes, a large collection of Arabidopsis lines containing enhancer-trap *Ds* transposons with a promoterless β-glucuronidase (GUS) reporter gene [101] was screened. One enhancer-trap line showed local GUS activity in the roots upon colonization by WCS417r (K. M. Léon-Kloosterziel, unpublished). The roots of this line showed a similar expression pattern after treatment with the ethylene precursor ACC, indicating that this line contains a transposon insertion in the vicinity of an ethylene-inducible gene that is upregulated in the roots upon colonization by WCS417r. There are several candidate genes in the vicinity of the enhancer-trap *Ds* transposon, one of which encodes a thaumatin-like protein with homology to tobacco PR-5 and tomato osmotin (K. M. Léon-Kloosterziel, unpublished). However, the role of this gene in ISR remains to be elucidated.

2. Potentiation of JA-Dependent Responses in Plants Expressing ISR

Potentiation is expressed as a faster and stronger activation of defense responses of induced plants after infection with a challenging pathogen and can make the plant react more effectively to an invading pathogen. This phenomenon has been observed in different plant systems. For instance, carnation plants expressing WCS417r-mediated ISR produce significantly more phytoalexins upon stem inoculation with *F. oxysporum* f.sp. *dianthi* [5]. This enhanced phytoalexin production may contribute to limiting ingress of the pathogen in ISR-expressing plants.

To investigate whether phytoalexins are involved in ISR in Arabidopsis, plant mutants impaired in the biosynthesis of the Arabidopsis phytoalexin camalexin were tested for their ability to express WCS417r-mediated ISR. Mutants *pad1-1*, *pad2-1*, *pad3-1*, and *pad4-1* showed wild-type levels of ISR against *P. syringae* pv. *tomato* after root bacterization with WCS417r (Fig. 3A), indicating that at least in this particular plant-pathogen combination phytoalexins do not contribute to the enhanced level of resistance (C. M. J. Pieterse, unpublished).

Potentiation has been demonstrated at the level of gene expression as well. For instance, tobacco plants exhibiting pathogen-induced SAR showed enhanced expression of *PR-10* and *Pal* genes upon challenge with a pathogen [102]. Similarly, SAR-expressing Arabidopsis plants accumulated enhanced levels of *PR-1*, *PR-2*, and *PR-5* transcripts after challenge [54]. Because rhizobacteria-mediated ISR is not associated with increased production of either JA or ethylene, potentiation of JA- or ethylene-dependent defense responses points to an important role in ISR. If the JA and ethylene dependency of ISR is based on enhanced sensitivity to these signal molecules, one would expect that ISR-expressing plants react faster or more strongly to pathogen-induced JA or ethylene production.

To clarify this point, van Wees et al. [54] studied the expression of the JA-responsive genes *Atvsp*, *Pdf1.2*, *Lox2*, and *Pal1* and the ethylene-responsive genes *Hel* and *ChiB* in control and ISR-expressing plants after challenge with *P. syringae* pv. *tomato*. In noninduced control plants, pathogen infection induced the expression of all genes tested. In challenged, ISR-expressing plants, only *Atvsp* displayed an enhanced level of expression in comparison to challenged control plants (Fig. 4). The expression of the other JA-responsive genes and the expression of the ethylene-responsive genes was not potentiated, suggesting that ISR is associated with the potentiation of a specific set of JA-responsive genes. The mechanism and the significance of this phenomenon in induced broad-spectrum resistance is still unknown.

3. Identification of a Novel Locus (*ISR1*) Controlling Rhizobacteria-Mediated ISR

Previously, Leeman et al. [73] and van Wees et al. [50] provided evidence that the expression of rhizobacteria-mediated ISR varies with the host/rhizobacterium combination by showing that radish and Arabidopsis plants respond differentially to a set of ISR-inducing *Pseudomonas* spp. strains: radish developed ISR against *F. oxysporum* f.sp. *raphani* after colonization of the roots by WCS417r or WCS374r, but not in response to WCS358r, whereas Arabidopsis was responsive to WCS417r and WCS358r, but not to WCS374r (see Fig. 3A). Although both radish and Arabidopsis possess the machinery to express rhizobacteria-mediated ISR and both WCS374r and WCS358r are able to trigger this response, they clearly fail to do so in the radish/WCS358r and the Arabidopsis/WCS374r combinations. Also within plant species, differential induction of ISR has become evident. In Arabidopsis most ecotypes, e.g., Columbia and Landsberg *erecta*, develop ISR in response to treatment with WCS417r, but ecotypes RLD and Wassilewskija (Ws) are nonresponsive [50,103]. This suggests that specific recognition between the plant and the ISR-inducing rhizobacterium is required for the elicitation of ISR.

In a genetic approach to identify ISR-related genes, 10 ecotypes of Arabidopsis were screened for their potential to express ISR and SAR against *P. syringae* pv. *tomato* [103]. All ecotypes tested developed SAR. However, of the 10 ecotypes RLD and Ws did not develop ISR after treatment of the roots with WCS417r. This WCS417r-nonresponsive phenotype was associated with a relatively high susceptibility to *P. syringae* pv. *tomato*, which was apparent as both a higher proliferation of the pathogen in the leaves and more severe disease symptoms. Genetic analysis of the F_1, F_2, and F_3 progeny of a cross between the WCS417r-responsive ecotype Col-0 and the WCS417r-nonresponsive ecotype

RLD revealed that both the potential to express ISR and the relatively high level of basal resistance against *P. syringae* pv. *tomato* are monogenic, dominant traits that are genetically linked. The corresponding locus, designated *ISR1*, was mapped between CAPS markers *B4* and *GL1* on chromosome III. Neither responsiveness to WCS417r nor the relatively high level of basal resistance against *P. syringae* pv. *tomato* was complemented in the F₁ progeny of crosses between RLD and Ws, indicating that both ecotypes are affected in the same locus.

Interestingly, mutants *jar1-1* and *etr1-1*, which are affected in their response to JA or ethylene, respectively, show the same phenotype as ecotypes RLD and Ws in that they are both unable to express WCS417r-mediated ISR and show enhanced susceptibility to infection by *P. syringae* pv. *tomato* [80]. Analysis of the ethylene responsiveness of RLD and Ws revealed that both ecotypes show a reduced sensitivity to ethylene that co-segregates with the recessive alleles of the *ISR1* locus [104]. Therefore, it was hypothesized that the Arabidopsis *ISR1* locus encodes a novel component of the ethylene-response pathway that plays an important role in disease-resistance signaling.

VI. PLANT PHYSIOLOGICAL AND ECOLOGICAL ASPECTS OF INDUCED RESISTANCE

A. Plant Growth and Defense Against Different Types of Attackers

Constitutive expression of pathogen-induced SAR and PRs in Arabidopsis *cpr* mutants has invariably been associated with reduced plant growth. This suggests that maintenance of the state of SAR imparts a metabolic burden on the plant. The induction of systemic resistance in tobacco by *P. fluorescens* CHA0, which is also associated with the accumulation of PRs, likewise reduces plant growth [42]. In contrast, many other rhizobacterial strains that can induce resistance promote plant growth and are hence called plant growth–promoting rhizobacteria (PGPR). Thus, *P. fluorescens* WCS417 was found to stimulate growth of Arabidopsis by 33% [105]. The mechanisms of plant growth promotion by rhizobacteria are only poorly understood [65] and may, or may not, be linked to disease-suppressing properties. However, the absence of major changes in gene expression of plants induced by most PGPR implies only minimal metabolic costs as compared to plants expressing SAR. The level of resistance induced by rhizobacteria is always quantitatively less than SAR induced by necrotizing pathogens [103]. The general absence of PRs in the former and their presence in the latter type of induced resistance, together with the established antipathogen activities of at least some among the PRs, indicates that PRs could be responsible for the higher level of induced resistance with SAR and tissue necrosis [99].

Evidence is accumulating that components from SA-, JA-, and ethylene-dependent defense pathways can affect each other's signaling. SA and its functional analogs 2,6-dichloroisonicotinic acid and benzothiadiazole suppress JA-dependent defense gene expression [54,106–110], possibly through the inhibition of JA synthesis and action [111]. SA inhibits the JA-dependent wound response, whereas JA has been reported to both stimulate and reduce SA-dependent resistance responses [110,112–114]. Cross-talk between different signal transduction pathways is thought to provide great regulatory potential for coordinating multiple resistance mechanisms in varying combinations and may help the plant to prioritize the activation of a particular defense pathway over another [15,105,115–117]. Such cross-talk between signaling pathways leads to conflicting outcomes with respect to induced resistance to pathogens, which depends primarily on the SA pathway, and to herbivorous insects, which activate the JA pathway [116,118,119].

Since ISR and SAR share the regulatory factor NPR1, the question was raised as to what extent the JA-dependent ISR pathway and the SA-dependent SAR pathway interact. Recently, van Wees et al. [120] investigated possible antagonistic interactions between both pathways. Simultaneous activation of both pathways resulted in an additive effect on the level of induced protection against *P. syringae* pv. *tomato*. No enhanced level of protection was evident in Arabidopsis genotypes that are blocked in either the ISR or the SAR response. Expression of the SAR marker gene *PR-1* was not altered in plants expressing both ISR and SAR compared to plants expressing SAR alone, indicating that the SAR and the ISR pathway are compatible and that there is no significant cross-talk between these signaling pathways. Furthermore, plants expressing both types of induced resistance did not show elevated levels of *Npr1* transcripts. Apparently, the constitutive level of NPR1 is sufficient to facilitate simultaneous expression of both SAR and ISR. It was hypothesized that the enhanced protection in plants expressing both types of induced systemic resistance is established through parallel activation of complementary, NPR1-dependent defense responses that are both active against *P. syringae* pv. *tomato*.

B. ISR and Crop Protection

The level of protection afforded by rhizobacteria-mediated ISR is seldomly more than 50% reduction in disease severity or the numbers of diseased plants, which corresponds to at most a 50% increase in crop yield. Chemical crop protection is usually cheaper and far more effective [121,122]. Therefore, for biocontrol to become economically competitive, its reliability and effectiveness must be improved. Such improvements can be envisaged when ISR is employed in combination with other strategies for suppressing disease. For instance, reducing the activity of a pathogen by microbial antagonism will weaken its pathogenic potential and inhibit it even more when it encounters a host plant in which the defensive capacity is enhanced through the induction of systemic resistance. In several cases, application of combinations of rhizobacterial strains with different mechanisms improves disease suppression. Moreover, under conditions in which single strains may fail to reduce disease, combinations are likely to afford at least protection by the other strain(s). In the protection of radish against *F. oxysporum* f.sp. *raphani*, strains *P. putida* RE8 and *P. fluorescens* RS111a each suppressed the percentage of wilted plants by about 50%. Both strains are able to induce systemic resistance in radish, but may also possess additional antagonistic mechanisms. Disease suppression by the combination of RE8 and RS111a was significantly better as compared to the single strains [123]. Similar results were obtained in the combination of the resistance-inducing strain RE8 and *P. putida* WCS358 [124]. Suppression of Fusarium wilt of radish by WCS358 does not involve ISR [73] but occurs through siderophore-mediated competition for iron [125]. Similar additive effects have been found in the suppression of damping-off of tobacco seedlings inoculated with *Pythium torulosum*, *Pythium aphanidermatum*, or *Phytophthora parasitica*, when treatment with *Bacillus cereus* strain UW85 was combined with SA-induced SAR [126]. Combination of ISR-inducing rhizobacteria with biocontrol fungi is another option for improving crop protection. Indeed, co-inoculations of rhizobacteria and antagonistic fungi suppressed Fusarium wilt in flax, tomato, and radish under conditions in which their efficacies or population densities were too low to do so on their own [127,128].

Relatively little attention has been given to combining biocontrol agents with low doses of chemical crop protectants. However, when a resistance-inducing biocontrol agent is not sensitive to the chemical compound used, in combination they may act synergistically, such that substantially lower doses of the chemical are needed to achieve a similar level of disease control. Such approaches may substantially decrease the inputs of toxic chemicals into the environment while maintaining adequate levels of crop protection.

REFERENCES

1. J Handelsman, EV Stabb. Biocontrol of soilborne plant pathogens. Plant Cell 8: 1855–1869, 1996.
2. LC van Loon, PAHM Bakker, CMJ Pieterse. Systemic resistance induced by rhizosphere bacteria. Annu Rev Phytopathol 36:453–483, 1998.
3. TFC Chin-A-Woeng, W de Priester, AJ van der Bij, BJJ Lugtenberg. Description of the colonization of a gnotobiotic tomato rhizosphere by *Pseudomonas fluorescens* biocontrol strain WCS365, using scanning electron microscopy. Mol Plant-Microbe Interact 10:79–86, 1997.
4. MF Turlier, A Eparvier, C Alabouvette. Early dynamic interactions between *Fusarium oxysporum* f.sp. *lini* and the roots of *Linum usitatissimum* as revealed by transgenic GUS-marked hyphae. Can J Bot 72:1605–1612, 1994.

5. R van Peer, GJ Niemann, B Schippers. Induced resistance and phytoalexin accumu-
    lation in biological control of fusarium wilt of carnation by *Pseudomonas* sp. strain
    WCS417r. Phytopathology 81:728–734, 1991.
6. G Wei, JW Kloepper, S Tuzun. Induction of systemic resistance of cucumber to
    Colletotrichum orbiculare by select strains of plant growth-promoting rhizobact-
    eria. Phytopathology 81:1508–1512, 1991.
7. MS Meera, MB Shivanna, K Kageyama, M Hyakumachi. Persistence of induced
    systemic resistance in cucumber in relation to root colonization by plant growth
    promoting fungal isolates. Crop Prot 14:123–130, 1995.
8. MS Meera, MB Shivanna, K Kageyama, M Hyakumachi. Responses of cucumber
    cultivars to induction of systemic resistance against anthracnose by plant growth
    promoting fungi. Eur J Plant Pathol 101:421–430, 1995.
9. RP Larkin, DL Hopkins, FN Martin. Suppression of fusarium wilt of watermelon
    by nonpathogenic *Fusarium oxysporum* and other microorganisms recovered from
    a disease-suppressive soil. Phytopathology 86:812–819, 1996.
10. A De Cal, S Pascual, P Melgarejo. Involvement of resistance induction by *Penicil-
    lium oxalicum* in the biocontrol of tomato wilt. Plant Pathol 46:72–79, 1997.
11. BJ Duijff, D Pouhair, C Olivain, C Alabouvette, P Lemanceau. Implication of sys-
    temic induced resistance in the suppression of fusarium wilt of tomato by *Pseudo-
    monas fluorescens* WCS417r and by nonpathogenic *Fusarium oxysporum* Fo47.
    Eur J Plant Pathol 104:903–910, 1998.
12. C Cordier, MJ Pozo, JM Barea, S Gianinazzi, V Gianinazzi-Pearson. Cell defense
    responses associated with localized and systemic resistance to *Phytophthora para-
    sitica* induced in tomato by an arbuscular mycorrhizal fungus. Mol Plant-Microbe
    Interact 11:1017–1028, 1998.
13. G De Meyer, J Bigirimana, Y Elad, M Höfte. Induced systemic resistance in *Tri-
    choderma harzianum* T39 biocontrol of *Botrytis cinerea*. Eur J Plant Pathol 104:
    279–286, 1998.
14. DY Han, DL Coplin, WD Bauer, HAJ Hoitink. A rapid bioassay for screening
    rhizosphere microorganisms for their ability to induce systemic resistance. Phytopa-
    thology 90:327–332, 2000.
15. K Maleck, RA Dietrich. Defense on multiple fronts: how do plants cope with di-
    verse enemies? Trends Plant Sci 4:215–219, 1999.
16. JA Ryals, UH Neuenschwander, MG Willits, A Molina, HY Steiner, MD Hunt.
    Systemic acquired resistance. Plant Cell 8:1809–1819, 1996.
17. C Chen, EM Bauske, G Musson, R Rodriguez-Cabana, J Kloepper. Biological con-
    trol of fusarium wilt on cotton by use of endophytic bacteria. Biol Control 5:83–
    91, 1995.
18. N Benhamou, JW Kloepper, A Quadt-Hallman, S Tuzun. Induction of defense-
    related ultrastructural modifications in pea root tissues inoculated with endophytic
    bacteria. Plant Physiol 112:919–929, 1996.
19. N Benhamou, RR Bélanger, TC Paulitz. Induction of differential host responses
    by *Pseudomonas fluorescens* in *Ri* T-DNA-transformed pea roots after challenge
    with *Fusarium oxysporum* f.sp. *pisi* and *Pythium ultimum*. Phytopathology 86:
    1174–1185, 1996.
20. N Benhamou, RR Bélanger, TC Paulitz. Pre-inoculation of *Ri* T-DNA-transformed

pea roots with *Pseudomonas fluorescens* inhibits colonization by *Pythium ultimum* Trow: an ultrastructural study. Planta 199:105–117, 1996.

21. P M'piga, RR Bélanger, TC Paulitz, N Benhamou. Increased resistance to *Fusarium oxysporum* f.sp. *radicis-lycopersici* in tomato plants treated with the endophytic bacterium *Pseudomonas fluorescens* strain 63-28. Physiol Mol Plant Pathol 50: 301–320, 1997.

22. N Benhamou, P Rey, M Chérif, J Hockenhull, Y Tirilly. Treatment with the myco-parasite *Pythium oligandrum* triggers induction of defense-related reactions in to-mato roots when challenged with *Fusarium oxysporum* f.sp. *radicis-lycopersici*. Phytopathology 87:108–122, 1997.

23. P Rey, N Benhamou, E Wulff, Y Tirilly. Interactions between tomato (*Lycopersicon esculentum*) root tissues and the mycoparasite *Pythium oligandrum*. Physiol Mol Plant Pathol 53:105–122, 1998.

24. N Benhamou, S Gagné, D Le Quéré, L Dehbi. Bacterial-mediated induced resistance in cucumber: beneficial effect of the endophytic bacterium *Serratia plymuthica* on the protection against infection by *Pythium ultimum*. Phytopathology 90: 45–56, 2000.

25. C Olivain, C Alabouvette. Colonization of tomato root by a non-pathogenic strain of *Fusarium oxysporum*. New Phytol 137:481–494, 1997.

26. C Chen, RR Bélanger, N Benhamou, TC Paulitz. Defense enzymes induced in cucumber roots by treatment with plant growth-promoting rhizobacteria (PGPR) and *Pythium aphanidermatum*. Physiol Mol Plant Pathol 56:13–23, 2000.

27. I Yedidia, N Benhamou, I Chet. Induction of defense responses in cucumber plants (*Cucumis sativus* L.) by the biocontrol agent *Trichoderma harzianum*. Appl Environ Microbiol 65:1061–1070, 1999.

28. A De Cal, R Garcia-Lepe, P Melgarejo. Induced resistance by *Penicillium oxalicum* against *Fusarium oxysporum* f.sp. *lycopersici*: histological studies of infected and induced tomato stems. Phytopathology 90:260–268, 2000.

29. RK Cameron, R Dixon, C Lamb. Biologically induced systemic acquired resistance in *Arabidopsis thaliana*. Plant J 5:715–725, 1994.

30. B Mauch-Mani, JP Métraux. Salicylic acid and systemic acquired resistance to pathogen attack. Ann Bot 82:535–540, 1998.

31. TP Delaney, S Uknes, B Vernooij, L Friedrich, K Weymann, D Negrotto, T Gaffney, M Gut-Rella, H Kessmann, E Ward, J Ryals. A central role of salicylic acid in plant disease resistance. Science 266:1247–1250, 1994.

32. N Yalpani, J León, MA Lawton, I Raskin. Pathway of salicylic acid biosynthesis in healthy and virus-inoculated tobacco. Plant Physiol 103:315–321, 1993.

33. L Serino, C Reimmann, H Baur, M Beyeler, P Visca, D Haas. Structural genes for salicylate biosynthesis from chorismate in *Pseudomonas aeruginosa*. Mol Gen Genet 249:217–228, 1995.

34. J Durner, J Shah, DF Klessig. Salicylic acid and disease resistance in plants. Trends Plant Sci 2:266–274, 1997.

35. JB Rasmussen, R Hammerschmidt, MN Zook. Systemic induction of salicylic acid accumulation in cucumber after inoculation with *Pseudomonas syringae* pv. *syringae*. Plant Physiol 97:1342–1347, 1991.

36. B Vernooij, L Friedrich, A Morse, R Reist, R Kolditz-Jawhar, E Ward, S Uknes,

H Kessmann, J Ryals. Salicylic acid is not the translocated signal responsible for inducing systemic acquired resistance but is required in signal transduction. Plant Cell 6:959–965, 1994.

37. LC van Loon, JF Antoniw. Comparison of the effects of salicylic acid and ethephon with virus-induced hypersensitivity and acquired resistance in tobacco. Neth J Plant Pathol 88:237–256, 1982.

38. M Leeman, FM den Ouden, JA van Pelt, FPM Dirkx, H Steijl, PAHM Bakker, B Schippers. Iron availability affects induction of systemic resistance to fusarium wilt of radish by *Pseudomonas fluorescens*. Phytopathology 86:149–155, 1996.

39. G De Meyer, K Capieau, K Audenaert, A Buchala, JP Métraux, M Höfte. Nanogram amounts of salicylic acid produced by the rhizobacterium *Pseudomonas aeruginosa* 7NSK2 activate the systemic acquired resistance pathway in bean. Mol Plant-Microbe Interact 12:450–458, 1999.

40. LC van Loon, WS Pierpoint, T Boller, V Conejero. Recommendations for naming plant pathogenesis-related proteins. Plant Mol Biol Reporter 12:245–264, 1994.

41. LC van Loon, EA van Strien. The families of pathogenesis-related proteins, their activities, and comparative analysis of PR-1 type proteins. Physiol Mol Plant Pathol 55:85–97, 1999.

42. M Maurhofer, C Hase, P Meuwly, JP Métraux, G Défago. Induction of systemic resistance of tobacco to tobacco necrosis virus by the root-colonizing *Pseudomonas fluorescens* strain CHA0: influence of the *gacA* gene and of pyoverdine production. Phytopathology 84:139–146, 1994.

43. KS Park, JW Kloepper. Activation of PR-1a promoter by rhizobacteria that induce systemic resistance in tobacco against *Pseudomonas syringae* pv. *tabaci*. Biol Control 18:2–9, 2000.

44. RE Zdor, AJ Anderson. Influence of root colonizing bacteria on the defense responses of bean. Plant Soil 140:99–107, 1992.

45. RK Hynes, J Hill, MS Reddy, G Lazarovits. Phytoalexin production by wounded white bean (*Phaseolus vulgaris*) cotyledons and hypocotyls in response to inoculation by rhizobacteria. Can J Microbiol 40:548–554, 1994.

46. L Xue, PM Charest, SH Jabaji-Hare. Systemic induction of peroxidases, 1,3-β-glucanases, chitinases, and resistance in bean plants by binucleate *Rhizoctonia* species. Phytopathology 88:359–365, 1998.

47. E Hoffland, CMJ Pieterse, L Bik, JA van Pelt. Induced systemic resistance in radish is not associated with accumulation of pathogenesis-related proteins. Physiol Mol Plant Pathol 46:309–320, 1995.

48. E Hoffland, J Hakulinen, JA van Pelt. Comparison of systemic resistance induced by avirulent and non-pathogenic *Pseudomonas* species. Phytopathology 86:757–762, 1996.

49. CMJ Pieterse, SCM van Wees, E Hoffland, JA van Pelt, LC van Loon. Systemic resistance in *Arabidopsis* induced by biocontrol bacteria is independent of salicylic acid accumulation and pathogenesis-related gene expression. Plant Cell 8:1225–1237, 1996.

50. SCM van Wees, CMJ Pieterse, A Trijssenaar, Y van 't Westende, F Hartog, LC van Loon. Differential induction of systemic resistance in Arabidopsis by biocontrol bacteria. Mol Plant-Microbe Interact 10:716–724, 1997.

51. G De Meyer, K Audenaert, M Höfte. *Pseudomonas aeruginosa* 7NSK2-induced systemic resistance in tobacco depends on in planta salicylic acid accumulation but is not associated with PR1a expression. Eur J Plant Pathol 105:513–517, 1999.

52. M Höfte, J Bigirimana, G De Meyer, K Audenaert. Induced systemic resistance in tomato, tobacco and bean by *Pseudomonas aeruginosa* 7NSK2: bacterial determinants, signal transduction pathways and role in host resistance. Proceedings of Fifth International PGPR Workshop, Córdoba, Argentina, 2000, pp 108–113.

53. CM Press, M Wilson, S Tuzun, JW Kloepper. Salicylic acid produced by *Serratia marcescens* 90-166 is not the primary determinant of induced systemic resistance in cucumber or tobacco. Mol Plant-Microbe Interact 10:761–768, 1997.

54. SCM van Wees, M Luijendijk, I Smoorenburg, LC van Loon, CMJ Pieterse. Rhizobacteria-mediated induced systemic resistance (ISR) in *Arabidopsis* is not associated with a direct effect on expression of known defense-related genes but stimulates the expression of the jasmonate-inducible gene *Atvsp* upon challenge. Plant Mol Biol 41:537–549, 1999.

55. C Chen, RR Bélanger, N Benhamou, TC Paulitz. Role of salicylic acid in systemic resistance induced by *Pseudomonas* spp. against *Pythium aphanidermatum* in cucumber roots. Eur J Plant Pathol 105:477–486, 1999.

56. R van Peer, B Schippers. Lipopolysaccharides of plant-growth promoting *Pseudomonas* sp. strain WCS417r induce resistance in carnation to fusarium wilt. Neth J Plant Pathol 98:129–139, 1992.

57. M Leeman, JA van Pelt, FM den Ouden, M Heinsbroek, PAHM Bakker, B Schippers. Induction of systemic resistance against fusarium wilt of radish by lipopolysaccharides of *Pseudomonas fluorescens*. Phytopathology 85:1021–1027, 1995.

58. BJ Duijff, V Gianinazzi-Pearson, P Lemanceau. Involvement of the outer membrane lipopolysaccharides in the endophytic colonization of tomato roots by biocontrol *Pseudomonas fluorescens* strain WCS417r. New Phytol 135:325–334, 1997.

59. MT Esquerré-Tugayé, G Boudart, B Dumas. Cell wall degrading enzymes, inhibitory proteins, and oligosaccharides participate in the molecular dialogue between plants and pathogens. Plant Physiol Biochem 38:157–163, 2000.

60. G Felix, JD Duran, S Volko, T Boller. Plants have a sensitive perception system for the most conserved domain of bacterial flagellin. Plant J 18:265–276, 1999.

61. L Gómez-Gómez, T Boller. FLS2: an LRR receptor-like kinase involved in the perception of the bacterial elicitor flagellin in *Arabidopsis*. Mol Cell 5:1003–1011, 2000.

62. Y Takeuchi, M Yoshikawa, G Takeba, K Tanaka, D Shibata, O Horino. Molecular cloning and ethylene induction of mRNA encoding a phytoalexin elicitor-releasing factor, β-1,3-endoglucanase, in soybean. Plant Physiol 93:673–682, 1990.

63. A Quadt-Hallmann, N Benhamou, JW Kloepper. Bacterial endophytes in cotton: mechanisms of entering the plant. Can J Microbiol 43:577–582, 1997.

64. M Maurhofer, C Keel, D Haas, G Défago. Influence of plant species on disease suppression by *Pseudomonas fluorescens* strain CHA0 with enhanced antibiotic production. Plant Pathol 44:40–50, 1995.

65. WT Frankenberger, M Arshad. Phytohormones in Soils; Microbial Production and Function. New York: Marcel Dekker, 1995.

66. BR Glick, C Liu, S Ghosh, EB Dumbroff. Early development of canola seedlings

in the presence of the plant growth-promoting rhizobacterium *Pseudomonas putida* GR12-2. Soil Biol Biochem 29:1233–1239, 1997.

67. T Boller. Ethylene in pathogenesis and disease resistance. In: AK Mattoo, JD Suttle, eds. The Plant Hormone Ethylene. Boca Raton, FL: CRC Press, 1991, pp. 293–314.

68. JM Meyer, P Azelvandre, C Georges. Iron metabolism in *Pseudomonas*: salicylic acid, a siderophore of *Pseudomonas fluorescens* CHA0. BioFactors 4:23–27, 1992.

69. P Visca, A Ciervo, V Sanfilippo, N Orsi. Iron-regulated salicylate synthesis by *Pseudomonas* spp. J Gen Microbiol 139:1995–2001, 1993.

70. M Maurhofer, C Reimmann, P Schmidli-Sacherer, S Heeb, D Haas, G Défago. Salicylic acid biosynthetic genes expressed in *Pseudomonas fluorescens* strain P3 improve the induction of systemic resistance in tobacco against tobacco necrosis virus. Phytopathology 88:678–684, 1998.

71. S Buysens, K Heungens, J Poppe, M Höfte. Involvement of pyochelin and pyoverdin in suppression of *Pythium*-induced damping-off of tomato by *Pseudomonas aeruginosa* 7NSK2. Appl Environ Microbiol 62:865–871, 1996.

72. G De Meyer, M Höfte. Salicylic acid produced by the rhizobacterium *Pseudomonas aeruginosa* 7NSK2 induces resistance to leaf infection by *Botrytis cinerea* on bean. Phytopathology 87:588–593, 1997.

73. M Leeman, JA van Pelt, FM den Ouden, M Heinsbroek, PAHM Bakker, B Schippers. Induction of systemic resistance by *Pseudomonas fluorescens* in radish cultivars differing in susceptibility to fusarium wilt, using a novel bioassay. Eur J Plant Pathol 101:655–664, 1995.

74. J Mercado-Blanco, KMGM van der Drift, PE Olsson, JE Thomas-Oates, LC van Loon, PAHM Bakker. Analysis of the *pmsCEAB* gene cluster involved in the biosynthesis of salicylic acid and the siderophore pseudomonine in the biocontrol strain *Pseudomonas fluorescens* WCS374. J Bacteriol 183:1909–1920, 2001.

75. CMJ Pieterse, JA van Pelt, J Ton, S Parchmann, MJ Mueller, AJ Buchala, JP Métraux, LC van Loon. Rhizobacteria-mediated induced systemic resistance (ISR) in *Arabidopsis* requires sensitivity to jasmonate and ethylene but is not accompanied by an increase in their production. Physiol Mol Plant Pathol 57:123–134, 2000.

76. C Nawrath, JP Métraux. Salicylic acid induction-deficient mutants of *Arabidopsis* express *PR-2* and *PR-5* and accumulate high levels of camalexin after pathogen inoculation. Plant Cell 11:1393–1404, 1999.

77. P Kachroo, K Yoshioka, J Shah, HK Dooner, DF Klessig. Resistance to turnip crinkle virus in Arabidopsis is regulated by two host genes and is salicylic acid dependent but *NPR1*, ethylene, and jasmonate independent. Plant Cell 12:677–690, 2000.

78. P Vijayan, J Shockey, CA Levesque, RJ Cook, J Browse. A role for jasmonate in pathogen defense of Arabidopsis. Proc Natl Acad Sci USA 95:7209–7214, 1998.

79. PE Staswick, GY Yuen, CC Lehman. Jasmonate signaling mutants of *Arabidopsis* are susceptible to the soil fungus *Pythium irregulare*. Plant J 15:747–754, 1998.

80. CMJ Pieterse, SCM van Wees, JA van Pelt, M Knoester, R Laan, H Gerrits, PJ Weisbeek, LC van Loon. A novel signaling pathway controlling induced systemic resistance in *Arabidopsis*. Plant Cell 10:1571–1580, 1998.

81. M McConn, RA Creelman, E Bell, JE Mullet, J Browse. Jasmonate is essential for insect defense in Arabidopsis. Proc Natl Acad Sci USA 94:5473–5477, 1997.

82. MJ Stout, AL Fidantsef, SS Duffey, RM Bostock. Signal interactions in pathogen and insect attack: systemic plant-mediated interactions between pathogens and herbivores. Physiol Mol Plant Pathol 54:115–130, 1999.

83. M Knoester, LC van Loon, J van den Heuvel, J Hennig, JF Bol, HJM Linthorst. Ethylene-insensitive tobacco lacks nonhost resistance against soil-borne fungi. Proc Natl Acad Sci USA 95:1933–1937, 1998.

84. T Hoffman, JS Schmidt, X Zheng, AF Bent. Isolation of ethylene-insensitive soybean mutants that are altered in pathogen susceptibility and gene-for-gene disease resistance. Plant Physiol 119:935–949, 1999.

85. BPHJ Thomma, K Eggermont, IAMA Penninckx, B Mauch-Mani, R Vogelsang, BPA Cammue, WF Broekaert. Separate jasmonate-dependent and salicylate-dependent defense-response pathways in Arabidopsis are essential for resistance to distinct microbial pathogens. Proc Natl Acad Sci USA 95:15107–15111, 1998.

86. BPHJ Thomma, K Eggermont, FMJ Tierens, WF Broekaert. Requirement of a functional *ethylene-insensitive 2* gene for efficient resistance of Arabidopsis to infection by *Botrytis cinerea*. Plant Physiol 121:1093–1101, 1999.

87. C Norman-Setterblad, S Vidal, TE Palva. Interacting signal pathways control defense gene expression in *Arabidopsis* in response to cell wall-degrading enzymes from *Erwinia carotovora*. Mol Plant-Microbe Interact 13:430–438, 2000.

88. PJ O'Donnell, C Calvert, R Atzorn, C Wasternack, HMO Leyser, DJ Bowles. Ethylene as a signal mediating the wound response of tomato plants. Science 274:1914–1917, 1996.

89. IAMA Penninckx, BPHJ Thomma, A Buchala, JP Métraux, JM Manners, WF Broekaert. Concomitant activation of jasmonate and ethylene response pathways is required for induction of a plant defensin gene in Arabidopsis. Plant Cell 10: 2103–2113, 1998.

90. M Knoester, CMJ Pieterse, JF Bol, LC van Loon. Systemic resistance in *Arabidopsis* induced by rhizobacteria requires ethylene-dependent signaling at the site of application. Mol Plant-Microbe Interact 12:720–727, 1999.

91. E Bell, RA Creelman, JE Mullet. A chloroplast lipoxygenase is required for wound-induced jasmonic acid accumulation in *Arabidopsis*. Proc Natl Acad Sci USA 92: 8675–8679, 1995.

92. H Cao, SA Bowling, AS Gordon, X Dong. Characterization of an *Arabidopsis* mutant that is nonresponsive to inducers of systemic acquired resistance. Plant Cell 6:1583–1592, 1994.

93. T Delaney, L Friedrich, J Ryals. *Arabidopsis* signal transduction mutant defective in plant disease resistance. Proc Natl Acad Sci USA 92:6602–6606, 1995.

94. J Shah, F Tsui, DF Klessig. Characterization of a salicylic acid-insensitive mutant (*sai1*) of *Arabidopsis thaliana* identified in a selective screen utilizing the SA-inducible expression of the *tms2* gene. Mol Plant-Microbe Interact 10:69–78, 1997.

95. Y Zhang, W Fan, M Kinkema, X Li, X Dong. Interaction of NPR1 with basic leucine zipper protein transcription factors that bind sequences required for salicylic acid induction of the *PR-1* gene. Proc Natl Acad Sci USA 96:6523–6528, 1999.

96. JM Zhou, Y Trifa, H Silva, D Pontier, E Lam, J Shah, DF Klessig. NPR1 differen-

tially interacts with members of the TGA/OBF family of transcription factors that bind an element of the *PR-1* gene required for induction by salicylic acid. Mol Plant-Microbe Interact 13:191–202, 2000.

97. C Després, C DeLong, S Glaze, E Liu, PR Fobert. The Arabidopsis NPR1/NIM1 protein enhances the DNA binding activity of a subgroup of the TGA family of bZIP transcription factors. Plant Cell 12:279–290, 2000.

98. K Lawton, K Weymann, L Friedrich, B Vernooij, S Uknes, J Ryals. Systemic acquired resistance in *Arabidopsis* requires salicylic acid but not ethylene. Mol Plant-Microbe Interact 8:863–870, 1995.

99. LC van Loon. Induced resistance in plants and the role of pathogenesis-related proteins. Euro J Plant Pathol 103:753–765, 1997.

100. SCM van Wees, CMJ Pieterse, R DeRose, T Rabilloud, LC van Loon. Attempted identification of genes and proteins associated with rhizobacteria-mediated induced systemic resistance (ISR) in *Arabidopsis*. In: SCM van Wees. Rhizobacteria-Mediated Induced Systemic Resistance in *Arabidopsis*: Signal Transduction and Expression. Ph.D. thesis, Utrecht University, 1999, pp. 95–103.

101. CW Vroemen, N Aarts, PMJ in der Rieden, A van Kammen, SC de Vries. Identification of genes expressed during *Arabidopsis thaliana* embryogenesis using enhancer trap and gene trap *Ds*-transposons. In: F LoSchavio, RL Last, G Morelli, NV Raikhel, eds. Cellular Integration of Signal Transduction Pathways. Berlin: Springer, 1998, pp. 207–232.

102. LAJ Mur, G Naylor, SAJ Warner, JM Sugars, RF White, J Draper. Salicylic acid potentiates defence gene expression in tissue exhibiting acquired resistance to pathogen attack. Plant J 9:559–571, 1996.

103. J Ton, CMJ Pieterse, LC van Loon. Identification of a locus in Arabidopsis controlling both the expression of rhizobacteria-mediated induced systemic resistance (ISR) and basal resistance against *Pseudomonas syringae* pv. *tomato*. Mol Plant-Microbe Interact 12:911–918, 1999.

104. J Ton, S Davison, SCM van Wees, LC van Loon, CMJ Pieterse. The Arabidopsis *ISR1* locus controlling rhizobacteria-mediated induced systemic resistance is involved in ethylene signaling. Plant Physiol 125:652–661, 2001.

105. CMJ Pieterse, LC van Loon. Salicylic acid-independent plant defence pathways. Trends Plant Sci 4:52–58, 1999.

106. HM Doherty, RR Selvendran, DJ Bowles. The wound response of tomato plants can be inhibited by aspirin and related hydroxy-benzoic acids. Physiol Mol Plant Pathol 33:377–384, 1988.

107. H Peña-Cortés, T Albrecht, S Prat, EW Weiler, L Willmitzer. Aspirin prevents wound-induced gene expression in tomato leaves by blocking jasmonic acid biosynthesis. Planta 191:123–128, 1993.

108. IAMA Penninckx, K Eggermont, FRG Terras, BPHJ Thomma, GW De Samblanx, A Buchala, JP Métraux, JM Manners, WF Broekaert. Pathogen-induced systemic activation of a plant defensin gene in *Arabidopsis* follows a salicylic acid-independent pathway. Plant Cell 8:2309–2323, 1996.

109. SA Bowling, JD Clarke, Y Liu, DF Klessig, X Dong. The *cpr5* mutant of Arabidopsis expresses both NPR1-dependent and NPR1-independent resistance. Plant Cell 9:1573–1584, 1997.

110. T Niki, I Mitsuhara, S Seo, N Ohtsubo, Y Ohashi. Antagonistic effect of salicylic acid and jasmonic acid on the expression of pathogenesis-related (PR) protein genes in wounded mature tobacco plants. Plant Cell Physiol 39:500–507, 1998.

111. SH Doares, J Narváez-Vásquez, A Conconi, C Ryan. Salicylic acid inhibits synthesis of proteinase inhibitors in tomato leaves induced by systemin and jasmonic acid. Plant Physiol 108:1741–1746, 1995.

112. KA Lawton, SL Potter, S Uknes, J Ryals. Acquired resistance signal transduction in *Arabidopsis* is ethylene independent. Plant Cell 6:581–588, 1994.

113. P Schweizer, A Buchala, JP Métraux. Gene expression patterns and levels of jasmonic acid in rice treated with the resistance inducer 2,6-dichloroisonicotinic acid. Plant Physiol 115:61–70, 1997.

114. Y Xu, PFL Chang, D Liu, ML Narasimhan, KG Raghothama, PM Hasegawa, RA Bressan. Plant defense genes are synergistically induced by ethylene and methyl jasmonate. Plant Cell 6:1077–1085, 1994.

115. P Reymond, EE Farmer. Jasmonate and salicylate as global signals for defense gene expression. Curr Opin Plant Biol 1:404–411, 1998.

116. RM Bostock. Signal conflicts and synergies in induced resistance to multiple attackers. Physiol Mol Plant Pathol 55:99–109, 1999.

117. J Glazebrook. Genes controlling expression of defense responses in Arabidopsis. Curr Opin Plant Biol 2:280–286, 1999.

118. GW Felton, KL Korth, JL Bi, SV Wesley, DV Huhman, MC Mathews, JB Murphy, C Lamb, RA Dixon. Inverse relationship between systemic resistance of plants to microorganisms and to insect herbivory. Curr Biol 9:317–320, 1999.

119. ND Paul, PE Hatcher, JE Taylor. Coping with multiple enemies: an integration of molecular and ecological perspectives. Trends Plant Sci 5:220–225, 2000.

120. SCM van Wees, EAM de Swart, JA van Pelt, LC van Loon, CMJ Pieterse. Enhancement of induced disease resistance by simultaneous activation of salicylate- and jasmonate-dependent defense pathways in *Arabidopsis thaliana*. Proc Natl Acad Sci USA 97:8711–8716, 2000.

121. DE Mathre, RJ Cook, NW Callan. From discovery to use: traversing the world of commercializing biocontrol agents for plant disease control. Plant Dis 83:972–983, 1999.

122. GD Lyon, AC Newton. Do resistance elicitors offer new opportunities in integrated disease control strategies? Plant Pathol 46:636–641, 1997.

123. M de Boer, I van der Sluis, LC van Loon, PAHM Bakker. Combining fluorescent *Pseudomonas* spp. strains to enhance suppression of fusarium wilt of radish. Eur J Plant Pathol 105:201–210, 1999.

124. M de Boer. Combining *Pseudomonas* strains to improve biological control of fusarium wilt in radish. Ph.D. thesis, Utrecht University, 2000.

125. JM Raaijmakers, M Leeman, MMP van Oorschot, I van der Sluis, B Schippers, PAHM Bakker. Dose-response relationships in biological control of fusarium wilt of radish by *Pseudomonas* spp. Phytopathology 85:1075–1081, 1995.

126. J Chen, LM Jacobson, J Handelsman, RM Goodman. Compatibility of systemic acquired resistance and microbial biocontrol for suppression of plant disease in a laboratory assay. Mol Ecol 5:73–80, 1996.

127. C Alabouvette, P Lemanceau, C Steinberg. Recent advances in the biological control of fusarium wilts. Pestic Sci 37:365–373, 1993.

128. M Leeman, FM den Ouden, JA van Pelt, C Cornelissen, A Matamala-Garros, PAHM Bakker, B Schippers. Suppression of fusarium wilt of radish by co-inoculation of fluorescent *Pseudomonas* spp. and root-colonizing fungi. Euro J Plant Pathol 102:21–31, 1996.

17
Comprehensive Testing of Biocontrol Agents

Suseelendra Desai
National Research Centre for Groundnut, Indian Council of Agricultural Research, Junagadh, Gujarat, India

Munagala S. Reddy and Joseph W. Kloepper
Auburn University, Auburn, Alabama

I. INTRODUCTION

Economically important agricultural, horticultural, and ornamental crop plants are attacked by various soilborne and foliar pathogenic fungi, resulting in billions of dollars in cumulative crop losses. Currently, the most widely used control measure for suppressing these diseases is the use of fungicides. However, problems encountered, such as development of resistance by pathogen to fungicides, inability of seed-treated fungicides to protect the roots of mature plants, rapid degradation of the chemicals, and a requirement for repeated applications, have given impetus to alternative disease-control measures. Other factors leading to increased interest in alternatives include the increasing cost of soil fumigation, lack of suitable replacements for methyl bromide, and public concerns over exposure to fungicides. Both the agriculture and agri-food sectors are now being expected to move toward environmentally sustainable development, while maintaining productivity. These concerns and expectations have led to renewed interest on the use of ''biologically based pest-management strategies.'' One approach to such biologically based strategies is the use of naturally occurring and environmentally safe biocontrol agents (BCAs) such as plant growth–promoting rhizobacteria (PGPR) and fungi, used alone or in conjunction with integrated pest-management (IPM) strategies.

Plant pathogens and some BCAs have co-evolved over time, establishing specific modes of co-existence. Biological control could be either to enrich the

Table 1 Commercial Biocontrol Products Used for Plant Disease Management Worldwide

Trade name	Biocontrol agent	Target pathogen/disease	Type of formulation	Delivery/Application	Manufacturer/ Distributor
AQ10 Biofungicide	*Ampelomyces quisqualis* M-10	Powdery mildew	Water-dispersible granule	Spray	Ecogen, Inc., 2005 Cabot Blvd. West, Langhorne, PA 19074 P.O. Box 4309, Jerusalem, Israel
Bio-Fungus (formerly Anti-Fungus)	*Trichoderma* spp.	*Sclerotinia, Phytophthora, Rhizoctonia solani, Pythium* spp., *Fusarium, Verticillium*	Granular, wettable powder, sticks, and crumbles	After fumigation incorporated in soil, sprayed or injected	De Cuester, Meststoffen NV, Belgium Forstsesteenweg 30, B-2860 St-Katelijne-Waver, Belgium
Aspire	*Candida oleophila* I-182	*Botrytis* spp., *Penicillium* spp.	Wettable powder	Postharvest to fruit as drench, drip, or spray	Ecogen, Inc., 2005 Cabot Blvd. West, Langhorne, PA 19074 P.O. Box 4309, Jerusalem, Israel
Binab T	*Trichoderma harzianum* (ATCC 20476) and *Trichoderma polysporum* (ATCC 20475)	Wilt, take-all, root rot, and internal decay of wood products and decay in tree wounds	Wettable powder and pellets	Spray, mixing with potting substrate, as paste painting on tree wounds, inserting pellets in holes drilled in wood	Bio-Innovation AB, Bredholmen, Box 56, S-545 02, ALGARAS, Sweden Henry Doubleday Research, Association Sales, Ltd., Ryton on Dunsmore, Coventry, CV8 3LG, UK

Product	Organism	Target	Formulation	Application	Manufacturer
Biofox C	*Fusarium oxysporum* (nonpathogenic)	*Fusarium oxysporum, Fusarium moniliforme*	Dust or alginate granule	Seed treatment or soil incorporation	S.I.A.P.A., Via Vitorio Veneto 1 Galliera, 40010, Bologna, Italy
Bio-save 100 Bio-save 1000	*Pseudomonas syringae* ESC-10	*Botrytis cinerea, Penicillium* spp., *Mucor pyroformis, Geotrichum candidum*	Frozen cell concentrated pellets	Pellets, postharvest, to fruit as drench, dip, or spray	EcoScience Corp., Produce Systems Div., P.O. Box 3228, Orlando, FL 32802
Bio-save 110	*Pseudomonas syringae* ESC-11	*Botrytis cinerea, Penicillium* spp., *Mucor pyroformis, Geotrichum candidum*	Frozen cell concentrated pellets	Pellets, postharvest to fruit as drench, dip, or spray	EcoScience Corp., Produce Systems Div., P.O. Box 3228, Orlando, FL 32802
BlightBan A506	*Pseudomonas fluorescens* A506	Frost damage, *Erwinia amylovora*, and russet-inducing bacteria	Wettable powder	Bloom time spray of the flower and fruit	Plant Health Technologies, 926 E. Santa Ana, Fresno, CA 93704
Cedomon	*Pseudomonas chlororaphis*	Leaf stripe, net blotch, *Fusarium* spp., spot blotch, leaf spot, and others	Seed treatment	Seed dressing	BioAgri AB, P.O. Box 914, Dag Hammarskjolds, 180 SE-751 09, Uppsala, Sweden
Companion	*Bacillus subtilis* GB03	*Rhizoctonia, Pythium, Fusarium*, and *Phytophthora*	Liquid	Drench at time of seeding and transplanting or as a spray for turf	Growth Products, P.O. Box 1259, Westmoreland Ave., White Plains, NY, 10602

Table 1 Continued

Trade name	Biocontrol agent	Target pathogen/disease	Type of formulation	Delivery/Application	Manufacturer/Distributor
Conquer	*Pseudomonas fluorescens*	*Pseudomonas tolassii*	Liquid	Spray	Mauri Foods, 67 Epping Rd., North Ryde, Australia Sylvan Spawn Laboratory, West Hills Industrial Park, Kittanning, PA 16201
Contans	*Coniothyrium minitans*	*Sclerotinia sclerotiorum* and *S. minor*	Water-dispersible granule	Spray	Prophyta Biologischer Pflanzenschutz GmbH, Inselstrasse 12, D-23999 Malchow/Poel, Germany
Deny (formerly Blue Circle, Precept)	*Burkholderia cepacia* (*Pseudomonas cepacia*) type Wisconsin	*Rhizoctonia, Pythium, Fusarium,* and disease caused by lesion, spiral, lance, and sting nematodes	Peat-based dried biomass from solid fermentation; aqueous suspension	Applied to seeds with a sticking agent in planter box (aqueous suspension formulation is for use in drip irrigation or as a seedling drench)	Stine Microbial Products, 6613 Haskins, Shawnee, KS 66216

Product	Organism	Target	Formulation	Application	Supplier
Epic	Bacillus subtilis	Rhizoctonia solani, Fusarium spp., Alternaria spp., and Aspergillus spp. that attack roots	Dry powder	Added to a slurry; mix with a chemical fungicide for commercial seed treatment	Gustafson, Inc., P.O. Box 660065, Dallas, TX 75266
Fusaclean	Fusarium oxysporum (nonpathogenic)	Fusarium oxysporum	Spores, microgranule	In drip to rock wool; incorporate in potting mix; in row	Natural Plant Protection, Route d'Artix B.P. 80,64150, Nogueres, France
Galltrol-A	Agrobacterium radiobacter Strain 84	Crown gall caused by Agrobacterium tumefaciens	Petri plates with pure culture grown on agar	Bacterial mass from one plate in one gallon of nonchlorinated water; suspension applied to seeds, seedlings, cuttings, roots, stems, and as soil drench	AgBioChem, Inc., 3 Fleetwood Ct., Orinda, CA 94563
Intercept	Pseudomonas cepacia	Rhizoctonia solani, Fusarium spp., Pythium spp.			Soil Technologies Corp., RR 4, Box 133, Fairfield, IA 52556
Kodiak, Kodiak HB, Kodiak AT	Bacillus subtilis	Rhizoctonia solani, Fusarium spp., Alternaria spp., and Aspergillus spp. that attack roots	Dry powder; usually applied with chemical fungicides	Added to a slurry mix for seed treatment; hopper box treatment	Gustafson, Inc., P.O. Box 660065, Dallas, TX 75266
KONI	Coniothyrium minitans	Sclerotinia sclerotiorum and S. minor	Granules	Incorporated into soil or soilless mix granule	BIOVED, Ltd., Ady Endre u. 10, 2310 Szigetszentmiklos, Hungary

Table 1 Continued

Trade name	Biocontrol agent	Target pathogen/disease	Type of formulation	Delivery/Application	Manufacturer/ Distributor
Monitor SD	*Trichoderma* spp.	Soilborne plant pathogens	Seed dresser	Seed dressing	M/s Agriland Biotech Pvt. Ltd.36, Industrial Estate, Savli, Baroda, India
Monitor WP	*Trichoderma* spp.	Soilborne plant pathogens	Wettable powder	Soil application	M/s Agriland Biotech Pvt. Ltd. 36, Industrial Estate, Savli, Baroda, India
Mycostop	*Streptomyces griseoviridis* strain K61	*Fusarium* spp., *Alternaria brassicola, Phomopsis* spp., *Botrytis* spp., *Pythium* spp., and *Phytophthora* spp. that cause seed, root, and stem rot, and wilt disease	Powder	Drench, spray or through irrigation system	Kemira Agro Oy, Porkkalankatu 3, P.O. Box 330, 00101 Helsinki, Finland
Nogall, Diegall	*Agrobacterium radiobacter*	*Agrobacterium tumifaciens*	Washed plates; culture suspensions	Root dips	Bio-Care Technology Pty. Ltd., RMB 1084, Pacific Highway, Somersby, NSW 2250, Australia

Norbac 84C	Agrobacterium radiobacter strain K84	Crown gall caused by Agrobacterium tumefaciens	Aqueous suspension containing bacterial cells, methyl cellulose, and phosphate buffer (refrigerate)	Root, stem, cutting dip, or spray	New BioProducts, Inc., 4737 N.W. Elmwood Dr., Corvallis, OR 97330
Paecil (also known as Bioact)	Paecilomyces lilacinus	Various nematode spp.	Dry spore concentrate	Seedling or soil drench	Technological Innovation Corporation Pvt. Ltd., Innovation House, 124 Gymnasium Dr., Macquarie University, Sydney NSW, 2109, Australia
Phagus	Bacteriophage	Pseudomonas tolaasii	Bacterial suspension		Natural Plant Protection, Route d' Artix B.P. 80, 64150 Nogueres, France
Polygandron	Pythium oligandrum	Pythium ultimum	Granule or powder	Seed treatment or soil incorporation	Vyskumny ustav rastlinnej [Plant Production Institute], Bratislavsk cesta 122, 921 68 Piestany, Slovak Republic
Primastop	Gliocladium catenulatum	Pythium spp., Rhizoctonia solani spp., Botrytis spp., Didymella spp.	Wettable powder	Drench and incorporation	Kemira Agro Oy, Porkkalankatu 3, P.O. Box 330, 00101 Helsinki, Finland

Table 1 Continued

Trade name	Biocontrol agent	Target pathogen/disease	Type of formulation	Delivery/Application	Manufacturer/Distributor
Protus WG	*Talaromyces flavus*, isolate V117b	*Verticillium daliae, V. albo-atrum*, and *Rhizoctonia solani*	Water-dispersible powder containing ascospores	Soil or seed treatment, soil drench, root dip application	Prophyta Biolgischer Pflanzenschutz GmbH, Inselstrasse 12, D-23999 Malchow/Poel, Germany
Rhizo-Plus, Rhizo-Plus Konz	*Bacillus subtilis* FZB24	*Rhizoctonia solani, Fusarium* spp., *Alternaria* spp., *Sclerotinia, Verticillium, Streptomyces scabies*	Water-dispersible granule	As suspension for seed treatment, soil drench, dip, and addition to nutrient solutions	KFZB Biotechnik GmbH, Glienicker Weg 185, D-12489 Berlin, Germany
Root Pro	*Trichoderma harzianum*	*Rhizoctonia solani, Pythium* spp., *Fusarium* spp., and *Sclerotium rolfsii*	Fungal spores mixed with peat and other organic material	Agent is mixed into growing media at time of seeding or transplanting	Mycontrol Ltd., Alon Hagalil M.P. Nazereth Elit 17920, Israel
RootShield (also sold as Bio-Trek T-22G)	*Trichoderma harzianum* Rifai strain KRL-AG2 (T-22)	*Pythium* spp., *Rhizoctonia solani, Fusarium* spp.	Granules or wettable powder	Granules mixed with soil or potting medium; powder mixed with water and added as a soil drench	Bioworks, Inc., 122 North Genesee St., Geneva, NY 14456
Rotstop, P.g. Suspension	*Phlebia gigantea*	*Heterobasidium annosum*	Spores in inert powder	Spray, chain saw oil	Kemira Agro Oy, Porkkalankatu 3, P.O. Box 330, 00101 Helsinki, Finland

Product	Organism	Target	Formulation	Application	Manufacturer/Address
Serenade	*Bacillus subtilis*	Powdery mildew, downy mildew, Cercospora leaf spot, early blight, late blight, brown rot, fire blight, and others	Wettable powder	Spray	AgraQuest, Inc., 1105 Kennedy Place, Davis, CA 95615
Shakti SD	Bacterial spp.	Soilborne diseases of field crops	Powder	Seed dressing	Sangam Bio-Dynamic, Panoli, Gujarat, India
Spot-Less	*Pseudomonas aureofaciens* strain Tx-1	Dollar spot, anthracnose, *Phythium aphanadermatium*, michrochium patch (pink snow mold)	Liquid	Overhead irrigation; can only be used with the BioJect Automatic Fermentation System	Eco Soil Systems, Inc., 10740 Thommint Rd., San Diego, CA 92127
SoilGard (formerly GlioGard)	*Gliocladium virens* GL-21	Damping-off and root rot pathogens especially *Rhizoctonia solani* and *Pythium* spp.	Granules	Granules are incorporated in soil or soil-less growing media prior to seeding	Thermo Trilogy, 9145 Guilford Road, Suite 175, Columbia, MD 21046
Star-T WP	*Trichoderma* spp.	Soilborne pathogens of field crops, wilt of sugarcane, gummosis of citrus	Wettable powder	Seed and soil	Sangam Biodynamic, 46, GIDC, Panoli, Gujarat, India
Supresivit	*Trichoderma harzianum*	Various fungi			Borregaard BioPlant, Helsingforsgade 27B, DK-8200 & Arhus N, Denmark

Table 1 Continued

Trade name	Biocontrol agent	Target pathogen/disease	Type of formulation	Delivery/Application	Manufacturer/ Distributor
System 3	*Bacillus subtilis* GB03 and chemical pesticides	Seedling pathogens	Dust	Seed treatment in planter box	Helena Chemical Co., 6075 Poplar Avenue, Suite 500, Memphis, TN 38119
T-22G, T-22 Planter Box (also sold as Bio-Trek)	*Trichoderma harzianum* Rifai strain KRL-AG2	*Pythium* spp., *Rhizoctonia solani, Fusarium* spp., and *Sclerotinia homeocarpa*	Granules or dry powder	Granules added in-furrow with granular applicator, by broadcast application to turf, mixed with greenhouse soil, or by mixing powder with seeds in planter box, or in commercial seed treatment slurry	Bioworks, Inc. (formerly TGT, Inc.), 122 North Genesee St., Geneva, NY 14456
Trieco	*Trichoderma viride*	*Rhizoctonia* spp., *Pythium* spp., *Fusarium* spp., gray mold	Powder	Dry or wet seed, tuber, or set dressing or soil drench	Ecosense Labs (I) Pvt. Ltd., 11/B Tiwari Industrial Estate, Ram Mandir Rd., Goregaon (W), Mumbai—400 104, India

Product	Organism	Target	Formulation	Application	Manufacturer
Trichodex	*Trichoderma harzianum*	Primarily *Botrytis cinerea*, also *Colletotrichum* spp., *Fulvia fulva*, *Monilia laxa*, *Plasmopara viticola*, *Pseudoperonospora cubensis*, *Rhizopus stolonifer*, *Sclerotinia sclerotiorum*	Wettable powder	Spray	Makhteshim Chemical Works, Ltd., P.O. Box 60, Beer Sheva, Israel
Trichopel, Trichoject, Trichodowels, Trichoseal	*Trichoderma harzianum* and *T. viride*	*Armillaria, Botryosphaeria, Chondrostereum, Fusarium, Nectria, Phytophthora, Pythium, Rhizoctonia*			Agrimm Technologies, Ltd., P.O. Box 13-245, Christchurch, New Zealand
Trichoderma 2000 (formerly "TY")	*Trichoderma* sp.	*Rhizoctonia solani, Sclerotium rolfsii, Pythium* spp., *Fusarium* spp.	Incorporated into soil or potting medium		Mycontrol, Ltd., Alon Hagalil M.P. Nazereth Elit 17920, Israel
Victus	*P. fluorescens* NCIB 12089		Aqueous suspension of fermenter broth		Sylvan Spawn, Kittanning, PA

Source: http://www.barc.usda.gov/psi/bpdl/bpdlprod/bioprod.html and personal contacts of the authors.

native BCAs by specific cultural practices such as the use of organic amendments or by using introduced BCAs. Introduction of BCAs for the control of plant diseases has been practiced in agricultural fields since at least 1927 [1]. Over the intervening years, from hundreds of biological control agents identified as potential candidates, only a few have been formulated for commercial use against various diseases (Table 1). Of these, only about 5% of BCAs have actually achieved their aim [2]. This large gap between research and successful industrial exploitation is due to several factors [3,4], one of which is the lack of logical comprehensive testing of BCAs, which is the focus of this chapter.

A comprehensive testing program helps in an in-depth analysis of a candidate BCA to bring out its strengths, weaknesses, and opportunities, thereby aiding in proper selection and deployment of BCAs for exploitation of their potential. It will also help to bring out the essential criteria for a BCA such as mode of action, environmental tolerance, speed of action, and persistence. The steps involved in the development of a successful BCA are depicted in Figure 1. Testing must begin with the identification of the BCA and continue up to the commercial product. In this chapter an effort has been made to give an overview of various aspects of BCA testing.

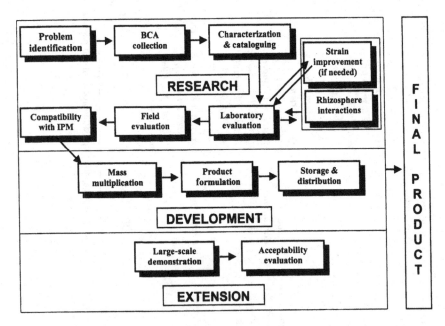

Figure 1   Steps involved in the development of a formulation of a biocontrol agent.

II. COMPREHENSIVE TESTING

For identification of a successful BCA, standard methodologies for isolation, screening, and mode of action of BCAs have been well documented [5]. Identification of an effective BCA requires thorough testing protocols at different stages. For convenience, testing at different stages has been spread over three interfaces: BCA-pathogen-crop, BCA-industry-extension, and BCA-regulatory. The critical tests that a BCA must undergo at each interface are discussed below.

A. BCA-Pathogen-Crop Interface

The BCAs exhibit different modes of action, and hence, a good testing program should elucidate all the mechanisms involved in the biocontrol activity of the BCA. Apart from biocontrol ability, the BCAs possess other traits such as rhizosphere competence, tolerance of fungicides, saprophytic competitive ability, ability to tolerate high and low temperatures, adaptability to different edaphic conditions, etc. These traits are useful for an otherwise good BCA as they help in the establishment of the BCA in a given agro-ecological region. A comprehensive testing program could be elaborative and cumbersome but is nevertheless crucial. Several protocols have been developed for the identification of a BCA, which can be broadly classified as in vitro and greenhouse and field tests.

1. In Vitro Testing

A compilation by Anjaiah [6] shows that the pseudomonads exhibit different modes of interactions such as competition, production of siderophores (pyoverdin), and antibiotics (hydrogen cyanide, oomycin A, pyoluteorin, pyrrolnitrin, 2,4-diacetyl phloroglucinol, phenazine 1-carboxylic acid, and pyocyanin) and induce resistance. Similarly, fungal BCAs also exhibit several modes of biocontrol ability such as mycoparasitism, competition, production of hydrolases, and antibiosis [7]. All the in vitro tests have been designed for identification/selection of potential BCAs and elucidate biocontrol mechanisms of known BCAs. For instance, methods such as dual culture, sclerotial parasitization, and soil tube methods are useful for identifying the mycoparasitism and competition phenomena. Under simulated conditions BCAs are checked for volatiles and nonvolatiles to comprehend the production of cell wall–degrading enzymes and antibiotics. Testing for rhizosphere competence and competitive saprophytic ability, tolerance of antibiotics, and adaptability to different edaphic conditions will help in characterizing the BCAs for their other desirable traits. These in vitro tests have been successfully used against all groups of BCAs. Some of the frequently used methods are briefly described here.

Dual Culture.   This technique, also known as biculture, cross culture, or paired culture [8], has been extensively used for preliminary screening of large populations of fungal, bacterial, and actinomycetous BCAs. In principle, the pathogen and the BCA should be allowed to interact in a petri dish under optimum conditions for both the pathogen and the BCA. The inhibition is recorded either in the form of the inhibition zone produced or the overgrowth of the pathogen by the BCA (Figs. 2 and 3). The antagonistic effects are scored in different patterns depending on the BCA to be evaluated against a pathogen [9–12]. To test the killing of the pathogen, agar discs are cut from the different zones of interaction between the BCA and the pathogen and are plated on differential media [13].

Production of Volatiles and Nonvolatile Inhibitory Compounds.   Many of BCAs produce chemicals that are inhibitory to the pathogens. These chemicals can either be volatile or be released into the medium (nonvolatile). Dennis and Webster [14,15] have developed methods for studying the production of volatile and nonvolatile inhibitory compounds by the BCA. Many individuals have used these methods with modifications wherever required [12,13]. While testing for the production of volatiles, the pathogen and the BCA are inoculated on individual petri dishes. The pathogen is then exposed to the BCA by removing the lid of the plate and inverting the dish over the petri dish inoculated with the BCA. Such plates are generally sealed with parafilm. Observations are recorded on the inhibition of the growth of the test pathogen. To test for nonvolatiles, the BCA is inoculated on sterile cellophane membrane spread over the medium in a petri dish. When BCA covers the cellophane membrane, the membrane is carefully removed and the test pathogen is introduced in the same petri dish. The inhibition of growth is a measure of the nonvolatile chemicals released by the BCA into the medium. Care should be taken to maintain both positive and negative controls to avoid erroneous conclusions.

Sclerotial Parasitization.   Often plant pathogens survive in the form of sclerotia to overcome adverse climatic conditions. Parasitization of sclerotia by a BCA would be quite useful in reducing potential inoculum. Tests could be conducted in the laboratory or in greenhouses under controlled conditions for assessing the ability of a BCA to successfully infect and kill sclerotia. In petri dishes, the sclerotia are first exposed to the BCA for a specified time and then the BCA is allowed to grow on the sclerotia. Subsequently, the BCA is differentially killed and the viability of sclerotia is checked. Alternatively, the sclerotia can be buried in sterilized sand in the petri dish and then inoculated with the suspensions of antagonists. Effective strains would macerate and kill the sclerotia and, hence, could be used as a measure of the efficacy of the BCA [16]. Köhl and Schlösser [17,18] observed variability in the efficacy among the isolates of *Trichoderma* spp. against the sclerotia of *Botrytis cinerea*. The efficacy also varied with the temperatures at which the tests were conducted. Desai and Schlösser

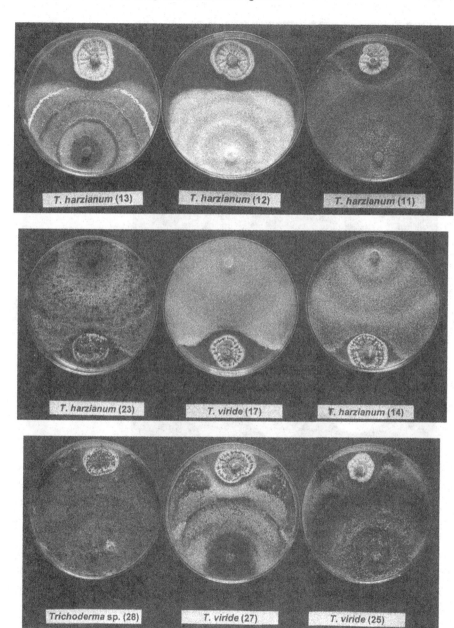

Figure 2 Dual culture technique for evaluation of isolates of *Trichoderma* against *Aspergillus flavus* causing aflatoxin contamination in peanut (*Arachis hypogaea*).

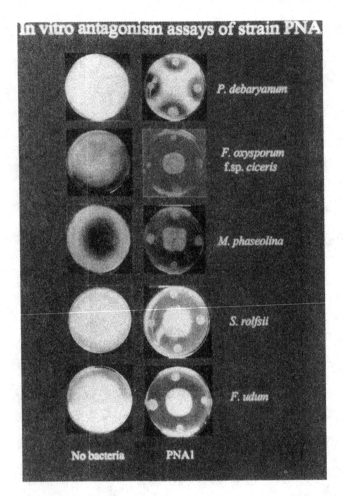

Figure 3 In vitro antagonism of *Psuedomonas fluorescens* (PNR1) against different soilborne plant pathogens. (From Ref. 6.)

[19] tested 44 isolates of *Trichoderma* spp. and found that only 14 could infect and kill all the sclerotia inoculated, 13 could kill only some of the sclerotia, and 17 did not affect the sclerotial viability. Sclerotia of *Rhizoctonia solani* on potato tubers were inactivated by the BCA *Verticillium biguttatum* [20].

Competitive Saprophytic Ability. Competitive saprophytic ability (CSA) is essential for survival of a BCA under varied climatic conditions, especially in the absence of a pathogen in and around rhizosphere. CSA can be tested by the colonization potential of the BCA of the organic matter available in the rhizosphere. The test is simulated in vitro by infesting sterilized and unsterilized soils with the BCA and incubating dried stem pieces of cotton or wheat, or other substrates are incubated in these soils. The colonization of these substrates by BCAs is assessed at regular intervals [21]. A CSA index can also be calculated to compare BCAs using the following formula:

$$\text{CSA index} = \sum_{N}^{i=1} [\ln(1/1 - C_i)/(t_i)(\log P_i)]/n$$

where C is the frequency of isolation of a specific strain of BCA from the segments of organic matter, t is the time of incubation, P is the population density of conidia added to the soil, and n is the number of treatments.

Köhl et al. [22] evaluated four BCAs for suppression of sporulation of *Botrytis cinerea* and found that *Ulocladium atrum* survived varied microclimatic conditions.

Rhizosphere Competence.   The competitiveness of BCAs with other soil inhabitants in colonization of root surfaces of the host plants is termed as rhizosphere competence (RC). An index for measuring RC of *Trichoderma* spp. as used by Ahmad and Baker [23]:

$$\text{RC index} = \sum_{N}^{i=1} [\log (P_i + 1)\ln(D_i + 1)]/n$$

where P is the population density per mg of rhizosphere soil, D is the root depth, and n is the total root length.

Recent studies show that RC is significantly influenced by the elucidation of specific recognition phenomena among the BCA, the pathogen, and species as well as the physiological stage of the crop plant. This phenomenon was demonstrated among antagonists against the phytopathogenic nematodes of velvet bean (*Mucuna deeringiana*), castor bean (*Ricinus communis*), sword bean (*Canavalia ensiformis*), and Abruzzi rye (*Secale cereale*) [24]. Significant variations in rhizosphere of these crops were reported in their total bacterial counts and in the populations of spore forming-bacteria, fungi, coryneform bacteria, and chitinolytic fungi and bacteria. Nautiyal [25] screened 256 bacterial strains, each representing different morphological types, for their biocontrol activity against *Fusarium oxysporum* f.sp. *ciceri*, *R. bataticola*, and *Pythium* spp. under greenhouse conditions. The purpose was to identify a superior biocontrol strain of *Pseudomonas* spp. (NBRI9926P3) coupled with RC. Occasionally, situations arose where there would be a very good BCA that lacked rhizosphere competence or there were highly rhizosphere competent strains with very poor biocontrol ability. In such cases, molecular biology tools, such as protoplast fusion, can be used to bring together desirable traits [26].

Sensitivity to Chemicals. Several fungicides, nematicides, and bactericides are used to manage various groups of pathogens either as seed treatments or through soil application. A BCA tolerant of these chemicals may integrate well into pest-management strategies for cost-effective plant protection. Here, the chemicals would provide good short-term seed protection, and the BCA would provide long-term root protection. Desai and Schlösser [27] observed that all 17 isolates of *Trichoderma* spp. belonging to nine species aggregates were sensitive to carbendazim and procloraz with the exception of *T. piluliferum*, which was able to tolerate procloraz. However, *T. piluliferum* did not possess good biocontrol ability against *S. rolfsii*. Nautiyal [25] was able to isolate a spontaneous rifampicin-resistant (Rifr) derivative of *Pseudomonas* spp. (NBRI9926), showing a growth rate and membrane protein composition comparable to the wild type coupled with rhizosphere competence in chickpea. Papavizas and Lewis [28] have induced stable benomyl tolerance through UV mutation in strains of *T. viride* that have also possessed a high level of biocontrol ability.

Production of Hydrolases and Antibiotics. Testing for production of hydrolases and antibiotics will help in the characterization of BCA and thus deploy them in a systematic way. Several fungal BCAs, including species of *Trichoderma* and *Verticillium*, are known to produce a battery of hydrolases such as chitinase, glucanase, polygalacturonase, protease, and mannase, which help in the maceration of cell walls of those plant pathogens causing death [29–31]. In the rhizosphere of wheat, strain 2-79 produces the antibiotic phenazine-1-carboxylic acid as its primary means of suppressing take-all disease [32]. Pleban et al. [33] found chitinolytic activity in a strain of *B. cereus* showing antibiosis against *Macrophomina phaseolina*, while Sela et al. [34] characterized a collagenolytic and proteolytic enzyme from *B. cereus* that damaged the cuticles of *Meloidogyne javanica*. Inhibitory furanone, identified as 3-(2-hydroxypropyl)-4-(2-hexadienyl)-2(5H)-furanone, was isolated from the culture filtrate of *T. harzianum* that inhibited the growth of microorganisms [35].

Induction of Resistance. Apart from the direct action against plant pathogens, many antagonists, especially bacterial BCAs, induce resistance in the plant system by signaling host defense mechanisms. Some of the enzyme systems were also common for production of phytoalexin in plants and antibiotic production in pseudomonads that suppressed take-all of wheat caused by *Gaeumannomyces graminis* var. *tritici*. The antibiotics (phenazine and phloroglucinol) produced by these bacteria played a significant role in both plant defense and ecological competence [36]. In an aseptic hydroponic system, *T. harzianum* (T-203) assisted in strengthening the epidermal and cortical cells of cucumber seedlings. Wall appositions contained large amounts of callose, while peroxidase and chitinase activities were triggered within 72 hours of infection. Since all these activities are related to defense mechanisms, it was concluded that the fungus helps in the induction of such responses to protect from pathogenic microbes [37].

Other Desirable Traits. In addition to the above traits, testing the BCAs for other desirable traits such as growth on different substrates, production of different survival structures (chlamydospores, sclerotia, spores, hyphal fragments), tolerance of different ambient and edaphic conditions (high and low temperatures; soil-moisture deficit stress, soil salinity, soil alkalinity, soil acidity, water-logged conditions) could be useful for not only their characterization but also their need-based deployment. Köhl [13] observed that only selected strains could tolerate extreme temperatures. BCAs may respond differentially to varied soil conditions. A soil-moisture deficit beyond -4.54 bars affected sporulation of *T. viride*, but not *T. harzianum* [38]. The antagonistic potential of *Trichoderma* spp. against *Fusarium udum* was not greatly altered by changing the environmental conditions. However, it was maximal at 35°C $\pm$ 2 and pH 6.5 over a wide range of C/N ratios [39].

2. Greenhouse and Field Testing

The performance of a BCA could be specific either to location or to a pathogen. Alternatively, a few BCAs can perform across locations and against more than one pathogen [40]. In both situations, tests must be conducted thoroughly to assess the actual potentiality of a BCA so that it can be maximally exploited, and several methods are available for such evaluation. Of several BCA isolates, only a few may qualify for greenhouse and field testing. The main objectives of greenhouse and field testing are to:

1. Select active BCAs that control a spectrum of plant pathogens
2. Evaluate selected BCAs under a set of environmental conditions
3. Evaluate test formulations and application methods

Once the objectives of screening are clear, screening in greenhouse and field should ensure enough and known plant pathogen population to induce sufficiently high disease pressure. These tests are also influenced by choice of BCA, testing environmental conditions, choice of cultivars, type of formulation, application method, and method and time of disease assessment. Simulation of testing situations gives an opportunity for the effective evaluation of a BCA. In all these cases, the BCA can be applied as seed treatments or directly to the soil. Soil application would be highly effective and cost effective in cases of transplanted crops where the nurseries would be protected and during which time the BCA could establish on the roots, thus protecting the crop even after transplanting [41]. The method of application of a BCA depends on the type of pathogen to be managed, the stage of crop to be protected, the nature and spread of disease in a region, and the climatic conditions of the region. An attempt has been made here to discuss the principle behind testing of a BCA.

Under greenhouse conditions, isolates of *Trichoderma* and *Gliocladium* were evaluated for the management of root rot and stem rot of groundnut caused by *R. solani* and *S. rolfsii* [42]. The BCA was applied as seed treatment delivered in a wheat-bran sawdust preparation. Similarly, for the management of root rot of pepper caused by *Phytophthora capsici*, Sid Ahmed et al. [43] used alveolar trays containing 2:1 mixture of peat and sand for the evaluation of efficacy of *T. harzianum*. These trays were planted with pepper plants and incubated in growth chambers. In greenhouse tests, of several geocarposphere bacteria, seven isolates significantly reduced colonization of peanut pods by *A. flavus* [44]. An inoculated flat test has been devised [45] to assess the suppressive effect of total soil microflora against *R. solani*. The seeds of test plants were surface-sterilized with sodium hypochlorite (0.5%) and densely sown into nontreated test soil filled in shallow trays. Test fungus pathogenic to test plant was introduced in one corner of the tray, and then the spread of the disease was monitored. Kay and Stewart [46] applied four BCAs as a soil amendment (sand:bran:fungal homogenate, 1:1:2) at the rate of 0.1% wheat bran/g dry soil in a soil-box method for the management of *S. cepivorum*, which causes white rot of onions. Although all BCAs provided various levels of disease control equivalent to the fungicide (procymidone 0.5 g a.i./100 g seed) treatment, *Cheatomium globosum* and *Trichoderma* spp. were more effective. Successful biocontrol of *S. rolfsii* with cultures of *Trichoderma* spp. has also been reported [42,47]. Dinakaran et al. [48] and Sankar and Jeyarajan [49] used BCAs as seed dressers for the management of root rot of sesamum caused by *R. bataticola*. The endophytic *Bacillus* spp. *B. cereus, B. subtilis,* and *B. pumilus* protected cotton seedlings against *R. solani* in the greenhouse by 51, 46, and 56%, respectively [50]. Similarly, 72, 79, and 26%, reduced incidence of *S. rolfsii* in bean seedlings was achieved when inoculated with *B. subtilis, B. cereus,* and *B. pumilus*, respectively. *P. chitinolytica* was found to be effective in greenhouse, screen house, and micro plot tests against *M. javanica* on tomatoes [51]. Biocompetitive and nontoxigenic *Aspergillus parasiticus* (NRRL 13539) was introduced into the soil by sprinkler to inhibit toxigenic *Aspergillus flavus* and thus reduce aflatoxin contamination in peanuts [52]. In field trials, two bacterial isolates, 91A-539 and 91A-599, suppressed colonization of peanut pods by *A. flavus* [44].

B. BCA-Industry-Extension Interface

The commercialization of a BCA is significantly influenced by the consistent performance, persistence, safety, stable formulation, application method, viable market size, preferably low costs of production, and preferably low capital costs.

At the BCA-industry-extension interface, testing is required for large-scale multiplication of the BCA, development of formulations, evaluation of their efficacy, demonstration of the benefits of using biopesticides to the end users, and assessment of the performance of the technology and its refinement back in the laboratory. For undertaking these activities, a strong network of scientist-industry-extension-farmer is required. A few efforts made so far in this direction are discussed below.

1. Formulation Development

For industries to commercialize such technologies, they need model studies for the large-scale multiplication of a BCA, which include a suitable and inexpensive medium, method of fermentation (solid or liquid), type of formulation (wettable powder, liquid, granular), nature of filler material, delivery systems, optimum storage conditions of the product, and information on shelf life. Often, dusts contain about 5–10% colony-forming units (cfu) of BCA by weight, wettable powders might have 50–80% cfu, granular formulations might contain 5–20% cfu, and liquid formulations about 10–40% cfu (S. Desai, personal communication). The testing protocols must consider the optimum formulation type as well as concentration of BCA. Certain specific conditions might increase the efficacy of a formulation. Addition of organic acids to *T. koningii* formulations and polysaccharides and polyhydroxyl alcohols to *T. harzianum* increased the activity of the BCAs [53].

Gangadharan et al. [54] found that tapioca rind, tapioca refuse, and well-decomposed farmyard manure formed good substrates for the mass production of *T. viride* and *T. harzianum*. A prototype fermentation system developed by Papavizas et al. [55] could be successfully used for fungal BCAs. This system, however, does not take care of all optimum conditions but can still help in the preliminary evaluation of certain conditions, such as finding a suitable medium for the mass multiplication of fungal BCAs. Vermiculite-based fermenter biomass of formulation with an initial population of 205×10^6 cfu/g stored in milky white bags showed an exponential phase up to 30 days (309×10^6 cfu/g). Further temperature of 20–30°C was optimum for the storage of the formulation at which, even after 75 days, the product contained $206{-}271 \times 10^6$ cfu/g [56]. Sankar and Jeyarajan [49] used similar techniques to develop seed dressing formulations of *T. viride, T. harzianum*, and *Gliocladium virens* for management of *Macrophomina phaseolina*, in sesamum. To overcome the barriers of the commercial use of phenazine-producing pseudomonads, Slininger et al. [32] optimized the culture conditions, physiological state, and associated metabolites on the biocontrol ability of pseudomonads. A fermenter biomass containing 24- to 48-hour-old cells had longer drying survival rate, but shorter shelf life. Similarly, methylcellulose-water formulations retained better viability.

2. Quality Control and Quality Assessment

One of the major bottlenecks for the widespread adaptation of the BCAs in production systems is development of reasonable quality control steps. Unlike the chemical formulations, there have been no stringent guidelines for maintaining good quality standards. Biological control is a very complicated natural phenomenon, and it requires a thorough understanding of the process for maintaining required order of quality and to design required regulations.

A number of criteria have to be fulfilled to develop a high-quality BCA product such as maintenance of a pure culture, good mass multiplication facilities, optimum formulation practices, and appropriate storage conditions. Contaminated cultures and bad fermentation and formulation conditions would lead to growth of other microbes, thus reducing the final optimal population counts of the BCA. For instance, solid-state fermentation, though simple and easy, is a potential source of cross-contamination as it is often difficult to maintain sterile conditions. Products formulated using such biomass fail to give expected results and also show variations between batches. On the contrary, liquid fermentation can offer production of pure biomass for formulation. Another concern as to the maintenance of quality involves the emergence of many entrepreneurs who do not possess the required facilities and so market spurious products in the name of biocontrol agents. Although biological control helps in managing some pathogens, such practices will lead to a loss of faith of farmers in biological control. These problems can be overcome if proper care is taken to establish good fermentation and formulation facilities and stringent regulatory systems are in place.

3. Technology Demonstration, Assessment, and Refinement

To develop a BCA through rigorous testing protocols as suggested above for each of the BCA formulations developed, its utility has to be demonstrated under field conditions. Such demonstrations need a good team of subject matter specialists who can interact with the growers and get feedback so that the technology is assessed as to its impact and refined to suit the exact needs. In general, there is reluctance to use any new technology. Hence, for its acceptance and widespread use, a proven BCA needs to be demonstrated among progressive farmers who are generally enthusiastic. They, in turn, become models for others and thus can spread the technology at a faster pace. Based on this principle, under the aegis of Indian Council of Agricultural Research, an innovative project, the Technology Assessment and Refinement-Institute Village Linkage Program, has been initiated. Each of the research centers involved adopts several villages and demonstrates the technologies as per the need of the farmers. Then the team interacts with the beneficiaries, gets feedback, refines the technology, and again brings the method to the farmers who will use it. Where the technology has been useful, it has been immediately accepted by the farmers. These beneficiaries are trained at the research centers and in their own fields. These farmers then transfer the technologies to their colleagues.

C. BCA-Regulatory Interface

With the growing demand for the use of biopesticides, several nations, including the United States, Canada, and India, are developing guidelines for the regulated use of the biopesticides. At the regulatory level, testing protocols should emphasize the criteria for the registration of BCA as biopesticides, the legal aspects of registration, safe handling of the antagonists, and mechanisms to monitor the introduced microbes. Strains of BCA would be treated as pesticides, and thus, while processing for registration, special care should be taken to generate need-based toxicological data to reduce costs. The image on BCAs has always been that they are eco-friendly. Hence, while considering the registration of biopesticides, care must be given to show that the formulations will not be harmful to mammals over time. For instance, *Trichoderma* has been considered to be an opportunistic human pathogen. Larson et al. [57] reported that *T. viride* spores at relatively high concentrations (0.1–2 mg/mL) triggered histamine release from bronchoalveolar lavage cells. *T. viride* has been on the biological warfare list of some countries in category C (S. Desai, personal communication).

1. Environmental Risk-Assessment Studies

The assessment of the environmental impact of a chemical molecule can take 7–8 years and cost several million dollars. Less information is required for the registration of BCAs because they are generally considered bio-safe, unlike chemicals. This evaluation cost is one of the major factors in the commercialization of BCAs. A well-designed regulatory process prevents the introduction of BCAs or BCA products that are potentially dangerous to the environment, thus allowing the release of all useful products without any unwarranted delays. The environmental impact studies vary depending on whether the BCA is indigenous or not, a natural isolate or genetically modified, a known pathogens of crops, humans, and other mammals, and according to its activity spectrum.

To assess the potential possible risks due to continuous usage of microbial pesticides or their products in a given environment, suitable risk-assessment studies are required. Once applied under field conditions, these BCAs are expected to establish in that region and may spread to nontarget regions over a period of time. The risks of microbial biocontrol agents to human health and to the environment should be identified as a basis for appropriate regulation of these beneficial organisms [58]. These guidelines were developed in the United States, Canada, and other countries after a series of deliberations. In India regulations were recently developed and are being further revised to suit the local federal requirements. In any case, the basic function of these regulations is to monitor and regulate the use of microbial pesticides to obviate any damage to mammals so that the basic purpose of the eco-friendly nature of these formulations is defeated. Consultation among technical specialists, manufacturers, users, academics, consultants, regulators, and public interest groups gives a balanced approach while formulating these procedures.

Agriculture and Agri-Food Canada have given a detailed account of the efforts being made in Canada for developing a four-tier testing procedure used for the assessment of risk of the microbial pesticides and their products.

2. Monitoring of BCA

After BCAs are successfully introduced and established in the soil, they should be monitored at regular intervals to understand what their population dynamics are and, in turn, their influence on the ecological niche. Such monitoring will also help regulatory authorities in making certain crucial policy decisions for either promotion or withdrawal of the formulations. While a good formulation strategy helps in the successful introduction of a BCA on plant surfaces, the ability of BCAs to efficiently occupy plant surfaces determines the method of monitoring. Production of pili and attachment of fluorescent pseudomonads to corn roots [59] and agglutination of P. putida and root colonization to suppress Fusarium wilt of cucumber [60] could be used for measuring indirectly the establishment of BCA on plant surfaces. Alternatively, monitoring could also be done with the help of marker traits that are either inherent to the BCA or induced by different mechanisms. Papavizas and Lewis [28] induced benomyl resistance in Trichoderma as their method of monitoring. With the help of rifampicin resistance, DalSoo et al. [61] monitored and studied the population dynamics of Bacillus spp. and P. fluorescens (antagonistic to G. graminis var. tritici) in the rhizosphere and spermosphere of wheat. Nautiyal [25] used a natural mutant of Pseudomonas with rifampicin resistance for similar purposes. Some of the commonly used marker genes may also be introduced into BCA so that they can be easily monitored over time. Cook et al [62] introduced the LacZY system into P. fluorescens 2–79 for purposes of monitoring the populations of the antagonist and take-all disease suppression in wheat. The highest population was observed 14 days after planting, declining thereafter by five orders of magnitude to 10^3/g root at harvesting 326 days after planting.

III. COMPATIBILITY TESTING

The success of a BCA depends on its compatibility with other disease-management systems. This requires holistic testing of BCA in combination with other disease-management practices in a systems approach. Once the BCA is found to be compatible, it can be successfully integrated with the integrated disease-management modules for each cropping system. Csinos et al. [63] evaluated the compatibility of *Trichoderma* spp. with fungicides for the management of *S. rolfsii* in groundnut. Similarly, in a pot study Sankar and Jeyarajan [64] found that *Gliocladium* and *Trichoderma* were compatible with *Azospirillum*. *T. harzianum*, *Rhizobium*, and carbendazim were successfully integrated for the management of stem rot of peanuts caused by *S. rolfsii* [65]. Addition of *T. harzianum* up to 5% resulted in 92% protection without affecting nodulation. Combination of either *Trichoderma* or *Gliocladium* with fungicides like carboxin or metalaxyl protected crop plants against *S. rolfsii*, *R. solani*, *F. oxysporum* f.sp. *ciceris*, and other soilborne pathogens [66–68]. Elad et al. [69] used a combination of *T. harzianum* and dicarboximide for successful control of *Botrytis cinerea* in cucumbers. However, the alternation of BCA with fungicides was shown to be more effective than mixtures. Integration of *T. harzianum* with a sublethal dose of methyl bromide (300 kg/ha) and soil solarization yielded a maximum control of Fusarium crown and root rot of tomato caused by *F. oxysporum* f.sp. *radicis-lycopersici* [70]. This also resulted in increased populations of BCA. Ordentlich et al. [71] integrated *T. harzianum* with captan to protect potato tubers against *V. dahliae* by reducing disease incidence and increasing potato yield by 15.7% under field conditions. Application of *Bacillus subtilis*, *Bradyrhizobium japonicum*, and *Glomus fasciculatum* used either alone or in combination increased shoot dry weight, number of nodules, phosphorus content, and reduced nematode multiplication and wilting index [72]. The BCAs earlier identified were integrated into an IDM module for the management of leafspots and rust [73] in which one spray of either colony-forming units of *Penicillium islandicum* or *Verticillium leccanii* or their culture filtrates was both beneficial and compatible with other components such as a spray each of fungicides and 2% neem leaf extract. More emphasis should be placed on such studies to explore the compatibility of BCAs with other options as it not only averts failure of a BCA in a given crop but will also increase the viability of the biocontrol option in crop production systems.

IV. INTRA- AND INTERNATIONAL NETWORK

The importance placed on biocontrol research in the recent past has been quite significant across the globe. However, there has been a wide gap in the quality of research among laboratories. Laboratories with good facilities have been able to make significant contributions, whereas those with minimum facilities have not been able to make a dent in this area of research. Biocontrol systems are often region or location specific, but to understand and sort out bottlenecks many basic studies need to be conducted. It is this situation that calls for strong intra-

and international collaboration, as a collaborative mode of work will help researchers from resource-poor laboratories to generate data required at well-equipped laboratories. With the large-scale adoption of information dissemination through superhighways, global communication has become cheap, quick, and fast. Several dedicated web sites are posting regularly updated information about biocontrol. At these web sites, facility has also been created for online discussions and posting of newsletters to the users. Professional societies can take the lead in organizing a network to properly manage and disseminate this information to the end users so that scarce resources are properly utilized to optimize returns.

V. CHARACTERIZATION AND CATALOGUING OF BCA STRAINS

For a successful biocontrol program, effective BCAs have to be identified depending on their efficacy, abundance, and ease of handling. A systematic characterization of morpho-physiological traits, molecular markers, biocontrol ability, and other desirable traits is a basic requirement for proper maintenance of the germplasm of BCAs. Other desirable traits may include rhizosphere competence, tolerance to commonly used pesticides, tolerance to high- and low-temperature stresses, sustainability to varied edaphic conditions, compatibility with other beneficial microorganisms used for crop productivity enhancement, and competitive saprophytic ability. Such characterization could be both qualitative and quantitative, depending on the trait. In recent years attempts have been made at detailed characterization of BCAs. Kloepper et al. [24] analyzed and characterized 50 randomly selected bacterial strains for various physiological traits associated with rhizosphere competence, including chitinolytic activity, gelatin hydrolysis, production of hydrogen cyanide, starch hydrolysis, phenol oxidation, siderophore production, and production of antifungal compounds. These strains were isolated from the seedlings and mature plants of soybean, velvet bean, rye, and castor that demonstrated antagonism toward phytopathogenic nematodes. The significant differences among the strains and crop species emphasize the need for such characterization in other crop-pathogen-BCA systems as well. A sterile hyaline basidiomycete (SHB) was found to be a good candidate BCA against *Thielaviopsis basicola*, the cause of black hull of groundnuts, exhibiting several modes of action [74]. Lorito et al. [75] attributed the highest level of antifungal activity to a combination of all four fractions of a 1,3-β-glucosidase and an *N*-acetyl-β-glucosaminidase, an endochitinase, and a 1,4-β-chitobiosidase, isolated from the culture filtrate of *T. harzianum*, against *B. cinerea*. ED_{50} (50% effective dose) values were as low as 1.6 μg/mL for the inhibition of conidial germination and 1.7 μg/mL for the inhibition of germ tube elongation of the surviving spores. The degree of antifungal activity also varied with combinations of enzymes. Similarly, a biomimetic system was used to demonstrate the molecular basis for the interaction between *S. rolfsii* and *T. harzianum* [76]. In this method, the presence of purified lectin, isolated from *S. rolfsii*, on the surface of inert nylon fibers specifically induced mycoparasitic behavior in *T. harzianum* forming tightly adhering coils, appressorium-like bodies, and hyphal loops. Numerical analysis of 15 isolates of *T. harzianum*, based on 82 morphological, physiological, and biochemical characters and 99 isoenzyme bands, revealed that the isolates formed four distinct groups. Representative sequences of the ITS 1 and ITS 2 regions in the ribosomal DNA gene cluster confirmed this distribution, and antagonist-specific populations displayed similarities between levels and specificities of biological activity and the numerical characterization groupings [77]. Schickler et al. [78] differentiated each of the three strains of *T. harzianum* based on a unique electrophoretic pattern of three to five different chitinases. Electrophoretic karyotyping using contourclamped homogeneous field (CHEF) electrophoresis and randomly amplified polymorphic DNA analysis was used for grouping 10 isolates of *T. harzianum* into six distinct groups according to their capacity for biocontrol of plant pathogens [79]. These groups further exhibited intragroup compatibility and intergroup incompatibility. Based on phylogenetic analysis of the ITS1 sequences, 17 strains of *Trichoderma* were classified into four groups, namely *T. harzianum–T. hamatum* complex, *T. longibrachiatum*, *T. asperellum*, and *T. atroviride–T. koningii* complex [80]. Further, the correlation between different genotypes and potential biocontrol activity was studied under dual culture of 17 BCAs in the presence of the phytopathogenic fungi *Phoma betae*, *Rosellinia necatrix*, *Botrytis cinerea*, and *Fusarium oxysporum* f. sp. *dianthi* on three different media types. Many of these findings were discovered in isolation and therefore may only apply to specific cases. If these findings are confirmed under different crop production systems, they could be generalized. This would help not only in optimizing scarce resources but also in avoiding duplication of efforts.

VI. REGISTRATION OF BCAs

A range of BCAs containing bacteria and fungi as active ingredients are now commercially available in many countries for the control of fungal and bacterial diseases (Table 1). The regulatory requirements have been generally favorable and less stringent for BCAs than chemicals. Registration of BCAs with a federal or central regulatory agency is mandatory before its release to end users (i.e., growers). In recent years, a number of countries have introduced their own individual registration requirements to address the nature and distinct characteristics of microorganisms. This has led to higher regulatory demands in individual countries. Europe follows the OECD definition of biopesticides, i.e., including pheromones, insect and plant growth regulators, plant extracts, transgenic plants, and microorganisms [81]. In the United States, the U.S. Environmental Protection Agency (EPA) regulates biological pesticides or biopesticides. The generic and product-specific data requirements for biological pesticides appear in Title 40, Part 158, of the Code of Federal Regulations (CFR). A complete description of all data requirements and study protocols for biological pesticides is presented in the Pesticide Assessment Guidelines, Subdivision M: Guidelines for Testing Biorational Pesticides [82]. In China, the Institute for the Control of Agrochemicals, Ministry of Agriculture (ICAMA), is the authority for biopesticides registration. The government of India, in its notification of 8-15/99-CIR, dated 02.08.1999, has allowed for the registration of two types of biopesticides under section 9(3) of the Insecticides Act of 1968, provisional and regular. Canadian and U.S. regulatory agencies have taken a common approach and interpretation of results on the harmonization of guidelines in semiochemicals and pheromones. This is a welcome approach, and extending this harmonization on a global scale would do much to foster implementation of BCAs. International harmonization and a uniform set of rules would not only lower costs, but also encourage researchers and industrial partners in the rapid development of innovative biocontrol approaches for the development of sustainable agriculture.

VII. FUTURE DIRECTIONS

The growing concerns in recent years about the environment, coupled with lack of viable solutions for pathogens such as *Sclerotium* and *Aspergillus*, which are taking heavy tolls on crops, offer the basic impetus for research and development of BCAs. For the development of one chemical pesticide, millions of dollars can be spent, whereas developing biopesticides is often treated as a small-scale industry. Such lack of parity is not only because of a lack of vision by industries but also because of the failure of the research and extension wings to work in a well-knit network mode for technology assessment and refinement. In the years to come, biological control will certainly be a mainstay in commercial agriculture. However, concerted input from all active participants will be required. Differences in effectiveness and reliability of strains may be encountered between field trials and investigators [83]. If such findings are neglected, the probability of success of a final product will be in jeopardy. Basic research on the identification of recognition phenomena [84] will definitely help in developing more precise testing protocols. The abundant information available through the internet has to be properly managed for deriving maximum benefits and optimizing resource management. Biocontrol has reached the limelight after a long gestation period. However, if this momentum is to be maintained, a few essential and immediate needs must be dealt with:

Development of protocols for rapid and differential detection of the plant pathogens and BCAs in the host system that facilitate reliable comprehensive testing of BCAs.

Establishment of a data bank of BCA strains and of a system to facilitate the exchange of information among workers

Encouraging interdisciplinary collaborations to pursue the process of formulation of a successful BCA

More inflow of funds for international collaborations to fill the existing wide gap and develop more efficient testing systems

Establishment of good, broad-based and functional network at micro- and macro-levels that enable all biocontrol scientists and others to work in harmony in a cropping systems mode and understand the intricacies of formulating the crop protection schedules that include BCA as one of the components

A regular survey and surveillance to help in the identification of natural BCAs for their exploitation in IPM modules

Encouraging farmers who adopt biocontrol and thus obtain better feedback on technology assessment and refinement

Development of stringent quality-control and quality-assessment protocols with hassle-free and easy-to-adopt registration requirements

ACKNOWLEDGMENTS

The authors are thankful to Dr. V. Raghunathan, Director, CIL, Government of India, for his kind help in providing all the information on registration requirements for biocontrol agents. The senior author is also thankful to Dr. A. Bandyopadhyay, Director, National Research Centre for Groundnut, for his kind support while preparing this manuscript.

REFERENCES

1. WA Millard, CB Taylor. Antagonism of microorganisms as the controlling factor in the inhibition of scab by green manuring. Ann Appl Biol 14:202, 1927.
2. AR Justum. Commercial application of biocontrol: status and prospects. Philosophical Trans Royal Soc London, B 318:357–373, 1988.
3. GE Templeton. Mycoherbicide research at the University of Arkansas, past, present and future. Weed Sci 34 (suppl 1):35–37, 1986.
4. N Gutterson, W Howie, T Suslow. Enhancing efficiencies of biocontrol agents by use of biotechnology. UCLA Symp Mol Cell Biol 112:749–765, 1990.

5. J Singh, JL Faull. Hyperparasitism and biological control. In: KG Mukherji, KL Garg, eds. Biological Control of Plant Pathogens. Vol. 2. Boca Raton, FL: CRC Press, 1988 pp 167–179.

6. V Anjaiah. Molecular analysis of biological control mechanisms of a fluorescent *Pseudomonas aeruginosa* strain PNA1, involved in the control of plant diseases. PhD dissertation, Vrije Univerteit, Brussels, Belgium, 1998.

7. GC Papavizas. *Trichoderma* and *Gliocladium*: biology, ecology and potential for biocontrol. Annu Rev Phytopathol 23:23–54, 1985.

8. KF Baker, RJ Cook. Biological Control of Plant Pathogens. San Francisco: WH Freeman, 1974, pp. 108–109.

9. JW Deacon. Studies on *Pythium oliogandrum*, an aggressive parasite of other fungi. Trans Br Mycol Soc 66:383, 1976.

10. DK Bell, HD Wells, CR Markham. In vitro antagonism of *Trichoderma* species against six fungal pathogens. Phytopathology 72:379–382, 1982.

11. R Selvarajan, R Jeyarajan. Inhibition of chickpea root rot pathogens, *Fusarium solani* and *Macrophomina phaseolina*, by antagonists. Indian J Mycol Pl Pathol 26: 248–251, 1996.

12. S Desai. Characterization of isolates of *Trichoderma* spp. for their biocontrol ability against *Sclerotium rolfsii*. International conference on integrated plant disease management for sustainable agriculture, Abstracts of Poster Sessions, New Delhi, India, 1997.

13. J Köhl. Eignung von Stämmen aus der Gattung *Trichoderma* für die biologische Bekämpfung phytopathogener Pilze. PhD dissertation, Justus-Liebig Universität, Giessen, Germany, 1989.

14. C Dennis, J Webster. Antagonistic properties of species groups of *Trichoderma* I. Trans Br Mycol Soc 57:25–39, 1971.

15. C Dennis, J Webster. Antagonistic properties of species groups of *Trichoderma* II. Trans Br Mycol Soc 57:41–48, 1971.

16. L Mohan, R Jeyarajan. An in vitro test for evaluating the efficacy of mycoparasites on the sclerotial germination of ergot (*Claviceps fusiformis* Lov.) of pearl millet. J Biol Control 4:75–76, 1990.

17. J Köhl, E Schlösser. Specificity in decay of sclerotia of *Botrytis cinerea* by species and isolates of *Trichoderma*. Med Fac Landbouww Rijksuniv Gent 53/2a:339–346, 1988.

18. J Köhl, E Schlösser. Decay of sclerotia of *Botrytis cinerea* by *Trichoderma* spp. at low temperatures. J Phytopathol 125:320–326, 1989.

19. S Desai, E Schlösser. Parasitisation of *Sclerotium rolfsii* Sacc. by *Trichoderma*. Indian Phytopathol 52:47–50, 1999.

20. G Jager, H Velvis. *Rhizoctonia solani* in potatoes and the use of the mycoparasite *Verticillium biguttatum* as a natural control agent. Gewasbescherming 23:33–41, 1992.

21. JS Ahmad, R Baker. Rhizosphere-competence of *Trichoderma harzianum*. Phytopathology 77:182–189, 1987.

22. J Köhl, WML Molhoek, CH-Van-der Plas, NJ Fokkema. Effect of *Ulocladium atrum* and other antagonists on sporulation of *Botrytis cinerea* on dead lily leaves exposed to field conditions. Phytopathology 85:393–401, 1995.

23. JS Ahmad, R Baker. Competitive saprophytic ability and cellulolytic activity of rhizosphere-competent mutants of *Trichoderma harzianum*. Phytopathology 77:358–362, 1987.
24. JW Kloepper, R Rodriguez-Kabana, JA McInroy, DJ Collins. Analysis of populations and physiological characterization of microorganisms in rhizospheres of plants with antagonistic properties to phytopathogenic nematodes. Plant Soil 136:95–102, 1991.
25. CS Nautiyal. Rhizosphere competence of *Pseudomonas* sp. NBRI9926 and *Rhizobium* sp. NBRI9513 involved in the suppression of chickpea (*Cicer arietinum* L.) pathogenic fungi. FEMS-Microbiol Ecol 23:145–158, 1997.
26. A Sivan, GE Harman. Improved rhizosphere competence in a protoplast fusion progeny of *Trichoderma harzianum*. J Gen Microbiol 137:23–29, 1991.
27. S Desai, E. Schlösser. Comparative sensitivity of isolates of *Trichoderma* spp. to selected fungicides in vitro. Med Fac Landbouwv Rijksuniv Gent 58/3b:1365–1372, 1993.
28. GC Papavizas, JA Lewis. Physiological and biocontrol characteristics of stable mutants of *Trichoderma viride* resistant to MBC fungicides. Phytopathology 73:407–411, 1983.
29. C Fanelli, F Cervone. Polygalacturonase and cellulase production by *Trichoderma koningii* and *Trichoderma pseudokonigii*. Trans Br Mycol Soc 68:291–294, 1977.
30. Y Elad, I Chet, Y Henis. Degradation of plant pathogenic fungi by *Trichoderma harzianum*. Can J Microbiol 28:719–725, 1982.
31. D Jacobs, O Kamoen. Role of cell wall degrading enzymes of *Trichoderma* antagonism. Med Fac Landbouww Rijksuniv 51:751–758, 1986.
32. PJ Slininger, RJ Bothast, DM Weller, LS Thomashow, RJ Cook, JE van-Cauwenberge. Effect of growth culture physiological state, metabolites, and formulation on the viability, phytotoxicity, and efficacy of the take-all biocontrol agent *Pseudomonas fluorescens* 2–79 stored encapsulated on wheat seeds. Appl Microbiol Biotechnol 45:391–398, 1996.
33. S Pleban, L Chernin, I Chet. Chitinolytic activity of an endophytic strain of *Bacillus cereus*. Lett Appl Microbiol 25:284–288, 1997.
34. S Sela, H Schickler, I Chet, Y Spiegel. Purification and characterization of a *Bacillus cereus* collagenolytic/proteolytic enzyme and its effect on *Meloidogyne javanica* cuticular proteins. Euro J Pl Pathol 104:59–67, 1998.
35. A Ordentlich, Z Wiesman, HE Gottlieb, M Cojocaru, I Chet. Inhibitory furanone produced by the biocontrol agent *Trichoderma harzianum*. Phytochemistry 31:485–486, 1992.
36. RJ Cook, LS Thomashow, DM Weller, D Fujimoto, M Mazzola, DS Kim. Molecular mechanisms of defense by rhizobacteria against root disease. Proc Natl Acad Sci 92:4197–4201, 1995.
37. I Yedidia, N Benhamou, I Chet. Induction of defense responses in cucumber plants (*Cucumis sativus* L.) by the biocontrol agent *Trichoderma harzianum*. Appl Environ Microbiol 65:1061–1070, 1999.
38. S Panicker, R Jeyarajan, S Panicker. Effect of osmotic water potential on growth and sporulation of *Trichoderma* spp. and *Rhizoctonia solani* Kuhn. Indian J Mycol Pl Pathol 21:254–256, 1991.

39. Himani Bhatnagar. Influence of environmental conditions on antagonistic activity of *Trichoderma* spp. against *Fusarium udum*. Indian J Mycol Pl Pathol 26:58–63, 1996.

40. K DalSoo, RJ Cook, DM Weller, DS Kim. *Bacillus* sp. L324–92 for biological control of three root diseases of wheat grown with reduced tillage. Phytopathology 87:551–558, 1997.

41. JS Cole, Z Zvenyika. Integrated control of *Rhizoctonia solani* and *Fusarium solani* in tobacco transplants with *Trichoderma harzianum* and tridimenol. Pl Pathol 37: 271–277, 1988.

42. S Sreenivasaprasad, K Manibhushanarao. Efficacy of *Gliocladium virens* and *Trichoderma longibrachiatum* as biocontrol agents of groundnut root and stem rot diseases. Int J Pest Manage 39:167–171, 1993.

43. A Sid Ahmed, C Pérez Sanchéz, C Egea, E Candela. Evaluation of *Trichoderma harzianum* for controlling root rot caused by *Phytophthora capsici* in pepper plants. Pl Pathol 48:58–65, 1999.

44. C Jan Mickler, KL Bowen, JW Kloepper. Evaluation of selected geocarposphere bacteria for biological control of *Aspergillus flavus* in peanut. Plant Soil 175:291–299, 1995.

45. J Ferguson. Reducing plant disease with fungicidal soil treatment, pathogen free-stock, and controlled microbial colonization. Ph.D. dissertation, University of California, Berkeley, 1958.

46. SJ Kay, A Stewart. Evaluation of fungal antagonists for control of onion white rot in soil box trials. Pl Pathol 43:371–377, 1994.

47. Y Elad, I Chet, J Katan. *Trichoderma harzianum*: a biocontrol agent effective against *Sclerotium rolfsii* and *Rhizoctonia solani*. Phytopathology 70:119–121, 1980.

48. D Dinakaran, R Sridhar, R Jeyarajan, G Ramakrishnan. Management of sesamum root rot with biocontrol agents. J Oilseeds Res 12:262–263, 1995.

49. P Sankar, R Jeyarajan. Seed treatment formulation of *Trichoderma* and *Gliocladium* for biological control of *Macrophomina phaseolina* in sesamum. Indian Phytopathol 49:148–151, 1996.

50. S Pleban, F Ingel, I Chet. Control of *Rhizoctonia solani* and *Sclerotium rolfsii* in the greenhouse using endophytic *Bacillus* spp. Euro J Pl Pathol 101:665–672, 1995.

51. Y Spiegel, E Cohn, S Galper, E Sharon, I Chet. Evaluation of a newly isolated bacterium, *Pseudomonas chitinolytica* sp. nov. for controlling the root-knot nematode *Meloidogyne javanica*. Biocontrol Sci Tech 1:115–125, 1991.

52. JW Dorner, RJ Cole, PD Blankenship. Use of biocompetitive agent to control preharvest aflatoxin in drought stressed peanuts. J Food Prot 55:888–892, 1992.

53. EB Nelson, GE Harman, GT Nash. Enhancement of *Trichoderma* induced biological control of Pythium seed rot and pre-emergence damping off of peas. Soil Biol Biochem 20:145–150, 1988.

54. K Gangadharan, R Jeyarajan, K Gangadharan. Mass multiplication of *Trichoderma* spp. J Biol Control 4:70–71, 1990.

55. GC Papavizas, MT Dunn, JA Lewis, J Beagle-Ristaino. Liquid fermentation technology for experimental production of biocontrol fungi. Phytopathology 74:1171–1175, 1984.

56. S Nakkeeran, P Sankar, R Jeyarajan. Standardization of storage conditions to increase the shelf life of *Trichoderma* formulations. J Mycol Pl Pathol 27:60–63, 1997.

57. FO Larsen, P Clementsen, M Hansen, N Maltbaek, S Gravesen, PS Skov, S Norn. The indoor microfungus *Trichoderma viride* potentiates histamine release from human bronchoalveolar cells. APMIS 104:673–679, 1996.

58. RJ Cook. Assuring the safe use of microbial biocontrol agents: a need for policy based on real rather than perceived risks. Can J Pl Pathol 8:439–445, 1996.

59. SJ Vesper. Production of pili (fimbriae) by *Pseudomonas fluorescens* and correlation with attachment with corn roots. Appl Environ Microbiol 53:1397–1405, 1988.

60. PH Tari, AJ Anderson. Fusarium wilt suppression and agglutinability of *Pseudomonas putida*. Appl Environ Microbiol 54:375–380, 1988.

61. K DalSoo, DM Weller, RJ Cook, DS Kim. Population dynamics of *Bacillus* sp. L324-92R12 and *Pseudomonas fluorescens* 2-79RN10 in the rhizosphere of wheat. Phytopathology 87:559–564, 1997.

62. RJ Cook, DM Weller, P Kovacevich, D Drahos, B Hemming, G Barnes, EL Pierson. Establishment, monitoring, and termination of field tests with genetically altered bacteria applied to wheat for biological control of take-all In: DR Mackenzie, SC Henry, eds. Proceedings of the Kiawah Island Conference Biological Monitoring of Genetically Engineered Plants and Microbes, Bethesda, Maryland, 1990, pp. 177–187.

63. AS Csinos, DK Bell, NA Minton, HD Wells. Evaluation of *Trichoderma* spp., fungicides and chemical combinations for control of southern stem rot on peanuts. Peanut Sci 10:75–79, 1983.

64. P Sankar, R Jeyarajan. Compatibility of antagonists with *Azospirillum* in sesamum. Indian Phytopathol 49:67–71, 1996.

65. M Muthamilan, R Jeyarajan. Integrated management of Sclerotium root rot of groundnut involving *Trichoderma harzianum*, *Rhizobium* and carbendazim. Indian J Mycol Pl Pathol 26:204–209, 1996.

66. IS Sawant, AN Mukhopadhyay. Integration of metalaxyl with *Trichoderma harzianum* for the control of Pythium damping-off in sugarbeet. Indian Phytopathol 43: 535–541, 1991.

67. AN Mukhopadhyay, SM Shreshtha, PK Mukharjee. Biological seed treatment for control of soil-borne plant pathogens. FAO Pl Prot Bull 40:21–30, 1992.

68. NP Kaur, AN Mukhopadhyay. Integrated control of 'chickpea wilt complex' by *Trichoderma* and chemical methods in India. Tropical Pest Manage 38:372–375, 1992.

69. Y Elad, G Zimand, Y Zaqs, S Zuriel, I Chet. Use of *Trichoderma harzianum* in combination or alternation with fungicides to control cucumber grey mould (*Botrytis cinerea*) under commercial greenhouse conditions. Plant Pathol 42:324–332, 1993.

70. A Sivan, I Chet. Integrated control of Fusarium crown and root rot of tomato with *Trichoderma harzianum* in combination with methyl bromide or soil solarization. Crop Prot 12:380–386, 1993.

71. A Ordentlich, A Nachmias, I Chet. Integrated control of *Verticillium dahliae* in potato by *Trichoderma harzianum* and captan. Crop Prot 9:363–366, 1990.

72. ZA Siddiqui, I Mahmood. Biological control of *Heterodera cajani* and *Fusarium udum* by *Bacillus subtilis*, *Bradyrhizobium japonicum* and *Glomus fasciculatum* on pigeonpea. Fund Appl Nematol 18:559–566, 1995.

73. MP Ghewande, S Desai, Prem Narayan; AP Ingle. Integrated management of foliar diseases of groundnut (*Arachis hypogaea* L.) in India. Int J Pest Manage 39:375–378, 1993.

74. SW Baard. Necrotrophic mycoparasitism of *Chalara elegans* (*Thielaviopsis basicola*) by a sterile basidiomycete. J Phytopathol 122:166–173, 1988.

75. M Lorito, CK Hayes, A di Pietro, SL Woo, GE Harman. Purification, characterization and synergistic activity of a glucan 1,3-beta-glucosidase and an N-acetyl-beta-glucosaminidase from *Trichoderma harzianum*. Phytopathology 84:398–405, 1994.

76. J Inbar, I Chet. A newly isolated lectin from the plant pathogenic fungus *Sclerotium rolfsii*: purification, characterization and role in mycoparasitism. Microbiol 140: 651–657, 1994.

77. I Grondona, R Hermosa, M Tejada, MD Gomis, PF Mateos, PD Bridge, E Monte, I Garcia-Acha. Physiological and biochemical characterization of *Trichoderma harzianum*, a biological control agent against soil borne fungal plant pathogens. Appl Environ Microbiol 63:1997, 3189–3198, 1997.

78. H Schickler, BC Danin Gehali, S Haran, I Chet. Electrophoretic characterization of chitinases as a tool for the identification of *Trichoderma harzianum* strains. Mycol Res 102:373–377, 1998.

79. I Gomez, I Chet, AH Estrella. Genetic diversity and vegetative compatibility among *Trichoderma harzianum* isolates. Mol Gen Genet 256:127–135, 1997.

80. MR Hermosa, I Grondona, EA Iturriaga, JM Diaz-Minguez, C Castro, E Monte, I Garcia-Acha. Molecular characterization and identification of biocontrol isolates of *Trichoderma* spp. Appl Environ Microbiol 66:1890–1898, 2000.

81. OECD Data requirements for registration of Biopesticides in OECD member countries: survey results. Environment Monograph No. 106, Paris, France. 1996.

82. U.S. Environmental Protection Agency, Office of Pesticides and Toxic Substances. Subdivision M of the Pesticide Testing Guidelines: Microbial and Biochemical Pest Control Agents. Document No. PB89-211676. National Technical Information Service, U.S. Department of Commerce, Springfield, VA. 1989.

83. D Hornby. Field testing putative biological controls of take-all: rationale and results. EPPO Bull 17:615–623, 1987.

84. S Fedi, E Tola, Y Moenne-Loccoz, DN Dowling, LM Smith, F O'Gara. Evidence for signaling between the phytopathogenic fungus *Pythium ultimum* and *Pseudomonas fluorescens* F113: *P. ultimum* represses the expression of genes in *P. fluorescens* F113, resulting in altered ecological fitness. Appl Environ Microbiol 63:4261–4266, 1997.

18

Formulation of Biological Control Agents for Pest and Disease Management

Prem Warrior
Valent BioSciences Corporation, Long Grove, Illinois

Krishnamurthy Konduru
Oklahoma State University, Stillwater, Oklahoma

Preeti Vasudevan .
Center for Advanced Studies in Botany, University of Madras–Guindy, Chennai, Tamil Nadu, India

I. INTRODUCTION

Effective biological control of pests or diseases relies on the successful establishment and maintenance of a threshold population of suppressive organisms on the planting material, the soil or more generically the matrix, below which their efficacy is impaired or insufficient. So far, few biological control agents have achieved success under field conditions. Among the hundreds of organisms identified as potential biological pest control agents, only very few have resulted in providing commercially acceptable control of pests/diseases. Varying degrees of efficacy have been achieved in the laboratory, greenhouse, or even small plot trials with different preparations and varying levels of performance have been noted. Reviews have attributed this variability in performance to factors such as stability or poor viability, sensitivity to UV light, desiccation, and fluctuating environmental conditions [1]. Scientific literature has also noted that some of these disadvantages were minimized or even overcome by the addition of selected ingredients and by preparing the final product in a form that is specific to the pest/disease-crop complex. Most of the successes in formulations of biological preparations and our current knowledge base has resulted from diligent studies on very few biologically active microorganisms such as *Bacillus thuringiensis*

(Bt), *Bacillus subtilis*, *Trichoderma harzianum*, and *Metarrhizium anisopliae* and are now being applied to the newer active ingredients.

The key objective of a formulation is to deliver an optimal dose of the agent at the optimal site and time. This apparently simple objective is, in principle, identical to the formulations of chemical active ingredients where significant advances in control release technologies have been developed. However, in the production of biological formulations, many additional obstacles have to be overcome, mainly due to the fact that the active ingredient itself, in most instances, is a living entity. In addition to the requirements for good physical properties for the final product, the biological pesticide is expected to maintain a functional living agent. Manufacturing of a biological product for commercial use necessitates large-scale production of the active organism mediating biocontrol, including fermentation. Unlike a chemical active ingredient, the development of a stable, active biological product begins from the selection of the production process and can be a fully integrated process. It is important to define the expectations, conceptualize the specific formulation, and identify multiple viable processes even while evaluating the biological manufacturing options. Unfortunately, these aspects have received very little attention, and only limited research has been carried out on biomass, fermentation, and delivery of microbes for the control of pathogens and pests. Although the stability and persistence of biocontrol agents is an essential attribute, the prolonged persistence of biocontrol agents could lead to selection for resistance in pathogen/pests and may be commercially unacceptable. Additionally, there are significant differences in the performance characteristics and desired attributes of a soil-applied vs. foliar biological product. The performance of biocontrol agents applied to the soil may depend largely on physical and chemical characteristics of soil, pH, moisture, temperature, as well as their ability to compete with the native microflora. Similarly, environmental factors such as temperature, moisture or dew period, protection against UV irradiation, and desiccation influence biological control in the phyllosphere [2]. The above-ground parts are often directly exposed to harsh climatic conditions that could be hostile to microorganisms and may be significantly different from the rhizosphere, which is generally considered more conductive to the survival of microbes, especially with regard to its structure, ecology, and nutrient status [3]. Also, the compatibility of the biocontrol agent with existing cultural practices and chemical control methods is an important criterion for successful use in formulation [4]. A good formulation should take into account all the above parameters and is expected to demonstrate its superiority over its nonformulated counterpart.

Devisetty et al. [5] provided an in-depth review of the various approaches to biopesticide formulations, with particular reference to Bt-based products. Burges [6] published a treatise on the formulation of microbial pesticides and succeeded in highlighting the importance of this area of research for commercializing biological products. This chapter will summarize the vast volume of work carried out in this important field and briefly review the use, development, and application of formulations of bacteria, fungi, virus, nematodes, and insects for disease and pest management from a biologist's perspective rather than that of a formulation chemist.

II. FACTORS INFLUENCING BIOCONTROL

The main function of a formulation is to facilitate the deployment of biocontrol agents on commercial scale and transfer its application from laboratory to field conditions. This may involve the optimization of factors such as stability upon storage, increasing persistence, protection from harmful environmental factors, and enhancing the activity at the target site [7] and is best determined by subjecting potential organisms to greenhouse or prefield trials.

The two most important abiotic factors that can affect the viability and consequently the efficacy of a living biological entity on the leaf surface are UV light and temperature. UV light can result in more than 50% damage in microbial agents as in the case of the crystalline toxins of the biological insecticide *B. thuringiensis* [8]. Bailey et al. [9] observed that the commercial Bt product used to control apple moth caused by *Epiphyas postvittana* lost more than half of its activity within a day on vine plants when the leaf surface was fully exposed to light. The same bacterium applied to shaded leaves showed survival of >60% after 2 days' exposure to sunlight. Stilbene-based optical brighteners have therefore been used as additives in formulations to protect microorganisms from the damage caused by UV exposure. Use of these additives was reported to also provide a 214-fold increase of virulence in baculoviruses [10]. Tinopal, Phorwite, Intrawhite, Leucophor, uric acid, folic acid, 2-hydroxy-4-methoxy-benzophenone, *p*-aminobenzoic acid, 2-phenylbenzimidazole-5-sulfonic acid, and dyes such as Congo red, methyl blue, safranin, brilliant yellow, and buffalo black have been used as UV protectants. In combination with high temperature, exposure to UV resulted in a drastic decrease in the activity of *Metarrhizium flavoviride*, causing up to 80% reduction in germination at 50°C [11,12]. Temperature plays a pivotal role in determining the metabolic activity of microbes.

The survival of a potential biological antagonist and its efficacy in mediating pest/disease control depends largely on the type of material used in the development of the formulation. Use of polymers for formulating biocontrol agents has been extensively explored [13]. Alginates are copolymers that contain 1,4-linked β-D-mannuronic and α-L-galuronic acid in different proportions [14]. Oil coating of alginate-formulated bacteria, fungi, and nematodes improved survival of biocontrol agents. The nematode *Subanguina picridis* survived in such a formulation for 9 months when stored at −20°C [15]. Additives may also contribute to increased survival of microbes in formulations. The loss in viability of bacteria associated with powdered formulations may be overcome by additions of sucrose and trehalose. These were reported to enhance viability by protecting membranes and proteins in intact bacteria during a freeze-drying process by replacing water molecules in the lipid bilayer [16].

Desiccation is a common problem encountered during the development of formulations, especially in the case of fungal antagonists such as *Metarrhizium* spp. and *Beauveria* spp. Tolerance to desiccation may be achieved by preparation of invert emulsions. Substances such as mineral oil, paraffin, and lecithin when added reduce evaporation of water from the formulations. Suspension in oil also excludes oxygen from the organisms and prevents respiration. Oils are more suitable for formulating lipophilic conidia and could eliminate the need for other wetting or spreading agents or stickers. Oil formulations may also help the performance of a product by facilitating adhesion of fungal spores to the insect cuticle and thus improve efficacy by aiding transport of spores into membranous folds on the insect body, where conditions are more conducive for spore germination and host infection [17]. Oil also seems to have cutinophilic properties that allow a larger number of conidia to penetrate the mouthparts of the insects. It is, however, important to note that some oils may have negative effects on formulations, as in the case of coconut oil, which seems to hinder the activity of *B. bassiana* [18]. Dried conidia of *M. flavoviride* stored in oil formulations, however, showed longer shelf life than dried powder–based formulations. This can be, in fact, prolonged by the incorporation of silica gel [19,12]. The addition of silica gel to oil formulations of *M. flavoviride* conidia also greatly increased temperature tolerance [20].

III. APPLICATION OF BIOLOGICAL FORMULATIONS TO THE PLANT SYSTEMS

An ideal formulation is expected to facilitate the delivery of the living biocontrol agent in its active state, at the right place, at the right time. While the formulated microbial products must be effective at the site of action and compatible with agronomic practices, they should be easy to apply to and adhere to plant parts such as seeds, tubers, cuttings, seedlings, transplants, and mature plants or be available in the soil medium.

Biological formulations applied to seeds greatly help deliver the agents to the spermosphere of plants, where, in general, extremely conducive environments prevail. The agents are therefore provided an excellent opportunity to survive, multiply, persist, and exercise control of soilborne pathogens [21]. With increased interest in limiting the use of pesticides and the need to deliver the active ingredients as close to the target as possible, this approach continues to receive consider-able attention from academic and commercial users [22]. Significant advances in seed treatment technology have been accomplished in the past few years, thanks to commercial organizations such as Gustafson, Inc. (Plano, TX), and the approach is an attractive means for introducing biological control agents into the soil-plant environment, as these introduced organisms are offered the selective advantage to be the first colonizers of plant roots. At the time of planting, the formulated product can be used directly (powders, liquids) without stickers. Com-mercial formulations of the actinomycete *Streptomyces griseoviridis* (Kemira Oy, Finland) used to control soilborne diseases is dusted onto the seeds prior to plant-ing. Much of the work on seed treatment has been carried out on *Bacillus subtilis* strains as in the case of the commercial product Kodiak®, used for control of *Rhizoctonia* and *Fusarium* in cotton and peanut. Several bacterial inoculants, as in the case of *Burkholderia* (formerly *Pseudomonas*) *cepaciae*, have been formu-lated as liquid inoculum preparations and applied to seeds prior to or at the time of planting. Several other gram-negative bacterial antagonists have been evalu-ated for potential commercial use (e.g., *Pseudomonas fluorescens* used in Dag-ger® G by Ecogen, Langhorn, PA); however, in many instances stability issues have impeded large-scale commercial acceptance of many of these strains. Seeds can also be precoated with dry powder or treated with liquid-based formulations of microbes. The microbes survive on such precoated seeds [23]. Additives are used to prolong the survival of microbial agents applied to seeds. The commonly used additives include gum arabic and xanthan gum, even though they were not able to provide an adequate degree of survival. Alginate hydrogel, used as a seed encapsulation material, maintains the entity in a viable state and protects it from other stresses. Soil treatment is preferred when biocontrol agents are too sensitive to desiccation. The antagonistic agents establish a high population in the soil, making them suppressive to the disease. Niche exclusion also becomes operative in such cases, as the increase in number of the introduced microbes renders essen-tial nutrients unavailable to soil pathogens and other less beneficial microflora [24]. Several species of *Trichoderma* have also been formulated extensively, us-ing cellulosic carriers and binders and modern thin-film coating techniques, in an attempt to introduce them into the rhizosphere regions of seedlings to protect them from diseases such as *Rhizoctonia solani* and *Pythium ultimum*. However, the major limitation of fungi as seed coatings remains that they do not colonize the rhizosphere as readily as the bacterial agents.

Liquid formulations can be applied to foliar parts of the plants for control of insect pests. Liquid-based formulation of *B. thuringiensis* has been used to control lepidopteran pests on forest, cotton, and several other agronomic and fruit/tree crops [25,26]. Ultra-low-volume (ULV) applications by air using high-potency formulations such as emulsifiable suspensions (ES) or soluble concentrates (SC) have been developed for forestry applications. Development of high-potency, cost-effective formulations with good suspension properties and good stability have contributed substantially to the successful global adaptation of biological insecticides in forestry. Additives such as stickers, spreaders, adjuvants, and emulsifiers in foliar sprays facilitate adhesion of microorganisms on plant tissues [27]. Such additives are essential for application to certain monocot plants (e.g., rice and sugarcane). The recent introduction of the dry flowable (DF) formulations for Bt strains is a significant step in this direction. Formulations of DiPel® DF and XenTari® DF overcome the stability limitations of liquid formulations and have enabled preparations of high-potency preparations with excellent suspension properties for control of lepidopteran pests.

The formulated organism or biological product may also be directly applied to plant roots in the form of a root dip, spray, drip, or flood application for control of soil pests and diseases. This is primarily applicable in the control of fungal diseases caused by pathogens such as *Fusarium, Pythium,* or *Rhizoctonia,* delivery of insect-parasitic nematodes, and for the management of plant parasitic nematode populations in agricultural soils. In all the above instances, the objective of the formulation is to stabilize and preserve the active ingredient and to distribute it evenly in the rhizosphere. Formulation ingredients targeted to protect the viable propagules such as spores may also be required. Uniform lateral and horizontal distribution in the soil matrix is the most important goal when the target is a soil pest, such as a nematode. Edaphic factors, such as soil type, pH, conductivity, cation exchange capacity (CEC), rainfall, etc., play a significant role in the efficacy of such a soil-applied product. In the case of the soil-applied nematicide DiTera®, emulsifiable suspension has been used successfully to deliver the killed-microbial product for commercial nematicide treatments on cole crops and grapes [28].

IV. CANDIDATES FOR BIOLOGICAL CONTROL

A. Bacteria

Bacteria have attracted enormous attention as agents for biocontrol, particularly since they are easy-to-handle, generally stable, aggressive colonizers of the rhizosphere or phyllosphere and inherently possess a quick generation time. They are also known to affect life cycles of different plant pathogens or pests by diverse mechanisms including the production of extracellular metabolites and intracellular proteinaceous toxins. In general, spore-forming bacteria (e.g., *Bacillus* spp.) survive to a greater extent even in harsh environments, compared to the non–spore-forming bacteria. Among the *Bacillus* spp., the ones that have attracted the most attention are *B. thuringiensis* and *B. subtilis.*

Bacteria may be formulated in either a dormant or a metabolically active state. They are easily mass-produced using a liquid fermentation process, although in some cases they may be more amenable to semisolid or solid-state fermentation. Components of the fermentation medium as well as the growth conditions are critical to both biomass and secondary metabolite production. Developing a final formulation usually requires processing of the fermentor broth and addition of components that stabilize and/or enhance the activity of the active ingredients. The end product can be a liquid or a solid formulation [23].

Liquid formulations may be oil, aqueous, or polymer-based. Aqueous formulations (AS) may require several additional components, such as stabilizers, stickers, surfactants, coloring agents, antifreeze compounds, and additional nutrients [24,29,30]. Oil-based formulations (ES) may involve blending a processed ferment with a mineral- or vegetable-based oil carrier and emulsifiers to allow dilution in water. Oil-based formulations reduce evaporation of droplets and allow for ultra-low-volume aerial application.

Dry formulations, which include wettable powders (WP), dry flowables (DF), granules (G) and wettable/water-dispersible granules (WG, WDG), can be produced through processes such as spray-drying, freeze-drying, or air-drying with or without the use of a fluidized bed. Wettable and dry granules are generally produced by adding binders, dispersants, wetting agents, and water to the dried, powdered fermented matter in a granulator. The extra processing steps in producing a dry formulation, however, increase manufacturing costs but could result in lower shipping costs due to reduced weight. Inert carriers such as fine clay, peat, vermiculite, alginate, and polyacrylamide beads may be used to develop dry formulations. The carrier facilitates delivery of the necessary concentration of viable cells in the most optimal physiological state. The carriers should be inexpensive, nontoxic, and contain an adequate number of viable propagules while protecting the biological control agent from adverse environmental conditions in order to be commercially viable. Other components such as diatomaceous earth, clay, talc, vermiculite, cellulose (carboxymethyl cellulose), and polymers (xanthan gum) have also been added to bacterial formulations [31]. Polyacrylamide and sodium alginate polymers are used for the immobilization of bacteria [32].

B. thuringiensis can be relatively easily produced in liquid fermenters, but production conditions strongly influence the quantitative and qualitative attributes of the final product. A few Bt-based products (DiPel®, XenTari®, BioBit®, Javelin®, Thuricide®) make up over 90% of the commercial biopesticides used today [33]. The bacterium produces delta-endotoxin protein, which is toxic to several insect species belonging to Lepidopera, Coleoptera, and Diptera. This bacterial active ingredient has received the most attention in terms of innovative approaches in formulations and has been formulated in liquids (both aqueous and oil-based), wettable powders, granules, and dusts [5,34]. In these products the additives help improve stability and enhance flowability characteristics. Since the proteinaceous toxin in the Bt cell needs to be ingested, attempts have been made to enhance the efficacy of Bt preparations using phagostimulants [35] and synergists of Bt toxins. Additionally, Bt genes encapsulated in nonliving *Pseudomonas fluorescens* (e.g., MVP® from Mycogen) have also been developed as commercial insecticides.

Many bacterial agents have been widely used in the area of plant disease suppression. *B. subtilis*–based seed treatment, such as Quantum-4000®, Kodiak®, and Epic®, have been available for use on legumes, vegetables, cotton, and ornamentals to control diseases caused by pathogens such as *Rhizoctonia* and *Fusarium*. A biological fungicide against gray mold based on *B. subtilis* (Serenade®) was recently approved for commercialization by the U.S. EPA. In China, *Bacillus* sp. has been used to enhance yield of rice, wheat, corn, sugar beet, cabbage, and rapeseed [36]. Several additional products based on *B. subtilis* are also being developed by commercial organizations around the world.

The formulation should provide a protective habitat for the introduced bacterium, thereby improving its potential for survival and successful colonization. While *Bacillus*-based organisms may be inherently more stable, some gram-negative bacterial species such as *Pseudomonas* need special preservative systems. Vidhyasekaran and Muthamilan [37] have developed a talc, peat, vermiculite, kaolinite, lignite, and farmyard manure–based formulation in which *P. fluorescens* strains survive for up to 240 days of storage. A methylcellulose:talc base–formulated *P. fluorescens* strain 7–14 survived for 10 months (at 4°C) and afforded 68.5% reduction in the rice blast disease; a similar formulation of *P. putida* V14i maintained the bacterium in a viable state for up to 10 months and afforded 60% suppression of rice sheath blight in the field [38,39].

Other bacterial products for disease control include *Agrobacterium radiobacter*, which has been commercially available in Australia, New Zealand, and the United States for the control of crown gall disease, as a concentrated liquid formulation or as a moist peat-based product [40]. These formulations are suspended in water before application to seeds and cuttings as dip, spray and/or drench. *Streptomyces griseoviridis* (Mycostop® produced by Kemira Oy, Finland) is used against many fungal pathogens and is available in several countries of the world as a wettable powder used for control of root diseases caused by *Pythium*, *Fusarium*, and *Phomopsis*. Liquid- or peat-based formulations of *Burkholderia cepacia* is also being used in the control of fungal pathogens like *Phytophthora*, *Pythium*, and *Fusarium*.

Bacteria are able to affect nematode life cycles in virtually all soils because of their constant association in the rhizosphere. A large number of rhizobacteria are known to reduce nematode populations, and important genera include *Agrobacterium*, *Alcaligenes*, *Bacillus*, *Clostridium*, *Desulfovibrio*, *Pseudomonas*, *Serratia*, and *Streptomyces* [41]. Obligate bacterial parasites such as *Pasteuria pene-*

trans, not commercially manufactured at this point, can reduce nematode reproduction by their parasitic behavior while the nonparasitic rhizobacterial species reduce nematode populations by preferentially colonizing the rhizosphere of the host plant or by producing nematoxic metabolites. Commercial formulations of *B. cepacia* prepared as wettable powders, water-dispersible granules, or emulsifiable suspensions are also being evaluated for control of nematodes like *Globodera rostochiensis*, *Meloidogyne incognita*, *Heterodera glycines*, and *Belonolaimus longicaudatus*.

B. Fungi

The major fungal species used in biocontrol are *Trichoderma harzianum*, *Gliocladium virens*, *Tilletiopsis pallescens*, and *Pseudozyma flocculose* [42]. Many root and foliar fungal pathogens such as *Rhizoctonia*, *Pythium*, *Botrytis*, and the powdery mildew fungus [40,42,43] are known to be controlled by these fungi.

    Different approaches exist for the development of fungal formulations. These include drying the fragmented mycelium [44], obtaining a mycelium pellet [45], encapsulation of spores or mycelia in starch or alginate [46,47], and coating of dry spores onto grain/bran [48]. Some fungal spores germinate rapidly in water. Also, aqueous preparations generally promote contaminant growth. Therefore, the use of liquid formulations is not preferred. Alternatively, dust and wettable powder formulations are more useful. *Trichoderma harzianum* exhibits better growth and delivery when formulated in diatomaceous earth with 10% molasses for the control of *Sclerotium rolfsii* in peanuts [49]. Considerable work has been focused on *Trichoderma* and *Gliocladium* as agents of biocontrol against other fungi, due to the relative ease of their isolation, culturing, and fermentation [24]. Commercial formulations of *T. harzianum* have been prepared by mixing the fungal biomass grown in solid medium with diatomaceous earth preparations such as Celite®.

    Environmental conditions such as temperature and moisture are important for the survival of fungal biocontrol agents. Several adjuvants and amendments are used to enhance fungal spore germination. Formulating fungal biocontrol agents using corn oil as an adjuvant significantly enhances bioactivity, reducing the dew period requirement from 12 to 2 hours and its spray volume requirement from 500 to 5 L/ha [50]. Surfactants are used in formulations to wet the plants by reducing surface tension. Surfactants also aid the dispersal of fungal spores in a spray-droplet mix. However, the selection of an appropriate surfactant is critical as should not inhibit spore germination. Tween®-20, nonoxyphenol, and sorbitol ester are common surfactants used with fungal agents [2]. The use of suitable inert emulsions for fungal formulations applied to leaf surfaces provides a favorable environment for germination [51]. The addition of vegetable oil as an emulsifier is reported to enhance the efficacy of fungal agents [52].

The commercial fungal formulations developed to date are dusts, granules, pellets, and wettable powders, ready to use as a suspension in water. Solid substrates commonly used to formulate fungi are sand, vermiculite, grain bran, cornmeal, and wheat kernels [2,51]. Fungi like *Colletotrichum truncatum*, *Alternaria cassiae*, and *Fusarium lateritium* have been formulated in liquid inoculum, wheat gluten, wheat flour, and kaolin [53]. The wheat matrix–based formulation "pesta" is applied to the soil. Alginate was used to develop formulations of *Gliocladium virens* for control of root-infecting fungi in pot culture plants [24]. Postharvest application of a shellac latex and a dissolved shellac ester formulation on grapefruits supported significantly high populations of the yeast *Candida oleophila* (a biocontrol agent of *Penicillium digitatum*) throughout a 4-month storage period at 13°C [54]. Field trials conducted over 4 years revealed that monthly applications of a peat-based granular formulation of *T. harzianum* reduced initial dollar spot disease severity by 71% and delayed disease development of *Sclerotinia homoeocarpa* by up to 30 days in creeping bent grass (turf grass) [55]. Pelletized alginate formulations of *Typhula phacorrhiza* containing kaolin clay supported over 70% survival of the organism at 4°C after 64 weeks. Field plots treated with this formulation were protected to a great extent against *Typhula incarnata* infecting turf grass [56]. *Gliocladium virens* applied to snap beans at planting state as a wheat bran alginate pellet formulation reduced preemergence and postemergence damping-off disease [57].

Fungal formulations approved for commercial use include those of *Ampelomyces quisqualis*, *G. virens*, *Fusarium oxysporum* (nonpathogenic), *Coniothyrium minitans*, *Candida oleophila*, *T. harzianum*, and *Phlebia gigantea*. Mycoinsectides are *Metarrhizium* spp., *Beauveria bassiana*, *B. brongniartii*, *Paecilomyces fumosorosens*, and *Verticillium lecanii*. The commercial alginate formulation of *G. virens* (GL-21) is available in the U.S. market as GlioGard®. More recently, a granular formulation named SoilGard® was introduced for the control of damping-off disease [58].

Fungal biocontrol agents, such as *Metarrhizium flavoviride* and *M. anisopliae* used in the control locusts, *Beauvaria bassiana* used to control whiteflies, beetles, and locusts, and *Verticillium lecanii* used against aphids [59] have also been well studied. Strains of *M. flavoviride* are known to be more virulent against African grasshoppers and locusts than those of *B. bassiana* [60]. These fungi applied as dusts, wettable powders, and emulsions amended to traps, baits, or soil are effective in bringing about pest control [61–64]. They are formulated with corn starch, oil, UV protectants, and sunlight blockers such as clay to improve and ensure control under field conditions [46,65,66].

Fungi or fungal metabolites have potential as biocontrol agents against nematodes. A large number of fungi are known to trap or prey on nematodes—the most important are *Paecilomyces*, *Verticillium*, *Hirsutella*, *Nematophthora*, *Arthrobotrys*, *Drechmeria*, *Fusarium*, and *Monacrosporium* [67]. The nematode trapping fungus *Arthrobotrys dactyloides* has been formulated in glucose corn steep with kaolin and vermiculite as carriers. Treatment of field soils with this granulated powder demonstrated a 57–96% reduction in nematode galls in tomato [68]. Novel formulations of the *Myrothecium verrucaria*–based nematicidal product DiTera® include a water-dispersible granule (WDG) for application to turf and ornamentals and an oil-based emulsifiable suspension (ES) commercially used in drip and flood irrigation application on grapes.

C. Viruses

Viruses are mainly used for the biological control of insect pests. Several viruses are pathogenic to insects and over the past few years have found commercial applications in several key markets. Among the 16 families of viruses are pathogenic to arthropods, the baculoviruses have generated the greatest interest as potential biocontrol agents against insect pests belonging to Lepidoptera, Hymenoptera, and Coleoptera. Baculoviruses are specific in their host range, do not infect beneficial insects, and persist in the environment, providing long-term control of insect pests. Some of the key products include Spod-X® (against beet armyworm, *Spodoptera exigua*) and Gemstar® (against cotton bollworm, *Helicoverpa zeae*). Baculoviruses are also used for foliar application against forest pests [69,70]. Although several baculoviruses have been registered for use as microbial control agents, only a few are currently used on a commercial basis. In addition to the fact that the utility of insect-specific viruses is primarily limited by the length of time they take to kill, their performance can be substantially influenced by their susceptibility to UV radiation. While the bacterial pathogens of insects (e.g., *B. thuringiensis*) cause cessation of feeding within a few hours of consumption, in the case of baculovirus, it may take 3–4 days after ingestion for mortality to occur. Additionally, in vivo manufacturing of these microbial pesticides, while quite feasible in small-scale "cottage" industries, raises considerable quality-control issues and batch-to-batch variability in large-scale commercial production. Also, considerable manual labor may be required in maintaining insect colonies and collecting the insect-derived viral particles.

Baculoviruses are most conveniently formulated in the form of concentrated wettable powders (WP). Since the viruses are obligate parasites of insects, the virus particles are produced in vivo in the insect body; the viral particles are collected from the insect cadavers, mixed with water and stored under refrigeration. These water-based formulations can be used in field trials and for noncommercial applications [69]). The viral liquid or solid base formulation products usually include stickers, spreaders, antioxidants, and UV-protectant dyes, such as acridine yellow, alkali blue, and mercurochrome [71]. The addition of stilbene brighteners to the formulation enhances its efficacy against gypsy moth (*Lymantria dispar*) [72]. Natural compounds such as skim milk powder are also used as surfactants.

In spite of the limitations, Baculoviruses in general have demonstrated high levels of insect control when properly used. However, commercial use of these biological agents has not gained popularity, primarily due to production and quality-control concerns. More recently, genetically modified Baculoviruses containing novel toxins have been developed and are being tested. These engineered products may help enhance their activity, making the insect more susceptible or reducing its feeding, thus reducing crop damage. Several successful field trials on cotton and lettuce have been conducted with a wettable powder formulation of *Autographa californica* nucleopolyhedrovirus (AcNPV), containing a recombinant gene of an insect-specific venom component from the Algerian scorpion (*Androctonus australis*, Hector). This strain could effect greater levels of control of the tobacco budworm pest *Heliothis virescens* and *Trichoplusia ni* pest in cabbage [73] than the chemical standards. Gypsy moth nuclear polyhedrosis virus (LdMNPV) formulated with an enhancing adjuvant, Blankophor® BBH, resulted in significant control of larval populations of gypsy moth (*Lymantria dispar* L) on foliar spray–treated trees [74]. Stilbene-derived brighteners greatly enhance the infectivity of a number of baculoviruses. The death of larvae exposed to virus formulated with Tinopal® LPW was significantly greater when compared to larvae inoculated with the virus alone. Analysis of the results of eight independent field trials in Mexico and Honduras revealed a significant positive relationship between log virus dose and percentage mortality observed in *S. frugiperda* larvae [75]. As newer strategies for genetically engineered baculoviruses emerge, newer approaches to releasing and preserving the recombinant organism and the toxins will have to be developed.

D. Nematodes

Most entomopathogenic nematodes have been isolated from naturally infected insects. Members of two families of nematodes, namely, Steinernematidae and Heterohabditidae, are thought to be important agents of biocontrol. These nematodes can be produced by fermentation, have a wide host range, and normally kill the host insects within 24–48 hours. Such nematodes are maintained either by repeated subcultures in susceptible insect larvae or by cryopreservation in liquid N_2 [76]. Nematodes may be mass-produced either within insect hosts or by liquid fermentation techniques. Significant progress in in vitro mass production and formulation of insect-parasitic nematodes has been accomplished in recent years, and several commercial products were introduced into the market-place. The nematode must be kept alive and in good condition before, during, and after application. Understanding the factors that limit nematode life cycle, the ecology of the target pest, and improved application technology have resulted in the development of products applicable to niche markets. Generally the nematode products are stored under refrigeration until used and applied either late in the afternoon, at night, or on overcast days to avoid drying and to minimize the harmful effects of ultraviolet radiation. One effective means of applying the nematode is through the irrigation system. It may be applied by spray applicator, and irrigated immediately after application.

Formulations of nematodes and their application have, met with limited success. They have been applied only on a small scale, in small gardens or glass-house crops; a few products have seen their utility in the control of mole crickets in the Florida Turf. Nematode formulations are usually developed in moist carrier substrates such as peat, vermiculite, polyether polyurethane sponge, and cedar shavings, which provide interstitial spaces to facilitate gas exchange. The sponge-based storage needs to be squeezed into water before application, while other carriers can be applied directly to the soil as mulch. Nematodes may also be formulated by encapsulation with calcium alginate [77]. Commercially the calcium alginate sheets spread a plastic screen to trap nematodes, and the nematodes are released from the alginate gel matrix by dissolving it in water with the help of sodium citrate [78]. Nematodes have also been formulated in polyacrylamide gels, flowable gels, and wheat flour [78–80], though these are less preferred than alginate gels. Water-dispersible granules composed of cellulose, lignin, silica, and clay–encased materials containing thick nematode suspension and dried by spraying onto roller pans to yield fine dry powders may be the ideal kind of formulation. Such a formulation ensures the availability of oxygen and prevents excessive water loss, facilitating the maintenance of nematodes in a viable state [81].

The easiest way to deliver nematodes for biocontrol is to apply them directly to the soil. They may also be applied via water channels, sprinklers, and drip irrigation channels. Nematodes can also be applied as a foliar treatment. However, they should be applied under stringent conditions such as high humidity (ideally early in the mornings or late in the evenings), as nematodes are very sensitive to desiccation and heat. Foliar applications of nematodes such as *Steinernema carpocapsae* have been successfully used in the control of beet army-worms and serpentine leaf miners [82,83]. They have been formulated in different materials, such as vermiculate, peat, clay, activated charcoal, alginate, and gel-forming polyacrylamide with additives like UV protectants and stored in anhydrobiotic form [78]. An enhanced nematode formulation containing an antidesiccant can be used in the control of a cotton foliage pest, *Earias insulana* [84].

E.   Insects and Mites

Isolation of potential insect enemies, their production, delivery, and application systems are important for the commercialization of biological control mediated by insects. Biological control requires a balance between predator and prey population for successful pest control. Many insects including predacious mites, wasps, and lacewings are known to be effective biocontrol agents against aphids, spider mites, European red mites, leafhoppers, whiteflies, bollworm, navel orangeworm, peach twig borer, and many other pests. Two major families of thrips—Aleolothripidae and Phlaeothripidae—are predacious on aphids, mites, thrips, and whiteflies.

Unfortunately, the use of insects as biocontrol agents has not gained commercial importance, mainly because of difficulties in mass production, inadequate packaging facilities, and the high frequency with which these insects must be released in order to effect significant levels of pest control. Rather than formulation for stability, commercial development efforts on this group of insects have focused more on the packaging/delivery mechanism. The release of adult insects for biocontrol is preferred, as they require less protection from environmental factors than do the other stages. Also, adults would be ready for reproduction immediately after release, and therefore their multiplication and sustenance in the field would be taken care of. Hence, optimal delivery devices such as sticky strips containing insect stages for release in greenhouses or field plots have been developed. Compared to nematodes and microorganisms, larger organisms like insects require more sophisticated methods of storage. Cold storage is ideal for storing insects for a few weeks to months. *Chrysoperla carnes* diapause adults can be stored with reduced mortality for more than 6 months at 5°C [85]. Safe shipment boxes with facility to control temperature and humidity to prevent desiccation and freezing may also be needed for insect shipment.

The immature stages of insects, especially eggs, may be ideal for use in formulations. The enclosure of eggs of the green lacewing *Chrysoperla* within a nonpoisonous adhesive was developed as an effective method of delivery [86]. Traps are perhaps best placed to release egg parasitoids or even adult insect for use in biocontrol programs. Both pheromone and visual traps are employed for the release of *Aphytis melinus* used against the citrus red scale pest *Aonidiella aurantii* [87]. *Epidinocarsis lopezi* was released to combat the cassava mealy bug in tropical Africa by dropping vials containing adult wasps from airplanes [88]. *Neoseiulus (Amblyseius) barkeri* mites were used to control broad mite and flower thrips and small insects on pepper [89].

The order Hymenoptera (wasps) includes more parasitoids than any other order of insects. *Trichogramma* spp. are commercially available and have been released for pest control in forests [90,91]. Predaceous mites like *Ipheseius degen-*

erans and *Euseius tularensis*, when released, survived well on citrus plants and reduced citrus thrips (*Scirtothrips citri*) populations in greenhouse and commercial nurseries. These treatments also improved citrus tree height and leaf number when compared to trees that received treatment with the insecticide abamectin [92]. Tydeid mite, *Orthotydeus lambi*, suppressed the development of grape powdery mildew on wild and cultivated grapes [93].

V. CONCLUSIONS

Biological control offers an environment-friendly alternative to the use of chemicals and pesticides for suppressing plant pests and diseases. Yet growers continue to prefer the use of chemicals to biological agents. While quick knock-down may be the immediate goal, it is widely believed that this trend is mainly due to the fact that the use of biological agents requires a thorough understanding of the organism and the environment in which the agent will function. The success of these agents in bringing about high levels disease suppression that match their chemical counterparts depends largely on the development of effective, stable, easy-to-handle yet economical formulations. Except in very unique circumstances, biological control or biotechnology-derived methods have not yet been able to completely replace the use of chemicals. Even though the U.S. EPA has approved more than 175 biological active ingredients, the biopesticides occupy a very small percentage of the global agrochemical market [94]. Biological control is a slow process and offers a narrow, specific spectrum of activity compared to the use of chemical. However, it should be noted that this weakness itself confers the primary advantage to biocontrol as a safer alternative by minimizing the possibilities of resistance development.

   The development of any formulation should take into account the economics, including the production and formulation costs. The formulation should be amenable for application to both phylloplane and rhizosphere, depending on the pathogen/disease to be controlled. Shelf life is a very important parameter to be considered in the development of a formulation, because most products will have to be stored for long periods of time before they can be marketed and later applied. The minimal requirement of one year of stability could be a challenge to most biocontrol agents. Biological safety, including effects on the environment and nontarget organisms, is also an important attribute.

   Though numerous organisms with the potential to be biological control agents are discovered each year, formulating these organisms effectively will be the key to their successful use. Living organisms must be handled carefully in order to maintain their viability throughout processing, storage, and application. Unlike chemical pesticides that begin to degrade after application, biocontrol agents need to survive and possibly proliferate at their point of application. Extensive research on the biology of control agents continues to be carried out by many academic groups; considerable efforts have been applied at the field research and commercial levels. In spite of this, in many instances their performance under field conditions may be diminished due to low viability and biological activity. Biological control of pests and diseases is still in its infancy and not yet accepted as a stand-alone treatment in many segments of the agriculture industry. Intensive precommercial formulation research and diligent assessments can avoid several pitfalls and minimize failures in the search for suitable organisms to be used in commercial products.

Along with specificity of the biological control agents comes the need for unique agents for controlling different pests or diseases. Different environments and ecological niches may require different strains or differing formulations. It becomes necessary to optimize biological control and to choose agents that operate via several modes and suppress a wide range of disease-controlling strains. The use of combinations of microbes with different mechanisms could be an approach to improve efficacy and consistency of biological control. The trend towards integrated pest management, increased concerns about food safety, and, more recently, the integration and combination of transgenic technology in modern agriculture require due consideration. Biological control is an option; formulation technology must be considered an essential tool that optimizes the activity, versatility, and utility of this option while maximizing the return to the grower.

Note: Mention of a commercial product, trade name, or manufacturer does not constitute an endorsement by the authors.

REFERENCES

1. KA Powell, AR Jutsum. Technical and commercial aspects of biocontrol products. Pesticide Sci 37:315–321, 1993.
2. CD Boyette, PC Quimby Jr., AJ Caesar, JL Birdsall, WJ Connick, Jr., DJ Daigle, MA Jackson, GH Egley, HK Abbas. Adjuvants, formulations, and spraying systems for improvement of mycoherbicides. Weed Technol 10:637–644, 1996.
3. JH Andrews. Biological control in the phyllosphere. Annu Rev Phytopathol 30:603–635, 1992.
4. BJ Jacobson, PA Backman. Biological and cultural plant disease controls: alternatives and supplements to chemicals in IPM systems. Plant Dis 77:311–315, 1993.
5. BN Devisetty, Y Wang, P Sudershan, BL Kirkpatrick, RJ Cibulsky, D Birkhold. Formulation and delivery systems for enhanced and extended activity of biopesticides. In: JD Nalewaja, GR Goss, RS Tann, eds. Pesticide Formulations and Application Systems, Vol. 18. ASTM STP 1347, 1998, pp. 242–272.

6. HD Burges, ed. Formulation of Microbial Biopesticides. Dordrecht Kluwer Academic Publishers, 1998.

7. J Lawrie, MP Greaves, VM Down, A Chassot. Some effects of spray droplet size on distribution, germination of and infection by mycoherbicide spores. Aspects Appl Biol 48:175–182, 1997.

8. M Pusztai, P Fast, H Kaplan, PR Carey. The effect of sunlight on the protein crystals from *Bacillus thuringiensis* var. *kurstaki* HD1 and NRD12: A Raman spectroscopic study. J Invertebrate Pathol 50:247–253, 1987.

9. P Bailey, G Baker, G Caon. Field efficacy and persistence of *Bacillus thuringiensis* var *kurstaki* against *Epiphyas postvittana* (walker) (Lepidoptera: Tortricidae) in relation to larval behaviour on grapevine leaves. Aust J Entomol 35:297–302, 1996.

10. EM Dougherty, KP Guthrie, M Shapiro. Optical brighteners provide baculovirus activity enhancement and UV radiation protection. Biol Control 7:71–74, 1996.

11. D Moore, OK Douro-Kpindou, NE Jenkins, CJ Lomer. Effects of moisture content and temperature on storage of *Metarrhizium flavoviride* conidia. Biocontrol Sci Technol 6:51–61, 1996.

12. D Moore, PM Higgins, CJ Lomer. Effects of simulated and natural sunlight on the germination of conidia of *Metarrhizium flavoviride* Gams and Rozsypal and interactions with temperature. Biocontrol Sci Technol 6:411–415, 1996.

13. GC Papavizas, DR Fravel, JA Lewis. Proliferation of *Talaromyces flavus* in soil and survival in alginate pellets. Phytopathology 77:131–136, 1987.

14. A Martinsen, G Skjak-Braek, O Smidsrod. Alginate as immobilization material: I. Correlation between chemical and physical properties of alginate gel beads. Biotechnol Bioeng 33:79–89, 1989.

15. TC Caesar-Tonthat, WE Dyer, PC Quimby, Jr., SS Posenthal. Formulation of an endoparasitic nematode, *Subanguina picridis* Brzeski, a biological control agent for Russian knapweed, *Acroptilon repens* (L) DC. Biol Control 5:262–266, 1995.

16. SB Leslie, E Israeli, B Lighthart, JH Crowe, LM Crowe. Trehalose and sucrose protect both membranes and proteins in intact bacteria during drying. Appl Environ Microbiol 61:3592–3597, 1995.

17. RP Bateman, M Carey, D Moore, C Prior. The enhanced infectivity of *Metarrhizium flavoviride* in oil formulations to desert locusts at low humidities. Ann Appl Biol 122:145–152, 1993.

18. S Lisansky. Biopesticides fall short of market projections. Performance Chem 16:387–396, 1989.

19. D Moore, RP Bateman, M Carey, C Prior. Long-term storage of *Metarrhizium flavoviride* conidia in oil formulations for the control of locusts and grasshoppers. Biocontrol Sci Technol 5:193–199, 1995.

20. GV McLatchie, D Moore, RP Bateman, C Prior. Effects of temperature on the viability of the conidia of *Metarrhizium flavoviride* in oil formulations. Mycol Res 98:749–756, 1994.

21. RJ Cook, KR Baker. The nature and practice of biological control of plant pathogens. St. Paul, MN: American Phytopathological Society, 1983.

22. MP McQuilken, P Halmer, D Rhodes. Application of microorganisms to seeds. In: HD Burges, ed. Formulations of Microbial Biopesticides. Dordrecht, The Netherlands: Kluwer Academic Publishers, 1998, pp. 255–285.

23. AS Paau. Formulations useful in applying beneficial microorganisms to see. TibTech 6:276–279, 1988.

24. RD Lumsden, JA Lewis, DR Fravel. Formulation and delivery of biocontrol agents for use against soil borne plant pathogens. In: FR Hall, JW Barry, eds. Biorational Pest Control Agents. Formulation and Delivery. Washington DC: ACS Symposium Series 595, 1995, pp. 166–182.

25. JE Bryant. Commercial production and formulation of *Bacillus thuringiensis*. Agric Ecosys Environ 49:31–35, 1994.

26. TR Shieh. Biopesticide formulations and their application. In: NN Ragsdale, PC Kearney, JR Plimmer, eds. Proceedings of American Chemical Society, 8th Int'l. Congress of Pesticide chemistry Options 2000, Washington DC, 1995, pp. 104–114.

27. LT Harvey. A Guide to Agricultural Spray Adjuvants Used in the United States. Fresno, CA: Thompson Publications, 1991.

28. P Warrior, L Rehberger, M Beach, PA Grau, GW Kirfman, JM Conley. Commercial development and introduction of DiTera, a new nematicide. Pestic Sci 55:376–379, 1999.

29. SM Boyetchko. Formulating bacteria for biological weed control. In: Proceedings of Expert Committee on Weeds, Victoria, Canada, 1996, pp. 85–87.

30. J Fages. An industrial view of *Azospirillum inoculans*: formulation and application technology. Symbiosis 13:15–26, 1992.

31. B Digat. Strategies for seed bacterization. Acta Hortic 253:121–130, 1989.

32. PK Jha, S Nair, S Babu. Encapsulation of seeds of *Sesbania sesban* with polyacrylamide and alginate gel entrapped rhizobia leads to effective symbiotic nitrogen fixation. Ind J Exp Biol 31:161–167, 1993.

33. TR Glare, M O'Callaghan. *Bacillus thuringiensis*: Biology, Ecology and Safety. Chichester, UK: John Wiley & Sons, Ltd., 2000.

34. RJC Cannon. Prospects and progress for *Bacillus thuringiensis*-based pesticides. Pestic Sci 37:331–335, 1993.

35. RR Farrar Jr., RL Ridgway. Enhancement of activity of *Bacillus thuringiensis* Berliner against four lepidopterous insect pests by nutrient-based phagostimulants. J Entomol Sci 30:29–42, 1995.

36. Z Shouan, X Weimin, Y Zhinong, M Ruhong. Research and commercialization of yield increasing bacteria (YIB) in China. In: T Wenhua, RJ Cook, A Rovira, eds. Advances of Biological Control of Plant Diseases. Proceedings of the International Workshop on Biological Control of Plant Diseases, Beijing, China, 1996, pp. 47–53.

37. P Vidhyasekaran, M Muthamilan. Development of formulations of *Pseudomonas fluorescens* for control of chickpea wilt. Plant Dis 79:782–786, 1995.

38. K Krishnamurthy, SS Gnanamanickam. Biological control of rice blast by *Pseudomonas fluorescens* strain Pf7-14: Evaluation of a marker gene and formulation. Biol Control 13:158–165, 1998.

39. K Krishnamurthy, SS Gnanamanickam. Biocontrol of rice sheath blight with formulated *Pseudomonas putida*. Indian Phytopathol 51:233–236, 1998.

40. Y Elad, I Chet. Practical approaches for biocontrol implementation. In: R Reuveni, ed. Novel Approaches to Integrated Pest Management. London: Lewis, 1995, pp. 323–338.

41. ZA Siddiqui, I Mahmood. Role of bacteria in the management of plant parasitic nematodes: a review. Bioresour Technol 69:167–179, 1999.

42. RR Belanger, M Benyagoub. Challenges and prospects for integrated control of powdery mildews in the greenhouse. Can J Plant Pathol 19:310–314, 1997.

43. ZK Punja. Comparative efficacy of bacteria, fungi, and yeasts as biological control agents for diseases of vegetable crops. Can J Plant Pathol 19:315–323, 1997.

44. MC Rombach, RM Aguda, DW Roberts. Production of *Beauveria bassiana* (Deuteromycotina: Hyphomycetes) in different liquid media and subsequent conidiation of dry mycelium. Entomophaga 33:315–324, 1988.

45. W Andersch, J Hartwig, P Reinecke, K Stenzel. Production of mycelial granules of the entomopathogenic fungus *Metarrhizium anisopliae* for biological control of soil pests. In: Proceedings of the 5th Int'l. colloquium on Invertebrate Pathology and Microbial Control, Adelaide, Australia, 1990, pp. 2–5.

46. RM Pereira, DW Roberts. Alginate and cornstarch mycelial formulations of entomopathogenic fungi, *Beauveria bassiana* and *Metarrhizium anisopliae*. J Econ Entomol 84:1657–1661, 1991.

47. GR Knudsen, JB Johnson, DJ Eschen. Alginate pellet formulation of a *Beauveria bassiana* (Fungi: Hyphomycetes) isolate pathogenic to cereal aphids. J Econ Entomol 83:2225–2228, 1990.

48. DL Johnson, MS Goettel. Reduction of grasshopper populations following field application of the fungus *Beauveria bassiana*. Bicontrol Sci Technol 3:165–175, 1993.

49. PA Backman, R Rodriquez-Kabana. A system for growth and delivery of biological control agents to the soil. Phytopathology 65:819–821, 1975.

50. CD Boyette. Unrefined corn oil improves the mycoherbicidal activity of *Colletotrichum truncatum* for hemp sesbania (*Sesbania exaltata*) control. Weed Technol 8:526–529, 1994.

51. BA Auld, L Morin. Constraints in the development of bioherbicides. Weed Technol 9:638–652, 1995.

52. BA Auld. Vegetable oil suspension emulsions reduce dew dependence of a mycoherbicide. Crop Prot 12:477–479, 1993.

53. WJ Connick Jr., DJ Daigle, PC Quimby Jr. An improved inert emulsion with high water retention for mycoherbicide delivery. Weed Technol 5:442–444, 1991.

54. RG McGuire, RD Hagenmaier. Shellac coatings for grapefruits that favor biological control of *Penicillium digitatum* by *Candida oleophila*. Biol Control 7:100–106, 1996.

55. EB Nelson, GE Harman. Biological control of turf grass disease with a rhizosphere competent strain of *Trichoderma harzianum*. Plant Dis 80:736–741, 1996.

56. C Wu, T Hsiang. Pathogenicity and formulation of *Typhaula phacorrhiza*, a biocontrol agent of gray snow mold. Plant Dis 82:1003–1006, 1998.

57. VL Smith. Enhancement of snap bean emergence by *Gliocladium virens*. Hurt Sci 31:984–985, 1996.

58. RD Lumsden, JF Walter, CP Baker. Development of *Gliocladium virens* for damping-off disease control. Can J Plant Pathol 18:463–468, 1996.

59. LA Lacey, MS Goettel. Current developments in microbial control of insect pests and prospects for the early 21st century. Entomophaga 40:3–27, 1995.

60. RM Nowierski, Z Zeng, S Jaronski, F Delgado, W Swearingen. Analysis and model-ing of time-dose-mortality of *Melanoplus sanguinipes*, *Locusta migratorioides*, and *Schistoicerca gregaria* (Orthoptera: Acrridiae) from *Beauveria*, *Metarrhizium*, and *Paecilomyces* isolates from Madagascar. J Invertebr Pathol 67:236–252, 1996.

61. MG Feng, TJ Poprawski, GG Khachatourians. Production, formulation and applica-tion of the entomopathogenic fungus *Beauveria bassiana* for insect control: current status. Bicontrol Sci Technol 4:2–34, 1994.

62. RW Caudwell, AG Gatehouse. Formulation of grasshopper and locust entomopatho-gens in baits using starch extrusion technology. Crop Prot 15:33–37, 1996.

63. MS Goettel, DL Johnson, GD Inglis. The role of fungi in the control of grasshoppers. Can J Bot 73:571–575, 1995.

64. GD Inglis, DL Johnson, MS Goettel. Effect of bait substrate and formulation on infection of grasshopper nymphs by *Beauveria bassiana*. Bicontrol Sci Technol 6: 35–50, 1996.

65. D Moore, PD Bridge, PM Higgins, RP Bateman, C Prior. Ultra-violet radiation dam-age to *Metarhizium flavoviride* conidia and the protection given by vegetable and mineral oils and chemicals sunscreens. Ann Appl Biol 122:605–615, 1993.

66. RW Caudwell, AG Gatehouse. Laboratory and field trial of bait formulations of the fungal pathogens, *Metarrhizium flavoviride*, against a topical grasshopper and locust. Biocontrol Sci Technol 6:561–567, 1996.

67. ZA Siddiqui, I Mahmood. Biological control of plant parasitic nematodes by fungi: a review. Bioresour Technol 58:229–239, 1996.

68. GR Stirling, LJ Smith, KA Licastro, LM Eden. Control of root-knot nematode with formulations of the nematode-trapping fungus *Arthrobotrys dactyloides*. Biol Con-trol 11:224–230, 1997.

69. JS Cory, DHL Bishop. Use of baculoviruses as biological insecticides. In: CD Rich-ardson, ed. Methods in Molecular Biology, Vol. 39: Baculovirus Expression Proto-cols. Totowa, NJ: Humana, 1995, pp. 277–294.

70. JC Cunningham. Baculoviruses as Microbial Insecticides. In: R Reuveni, ed. Novel Approaches to Integrated Pest Management. Boca Raton, FL: CRC Press, Inc., 1995, pp. 261–292.

71. M Shapiro. Radiation protection and activity enhancement of viruses. In: FR Hall, JW Barry, eds. Biorational Pest Control Agents. Formulation and Delivery. Wash-ington, DC: ACS Symposium Series 595, 1995, pp. 153–164.

72. JR Adams, CA Shepard, M Shapiro, GJ Tompkins. Light and electron microscopy of the histopathology of the midgut of gypsy moth larvae infected with ldMNPV plus a fluorescent brightener. J Invertebr Pathol 64:156–159, 1994.

73. MF Treacy. Recombinant baculoviruses. In: FR Hall, JJ Menn, eds. Biopesticides Use and Delivery. Totowa, NJ: Humana Press, 1999, pp. 321–340.

74. RE Webb, NH Dill, JM McLaughlin, LS Kershaw, JD Podgwaite, SP Cook, KW Thorpe, RR Farrar Jr., RL Ridgway, RW Fuester. Blankophor BBH as an enhancer of nuclear polyhedrosis virus in arborist treatments against the gypsy moth (Lepidop-tera: Lymantriidae). J Econ Entomol 89:957–962, 1996.

75. AM Martinez, D Goulson, JW Chapman, P Caballero, RD Cave, T Williams. Is it feasible to use optical brightener technology with a baculovirus bioinsecticide for resource-poor maize farmers in Mesoamerica? Biol Control 17:174–181, 2000.

76. J Curran, C Gilbert, K Butler. Routine cryopreservation of isolates of *Steinernema* and *Heterorhabditis* spp. J Nematol 24:1–2, 1992.

77. HK Kaya, CE Nelsen. Encapsulation of steinernematid and heterorhabditid nematodes with calcium alginate: a new approach for insect control and other applications. Environ Entomol 14:572–574, 1985.

78. R Georgis. Formulation and application technology. In: R Gaugler, HK Kaya, eds. Entomopathogenic Nematodes in Biological Control. Boca Raton, FL: CRC Press, 1990, pp. 173–191.

79. R Georgis, SA Manweiler. Entomopathogenic nematodes: a developing biological control technology. Agric Zool Rev 6:63–94, 1994.

80. WJ Connick Jr., WR Nickle, BJ Vinyarad. 'Pasta': new granular formulations for *Steinernema carpocapsae*. J Nematol 25:198–203, 1993.

81. SC Silver, DB Dunlop, DI Grove. Granular formulation of biological entities with improved storage stability. Int. Patent WO 95/05077 (1995).

82. I Glazer. Survival and efficacy of Steinernema carpocapsae in an exposed environment. Biocontrol Sci Technol 2:101–107, 1992.

83. AB Broadbent, THA Olthof. Foliar application of *Steinernema carpocapsae* (Rhabditida: Steinernematidae) to *Liriomyza trifolii* (Diptera: Agromyzidae) larvae in chrysanthemums. Environ Entomol 24:431–435, 1995.

84. I Glazer, M Klein, A Navon, and Y Nakache. Comparison of efficacy of entomopathogenic nematodes combined with antidesiccants applied by canopy sprays against three cotton pests (Lepidoptera: Noctuidae). J Econ Entomol 85:1636–1641, 1992.

85. MJ Tauber, CA Tauber, S Gardescu. Prolonged storage of *Chrysoperla carnea* (Neuroptera: Chrysopidae). Environ Entomol 22:843–848, 1993.

86. LR Wunderlich, DK Giles. Field assessment of adhesion and hatch of *Chrysoperla* eggs mechanically applied in liquid carriers. Biol Control 14:159–167, 1999.

87. P Phillips. Timing *Aphytis* release in coastal citrus. Citrograpy 72:128–131, 1987.

88. HR Herren. Africa-wide biological control project of Cassava pests. A review of objectives and achievements. Insect Sci Applic 8:837–840, 1987.

89. Y Fan, FL Petitt. Biological control of broad mire, *Polyphagotarsonemus latus* (Banks), by *Neoseiulus barkeri* Hughes on pepper. Biol Control 4:390–395, 1994.

90. B Bai, C Cobanoglu, SM Smith. Assessment of *Trichogramma* species for biological control of forest lepidopteran defoliators. Entomol Exp Applic 75:135–143, 1995.

91. SM Smith. Biological control with *Trichogramma*: advances, success, and potential of their use. Annu Rev Entomol 41:375–406, 1996.

92. EE Grafton-Cardwell, Y Ouyang, RA Striggow. Predacious mites for control of citrus thrips, *Scirtothrips citri* (Thysanoptera: Thripidae) in nursery citrus. Biol Control 14:29–36, 1999.

93. G English-Loeb, AP Norton, DM Gadoury, RC Seem, WF Wilcox. Control of powdery mildew in wild and cultivated grapes by a Tydeid mite. Biol Control 14:97–103, 1999.

94. P Warrior. Living systems as natural crop protection agents. Pest Manage Sci 56: 681–687, 2000.

19
Future Trends in Biocontrol

Jo Handelsman
University of Wisconsin–Madison, Madison, Wisconsin

I. INTRODUCTION

Historically, research in biocontrol has yielded discoveries of both fundamental biology and solutions for practical problems in agriculture. Advances on both fronts will be furthered by a better understanding of the complex ecology that surrounds the disease-retardant interactions of microorganisms and plants. The emerging tools of modern biology afford increasingly sophisticated approaches to dissect the multichannel dialogue among the plants, pathogens, biological control agents, and microbial communities that provide the biological context for disease and its suppression. As these research avenues are pursued, new principles of organismal interactions and community function and new strategies for deployment of biocontrol agents will emerge.

To realize the practical potential of biocontrol for agricultural production, it will be imperative to unite knowledge of mechanistic interactions with an appreciation of the complexity of the agroecosystem. Understanding the recognition, signaling, and cooperative and antagonistic interplay between the biological partners will lead to strategies to direct the outcome of the interaction more precisely and consistently. Knowledge of the events leading to disease control may suggest modifications in the timing, placement, and formulation of the biocontrol agent to achieve maximum disease control. Similarly, that knowledge may indicate situations in which certain biocontrol agents will not be successful and may thus lead to wiser choices to tailor the agent to the physical, chemical, and biological context into which it will be introduced.

The future research challenges in biocontrol move beyond the suppression of plant disease to include impacts on human health and registration processes.

Research will be needed to address the safety of biocontrol agents, as well as their efficacy, to avoid public health disasters, allay concerns of the general public, satisfy regulatory agencies, and promote commercial acceptance.

II. FUNDAMENTAL BIOLOGY

Research on biocontrol has been an important vehicle for expanding our knowledge of environmental microbiology. Through biocontrol research, new antibiotics have been identified, the basis for mycoparasitism has been elucidated, and mechanisms of nutrient competition among microbes have been empirically substantiated. Future research in biocontrol is likely to contribute to our knowledge of fundamental biology in microbial community function. Specifically, the structure of microbial communities, signaling among members of the community and from the community to plants that serve as hosts for them, and the basis for mutual dependence among members of microbial consortia will be key areas for study.

In the field, biocontrol agents must be effective in a complex biological milieu. The leaves, flowers, seeds, and roots on which biocontrol agents must suppress the action of pathogens are teeming with other microorganisms. The communities on these surfaces are in flux, influenced by cycles of moisture, temperature, light, and jetties of air or water. Thus, the biocontrol agent must contend with a complex physical environment that is constantly changing the biological environment that provides both assistance and competition to the biocontrol agent.

Despite the importance of microbial communities on plant surfaces, much biocontrol research has ignored their influence by studying biocontrol in laboratory settings with much simpler microbial communities than occur in the field. While this choice has led to the precise dissection of mechanisms of biocontrol that could not have been delineated so clearly in a more complex environment, the cleanliness of the lab setting is certainly not a reasonable approximation of the field environment. It is no surprise, therefore, that many biocontrol experiments have produced divergent results in the lab and field. The microbial communities on plant surfaces may be a key to understanding the discrepancies between lab and field results. If we are to understand the network of events that cooperate to effect disease suppression, then we must begin to understand the community context in which the key events occur. This context is difficult to study, and improved methods are needed. Thus, the future of biocontrol research must include experimental strategies to understand the interaction of the biocontrol agent and the community into which it is introduced and in which it is expected to function.

A. Community Dynamics

The first step in understanding community function is to describe the membership of the community. Some studies have provided an inventory of rhizosphere or phyllosphere communities in the presence and absence of a biocontrol agent. However, no study to date has followed community dynamics over the time scale that is relevant to a microorganism—minutes or hours. Such detailed studies are largely beyond today's technology for tracking either cultured organisms or molecular markers for uncultured communities. However, it may be the community fluctuations that occur over short intervals that determine the outcome of biocontrol applications. Consequently, it is imperative that the next phase of innovation in microbial detection includes developing the capacity to monitor community dynamics on a bacterial time scale.

B. Community Signaling

A second key area for future study is signaling among members of the microbial community and between the community and the plant host. The last decade has cracked open the fascinating arsenal of molecules that constitutes the language of microbes. Studies of biocontrol agents revealed that microbes use a variety of molecules to carry on functions as diverse as inciting defense responses in plants and sensing their own population densities. Perhaps one of the notable surprises that provides a directional signal for future research is that quorum-sensing molecules can be shared among members of different species. Communication among microbes within a community with small molecule signals could be the unifying element that makes the community an entity that enables the members of the community to know their place and their jobs. A substantial research effort in chemical communication in communities is likely to reveal that a cornucopia of chemical messages play a role in biocontrol by enabling the biocontrol agent to sense its surrounding, change the behavior of its competitors or cooperators, or alter the plant's defenses (see Chapter 16) or its contribution of carbon or other nutrients to the nutritional base available to the community.

C. Microbial Consortia

Biocontrol usually involves the isolation, culturing, and application of a single microorganism. And yet, most organisms live in close association with other species that provide services for them, including nutrient production and waste removal. It may, therefore, be unrealistic to expect many organisms to perform optimally in terms of growth, spread, and antagonism against a pathogen without providing the members of the community on which they depend. Biculture or dual culture (see Chapter 17) has often been thought to be successful because of the combined effects of the biocontrol agents on the pathogen, but it is also possible that the two microorganisms provide direct benefit to each other. An understanding of community functions in terms of interdependence of microorganisms may provide insight into all microbial communities and may suggest mixed inocula for biocontrol that will be more effective than single cultures.

III. APPLICATIONS

The yield of commercial products spawned by the modest amount of biocontrol research supported by public and private agencies is impressive compared to the number of fungicides that have resulted from the massive, multibillion dollar investment in fungicide research led largely by the agrichemical companies. The reputation of biocontrol, however, is that few agents work reliably or as well as their synthetic chemical counterparts. It is likely that the frequency of failure of biological versus chemical experimental agents is no different. The difference may be that chemical agents have been tested largely by private companies and many of their results have remained confidential; their reports to the scientific community mostly deal with the successful chemicals that become products. In contrast, biological agents have been tested mostly by public sector scientists who publish or discuss with colleagues the results of both successful and unsuccessful trials. A record of success and safety is the only means to reverse the reputation of biocontrol, and to achieve this record the practical issues discussed here must be addressed.

A. Formulation Challenges

Even among those biocontrol agents that have successfully made the difficult transition from lab to field conditions, the record for transfer from experimental field conditions to on-farm use has been abysmal. Many agents that perform spectacularly, in some cases as well as the best synthetic pesticide, under controlled conditions demonstrate little or no efficacy under agricultural production conditions. A significant barrier appears to be survival of the biocontrol agents when they are fermented, formulated, and applied in scale. Under experimental conditions, it is feasible to prepare the inoculum within 24–48 hours of planting, whereas under production conditions the inoculum often must survive transportation on seed or in packages. An area of future research that will have important implications is the formulation of biocontrol agents to facilitate storage and transport to the site of use. Gains have been made in this area by providing stabilizing and nutritional agents for fungi and bacteria in the formulations (see Chapters 17 and 18) as well as in the use of species that produce hearty resting structures such as spores, but more progress in this area is needed to expand the acreage and types of crops in which biocontrol can be used reliably.

B. Safety

Biocontrol has long been touted as a safe alternative to synthetic pesticides. Recent developments in human health challenge this assertion. The elevation of importance of opportunistic human pathogens in recent years has led to public concern about the widespread use of certain bacteria and fungi in agriculture. The emergence of immune-compromising infectious diseases and the increase in organ transplants, which are accompanied by temporary or long-term immune suppression to prevent tissue rejection, have made opportunistic pathogens a more visible threat to human health. Many biocontrol agents are—or are closely related to—opportunistic pathogens. Examples of biocontrol organisms of questionable safety abound. *Pseudomonas aeruginosa*, a biocontrol agent of gray leaf spot on turf (see Chapter 14) is a virulent opportunistic pathogen infecting surgical wounds and severe burns. *Burkholderia cepacia*, a highly successful biocontrol agent of pea root rot and other diseases, is associated with opportunistic lung infections of patients with cystic fibrosis. *Trichoderma viride* is an opportunistic human pathogen and is on the biological warfare list in some countries (see Chapter 17). *Bacillus cereus*, a biocontrol agent of soybean damping-off and root rot, is a known food toxicant and is closely related to *B. anthracis*, the causal agent of anthrax and a focus of biological warfare threats. The microbiological question that emerges in each of these cases is whether the strains used for biocontrol are in fact pathogenic to humans or whether they simply fall in the same species as a known pathogen. Given the variation of strains within a species and the state of confusion in microbial taxonomy today, this question is not easy to answer. The epidemiological questions that require attention center around the significance of agricultural use of these organisms in exposure of immunocompromised people to these agents. Some of the bacteria are so abundant in the environment that it is possible that exposure to natural populations, which cannot be controlled, far exceeds the exposure likely to result from agricultural applications except to people working in production facilities or on farms that use these products.

The human safety issue may present some of the most important unaddressed questions in biocontrol research. If left unanswered, the concerns may block public acceptance, registration, and adoption of biocontrol agents. If a serious threat exists, even with one of the many biocontrol agents on or near the marketplace, ignoring these risks could lead to highly visible human infections associated with the use of biocontrol agents. The results of such a disaster would be personal tragedy as well as indelible damage to the reputation of all biocontrol agents. It is in the interest of public safety as well as the continuance of biocontrol research for researchers in the area to take an interest in and encourage the environmental/medical/epidemiological studies that will answer the critical safety questions.

C. Cost

The challenges of formulation and safety testing generate substantial costs associated with the development of biocontrol agents for the marketplace. This presents a paradox for the industry: while biocontrol has long been lauded as more specific and targeted than synthetic pesticides, it is difficult to justify the costs of development for a narrow market, which is the natural outcome of a finely targeted agent. Therefore, the market pressures push the industry toward broad-spectrum biocontrol agents that suppress a spectrum of diseases on many large acreage crops. Certainly not all biocontrol agents on or near the market meet these criteria, but as regulations become more stringent, performance expectations higher, and safety issues more visible, the economic pressures will increase, driving the industry further in this direction. Interesting questions arise from the predicted movement of development of biocontrol agents toward common diseases of widely grown crops. If biocontrol agents are used on massive acreages, will the selection pressure for pathogens that are resistant to, or overcome the effects of, the biocontrol agent be increased, thereby shortening the effective life span of the product? Is it biologically feasible to find single organisms that are adapted to diverse locations, agroecosystems, and environmental conditions? Are there ecological events unique to large-scale application of microorganisms that might lead to concerns for environmental safety? The answers to these questions will reveal important biological principles as well as provide guidance for the development of biocontrol as a significant aspect of modern agricultural practice.

IV. CONCLUSION

By most definitions, biocontrol is an applied field of research involving the human use of a microorganism to enhance crop productivity. But the research generated by the desire to use this practice in agriculture has revealed fascinating biology and led to copious fundamental discoveries. The opportunity to conduct such basic research is coupled with a responsibility to solve the practical problems that prevent the successful deployment of many biocontrol agents. Future trends in biocontrol research will unite fundamental biology with the quest for solutions that will make biocontrol integral to the safe and wise management of every agroecosystem.

Index

Printed in the United States
By Bookmasters